Chemistry: An Analytical Approach

Chemistry: An Analytical Approach

Edited by
Johana Meyer

WILLFORD PRESS

www.willfordpress.com

Published by Willford Press,
118-35 Queens Blvd., Suite 400,
Forest Hills, NY 11375, USA

ISBN: 978-1-68285-761-8

Cataloging-in-Publication Data

Chemistry : an analytical approach / edited by Johana Meyer.
    p. cm.
Includes bibliographical references and index.
ISBN 978-1-68285-761-8
1. Chemistry, Analytic. 2. Chemistry. I. Meyer, Johana.
QD75.22 .C44 2020
543--dc23

For information on all Willford Press publications
visit our website at www.willfordpress.com

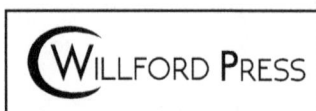

WILLFORD PRESS

Contents

Preface

The field of chemistry concerned with the study and use of instrumental and classical methods for the analysis of matter is known as analytical chemistry. It encompasses the separation, identification and quantification of matter. Separation processes include classical methods, such as distillation, precipitation and extraction, as well as instrumental methods of chromatography, field flow fractionation and electrophoresis. Modern analytical chemistry builds on the classical techniques of qualitative and quantitative analyses. These include chemical and flame tests, gravimetric analysis and volumetric analysis. This discipline is aided by several instrumental methods, which include spectroscopy, thermal analysis, electrochemical analysis, etc. This book presents the complex subject of analytical chemistry in the most comprehensible and easy to understand language. From theories to research to practical applications, case studies related to all contemporary topics of relevance to this field have been included herein. It will prove to be immensely beneficial to students and researchers in this field.

Various studies have approached the subject by analyzing it with a single perspective, but the present book provides diverse methodologies and techniques to address this field. This book contains theories and applications needed for understanding the subject from different perspectives. The aim is to keep the readers informed about the progresses in the field; therefore, the contributions were carefully examined to compile novel researches by specialists from across the globe.

Indeed, the job of the editor is the most crucial and challenging in compiling all chapters into a single book. In the end, I would extend my sincere thanks to the chapter authors for their profound work. I am also thankful for the support provided by my family and colleagues during the compilation of this book.

Editor

Inexpensive Bismuth-Film Electrode Supported on Pencil-Lead Graphite for Determination of Pb(II) and Cd(II) Ions by Anodic Stripping Voltammetry

Karen C. Bedin, Edson Y. Mitsuyasu, Amanda Ronix,
André L. Cazetta ⓘ, Osvaldo Pezoti, and Vitor C. Almeida ⓘ

Department of Chemistry, State University of Maringá, Av. Colombo 5790, CEP 87020-900 Maringá, PR, Brazil

Correspondence should be addressed to Vitor C. Almeida; vcalmeida@uem.br

Academic Editor: Valentina Venuti

The present work reports the development and application of bismuth-film electrode (BiFE), obtained by in situ method on the pencil-lead graphite surface, for simultaneous Cd(II) and Pb(II) determination at trace levels, as alternative to replace the mercury-film electrodes. Experimental factors, deposition time (t_d), deposition potential (E_d), and Bi(III) concentration (C_{Bi}), were investigated by applying a 2^3 factorial design using 0.10 mol/L acetate buffer solution (pH 4.5) as supporting electrolyte. The analysis conditions of the differential pulse technique were $t_d = 250$ s, $E_d = -1.40$ V, and $C_{Bi} = 250$ mg L^{-1}. The validation of the method employing BiFE was accomplished by determination of merit figures. The detection limits were of 11.0 μg L^{-1} for Cd(II) and 11.5 μg L^{-1} for Pb(II), confirming that proposed method is attractive and suitable for heavy metals determination. Additionally, the BiFE developed was successfully applied for the Cd(II) and Pb(II) determination in wastewater sample of battery industry.

1. Introduction

Trace elements, such as Cu, Mo, Mn, and Zn, are considered essential for human health, while others as Hg, Pb, and Cd may accumulate in body tissues causing problems due to their toxicity [1, 2]. This accumulation is due to anthropogenic activities, associated with disposal of solid and liquid waste without proper treatment [3]. Considering the increasing industrial use of these heavy metals and their serious environmental and toxicological impacts, it is necessary to develop new analytical methods to determine them at trace levels [4, 5].

The analytical methods used for heavy metals determination include spectrometry and electroanalytical techniques. Atomic absorption spectrometry (AAS) is the most applied technique for metals determination, since it provides satisfactory sensitivity, high selectivity, and relatively low cost equipment. However, this technique has the disadvantage of not allowing simultaneous determination of chemical elements [6]. The inductively coupled plasma optical emission spectrometry (ICP-OES) and inductively coupled plasma mass spectrometry (ICP-MS) are techniques which allows multielement analysis; however, the high cost of installation and maintenance has restricted their use in research and routine analysis [7–10].

Anodic stripping voltammetry (ASV) is a versatile electroanalytical technique for the trace metals determination in various environmental, clinical and industrial samples [6]. Stripping methods are important in trace analysis because the electrodeposition step concentrates the analyte on the electrode surface, enabling its determination even in extremely low quantities with reasonable accuracy [11, 12].

During the past five decades, mercury-based electrodes were often applied in ASV for presenting an excellent analytical performance [13, 14]. However, due to the high toxicity of mercury and its compounds, the use of this metal in electrodes was restricted. Consequently, an intense research for less toxic and environmentally friendly materials than mercury has been promoted [15–17].

Bismuth-film electrodes were introduced in the last decade to replace the mercury-based electrodes used for the determination of trace heavy metals and organic compounds, due to very similar electrochemical properties of Bi and Hg, including alloying heavy metals with bismuth as an analogy to amalgamation at mercury [10, 18]. The bismuth-film electrode consists of a bismuth film deposited on a substrate, which can be Au [19], Pt [20], carbon paste [21], glassy carbon [22], carbon fiber [23], pyrolytic graphite [24], and oxides [3, 25, 26]. Moreover, different combinations also are possible as Bi_2Te_3-graphene oxide hybrid film [27] and bismuth-dispersed xerogel-based composite film [28], besides new methods for Bi_2O_3-electrode modification as spark discharge [29], proving that there are still issues to be explored in the development of bismuth-based electrodes. Performance of bismuth-film electrodes is similar to traditional mercury electrodes, in addition to the advantage of having negligible toxicity compared to them [30]. The bismuth-film electrode has been applied in voltammetric studies from the ex situ or in situ method, at potential range of -1.2 to 0 V. Positives potentials are not applied, since bismuth is completely removed from the substrate surface by oxidation under these conditions [14].

Carbon electrodes are used extensively as voltammetric sensors in various applications [31]. Mechanical pencils are inexpensive and alternative material to produce carbon electrodes, since they are commercially available with different diameters and hardness. This type of graphite has the advantage of not being fragile as the pyrolytic or paste carbon electrodes but is not hard as the glassy carbon electrode. In addition, graphite has good characteristics such as high electrical conductivity, quick and easy pretreatment, low cost, wide availability, minimum trace metals residue in its composition, and low background current [18].

Bond et al. [12] reported the use of a graphite electrode, which was optimized to anodic stripping analysis using mercury thin-film for Cd and Pb determination and the results were consistent with those obtained by analysis with glassy carbon electrode. The use of bismuth-film electrode supported on pencil graphite (BiFE) has been reported in the literature [3, 15, 32, 33]; however, investigation of manufacturing settings can provide better responses for analytical determination of metal ions.

The present work aimed to prepare a BiFE, investigating the experimental parameters by 2^3 fractional factorial design, for the simultaneous determination of Pb(II) and Cd(II) ions by differential pulse anodic stripping voltammetry (DPASV). Additionally, the applicability of BiFE in DPASV was evaluated by the analytical figures of merit.

2. Materials and Methods

2.1. Reagents, Chemicals, and Samples.
All chemicals used were of analytical grade and milli-Q water was used in the preparation of solutions. The Bi(III), Pb(II), and Cd(II) stock solutions (1000 mg L^{-1}) were prepared from $Bi(NO_3)_2.4H_2O$, $Pb(NO_3)_2$ and $Cd(NO_3)_2.4H_2O$ acquired from Sigma-Aldrich. The powder graphite, H_2SO_4, HNO_3, $K_4Fe(CN)_6$, and CH_3COOH were purchased from Merck. A 0.20 mol L^{-1} H_2SO_4 solution and a 0.10 mol L^{-1} acetate buffer (pH 4.50) were prepared as substrates electrolytes for the cyclic voltammetry studies of the prepared electrode and for

Cd(II) and Pb(II) determination from DPASV, respectively. To evaluate the applicability of the electrode, wastewater samples of battery industry without any previous preparation were analyzed.

2.2. Apparatus.
Voltammetric measurements were performed using a potentiostat (Autolab potentiostat/galvanostat GPES IME 663, PGSTAT302n 247V 50/60 Hz). The electrochemical cell coupled to potentiostat was composed of pencil-lead graphite working electrode (GE) or bismuth-film electrode supported on pencil-lead graphite (BiFE), Ag/AgCl as reference electrode and counter electrode of Pt wire.

2.3. Preparation of BiFE.
Pencil-lead rods (Pentel Super, 2B, 0.7 mm in diameter) were used to prepare the GE and its assembly scheme is shown in Figure 1. Pencil-lead pieces of 2.5 cm were fitted to the micropipette tips, which were filled with carbon paste prepared by mixing graphite powder and mineral oil (Nujol) to promote the electric contact between the GE and copper wire. This wire was fixed in a glass tube that was connected to the micropipette tip, forming the electrode body. All connections were isolated using paste (cyanoacrylate), Teflon tape, and nonconductive epoxy resin. Due to this assembly, only the lower extremity of the pencil-lead rod (approximately 3.0 mm) is out of the micropipette tip and then available to act as the contact surface for film formation and determination of cations of interest. Since this surface was completely isolated with cyanoacrylate, a polishing step was necessary prior the experiments. Then, the polishing and the surface renewal of the electrodes were done by polishing them on a silk paper, until the base of the electrodes remained with metallic appearance.

The BiFE was prepared from the Bi(III) film formation on the GE surface by the cathode potential application (versus Ag/AgCl$_{sat}$) by the in situ method. In this, the bismuth deposition occurred simultaneously with the electrochemical deposition of analytes. Moreover, the in situ procedure was chosen because it allows better adhesion of the Bi film to the GE surface, besides the obtaining of higher and better resolved peaks in the voltammograms, along with the higher sensitivity to Cd(II) and Pb(II) when compared to the ex situ method [34, 35].

2.4. Procedure.
The cyclic voltammetry analysis was performed as a performance test of the GE prepared as working electrode, using 2.0 mol L^{-1} H_2SO_4 as the supporting electrolyte. Differential pulse anodic stripping voltammetry (DPASV) was carried out using standard Bi(III) solution (150 to 250 mg L^{-1}) in the electrochemical cell to form the Bi film. The deposition time, deposition potential and Bi(III) concentration were investigated using acetate buffer solution (0.1 mol L^{-1}) as supporting electrolyte and potential step of 1.95 mV, while the solution was stirred. Before each cycle a 30 s conditioning step at 300 mV (under stirring) was used to remove the bismuth excess and/or the target metals on the electrode surface. The DPASV measurements were performed from equilibrium time of 15 s, time modulation of 0.05 s, pulse interval time of 0.25 s, amplitude modulation of

TABLE 1: Factors and respective lower and superior levels values.

Factors	Lower level (−)	Superior level (+)
Bi(III) concentration (mg L^{-1})	150	250
Deposition potential (V)	−1.50	−1.10
Deposition time (s)	100	250

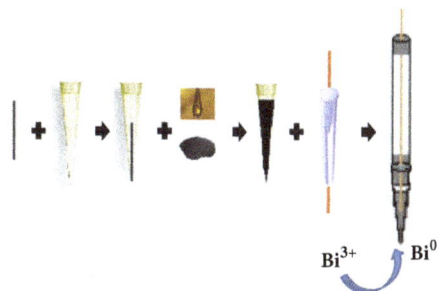

FIGURE 1: The assembly of pencil-lead graphite working electrode and bismuth-film formation.

25 mV, scan rate of 10 mV s^{-1}, and being with different Cd(II) and Pb(II) concentrations.

2.5. Experimental Design.

The 2^3 factorial design was applied to investigate the factors: deposition time (t_d), Bi(III) solution concentration (C_{Bi}), and deposition potential (E_d), represented in Table 1. The responses to the proposed design were the current peak areas of the ASV for Cd(II) and Pb(II) at concentrations of 48.3 to 233 μg L^{-1}. The selection of the experimental domain for each factor was determined by previous experiences and literature. Experimental design and data processing were performed using Design Expert 7.1.3 software.

3. Results and Discussion

3.1. Cyclic Voltammetry.

The GE electrochemical performance was investigated from cyclic voltammetry measurements of the potassium ferroferricyanide ($K_4[Fe(CN)_6]$ / $K_3[Fe(CN)_6]$) redox process. Figure 2 shows the cyclic voltammograms of different $K_4[Fe(CN)_6]$ concentrations. It can be seen that cathodic and anodic peaks occur in a short potential interval, indicating the reversibility of redox system.

Current values for anodic and cathodic peaks were proportional to the increase of the $K_4[Fe(CN)_6]$ concentration. From the voltammograms of the cyclic voltammetry, it can be seen that the performance of the developed GE is comparable with the glassy carbon [12]. Additionally, a linear relationship between the peak current intensity and $K_4[Fe(CN)_6]$ concentration was obtained and showed a linear regression equation of $I_d = 3.16 \times 10^{-7} + 4.18 \times 10^{-9} \, C_{K4[Fe(CN)6]}$ and determination coefficient (R^2) of 0.9992.

3.2. Anodic Stripping Voltammetry (ASV) with BiFE.

During the electrodeposition step in ASV method with BiFE, the codeposition of metallic ions present in the solution along with Bi^0 on the graphite surface is possible [36]. Therefore,

the film formation on the GE surface contributes to the improvement of the analytical signal corresponding to such ions, as shown in the Figure 3. It can be seen that the measurement performed with GE showed low current signals at -0.77 and -0.57 V, which are corresponding to Cd(II) and Pb(II), respectively. Applying BiFE, an increase in the analytic signal for the metal species present in solution can be observed, indicating a high sensitivity in the analysis. The highest current peak observed at -0.10 V corresponds to oxidation of bismuth film. According to the results, it can be inferred that other species with characteristic potential values higher than -0.10 V can be analyzed by the method.

3.3. Effect of the Factors Time (t_d), Deposition Potential (E_d), and Bi(III) Concentration (C_{Bi}) on ASV Analyses.

In order to investigate the effects of the factors t_p, E_d, and C_{Bi} on current measured for Cd(II) and Pb(II) determination using BiFE, a 2^3 factorial design was carried out. Table 2 shows the lower and superior levels values of the factors and its responses (current) obtained from the 8 ASV experiments.

Current measurements were made from the voltammograms obtained in potential values of -0.70 and -0.49 V for Cd(II) and Pb(II), respectively. As can be seen in Table 2, the lowest current values were observed for Cd(II), which ranged from 18.4 to 1.79 nA. The current peaks for Pb(II) determination ranged from 3.23 to 29.3 nA. Effects analyses were investigated by Pareto chart (Figure 4), showing the statistic t-test for each effect, where each bar represents the standard effect, *i.e.*, the estimated effect divided by its standard error and the t-critical value [37].

The factors represented by the bars that extend beyond the line are considered significant. According to the results, the t_d factor was significant for both metals causing a positive effect on the response. The other factors, E_d and C_{Bi}, as well as the interaction effects between the factors, were not significant. This indicates that metal and film saturation on the electrode surface is not achieved at low time values. Therefore, t_d is the most important factor for the sensitivity of the technique

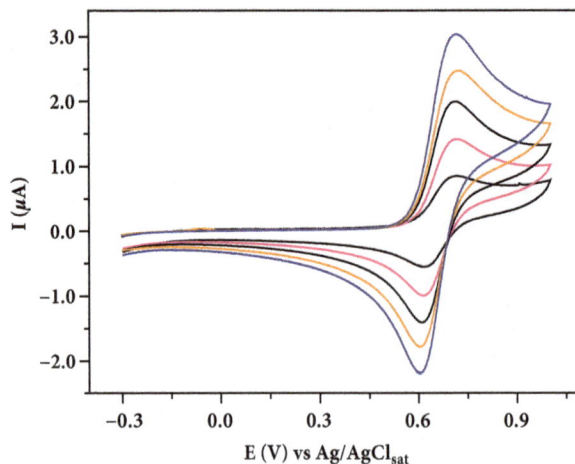

FIGURE 2: Cyclic voltammograms of potassium ferroferricyanide in 2.0 mol L^{-1} H$_2$SO$_4$ using GE (scan rate: 100 mV s^{-1}).

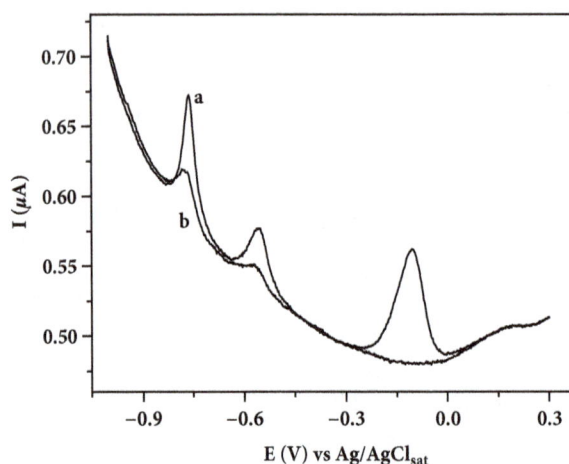

FIGURE 3: Differential pulse anodic stripping voltammograms obtained for Cd(II) and Pb(II) in 0.10 mol L^{-1} acetate buffer solution (pH 4.5) as supporting electrolyte; (a) 250 mg L^{-1} of Bi(III) dissolved in supporting electrolyte, (b) without Bi(III), determined with GE containing 142 μg L^{-1} of Cd(II) and Pb(II) (E_{dp} = -1.40 V; t_{dp} = 250 s; potential step =1.95 mV; amplitude = 25 mV).

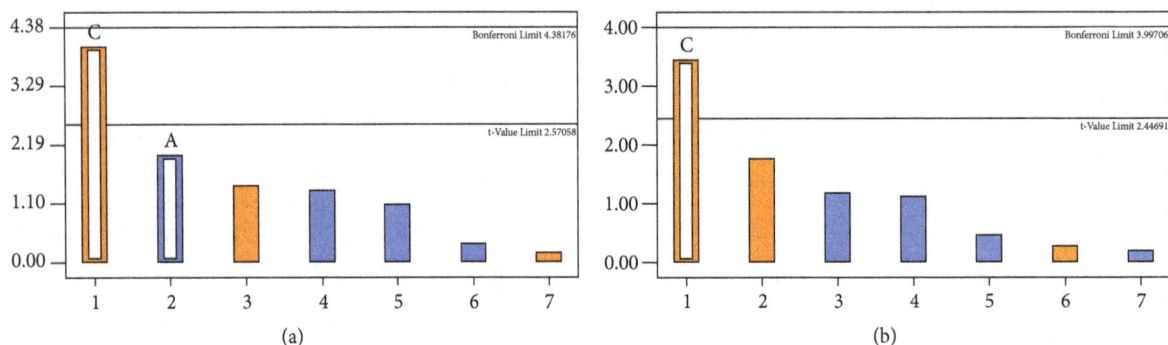

FIGURE 4: Pareto chart to evaluate the peaks area for Cd(II) (a) and Pb(II) (b), where C_{Bi} (A) and t_d (C).

using BiFE, which is in agreement with other works that reported the development of bismuth electrodes [18, 37, 38]. Thus, the optimized conditions selected for subsequent studies were those frequently reported in the literature: 250 s deposition time, Bi(III) concentration of 250 mg L^{-1}, and -1.40 V deposition potential [3, 26, 38–40].

3.4. Analytical Method Development. Analytical figures of merit were determined using the selected conditions (t_d, E_d, and C_{Bi}) to evaluate the applicability of BiFE (film obtained by the in situ method) on the Cd(II) and Pb(II) determination. Figure 5 shows differential pulse anodic stripping voltammograms for simultaneous determination of these ions using the

TABLE 2: Experiments of 2^3 factorial design and response values.

Order	C_{Bi} (mg L^{-1})	E_d (V)	t_d (s)	$I_{Cd(II)}$ (nA)	$I_{Pb(II)}$ (nA)
1	150	-1.50	100	5.31	7.92
2	150	-1.50	250	16.8	29.3
3	250	-1.50	100	4.71	9.26
4	150	-1.10	250	18.4	29.3
5	250	-1.50	250	7.02	14.9
6	250	-1.10	250	6.97	14.9
7	150	-1.10	100	2.75	3.23
8	250	-1.10	100	1.79	6.37

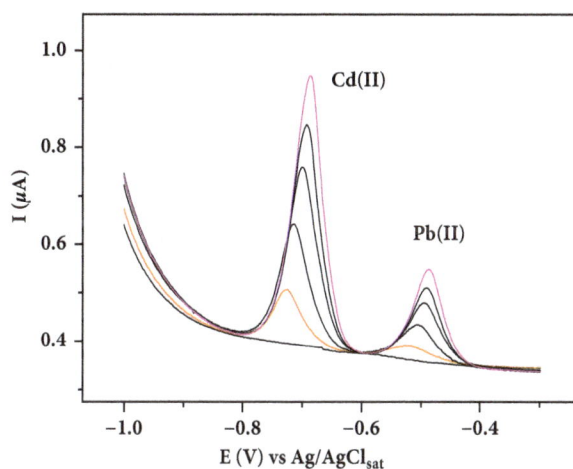

FIGURE 5: Differential pulse anodic stripping voltammograms for increasing concentrations (ranging from 48.3 to 233 μg L^{-1}) of Cd(II) and Pb(II) in 0.10 mol L^{-1} acetate buffer solution (pH 4.5) as supporting electrolyte using BiFE.

BiFE. According to the figure, the two peaks corresponding to metal ions are well resolved and increase with increasing concentrations.

From the calibration curve of each chemical species, it can be demonstrated linearly between the current values and concentrations within the range of 48.3 to 233 μg L^{-1} by the linear regression equation and determination coefficient of experimental data: $I_d = 2.44 \times 10^{-3} C_{Cd} + 4.29 \times 10^{-3}$ and R^2 = 0.9964 for Cd(II) and $I_d = 8.35 \times 10^{-4} C_{Pb} - 5.02 \times 10^{-3}$ and R^2 = 0.9954 for Pb(II).

3.4.1. Detection and Quantification Limits. To evaluate the limit of detection (LOD) and limit of quantification (LOQ), five different blank solutions were used, which were analyzed in triplicate and the standard deviation (S_B) of the means was calculated. The measurements were performed in 5.0 mL of blank solutions with 5.0 mL of acetate buffer solution, under the optimized conditions. The LOD and LOQ values were determined from the $3S_b/m$ and $10S_b/m$ relations, respectively, where m is the slope of the calibration curve of each chemical species. For Cd(II), the calculated values of LOD and LOQ were of 11.0 μg L^{-1} and 36.8 μg L^{-1}, respectively. For Pb(II), the LOD was 11.5 μg L^{-1} and LOQ was 38.2 μg L^{-1}. The

LOD values obtained at levels of μg L^{-1} are comparable with other studies involving modified electrodes. Serrano et al. [41] prepared modified graphite-epoxy composite electrodes which showed LOD of 2.40 and 4.70 μg L^{-1} for Cd(II) and of 1.50 and 3.30 μg L^{-1} for Pb(II), simultaneously determined. Kadara and Tothill [42] found LOD of 8.00 and 16.0 μg L^{-1} for Pb(II) and Cd(II), respectively, using a screen-printed Bi$_2$O$_3$-modified electrode. Lezi et al. [43] employed screen-printed electrodes modified with five bismuth precursor compounds in Pb(II) and Cd(II) determination, obtaining LOD values of 0.90-1.40 and 1.10-3.20 μg L^{-1}, respectively. Therefore, low LOD values could be achieved by increasing the deposition time during the Bi film formation in the proposed method, once the deposition time showed a positive effect on the response by increasing the current intensity, as demonstrated in Table 2. Some studies in the literature corroborate this fact, as an example: Zhang et al. [35] employed a deposition time of 300 s during the in situ Bi deposition on working electrode and LOD of 0.02 μg L^{-1} for Pb(II) and 0.01 μg L^{-1} for Cd(II) were obtained; Demetriades et al. [18] prepared a bismuth-film electrode supported on pencil graphite with Bi deposition time of 600 s and LOD of 0.30 μg L^{-1} for Cd(II) and 0.40 μg L^{-1} for Pb(II) were determined.

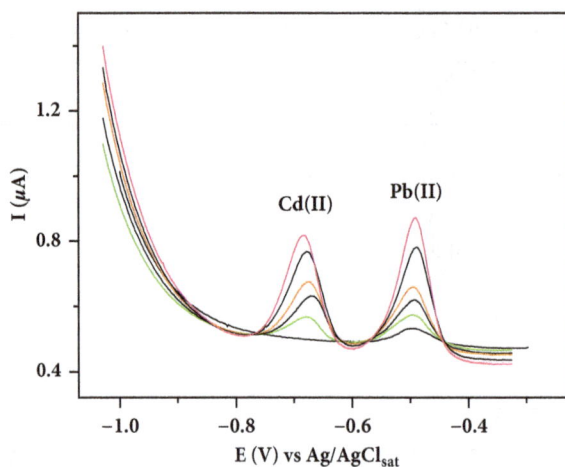

FIGURE 6: Differential pulse anodic stripping voltammograms for 1.00 mL wastewater sample in 9.00 mL acetate buffer solution (pH 4.5) with increasing Cd(II) and Pb(II) concentrations (48.3 to 233 μg L^{-1}) using BiFE.

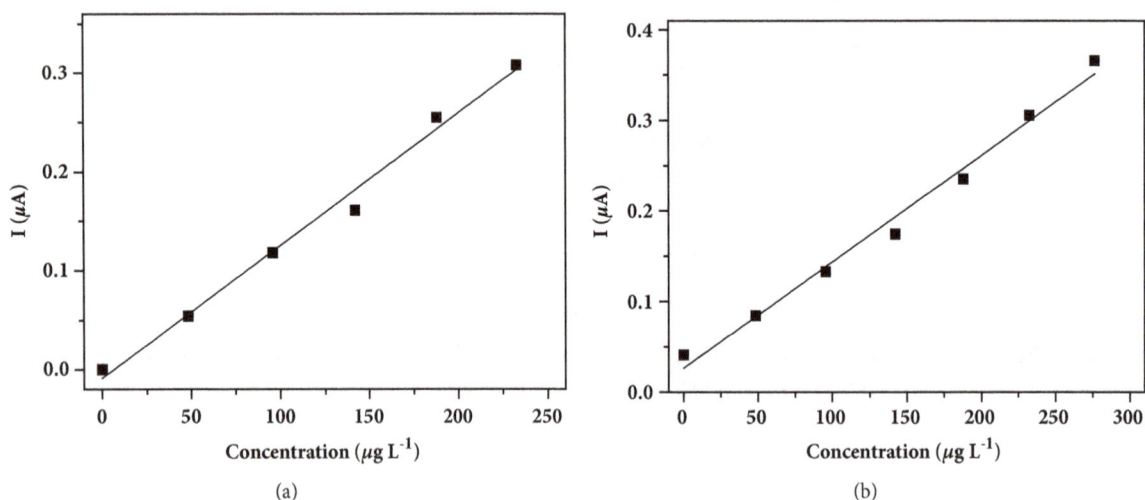

(a)

(b)

FIGURE 7: Calibration curves of standard addition method for Cd(II) (a) and Pb(II) (b).

3.4.2. Repeatability. The repeatability (intraday) was evaluated from five measurements (in triplicate) of solutions containing 150 μg L^{-1} Cd(II) or 150 μg L^{-1} Pb(II) with 10.0 mL of acetate buffer solution, with just one BiFE. The mean of triplicates and the corresponding relative standard deviation (RSD) were calculated to obtain the analytical parameter. The repeatability values obtained, expressed as % RSD, were of 9.07% for Cd(II) and 11.6% for Pb(II), which are acceptable with respect to trace level determination of heavy metals.

3.4.3. Reproducibility. The reproducibility was determined from measurements (in triplicate) of five different BiFEs for solutions of 150 μg L^{-1} Cd(II) or 150 μg L^{-1} Pb(II) with 10.0 mL of acetate buffer solution. This analytical parameter was calculated from RSD obtained from the mean value of measurements. The reproducibility values (expressed as the % RSD) achieved in this work were of 9.72% for Cd(II) and 7.87% for Pb(II) determination.

3.4.4. Standard Addition Method. The BiFE was applied in the determination of Cd(II) and Pb(II) in a wastewater sample from a battery industry. The sample was analyzed without any previous preparation and ions determination was performed by DPASV under optimized conditions. Figure 6 shows the voltammograms obtained by the standard addition method for Cd(II) and Pb(II) determination in the sample. The curve was constructed using concentrations of Cd(II) and Pb(II) ranging from 48.3 to 233 μg L^{-1}.

The calibration curves for the standard addition method of each chemical species are shown in Figure 7. According to the figure, a good linearity between current signals and analytes concentrations can be observed. The linear regression equation and determination coefficient obtained were $I_d = 1.36 \times 10^{-3} C_{Cd} - 1.05 \times 10^{-2}$ and R^2 = 0.9873 for Cd(II) and $I_d = 1.35 \times 10^{-3} C_{Pb} + 1.79 \times 10^{-2}$ and R^2 = 0.9706 for Pb(II). The Cd(II) was not detected in the sample, while the Pb(II) determined concentration was of 48.8 μg L^{-1}. To

TABLE 3: Recovery of Cd(II) and Pb(II) in battery industry wastewater sample fortified in two concentration levels: (I) 190 μg L^{-1} and (II) 235 μg L^{-1}.

| | Cd(II) (μg L^{-1}) | | Recovery (%) | | Pb(II) (μg L^{-1}) | | | Recovery (%) | |
Blank	(I)	(II)	(I)	(II)	Blank	(I)	(II)	(I)	(II)
ND	205 ± 7.4	228 ± 8.5	108 ± 3.9	96.9 ± 3.6	15.1	190 ± 10.4	239 ± 12.2	100 ± 5.5	102 ± 5.2

compare these results with a reference method the wastewater sample was analyzed by graphite furnace atomic absorption spectrometry (GFAAS); Cd(II) was not detected and the Pb(II) determined concentration was 50.1 μg L^{-1}, consistent with the measurements using BiFE.

3.4.5. Recovery Test. The recovery method was used in the battery industry wastewater sample, which was fortified with 190 μg L^{-1} Cd(II) and 235 μg L^{-1} Pb(II) solutions. Table 3 shows the results of the recovery study in wastewater sample using BiFE and optimized experimental conditions. From the table, it can be noticed that the recovery for both metals was close to 100%, ranging from 96.9 to 108% for the addition levels studied. This indicates adequate precision and accuracy of the method employing BiFE for determination of heavy metals at trace levels.

4. Conclusions

The BiFE developed by in situ method for Pb(II) and Cd(II) simultaneous determination at trace levels showed an efficient analytical performance. Among the parameters investigated for differential pulse anodic stripping voltammetry, the deposition time was the most important factor for the technique sensitivity. From the optimized conditions, the BiFE exhibited a linear response in the range between 48.3 and 233 μg L^{-1} with a detection limit of 11.0 μg L^{-1} for Cd(II) and 11.5 μg L^{-1} for Pb(II). Repeatability and reproducibility showed acceptable values for trace level determination of heavy metals. A wastewater sample of battery industry was used in the recovery method to evaluate the analytical application of BiFE on real samples. The recovery for both metals was close to 100%, demonstrating the precision and accuracy of the electrode developed. In this way, the results imply that BiFE can be useful in trace level determination of Cd(II) and Pb(II) in heavily polluted samples (industrial waste waters, aqueous sources from mining areas, etc.), thanks to its analytical performance, low cost, wide availability, easy preparation, and less time consuming.

Acknowledgments

The authors acknowledge CAPES and CNPq for the financial support (this research did not receive specific funding).

References

[1] W. Wonsawat, S. Chuanuwatanakul, W. Dungchai, E. Punrat, S. Motomizu, and O. Chailapakul, "Graphene-carbon paste electrode for cadmium and lead ion monitoring in a flow-based system," *Talanta*, vol. 100, pp. 282–289, 2012.

[2] N. Serrano, J. M. Díaz-Cruz, C. Ariño, and M. Esteban, "Stripping analysis of heavy metals in tap water using the bismuth film electrode," *Analytical and Bioanalytical Chemistry*, vol. 396, no. 3, pp. 1365–1369, 2010.

[3] K. Pokpas, S. Zbeda, N. Jahed, N. Mohamed, P. G. Baker, and E. I. Iwuoha, "Electrochemically reduced graphene oxide pencil-graphite in situ plated bismuth-film electrode for the determination of trace metals by anodic stripping voltammetry," *International Journal of Electrochemical Science*, vol. 9, no. 2, pp. 736–759, 2014.

[4] J. G. Ayenimo and S. B. Adeloju, "Rapid amperometric detection of trace metals by inhibition of an ultrathin polypyrrole-based glucose biosensor," *Talanta*, vol. 148, pp. 502–510, 2016.

[5] L. Fu, X. Li, X. Yu, and J. Ye, "Facile and Simultaneous Stripping Determination of Zinc, Cadmium and Lead on Disposable Multiwalled Carbon Nanotubes Modified Screen-Printed Electrode," *Electroanalysis*, vol. 25, no. 2, pp. 567–572, 2013.

[6] J. Ping, J. Wu, Y. Ying, M. Wang, G. Liu, and M. Zhang, "Evaluation of trace heavy metal levels in soil samples using an ionic liquid modified carbon paste electrode," *Journal of Agricultural and Food Chemistry*, vol. 59, no. 9, pp. 4418–4423, 2011.

[7] F. Fan, J. Dou, A. Ding, K. Zhang, and Y. Wang, "Determination of lead by square wave anodic stripping voltammetry using an electrochemical sensor," *Analytical Sciences*, vol. 29, no. 5, pp. 571–577, 2013.

[8] T. Ndlovu, O. A. Arotiba, S. Sampath, R. W. Krause, and B. B. Mamba, "Electroanalysis of copper as a heavy metal pollutant in water using cobalt oxide modified exfoliated graphite electrode," *Physics and Chemistry of the Earth*, vol. 50-52, pp. 127–131, 2012.

[9] S. Prakash and V. K. Shahi, "Improved sensitive detection of Pb2+ and Cd2+ in water samples at electrodeposited silver nanonuts on a glassy carbon electrode," *Analytical Methods*, vol. 3, no. 9, pp. 2134–2139, 2011.

[10] H. Xu, L. Zeng, D. Huang, Y. Xian, and L. Jin, "A Nafion-coated bismuth film electrode for the determination of heavy metals in vegetable using differential pulse anodic stripping voltammetry: An alternative to mercury-based electrodes," *Food Chemistry*, vol. 109, no. 4, pp. 834–839, 2008.

[11] K. Keawkim, S. Chuanuwatanakul, O. Chailapakul, and S. Motomizu, "Determination of lead and cadmium in rice samples by sequential injection/anodic stripping voltammetry using a bismuth film/crown ether/Nafion modified screen-printed carbon electrode," *Food Control*, vol. 31, no. 1, pp. 14–21, 2013.

[12] A. M. Bond, P. J. Mahon, J. Schiewe, and V. Vicente-Beckett, "An inexpensive and renewable pencil electrode for use in field-based stripping voltammetry," *Analytical Chemistry Acta*, vol. 345, no. 1–3, pp. 67–74, 1997.

[13] U. Injang, P. Noyrod, W. Siangproh, W. Dungchai, S. Motomizu, and O. Chailapakul, "Determination of trace heavy metals in herbs by sequential injection analysis-anodic stripping voltammetry using screen-printed carbon nanotubes electrodes," *Analytica Chimica Acta*, vol. 668, no. 1, pp. 54–60, 2010.

[14] R. M. Dornellas, R. A. A. Franchini, and R. Q. Aucelio, "Determination of the fungicide picoxystrobin using anodic stripping voltammetry on a metal film modified glassy carbon electrode," *Electrochimica Acta*, vol. 97, pp. 202–209, 2013.

[15] C. W. Foster, A. P. de Souza, J. P. Metters, M. Bertotti, and C. E. Banks, "Metallic modified (bismuth, antimony, tin and combinations thereof) film carbon electrodes," *Analyst*, vol. 140, no. 22, pp. 7598–7612, 2015.

[16] J. K. B. Bernardelli, F. R. Lapolli, C. M. G. Da Silva Cruz, and J. B. Floriano, "Determination of zinc and cadmium with characterized electrodes of carbon and polyurethane modified by a bismuth film," *Materials Research*, vol. 14, no. 3, pp. 366–371, 2011.

[17] V. Rehacek, I. Hotovy, M. Vojs, and F. Mika, "Bismuth film electrodes for heavy metals determination," *Microsystem Technologies*, vol. 14, no. 4-5, pp. 491–498, 2008.

[18] D. Demetriades, A. Economou, and A. Voulgaropoulos, "A study of pencil-lead bismuth-film electrodes for the determination of trace metals by anodic stripping voltammetry," *Analytica Chimica Acta*, vol. 519, no. 2, pp. 167–172, 2004.

[19] M. O. Salles, A. P. R. De Souza, J. Naozuka, P. V. De Oliveira, and M. Bertotti, "Bismuth modified gold microelectrode for Pb(II) determination in wine using alkaline medium," *Electroanalysis*, vol. 21, no. 12, pp. 1439–1442, 2009.

[20] I. Švancara, L. Baldrianová, E. Tesařová et al., "Recent advances in anodic stripping voltammetry with bismuth-modified carbon paste electrodes," *Electroanalysis*, vol. 18, no. 2, pp. 177–185, 2006.

[21] J. Wang, J.-W. Mo, S. Li, and J. Porter, "Comparison of oxygen-rich and mediator-based glucose-oxidase carbon-paste electrodes," *Analytica Chimica Acta*, vol. 441, no. 2, pp. 183–189, 2001.

[22] L. Liu, Z. Ma, X. Zhu, R. Zeng, S. Tie, and J. Nan, "Electrochemical behavior and simultaneous determination of catechol, resorcinol, and hydroquinone using thermally reduced carbon nano-fragment modified glassy carbon electrode," *Analytical Methods*, vol. 8, no. 3, pp. 605–613, 2016.

[23] E. A. Hutton, S. B. Hočevar, and B. Ogorevc, "Ex situ preparation of bismuth film microelectrode for use in electrochemical stripping microanalysis," *Analytica Chimica Acta*, vol. 537, no. 1-2, pp. 285–292, 2005.

[24] M. Lu, N. V. Rees, and R. G. Compton, "Determination of Sb(V) Using Differential Pulse Anodic Stripping Voltammetry at an Unmodified Edge Plane Pyrolytic Graphite Electrode," *Electroanalysis*, vol. 24, no. 6, pp. 1306–1310, 2012.

[25] E. Hull, R. Piech, and W. W. Kubiak, "Iridium oxide film electrodes for anodic stripping voltammetry," *Electroanalysis*, vol. 20, no. 19, pp. 2070–2075, 2008.

[26] G. H. Hwang, W. K. Han, J. S. Park, and S. G. Kang, "Determination of trace metals by anodic stripping voltammetry using a bismuth-modified carbon nanotube electrode," *Talanta*, vol. 76, no. 2, pp. 301–308, 2008.

[27] F. Tseliou, A. Avgeropoulos, P. Falaras, and M. I. Prodromidis, "Low dimensional Bi2Te3-graphene oxide hybrid film-modified electrodes for ultra-sensitive stripping voltammetric detection of Pb(II) and Cd(II)," *Electrochimica Acta*, vol. 231, pp. 230–237, 2017.

[28] P. A. Dimovasilis and M. I. Prodromidis, "Bismuth-dispersed xerogel-based composite films for trace Pb(II) and Cd(II) voltammetric determination," *Analytica Chimica Acta*, vol. 769, pp. 49–55, 2013.

[29] D. Riman, D. Jirovsky, J. Hrbac, and M. I. Prodromidis, "Green and facile electrode modification by spark discharge: Bismuth oxide-screen printed electrodes for the screening of ultra-trace Cd(II) and Pb(II)," *Electrochemistry Communications*, vol. 50, pp. 20–23, 2015.

[30] M. Frena, I. Campestrini, O. C. De Braga, and A. Spinelli, "In situ bismuth-film electrode for square-wave anodic stripping voltammetric determination of tin in biodiesel," *Electrochimica Acta*, vol. 56, no. 12, pp. 4678–4684, 2011.

[31] P. H. C. P. Tavares and P. J. S. Barbeira, "Influence of pencil lead hardness on voltammetric response of graphite reinforcement carbon electrodes," *Journal of Applied Electrochemistry*, vol. 38, no. 6, pp. 827–832, 2008.

[32] G. D. Pierini, M. F. Pistonesi, M. S. Di Nezio, and M. E. Centurión, "A pencil-lead bismuth film electrode and chemometric tools for simultaneous determination of heavy metals in propolis samples," *Microchemical Journal*, vol. 125, pp. 266–272, 2016.

[33] K. Asadpour-Zeynali and P. Najafi-Marandi, "Bismuth modified disposable pencil-lead electrode for simultaneous determination of 2-nitrophenol and 4-nitrophenol by net analyte signal standard addition method," *Electroanalysis*, vol. 23, no. 9, pp. 2241–2247, 2011.

[34] F. Arduini, J. Q. Calvo, G. Palleschi, D. Moscone, and A. Amine, "Bismuth-modified electrodes for lead detection," *TrAC - Trends in Analytical Chemistry*, vol. 29, no. 11, pp. 1295–1304, 2010.

[35] X. Zhang, Y. Zhang, D. Ding et al., "On-site determination of Pb2+ and Cd2+ in seawater by double stripping voltammetry with bismuth-modified working electrodes," *Microchemical Journal*, vol. 126, pp. 280–286, 2016.

[36] D. Li, J. Jia, and J. Wang, "A study on the electroanalytical performance of a bismuth film-coated and nafion-coated glassy carbon electrode in alkaline solutions," *Microchimica Acta*, vol. 169, no. 3, pp. 221–225, 2010.

[37] L. Pinto and S. G. Lemos, "Multivariate optimization of the voltammetric determination of Cd, Cu, Pb and Zn at bismuth film. Application to analysis of biodiesel," *Microchemical Journal*, vol. 110, pp. 417–424, 2013.

[38] C. Chen, X. Niu, Y. Chai et al., "Determination of lead(II) using screen-printed bismuth-antimony film electrode," *Electroanalysis*, vol. 25, no. 6, pp. 1446–1452, 2013.

[39] R. T. Kachoosangi, C. E. Banks, X. Ji, and R. G. Compton, "Electroanalytical determination of cadmium(II) and lead(II) using an in-situ bismuth film modified edge plane pyrolytic graphite electrode," *Analytical Sciences*, vol. 23, no. 3, pp. 283–

289, 2007.

[40] K. C. Armstrong, C. E. Tatum, R. N. Dansby-Sparks, J. Q. Chambers, and Z.-L. Xue, "Individual and simultaneous determination of lead, cadmium, and zinc by anodic stripping voltammetry at a bismuth bulk electrode," *Talanta*, vol. 82, no. 2, pp. 675–680, 2010.

[41] N. Serrano, A. González-Calabuig, and M. Del Valle, "Crown ether-modified electrodes for the simultaneous stripping voltammetric determination of Cd(II), Pb(II) and Cu(II)," *Talanta*, vol. 138, pp. 130–137, 2015.

[42] R. O. Kadara and I. E. Tothill, "Development of disposable bulk-modified screen-printed electrode based on bismuth oxide for stripping chronopotentiometric analysis of lead (II) and cadmium (II) in soil and water samples," *Analytica Chimica Acta*, vol. 623, no. 1, pp. 76–81, 2008.

[43] N. Lezi, A. Economou, P. A. Dimovasilis, P. N. Trikalitis, and M. I. Prodromidis, "Disposable screen-printed sensors modified with bismuth precursor compounds for the rapid voltammetric screening of trace Pb(II) and Cd(II)," *Analytica Chimica Acta*, vol. 728, pp. 1–8, 2012.

Quantitative Determination of Chlormequat Chloride Residue in Wheat using Surface-Enhanced Raman Spectroscopy

Shizhuang Weng,[1,2] Mengqing Qiu ⓘ,[1,2] Ronglu Dong,[3] Fang Wang,[1,2] Jinling Zhao ⓘ,[1,2] Linsheng Huang,[1,2] and Dongyan Zhang[1,2]

[1]Anhui Engineering Laboratory of Agro-Ecological Big Data, Anhui University, 111 Jiulong Road, Hefei 230601, China
[2]Science and Technology on Communication Networks Laboratory, Shijiazhuang 050000, China
[3]Hefei Institute of Physical Science, Chinese Academy of Sciences, 350 Shushanhu Road, Hefei 230031, China

Correspondence should be addressed to Jinling Zhao; apcomm_2010@163.com

Academic Editor: Richard G. Brereton

A simple and sensitive method for detection of chlormequat chloride residue in wheat was developed using surface-enhanced Raman spectroscopy (SERS) coupled with chemometric methods on a portable Raman spectrometer. Pretreatment of wheat samples was performed using a two-step extraction procedure. Effective and uniform active substrate (gold nanorods) was prepared and mixed with the sample extraction solution for SERS measurement. The limit of detection for chlormequat chloride in wheat extracting solutions and wheat samples was 0.25 mg/L and 0.25 μg/g, which was far below the maximum residual value in wheat of China. Then, support vector regression (SVR) and kernel principal component analysis (KPCA), multiple linear regression, and partial least squares regression were employed to develop the regression models for quantitative analysis of chlormequat chloride residue with spectra around the characteristic peaks at 666, 713, and 853 cm^{-1}. As for the residue in wheat, the predicted recovery of established optimal model was in the range of 94.7% to 104.6%, and the standard deviation was about 0.007 mg/L to 0.066 mg/L. The results demonstrated that SERS, SVR, and KPCA can provide the accurate and quantitative determination for chlormequat chloride residue in wheat.

1. Introduction

Plant growth regulator can increase crop production, improve quality, and enhance stress resistance [1, 2] through regulating cell growth, cell division, rhizogenesis, germination, blossom and maturation. Chlormequat chloride [3, 4] is an excellent plant growth regulator which is widely applied in wheat, rice, cotton, tobacco, corn, and tomato. Taking advantage of inhibition effect of cell elongation, chlormequat chloride improves the resistance of crops to drought, waterlogging, saline-alkali soil, or lodging. However, the residue in agricultural grains induced by the unreasonable and excessive application is health hazard for humans and animals, leading to economic losses to agricultural trade [5–7].

Chlormequat chloride detection methods as gas [8] or liquid chromatography [9, 10] coupled with mass spectrometry are highly selective and very accurate; thus, they are considered as standard detection methods. Nevertheless, these methods are not practical and convenient due to expensive and toxic reagent consumption, high analysis complexity, and requirement of large laboratory instruments and trained personnel [11, 12]. Meanwhile, considering the large-scale residue, the detection process must be simple and rapid. Therefore, a method employing portable equipment and simple detection procedure would be more suitable for detection of chlormequat chloride in grains.

Spectroscopic methods are promising tools for detection of farm chemical residue because they are simple, rapid, specific, and partially or completely automatic. The commonly used spectroscopic methods include near-infrared spectroscopy (NIR) [13, 14], Fourier-transform infrared spectroscopy (FTIR) [15, 16], Raman spectroscopy (RS) [17, 18], and surface-enhanced Raman spectroscopy (SERS) [19, 20]. NIR and FTIR are unsuitable for residue detection in grains because of severe interference from aqueous environments. RS can provide comprehensive and fingerprint information

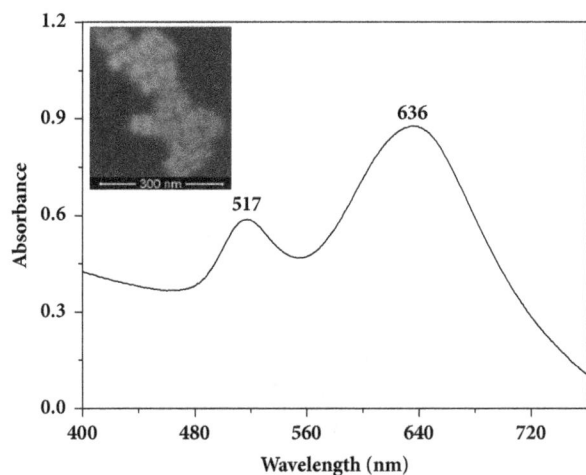

FIGURE 1: Ultraviolet-visible absorption spectrum of the prepared GNRs colloid; the inset is SEM image of GNRs.

of analyte without impact from aqueous phase, but its application is also limited by the low sensitivity for small cross-section of Raman scattering. By contrast, SERS is the most promising technique in trace detection because it greatly improved the sensitivity of RS through large enhancement of inelastic Raman scattering of molecule absorbed to the surface of nanoscale noble metal like silver, gold, and copper [11, 21]. Meanwhile, SERS inherits the advantage of RS with rapidity to provide fingerprint and comprehensive information. Due to these advantages, SERS is broadly applied in detection and discrimination of farm chemicals [22, 23], toxins [24], additives [25], drugs [26], and biomacromolecules [27]. Particularly, for farm chemicals, SERS has been used to detect isofenphos-methyl [12], chlorpyrifos [28], thiram [23], fenthion [29], triazophos [30], and so on. In addition, the intelligent analysis of spectra using chemometric methods can initiate acquisition of analyte information independent of professionals, which makes SERS easy and simple for popularization and application in detection. Polynomial fitting [31], derivative transformation [12], and asymmetric least squares [32] are often used to deduct baseline shift caused by the fluorescent background and other interference effects. Principal component analysis [12] and nonnegative factorization [33] are commonly adopted for extracting main information and reducing data dimensions. Some other methods such as multiple linear regression (MLR) [34], partial least squares regression (PLSR) [12], artificial neural network [35], and support vector regression (SVR) [36] are usually employed to develop the regression models for quantitative determination of substances with excellent predictability.

The objective of this study is to develop a simple and sensitive SERS method for quantitative determination of chlormequat chloride in wheat coupled with some chemometric methods on a portable Raman spectrometer, in which pretreatment of wheat samples is performed using a two-step extraction procedure. To the best of our knowledge, this paper is the first to report detection chlormequat chloride in grains using SERS technique.

2. Materials and Methods

2.1. Materials. Wheat samples were purchased from Hefei Zhougudui market. Chlormequat chloride powder (99.6%) was obtained from Beijing Puxi Technology Co., Ltd. Anhydrous methanol was acquired from Sinopharm Chemical Reagent Co., Ltd. Cetyltrimethylammonium bromide (CTAB), hydrogen tetrachloroaurate, trisodium citrate, L-ascorbic acid, sodium borohydride, and silver nitrite were purchased from Aladdin Industrial Corporation.

2.2. Sample Preparation. The pretreatment method for wheat was developed based on the extraction method in gas chromatography (GC). Wheat was first grinded using a pulverizer (Xinrui DFT-150, Changzhou, China) and filtered through 10-mesh sieves. Wheat powder of 5.00 g was added with 15 mL of methanol in 50 mL centrifuge tube and then vibrated for 10 min. The mixture was centrifuged at 4000 rpm for 3 min, and the supernatant was moved to the concentrated bottle. Wheat residue was extracted using 10 mL of methanol again, and the supernatant was also moved to the concentrated bottle. The supernatant was evaporated to dry on a Rotavapor (Yarong RE-52A, Shanghai, China) and redissolved in 5 ml of methanol.

Wheat extracting solutions containing different chlormequat chloride were then prepared. The obtained extraction solution was used to dissolve chlormequat chloride powder for getting the solution of 20, 10, 5, 2.5, 1, 0.5, and 0.25 mg/L. Additionally, to simulate actual residue, wheat powder was spiked with chlormequat chloride to yield final residue at 10, 5, 2.5, 1, 0.5, and 0.25 μg/g. The contaminated samples were extracted using the above pretreatment method.

2.3. SERS Measurement. The synthesis of gold nanorods (GNRs) was performed using a seed-mediated growth method previously developed by El-Sayed [37]. GNRs sol-solution was centrifuged at 8000 rpm for 10 min to get gray colloid, and 2 μL of GNRs colloid was dropped on silicon chip. After the droplet became dry, 2 μL of testing solution was dropped on the GNRs film. When the solvent was evaporated to dry, spectra were collected on a portable Raman spectrometer (B&WTEK, i-Raman785® Plus, USA) equipped with a 785 nm laser of 150 mW. The measurement was performed with 3 scans and exposure time of 5 s, and the spectral resolution was 2 cm^{-1} in the Raman shift range of 600 cm^{-1} to 1800 cm^{-1}. The spectra of 10 from five different spots on each sample were collected as the representative spectra. Five samples were measured for chlormequat chloride residue of each concentration.

Absorption spectra of the GNRs were recorded on an ultraviolet-visible (UV–Vis) spectrometer (UV-2600, Shimadzu, Japan). Morphologies of the GNRs were surveyed using the scanning electron microscope (SEM) image on a JSM 7500F microscope (JEOL Ltd., Tokyo, Japan). As shown in Figure 1, the GNRs exhibited two plasmon resonance bands of 517 and 636 nm which correspond to electron oscillations along the short and long axes of nanorods. SEM images revealed that the GNRs are ordered and uniform.

2.4. Chemometric Methods. The obtained spectra were first baseline-corrected using asymmetric least squares method to eliminate baseline and linear slope effects [32]. Subsequently, kernel principal component analysis (KPCA) was applied for the feature extraction of spectral data to obtain main information and reduce the dimension. KPCA is a nonlinear PCA developed with the kernel method. Concretely, KPCA projects spectra to the high-dimensional space and achieves separable data. Then, the obtained high-dimensional data is transformed into many principal variables (scores) using principle of principal component analysis. These variables are used to describe and replace the original spectra with advantage of weakening noise interference [38]. Radial basis function (RBF) was selected as the kernel function in KPCA for its high effectiveness in training process, and the effect of different width of kernel function (σ) on feature extraction was discussed.

To examine accurate and quantitative determination of analyte further, MLR, PLSR, and SVR were used to develop the regression models. MLR is a regression algorithm that is very efficient in building calibration models when the number of samples is more than that of variables. PLSR is one of the most robust and reliable tools in the development of a multivariate calibration model. Based on the linear algorithm, PLSR is often applied to predict a set of dependent variables from a large set of independent variables. PLSR decomposes the spectral array and concentration array with considering their relationships. Corresponding calculation relationships are strengthened for the better correction model. SVR is a variation of support vector machine with introduction of insensitive loss function. Despite finite sample, SVR still possesses excellent robustness and high sensitivity through balancing complexity and learning ability of model. Meanwhile, with the aid of kernel function, SVR can project data into the high-dimensional space for obtaining higher analysis accuracy. Moreover, RBF was also selected as the kernel function of SVR. Considering the performance of obtained regression models highly depends on penalty coefficient (C) in loss function and width of kernel function (σ), the optimal values of C and σ are obtained by traversing their empirical values [36]. The performances of the above models were quantitatively evaluated using 5-fold cross-validation method with root mean square error of cross-validation (RMSECV). All data analyses and validation of chemometric methods were performed in MATLAB 2013a (MathWorks Inc., Natick, MA, USA).

3. Results and Discussion

3.1. SERS of Chlormequat Chloride. The characteristic peaks reflect the information of molecular vibration and rotation, and these peaks are the basis for analysis and detection of substance using Raman or SERS technique. To determine the characteristic peaks of chlormequat chloride, pure chlormequat chloride powder was placed on the silicon wafer, and then Raman spectra were obtained through direct laser irradiation on it. The main Raman peaks of chlormequat chloride at 666, 713, 765, 853, and 1447 cm^{-1} were observed in Figure 2. According to relevant Raman peak assignment

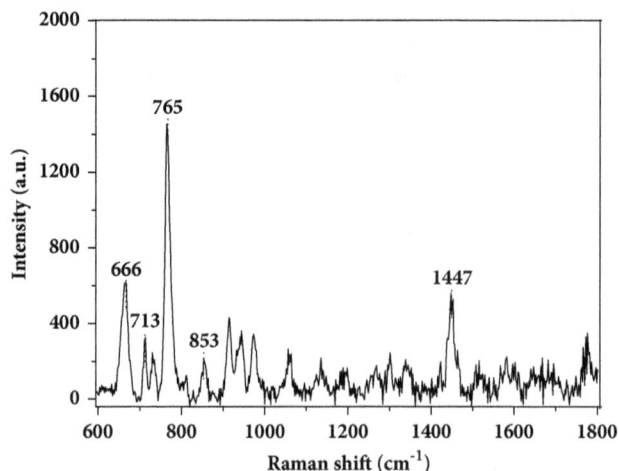

FIGURE 2: Raman spectra of pure chlormequat chloride powder.

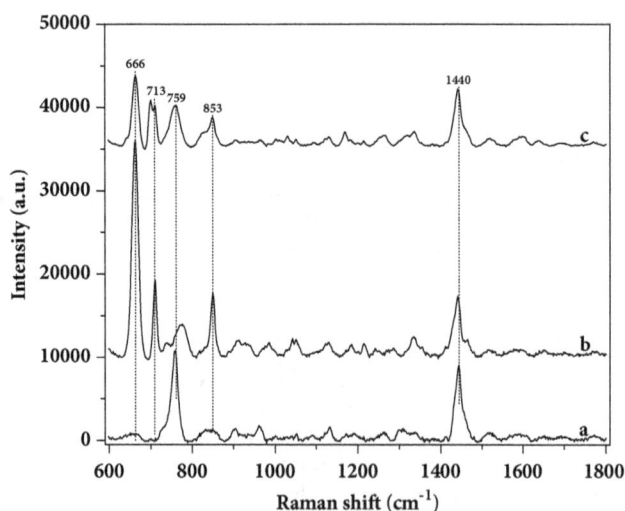

FIGURE 3: SERS spectra of gold nanorods (a), 100 mg/L of chlormequat chloride in methanol (b), and 20 mg/L of chlormequat chloride in wheat extraction solution (c).

and structure of chlormequat chloride molecule, the peaks at 666 and 765 cm^{-1} were attributed to C-Cl stretching vibration in synclinal and synperiplanar conformation. The bands at 713 and 1447 cm^{-1} can be associated with CH$_2$ oscillating vibration and CH$_2$-Cl bending vibration (in plane), respectively. Furthermore, the peak at 853 cm^{-1} was assigned to C-N symmetric stretching vibration.

However, the characteristic bands of SERS of molecule in complex media may have changes, which is mainly due to influence of Raman active substrate and background signals of complex media. Then, SERS spectra of GNRs, 100 mg/L of chlormequat chloride in methanol, and 20 mg/L of chlormequat chloride in wheat extraction solution were measured and shown in Figure 3. As shown in figure, the bands at 759 and 1440 cm^{-1} were for SERS of GNRs, which were attributed to CTAB residue. Furthermore, the two bands influenced the appearance of peaks at 765 and 1447 cm^{-1} of chlormequat chloride. Conversely, the other peaks at 666, 713,

TABLE 1: Predicted results of the model developed using chemometric methods.

Data	MLR	PLSR	KPCA+SVR	
	RMSECV (mg/L)	RMSECV (mg/L)	σ in KPCA	RMSECV (mg/L)
Spectra of 653-683, 705-728, and 847-872 cm^{-1}	0.3757	0.3758	1000	4.235
			5000	0.0299
			8000	0.0268
			10000	0.1131

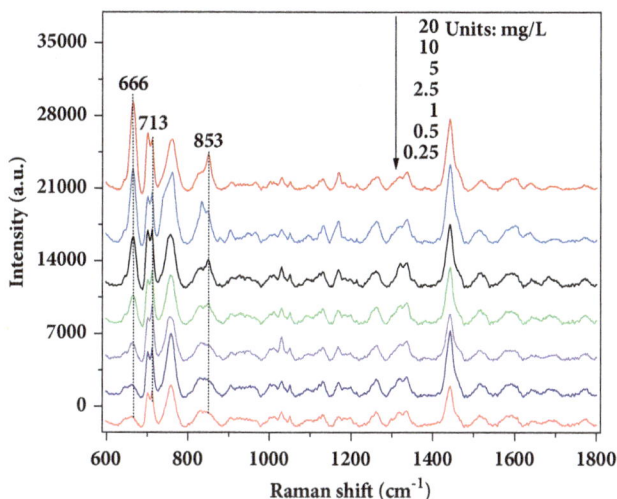

FIGURE 4: Spectra of 20, 10, 5, 2.5, 1, 0.5, or 0.25 mg/L of chlormequat chloride in wheat extraction solution.

and 853 cm^{-1} were not affected by the substrate and complex media. Therefore, the peaks at 666, 713, and 853 cm^{-1} were the key features for detection of chlormequat chloride residue in wheat extraction solution using SERS.

Then, SERS spectra of wheat extraction solution with 20, 10, 5, 2.5, 1, 0.5, or 0.25 mg/L of chlormequat chloride were measured with the uniform GNRs (Figure 4). As seen in the figure, the intensity of characteristic peaks weakened concomitantly with the decrease in concentration, which suggests SERS have potential for quantitative analysis. When the concentration of chlormequat chloride in solution was 0.25 mg/L, the peak at 713 cm^{-1} was still obvious, but the peaks at 666 and 853 cm^{-1} were just dimly visible. The phenomenon indicated that SERS technique with GNRs can detect the limit of detection for 0.25 mg/L of chlormequat chloride in wheat extraction solution.

3.2. Spectral Analysis Using Chemometric Methods. Intelligent analysis of spectra using chemometric methods can automatically obtain the information of substances without intervention of professionals, and this process is of significance for simple and rapid detection. SERS spectra are of high dimension and carry useless information for target analyte. The appropriate variable selection and feature extraction can improve the analysis results. Considering the fingerprint properties of SERS, the spectra around characteristic peaks were selected for the intelligent analysis, and the interference

can be avoided from the irrelevant information in spectra of other ranges. In particular, for SERS of chlormequat chloride in wheat extraction, the spectra of 653–683, 705–728, and 847–872 cm^{-1} were selected for the subsequent analysis. Then, KPCA with RBF was adopted to extract the principal feature of processed spectra. Feature extraction was highly dependent on σ in RBF, and the effects of different σ were discussed. Figure 5 shows the scatter of first two principal component scores obtained by KPCA with σ of 1000, 5000, 8000, and 10000. When σ was 1000, the corresponding score of wheat extraction with 20, 10, and 5 mg/L chlormequat chloride overlapped one another, which suggests bad subsequent results. However, for other larger values, the scatter of each category was separated well. Afterward, the obtained first two principal component scores were employed to develop the regression models for quantitative determination, and the model performance was evaluated with RMSECV (Table 1). As can be seen from Table 1, the RMSECV value of the linear model established with MLR and PLSR is relatively large, which may lead to the inaccurate prediction results. Meanwhile, the model obtained by SVR and KPCA with σ of 1000 was worse than others and consistent with the above assumption. Too large σ (10000) also caused the quantitative analysis results to yield poor results [38]. Accordingly, SVR and KPCA with σ of 8000 was used for quantitative determination of chlormequat chloride, and the predicted error of the optimal model is shown in Figure 6. From the figure, the concentration of all samples was accurately predicted with small error, and RMSECV of the model was 0.0268 mg/L. In conclusion, SERS and SVR with KPCA can provide an accurate detection method for chlormequat chloride residue in wheat solution. Subsequently, the optimal established model would be used to predict concentration of residue in the real case.

In addition, an unbiased estimation for generalization of the model was conducted with an independent testing set. The independent testing set was spectra of wheat extraction with 15, 8, 4, and 2 mg/L obtained through remeasurement, and the representative spectra were shown in Figure 7(a). From figure it is known that the representative spectra were almost identical to the previously measured spectra. The predicted error (RMSECV) for the new testing set with the optimal model was about 0.2110 mg/L (Figure 7(b)). Meanwhile, standard deviation was from 0.052 mg/L to 0.102 mg/L, and the predicted recovery was 97.4 %–110.3 % in Table 2. The above results indicated that concentration of residue can be accurately predicted, and the model established by SVR and KPCA had good generality.

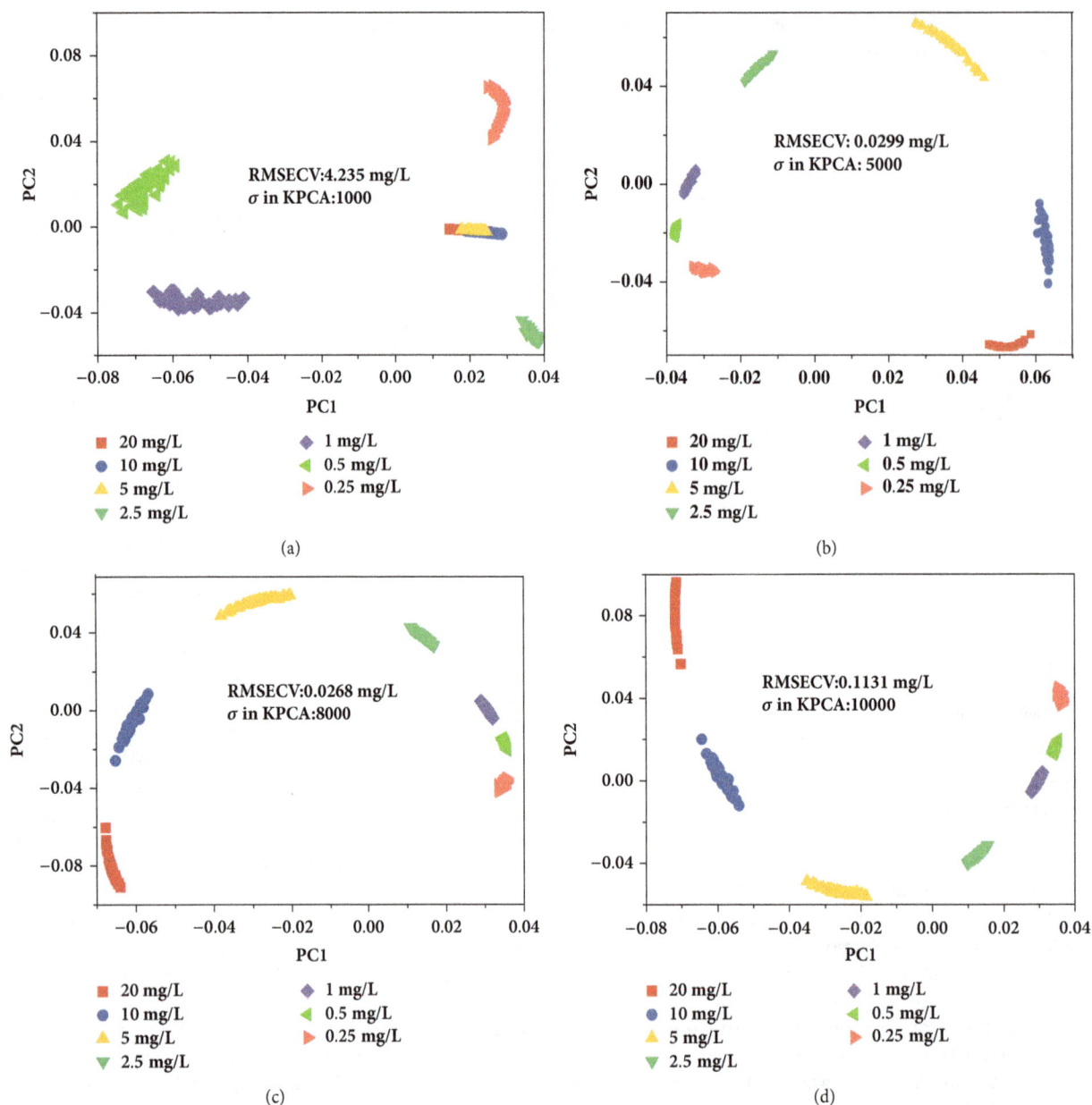

FIGURE 5: Scatter plot of first two principal component scores obtained by KPCA with σ of 1000 (a), 5000 (b), 8000 (c), and 10000 (d); PC1: the first principal component. PC2: the second principal component.

TABLE 2: Predicted results of 15, 8, 4, and 2 mg/L of chlormequat chloride in wheat extraction solution using SERS, SVR, and KPCA.

Spiked value (μg/g)	Mean predicted value (mg/L)	Standard deviation (mg/L)	Recovery (%)
15	14.87	0.066	99.1
8	8.18	0.091	102.3
4	3.90	0.052	97.4
2	2.21	0.102	110.3

3.3. Quantification of Chlormequat Chloride Residue in Wheat Samples. To simulate actual residue, wheat powder was spiked with chlormequat chloride to yield final residue at 10, 5, 2.5, 1, 0.5, and 0.25 μg/g. The contaminated samples were extracted using the two-step procedure, and the obtained

extraction solution was directly used for SERS measurement. For the residue of each concentration, 50 spectra were collected from five samples, respectively. The representative spectra are shown in Figure 8. From figure, spectra of chlormequat chloride residue in wheat were highly consistent

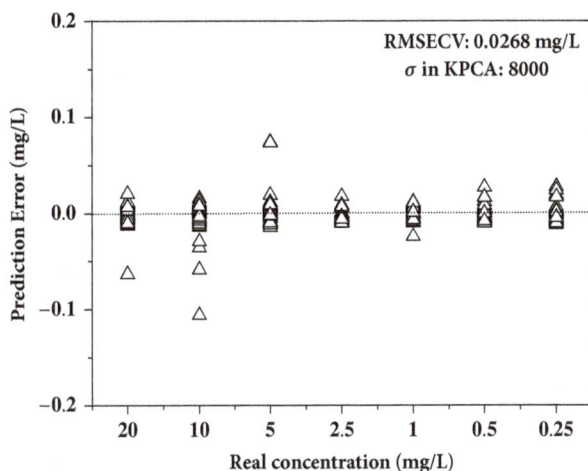

FIGURE 6: Predicted error of the optimal model built using SVR and KPCA with σ of 8000.

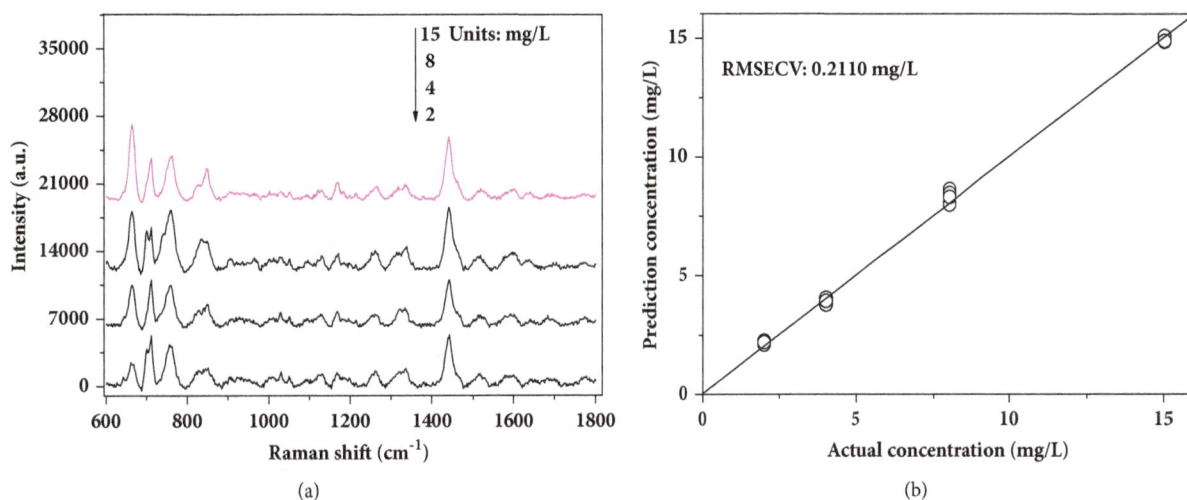

(a)

(b)

FIGURE 7: Spectra of 15, 8, 4, and 2 mg/L of chlormequat chloride in wheat extraction solution (a), predicted results by using SVR and KPCA (b).

with the spectra of residue in extraction solution, and the characteristic peaks at 713, 666, and 853 cm^{-1} were still obvious and feasible for quantification of analyte. However, a small difference in spectral intensity can be observed, which depended on the extraction efficiency of pretreatment methods for residue in wheat samples.

Afterward, all the spectra were processed using KPCA, and the first two principal component scores were used to predict the sample concentration combining with the established model. The experiment results are shown in Table 3. The predicted recovery was in the range of 94.7 % to 104.6 %, and standard deviation was from 0.007 mg/L to 0.066 mg/L. Results proved that the proposed pretreatment method was feasible and effective for extraction of chlormequat chloride residue in wheat. In addition, the lowest tested concentration of 0.25 μg/g was far below maximum residue limit of chlormequat chloride in wheat (5 μg/g). These results also demonstrated SERS and SVR with KPCA can realize accurate quantification of residue with good repeatability and high

sensitivity. Furthermore, the portable Raman spectrometer made the quantitative determination easy and efficient to be performed. In the future, the presented method can be applied for the detection of various farm chemicals in other grains.

4. Conclusions

In this work, a method for detection of chlormequat chloride in wheat was developed using SERS and chemometric methods on a portable Raman spectrometer. The extraction of residue in wheat was performed using a two-step procedure originated from GC detection. As for the spiked wheat samples, the optimal predicted recovery was in the range of 94.7 % to 104.6 %, and standard deviation was from 0.007 mg/L to 0.066 mg/L. These results indicated that the present method is an effective and feasible approach for determination of chlormequat chloride residue in wheat. Meanwhile, with aid of a portable Raman spectrometer, the present method could

TABLE 3: Predicted results of chlormequat chloride in wheat using SERS, SVR, and KPCA.

Spiked value (μg/g)	Mean predicted value (mg/L)	Standard deviation (mg/L)	Recovery (%)
10	9.96	0.066	99.6
5	4.74	0.042	94.7
2.5	2.614	0.064	104.6
1	1.012	0.014	101.2
0.5	0.479	0.007	95.8
0.25	0.242	0.025	96.8

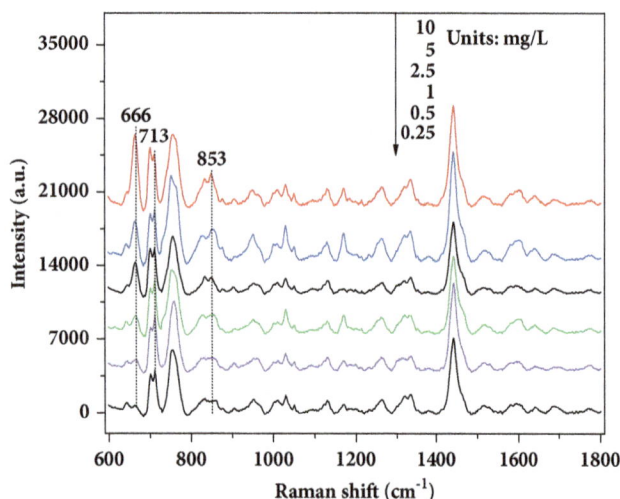

FIGURE 8: Spectra of wheat samples spiked with chlormequat chloride at 10, 5, 2.5, 1, 0.5, or 0.25 μg/g.

be executed onsite, which is suitable for rapid residue analysis in grains. However, spectral variation induced by instability of substrate and differences in sample pretreatment should be avoided and resolved prior to application of SERS. In conclusion, SERS with chemometric methods is a potentially powerful approach for detecting chlormequat chloride or other toxic residues in grains which can greatly help improve the safety and quality of agricultural products.

Acknowledgments

This study is supported by Natural Science Foundation of Anhui Province (nos. 1708085QF134 and 1604a0702016), Natural Science Research Project of Anhui Provincial Education Department (no. KJ2017A006), National Natural Science Foundation of China (nos. 31401285 and 61475163), National Key Research and Development Program (no. 4014YFD0800904), Anhui Provincial Science and Technology Project (no. 17030710162), and Open Foundation of Science and Technology on Communication Networks Laboratory (no. XX17641X011-02).

References

[1] M. B. Arnao and J. Hernández-Ruiz, "Melatonin: Plant growth regulator and/or biostimulator during stress?" *Trends in Plant Science*, vol. 19, no. 12, pp. 789–797, 2014.

[2] B. Guo, B. H. Abbasi, A. Zeb, L. L. Xu, and Y. H. Wei, "Thidiazuron: a multi-dimensional plant growth regulator," *African Journal of Biotechnology*, vol. 10, no. 45, pp. 8984–9000, 2011.

[3] H. P. Anosheh, Y. Emam, M. Ashraf, and M. R. Foolad, "Exogenous application of salicylic acid and chlormequat chloride alleviates negative effects of drought stress in wheat," *Advanced Studies in Biology*, vol. 4, no. 11, pp. 501–520, 2012.

[4] A. R. Gurmani, A. Bano, S. U. Khan, J. Din, and J. L. Zhang, "Alleviation of salt stress by seed treatment with abscisic acid (ABA), 6-benzylaminopurine (BA) and chlormequat chloride (CCC) optimizes ion and organic matter accumulation and increases yield of rice (oryza sativa L.)," *Australian Journal of Crop Science*, vol. 5, no. 10, pp. 1278–1285, 2011.

[5] D. Huang, S. Wu, X. Hou et al., "The skeletal developmental toxicity of chlormequat chloride and its underlying mechanisms," *Toxicology*, vol. 381, pp. 1–9, 2017.

[6] B. Xiagedeer, S. Wu, Y. Liu, and W. Hao, "Chlormequat chloride retards rat embryo growth in vitro," *Toxicology in Vitro*, vol. 34, pp. 274–282, 2016.

[7] P. Nisse, R. Majchrzak, J. P. Kahn, P. A. Mielcarek, and M. Mathieu-Nolf, "Chlormequat poisoning is not without risk: Examination of seven fatal cases," *Journal of Forensic and Legal Medicine*, vol. 36, pp. 1–3, 2015.

[8] Y. Xu, S. Jiang, H. Fu et al., "Determination of 7 plant growth regulators in apple, tomato, maize by gas chromatography-tandem mass spectrometry," *Agrochemicals*, vol. 53, no. 2, pp. 113–115, 2014.

[9] L. Ma, H. Zhang, W. Xu et al., "Simultaneous Determination of 15 Plant Growth Regulators in Bean Sprout and Tomato with Liquid Chromatography-Triple Quadrupole Tandem Mass Spectrometry," *Food Analytical Methods*, vol. 6, no. 3, pp. 941–951, 2013.

[10] Y. Zhou, Y. Han, H. Tian, and B. Chen, "Determination of Chlormequat Residues in Vegetables by High Performance Liquid Chromatography-Mass Spectrometry," *Journal of Food Science*, vol. 14, p. 46, 2010.

[11] K.-M. Lee, T. J. Herrman, Y. Bisrat, and S. C. Murray, "Feasibility of surface-enhanced raman spectroscopy for rapid detection of aflatoxins in maize," *Journal of Agricultural and Food Chemistry*, vol. 62, no. 19, pp. 4466–4474, 2014.

[12] D. Liu, Y. Han, L. Zhu et al., "Quantitative Detection of Isofenphos-Methyl in Corns Using Surface-Enhanced Raman Spectroscopy (SERS) with Chemometric Methods," *Food Analytical Methods*, vol. 10, no. 5, pp. 1202–1208, 2017.

[13] M.-T. Sánchez, K. Flores-Rojas, J. E. Guerrero, A. Garrido-Varo, and D. Pérez-Marín, "Measurement of pesticide residues in peppers by near-infrared reflectance spectroscopy," *Pest Management Science*, vol. 66, no. 6, pp. 580–586, 2010.

[14] B. Stenberg, R. A. V. Rossel, A. M. Mouazen, and J. Wetterlind, "Chapter five-visible and near infrared spectroscopy in soil science," *Elsevier Science Technology*, vol. 107, no. 107, pp. 163–215, 2010.

[15] G. Xiao, D. Dong, T. Liao et al., "Detection of Pesticide (Chlorpyrifos) Residues on Fruit Peels Through Spectra of Volatiles by FTIR," *Food Analytical Methods*, vol. 8, no. 5, pp. 1341–1346, 2015.

[16] S. Jawaid, F. N. Talpur, S. T. H. Sherazi, S. M. Nizamani, and A. A. Khaskheli, "Rapid detection of melamine adulteration in dairy milk by SB-ATR-Fourier transform infrared spectroscopy," *Food Chemistry*, vol. 141, no. 3, pp. 3066–3071, 2013.

[17] J. Moros, J. A. Lorenzo, P. Lucena, L. M. Tobaria, and J. J. Laserna, "Simultaneous Raman spectroscopy-laser-induced breakdown spectroscopy for instant standoff analysis of explosives using a mobile integrated sensor platform," *Analytical Chemistry*, vol. 82, no. 4, pp. 1389–1400, 2010.

[18] S. Dhakal, Y. Li, Y. Peng, K. Chao, J. Qin, and L. Guo, "Prototype instrument development for non-destructive detection of pesticide residue in apple surface using Raman technology," *Journal of Food Engineering*, vol. 123, pp. 94–103, 2014.

[19] Y. Y. Zhang, Y. Q. Huang, F. L. Zhai, R. Du, Y. D. Liu, and K. Q. Lai, "Analyses of enrofloxacin, furazolidone and malachite green in fish products with surface-enhanced Raman spectroscopy," *Food Chemistry*, vol. 135, no. 2, pp. 845–850, 2012.

[20] J. F. Li, Y. F. Huang, Y. Ding et al., "Shell-isolated nanoparticle-enhanced Raman spectroscopy," *Nature*, vol. 464, no. 7287, pp. 392–395, 2010.

[21] P. Wang, L. Wu, Z. Lu et al., "Gecko-Inspired Nanotentacle Surface-Enhanced Raman Spectroscopy Substrate for Sampling and Reliable Detection of Pesticide Residues in Fruits and Vegetables," *Analytical Chemistry*, vol. 89, no. 4, pp. 2424–2431, 2017.

[22] J. Kubackova, G. Fabriciova, P. Miskovsky, D. Jancura, and S. Sanchez-Cortes, "Sensitive surface-enhanced Raman spectroscopy (SERS) detection of organochlorine pesticides by alkyl dithiol-functionalized metal nanoparticles-induced plasmonic hot spots," *Analytical Chemistry*, vol. 87, no. 1, pp. 663–669, 2015.

[23] P. Guo, D. Sikdar, X. Huang et al., "Plasmonic core-shell nanoparticles for SERS detection of the pesticide thiram: Size-and shape-dependent Raman enhancement," *Nanoscale*, vol. 7, no. 7, pp. 2862–2868, 2015.

[24] G. C. Phan-Quang, H. K. Lee, I. Y. Phang, and X. Y. Ling, "Plasmonic Colloidosomes as Three-Dimensional SERS Platforms with Enhanced Surface Area for Multiphase Sub-Microliter Toxin Sensing," *Angewandte Chemie International Edition*, vol. 54, no. 33, pp. 9691–9695, 2015.

[25] V. Peksa, M. Jahn, L. Štolcová et al., "Quantitative SERS analysis of azorubine (E 122) in sweet drinks," *Analytical Chemistry*, vol. 87, no. 5, pp. 2840–2844, 2015.

[26] R. Dong, S. Weng, L. Yang, and J. Liu, "Detection and direct readout of drugs in human urine using dynamic surface-enhanced Raman spectroscopy and support vector machines," *Analytical Chemistry*, vol. 87, no. 5, pp. 2937–2944, 2015.

[27] C. Guo, G. N. Hall, J. B. Addison, and J. L. Yarger, "Gold nanoparticle-doped silk film as biocompatible SERS substrate," *RSC Advances*, vol. 5, no. 3, pp. 1937–1942, 2015.

[28] Q. Xu, X. Guo, L. Xu et al., "Template-free synthesis of SERS-active gold nanopopcorn for rapid detection of chlorpyrifos residues," *Sensors and Actuators B: Chemical*, vol. 241, pp. 1008–1013, 2017.

[29] X. Li, S. Zhang, Z. Yu, and T. Yang, "Surface-enhanced Raman spectroscopic analysis of phorate and fenthion pesticide in apple skin using silver nanoparticles," *Applied Spectroscopy*, vol. 68, no. 4, pp. 483–487, 2014.

[30] M. Fan, F. Cheng, C. Wang et al., "SERS optrode as a "fishing rod" to direct pre-concentrate analytes from superhydrophobic surfaces," *Chemical Communications*, vol. 51, no. 10, pp. 1965–1968, 2015.

[31] S. He, W. Xie, W. Zhang et al., "Multivariate qualitative analysis of banned additives in food safety using surface enhanced Raman scattering spectroscopy," *Spectrochimica Acta Part A: Molecular and Biomolecular Spectroscopy*, vol. 137, pp. 1092–1099, 2015.

[32] S. Baek, A. Park, Y. Ahn, and J. Choo, "Baseline correction using asymmetrically reweighted penalized least squares smoothing," *Analyst*, vol. 140, no. 1, pp. 250–257, 2015.

[33] X. Luo, M. Zhou, Y. Xia, and Q. Zhu, "An efficient non-negative matrix-factorization-based approach to collaborative filtering for recommender systems," *IEEE Transactions on Industrial Informatics*, vol. 10, no. 2, pp. 1273–1284, 2014.

[34] T. Janči, D. Valinger, J. Gajdoš Kljusurić, L. Mikac, S. Vidaček, and M. Ivanda, "Determination of histamine in fish by Surface Enhanced Raman Spectroscopy using silver colloid SERS substrates," *Food Chemistry*, vol. 224, pp. 48–54, 2015.

[35] S. Seifert, V. Merk, and J. Kneipp, "Identification of aqueous pollen extracts using surface enhanced Raman scattering (SERS) and pattern recognition methods," *Journal of Biophotonics*, vol. 9, no. 1-2, pp. 181–189, 2016.

[36] S. Li, Y. Zhang, J. Xu et al., "Noninvasive prostate cancer screening based on serum surface-enhanced Raman spectroscopy and support vector machine," *Applied Physics Letters*, vol. 105, no. 9, p. 091104, 2014.

[37] B. Nikoobakht and M. A. El-Sayed, "Preparation and growth mechanism of gold nanorods (NRs) using seed-mediated growth method," *Chemistry of Materials*, vol. 15, no. 10, pp. 1957–1962, 2003.

[38] L. J. Cao, K. S. Chua, W. K. Chong, H. P. Lee, and Q. M. Gu, "A comparison of PCA, KPCA and ICA for dimensionality reduction in support vector machine," *Neurocomputing*, vol. 55, no. 1-2, pp. 321–336, 2003.

3

Host–Guest Extraction of Heavy Metal Ions with p-t-Butylcalix[8]arene from Ammonia or Amine Solutions

Md. Hasan Zahir [iD],[1] Shakhawat Chowdhury,[2]
Md. Abdul Aziz,[3] and Mohammad Mizanur Rahman[4]

[1]Center of Research Excellence in Renewable Energy, Research Institute, King Fahd University of Petroleum and Minerals, Dhahran 31261, Saudi Arabia
[2]Department of Civil and Environmental Engineering, Water Research Group, King Fahd University of Petroleum and Minerals, Dhahran 31261, Saudi Arabia
[3]Center of Research Excellence in Nanotechnology, King Fahd University of Petroleum and Minerals, Dhahran 31261, Saudi Arabia
[4]Center of Research Excellence in Corrosion, King Fahd University of Petroleum and Minerals, Dhahran 31261, Saudi Arabia

Correspondence should be addressed to Md. Hasan Zahir; hzahir@kfupm.edu.sa

Academic Editor: Seyyed E. Moradi

The capacities of the p-t-butylcalix[8]arene (abbreviated as H_8L) host to extract toxic divalent heavy metal ions and silver from aqueous solution phases containing ammonia or ethylene diamine to an organic phase (nitrobenzene, dichloromethane, or chloroform) were carried out. When the metal ions were extracted from an aqueous ammonia solution, the metal ion selectivity for extraction was found to decrease in the order $Cd^{2+} > Ni^{2+} > Cu^{2+} > Ag^+ > Co^{2+} > Zn^{2+}$. When the aqueous phase contained ethylene diamine, excellent extraction efficiencies of 97% and 90% were observed for the heavy metal ions Cu^{2+} and Cd^{2+}, respectively. Under the same conditions the extraction of octahedral type metal ions, namely, Co^{2+} and Ni^{2+}, was suppressed. The extraction of transition metal cations by H_8L in ammonia and/or amine was found to be pH dependent. Detailed analysis of extraction behavior was investigated by slope analysis, the continuous variation method, and by loading tests.

1. Introduction

Host–guest chemistry has attracted great attention in the field of separation and/or extraction of alkaline, rare earth, and divalent heavy metal ions with calixarenes and their ester derivatives. Calixarenes have unusual capabilities to identify and distinguish between different ions and molecules, which makes them appropriate to use as specific receptors [1–3]. Calixarenes are macrocyclic phenolic oligomers with phenolic hydroxyl groups, which are able to coordinate the metal ions very tightly. As a result, the aromatic phenolic rings can form a cavity to integrate the guest metal ion. Recently, Yusof et al. reported that the heavy metal ions could be included into the host calix[4]resorcinarenes cavity in water-chloroform extraction systems [4, 5]. Calix[6]arenes can be also modified with a carboxylic acid as host in the

host–guest extraction of immunoglobulin G (IgG) [6]. In fact, a precise affinity can be created for specific ions and/or molecules by modification the hydroxyl functional group and/or by creating a new cavity size [7]. In calixarenes, the cavity size, position and type of donor groups, and molecular flexibility lead to their high potential for the complexation and extraction of metal ions. Ludwig et al. reported the impact of calixarenes in analytical chemistry and chemical separation technology in a review article [8].

The modification of p-t-butylcalix[n]arenes (n = 4, 6, 8) has been extensively investigated, facilitating the synthesis of a variety of host compounds with varying shapes and sizes that have been shown to be valuable in ion or molecular recognition studies [9, 10]. In particular, p-t-butylcalix[8]arene (abbreviated as H_8L) has shown interesting complexing properties towards C_{60}-fullerene, cesium, or strontium cations. This means that new synthetic routes with different functionalized derivatives of H_8L could be more versatile [11].

H_8L can form host–guest complexes through hydrophobic and π–π interactions within the cavities of π-donors composed of benzene rings, polycyclic aromatic hydrocarbons, anthraquinones, phenol regioisomers, and fullerene π-systems, due to their considerable electron affinity [12]. From a coordination standpoint, the reaction of H_8L with metal ions is more complex than that of H_6L; however, already few calix[8]crowns have been reported in the open literature [13, 14]. Derivatives of calix[n]arenes could be formed by inserting binding groups such as amines and/or alcohols into the upper and lower rim positions of the calixarene moiety [15]. It has been reported that proton transfer from an OH-group of the parent calixarene to an amine might occur during the reaction period, eventually resulting in a new compound from the association and inclusion of the amine into the cavity of the calixarene [16]. Finally, the abovementioned studies have established that calixarenes can effectively react with most metal ions and the new organometallic compounds can be obtained in good yields [8–10, 17].

H_8L has been utilized in the separation of metallic cations [18, 19], the extraction of methyl esters of some amino-acids [20], uranium (VI) preconcentration [21], lanthanide complexes [22], and molecular recognition of 1,5-diaminoantraquinone [23]. Erdemir et al. investigated the extraction abilities of carboxylic acid and methyl ester derivatives of p-t-butylcalix[n]arenes (n = 6, 8) for carcinogenic aromatic amines [24]. Other calixarenes and their derivatives have been utilized in the fields of complexation, separation, electroanalysis, spectroscopy, and chemometrics [25].

A larger ligand, i.e., H_8L, can act as a ditopic receptor for lanthanide and transition metal ions [26] and hence in principle may bind to a single metal ion in various ways. It is important to mention that solvent extraction of transition metal ions, particularly toxic metal ions, using H_8L alone has scarcely been reported in the open literature. Makrlik et al. studied the liquid–liquid extraction of Eu^{3+} trifluoromethanesulfonate into nitrobenzene in the presence of H_6L and H_8L [27]. Sansone et al. reported the separation of An^{3+}/Ln^{3+} from radioactive waste using CMPO (carbamoylmethyl-phosphine oxide) substituted H_6L and H_8L [28]. Gutsche et al. synthesized aminocalixarenes complexes of Ni^{2+}, Cu^{2+}, Pd^{2+}, Co^{2+}, and Fe^{2+} and determined their spectral and chemical characteristics [10]. Their findings indicated that metal-aminocalixarenes are more flexible than had previously been thought. It has also been reported that H_8L can be combined into a polymeric medium to produce a material that shows a high sorption ability towards transition metal ions (Cu^{2+}, Fe^{2+}, Zn^{2+}, Ni^{2+}, Co^{2+}, Cd^{2+}, and Pb^{2+}) in aqueous solution [29]. The authors also performed research on the extraction behavior of transition metal ions with H_4L and H_6L [30, 31].

An effective extractant with high selectivity for metal ions is in high demand for analytical applications, recycling of resources, and waste treatment purposes. Heavy metals such as lead (Pb), copper (Cu), and nickel (Ni) are harmful to humans. Obviously, the harmful impact of some ions, for example, Cd^{2+} and As^{2+}, is of great concern in such research. Cd^{2+} is one of the most toxic elements for humans. At high concentrations, it causes various debilitating conditions such as painful bone disease, bone marrow disorders, kidney problems, and *"Itaiitai"* or *"ouch-ouch"* disease [32]. However, calixarene derivatives may be useful binders for these cations.

In this study, H_8L has been investigated as host extractant for divalent heavy metal ions and silver from ammonia and amine solutions into various types of organic solvents. We also synthesized H_8L ethyl ester derivatives and the extraction behavior of transition metal cations from aqueous solution was also investigated.

2. Experimental Section

2.1. Materials. H_8L was purchased from the Sigma Aldrich Chemical Company, USA. All transition metal nitrate solutions were prepared according to the method in [30, 31]. Other reagents such as chloroform, dichloromethane and nitrobenzene, and ethanol were purchased from Carlo Erba Reagents, France. Ammonia, ethylenediamine and trimethylene diamine, ethyl bromoacetate, NaH, and THF/DMF were bought from Acros Organic, Belgium. The water was deionized. All the remaining reagents were as pure as is commercially available. Stock solutions were standardized by potentiometric and EDTA titrations. Other metal salts were guaranteed to be reagent grade.

2.2. Extraction Procedure. The hosts H_8L and H_8L-ester were prepared by dissolving the appropriate amount in various organic solvents followed by the dilution, typically to 1×10^{-4} M working solution; aqueous solutions of metal solutions were made from analytical purity nitrates of Cd^{2+}, Ni^{2+}, Cu^{2+}, Ag^{2+}, Co^{2+}, and Zn^{2+}. Extraction experiments typically performed by equilibrating 8 mL of a 5×10^{-5} M solution of the metal ions, 1 ml succinic acid (0.01M), and 1 mL of a buffer solution with 10 mL of a 5×10^{-4} M solution of the H_8L in 1,2-dichloroethane. The mixture was placed in a stoppered 50 mL glass tube at a volume ratio of 10:10 mL (organic phase to aqueous phase). The pH was adjusted by three types of buffer solution: CH_3COOH-CH_3COONa (acidic region), H_3BO_3-NaOH (neutral), and NH_3-NH_4Cl (alkali). In the case of H_8L-ester extractant, picric acid [2.5×10^{-5} M] was used as the counter anion. The extraction equilibrium was attained within 40 min of shaking with nitrobenzene; the extraction into chloroform reached equilibrium within 20 h of shaking. Therefore, the shaking time was fixed at 2 h for nitrobenzene and at 20 h for chloroform. The extractability was not affected by further shaking, indicating that the equilibrium was attained within 12 h. All the experiments were performed in presence of succinic acid to avoid emulsification during the extraction process. The distribution experiments were performed at room temperature.

Before shaking, the samples were left standing in the water bath at 25°C for 15 min to ensure that the extraction solutions were maintained at the same temperature. Then, a shaker at 200 stroke min^{-1} at 25.0 ± 0.1°C mixed the two phases; a shaking time of 12 h was sufficient to reach the extraction equilibrium. After shaking, the two phases were centrifuged at 2000 rpm. for 10 min, which was sufficient for complete separation. Before the measurement, the pH of the aqueous and organic phases was adjusted to 2.5 using 5 M HNO$_3$ and 3 M LiOH. The amount of extracted metal ions was calculated from the difference between the metal concentrations in the aqueous phase before and after the equilibration. The concentration of the metal ion in the organic phase was determined by the back-extraction method; 5 cm^3 of the organic phase was transferred into another glass-stoppered tube and shaken with 4 M hydrochloric acid. After phase separation, the equilibrium concentrations of metal cations in the aqueous phase were measured by an inductively coupled plasma atomic emission spectrometer (Seiko model SPS 1200AR). The equilibrium pH in the aqueous solutions was measured by a pH meter (Beckmann model ø45).

The extractability (Ex%) was determined from the decrease in the metal concentration in the aqueous phase:

$$Ex\% = \left\{ \frac{(Metal)]_{blank} - [Metal]_{water)}}{(Metal)]_{blank}} \right\} \times 100, \quad (1)$$

where [Metal]$_{blank}$ and [Metal]$_{water}$ denote the metal concentrations in the aqueous phase after extraction with nitrobenzene and with the nitrobenzene solution containing extractants, respectively, and [Metal]$_{or}$ denotes the metal concentration extracted into the organic phase.

2.3. Analysis. Morphology of the product particles was examined using scanning electron microscopy (SEM, JEOL JSM6330F). Fourier transform infrared (FT-IR) spectra were collected on a Bruker FT-IR spectrometer by using the KBr pellet technique. ^{1}H-NMR data were recorded on a JEOL JNM-GX 61D FT-NMR spectrometer operating at 400 MHz in CDCl$_3$, using TMS as internal standard.

3. Results and Discussion

3.1. Effect of H$_8$L Host Concentration. Experiments were performed using H$_8$L concentrations of 1×10^{-3} - 5×10^{-4} M and a metal ion concentration of 5×10^{-5} M; all other conditions were kept the same. We found that the extraction percentage increased with increasing H$_8$L concentration. The best extraction was achieved when 5×10^{-4} M H$_8$L was used. As H$_8$L was soluble in nitrobenzene up to 4×10^{-2} M and up to 1×10^{-2} M in 1,2-dichloroethane at room temperature, the saturated solution was used as the stock solution.

3.2. Role of Extractant. Three organic solvents (nitrobenzene, dichloromethane, and chloroform) were also tested as inert diluents at a fixed pH for solutions containing an equal amount of metal ions and H$_8$L. The phase volume ratio was maintained at 1:1 to avoid emulsion formation; this was found to be the most effective ratio. This means that the tendency for association is, in general, greater when the solvent–solute interactions are weaker; however, chloroform is the least effective. The exact cause of this type of behavior

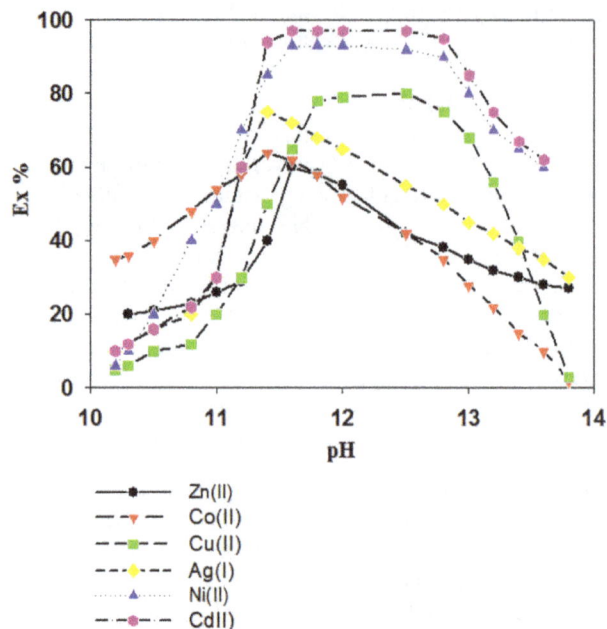

FIGURE 1: Extractability of all tested transition metal ions as a function of pH. Organic phase: $H_8L = 5 \times 10^{-4}$ M; aqueous phase: = (●) 5×10^{-5} M; succinic acid = 0.01 M; and buffer solution: 0.01 M; MES-NaOH (pH 5.0–7.0) and 0.1 M Tris-HClO$_4$ (pH 7.0–9.0). [Cd^{2+}] = (■)1×10^{-4} M, 0.2 M NaClO$_4$. O/A = 1; T = 25°C. O:A represents the ratio between organic (O) and aqueous (A) volumes in the experiments.

is not known. It was observed that the extraction percentage increased with the diluent type in the order of chloroform > dichloromethane > nitrobenzene.

3.3. Choice of Stripping Agent. After extraction of Cd^{2+} with H$_8$L, the metal ions were stripped with 7 mL of various concentrations of mineral acid reagents, specifically 4 M HCl, 0.1–5 M HNO$_3$, or 2 M H$_2$SO$_4$; lower concentrations of nitric acid (< 4.5 M) were not suitable as Cd^{2+} forms a stable complex. Finally, it was observed that 4 M HCl was suitable as a stripping agent.

3.4. Effect of Succinic Acid. Very small amounts of succinic was added to the reaction media to inhibit emulsification particularly in the case of Cd^{2+}, Cu^{2+}, and Cr^{3+} transition metal cations. By keeping all other parameters the same, experiments were performed using different concentrations of succinic acid. The best extraction was achieved when 0.01M succinic was used. It is noteworthy to mention that upon addition of the succinic acid into the highly basic buffer solution, the extraction percentage was stabilized probably due to the pH control.

3.5. Nature of the Extracted Species. At first, we studied the effect of pH on the extraction of several transition metal ions with H$_8$L in nitrobenzene. Specifically, the solvent extraction percentages of Cd^{2+}, Ni^{2+}, Cu^{2+}, Ag$^+$, Co^{2+}, Zn^{2+}, Cr^{3+}, and Mn^{2+} transition metal cations were examined. Figure 1

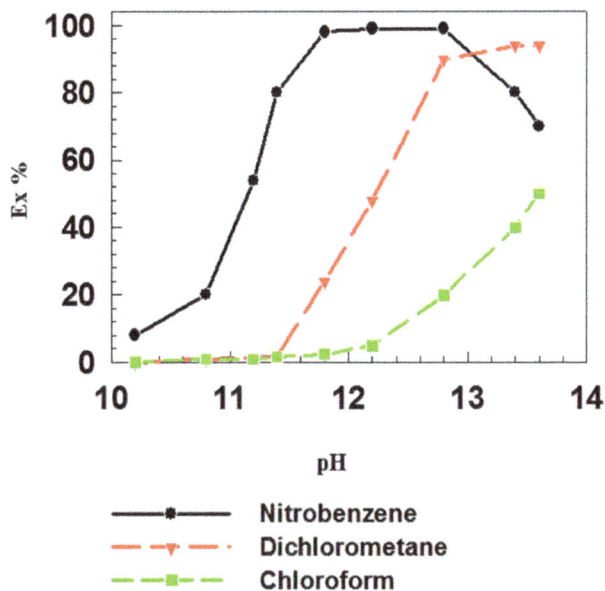

FIGURE 2: Effect of the organic solvent (nitrobenzene, dichloromethane, or chloroform) on the extraction percentage of Cd^{2+} with H_8L. $H_8L = 5 \times 10^{-4}$ M; aqueous phase [metal ion]: = (●) 5×10^{-5} M. O/A = 1; succinic acid = 0.01 M; and buffer solution, T = 25°C.

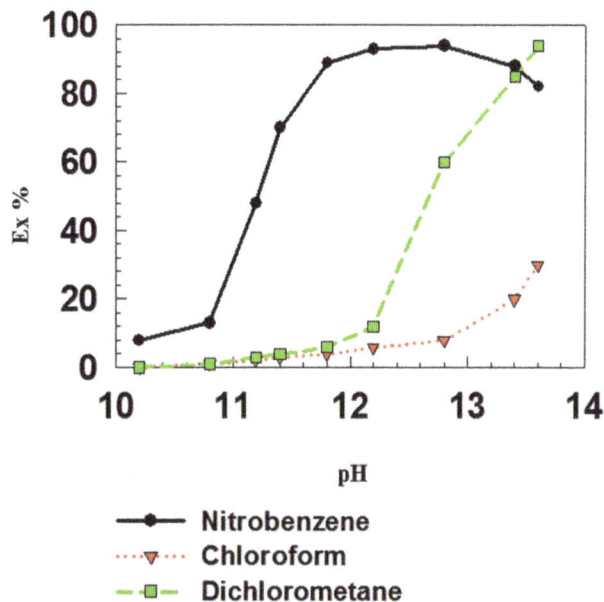

FIGURE 3: Effect of the organic solvent (nitrobenzene, dichloromethane, or chloroform) on the extraction percentage of Ni^{2+} with H_8L. p-calix[8] = 5×10^{-4} M; aqueous phase [metal ion]: = (●) 5×10^{-5} M. succinic acid = 0.01 M and buffer solution, O/A = 1; T = 25°C.

shows that the percentage of metal ions extracted increases as the pH increases from 10.0 to 13.0. In the acidic or neutral pH region, metal ions were not extracted with H_8L, whereas at pH 11.3 more than 60% of metal ions were extracted with H_8L. The maximum extraction percentage was observed in the pH range of the 11.5 to 13.00 for all tested samples. The extractability order is Cd^{2+} > Ni^{2+} > Cu^{2+} > Ag^+ > Co^{2+} > Zn^{2+}. H_8L has phenolic hydroxyl groups which deprotonate at pH > ca.10 and deprotonated H_8L can extract the metal ions. Among the cations tested, almost the same percentages of Cd^{2+} and Ni^{2+} ions were extracted under the experimental conditions. By contrast, no amounts of Cr^{3+} and Mn^{2+} were extracted, possibly due to the formation of a precipitate with ammonia solution. The results indicate that H_8L has a good affinity for complexation with transition metal ions. The study of pH effect indicates that the extraction mechanism depends on a proton exchange mechanism together with hydrogen bonding. Petit et al. also reported a dinuclear cobalt(II) complex of calix[8]arenes compound, prepared by solvothermal reaction of cobalt(II) acetate with p-t-butylcalix[8]arene and trimethylamine; the compound was formed by hydrogen bond bridging [33].

The effects of three organic solvents, nitrobenzene, dichloromethane, and chloroform, were examined for both Cd^{2+} (Figure 2) and Ni^{2+} (Figure 3) in the pH range of 10-13.5. Nitrobenzene was very effective as an extractant for both Cd^{2+} and Ni^{2+} metal ions, extracting 96% of these metal ions in the pH range of 11.50-12.70. Dichloromethane was effective in the pH range 11.4-13 for Cd^{2+}, whereas the effective pH range for Ni^{2+} was 12–13. Chloroform was less effective as an extractant at all pH values tested, showing

only 50% extraction of Cd^{2+} and Ni^{2+} combined and similar extraction percentages for Cd^{2+} and Ni^{2+} individually. It has been reported that the H_6L complex could be obtained as an adduct with chloroform (1 M); however, chloroform could not be removed from the adduct after calcination at 130°C for 3 days under reduced pressure. This indicates that the chloroform molecule might be encapsulated within the cavity of the calixarene and particularly for p-t-butyl calix[6]arene [H_6L] molecule, which may account for the low extraction percentage obtained in the chloroform solution [25]. Three types of organic solvents were tested and the nitrobenzene is found to be the most efficient. Two other solvents, dichloromethane and chloroform, are least effective. The exact cause of this type of behavior is not known. Masuda et al. observed that the relative descending order of extraction with other solvents is not the same with H_6L and the same order does not necessarily accord the order of their dielectric constants [31]. Actually the extraction percentages of Ce(III) with H_6L were 95%, 27%, and 18% at pH 11.85 in the case of organic solvents nitrobenzene, dichloromethane, and chloroform. The dielectric contestants of nitrobenzene, dichloromethane, and chloroform are 34.82, 7.77, and 4.80, respectively. Therefore, it can be assumed that the dielectric contestant of the medium has some contribution in the extraction process. However, the main factor determining the extraction efficiency in the extraction process must be taken into account and a better term correlating the relative extraction order is solubility parameter. Moreover, Thuéry et al. reported that the reaction at room temperature between H_8L and an excess of trimethylamine in chloroform provided a solid that could be recrystallized in methanol

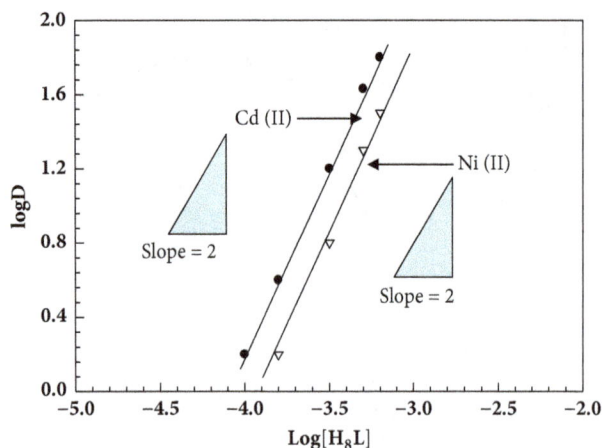

FIGURE 4: Plot of logD vs. log[H_nL] for Cd^{2+} and Ni^{3+} under the same reaction condition of Figure 3.

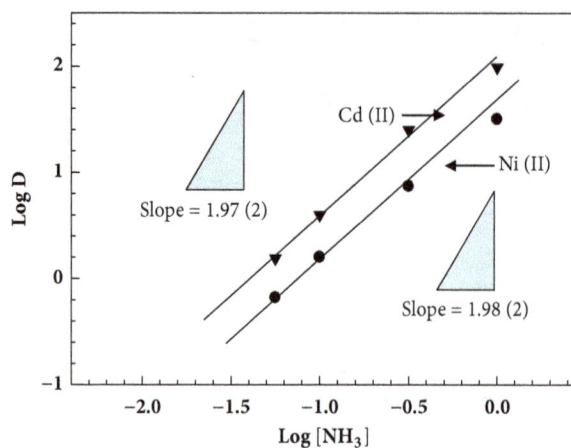

FIGURE 5: Plot of logD vs. log[NH_3] for Cd^{2+} and Ni^{3+} with H_8L under the same reaction condition of Figure 3.

to yield dark red crystals of new compound suitable for X-ray crystallography. Therefore, we can conclude that the chloroform solvents have some affinity and/or suitable for complex formation with H_8L ligand [34].

3.6. Slope Analysis. A traditional and effective means of obtaining both stoichiometric and equilibrium constant information about extraction processes, slope analysis, are based on an examination of the logarithmic variation of the distribution ratio, D, with relevant experimental variables. The log-log plots of the extraction in the form of D vs. a concentration variable indicate the stoichiometry of the formation of the extractable complex and thus lead to the derivation of a suitable equilibrium expression and then to the calculation of equilibrium constants.

Since the extraction reagents exhibited high selectivity for Cd^{2+} and Ni^{2+}, detailed extraction behavior for Cd^{2+} and Ni^{2+} was investigated by slope analysis, the continuous variation method, and by loading tests. The extraction mechanism was studied by evaluating the composition of the extracted Cd^{2+} and Ni^{2+} species. Figure 4 shows the effect of the initial concentration of H_8L in the organic phase on the extraction of Cd^{2+} and Ni^{2+} from an aqueous ammonia solution. With increasing initial concentration of H_8L, straight lines with a slope of 2 were obtained in the logD vs. log[H_8L] plots for Cd^{2+} and Ni^{2+}, indicating that the binding ratio of H_8L with Cd^{2+} and Ni^{2+} is 2:1. Figure 5 shows the effect of the initial concentration of ammonia in the aqueous phase on the extraction of Cd^{2+} and Ni^{2+}. The logD value increases linearly with an increasing initial concentration of ammonia with a slope of 2. These findings suggest that a 2:1 (H_8L: metal) complex was extracted into the organic phase by releasing an equimolar amount of protons from H_8L along with two ammonia molecules.

Distribution experiments were carried out in order to obtain information on the viability of the extraction process, the stoichiometry and distribution equilibrium of the extracted metal ions between phases, and the extent of Cd^{2+} and Ni^{2+} extraction. During this experiment, the

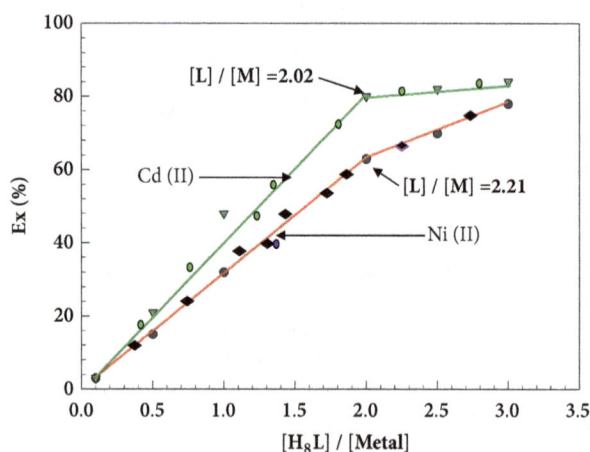

FIGURE 6: Plots of percentage extraction vs. molar ratio [L]/[M] for the extraction of Cd^{2+} and Ni^{3+} with H_8L under the same reaction condition of Figure 3. O:A = 1:1. Symbol (empty green inverted triangle), (filled green circle) (repeated) (filled red diamond) (filled red circle) (repeated)-H_8L; O/A = 1; T = 25°C.

concentrations of Cd^{2+} and Ni^{2+} were kept constant at 1 × 10^{-3} M, while the concentration of the extractant was varied from 1 × 10^{-4} to 2 × 10^{-3} M. This means that the relative concentration of the extractant H_8L (defined as the molar ratio of the initial extractant concentration and the concentration of the extracted metal) changed from 0.1 to 3.4 (Figure 6). A plot of the residual concentration of Cd^{2+} and Ni^{2+} in the aqueous phase, against the relative concentration of extractant, is presented in Figure 6. Initially, experiments were completed three times to examine their reproducibility. These results are also presented in Figure 6. As can be seen, the extraction method showed good reproducibility. An extraction percentage higher than 75% was obtained using H_8L at low concentrations of Cd^{2+} and Ni^{2+}. For Cd^{2+} and Ni^{2+} extracted by the same reaction condition, the higher Ex % of Cd^{2+} must have a high extraction equilibrium constant. That is why the slope of straight line after the inflection point in Figure 6 becomes more horizontal in green line (Cd^{2+}) than red line (Ni^{2+}).

As mentioned above the molar ratio of H_8L: metal was 2:1. These extraction studies suggest that the 2:1 (H_8L: metal) complex was extracted into the organic phase releasing 1 M protons from 1 M H_8L, accompanying two ammonia molecules [Figures 4 and 5]. The extraction equilibrium and the extraction equilibrium constant K_{ex} can be expressed as

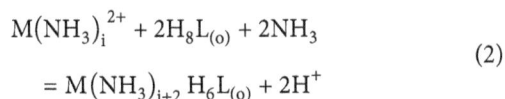

$$M(NH_3)_i^{2+} + 2H_8L_{(o)} + 2NH_3$$
$$= M(NH_3)_{i+2}H_6L_{(o)} + 2H^+ \qquad (2)$$

$$K_{ex} = \frac{\left[M(NH_3)_{i+2}H_6L\right]_o \left[H^+\right]^{2+}}{\left[M(NH_3)_i^{2+}\right]\left[H_8L\right]_o^2 \left[NH_3\right]^2} \qquad (3)$$

where (o) indicates the species in the organic phase. Equation (3) can be simplified as follows using the distribution ratio of the metal ($D = [M(NH_3)_{i+2}H_8L]_o / [M(NH_3)_i^{2+}]$):

$$K_{ex}' = \frac{D\,[H]^2}{[H_6L]_o^2 \,[NH_3]^2} \qquad (4)$$

$$\log D = \log K_{ex}' + 2\log[H_8L]_o + 2pH + 2\log[NH_3] \qquad (5)$$

From the above observations, it is suggested that the ammonia molecules and amine complex participate in the extraction of transition metal ions with H_8L.

The extraction of transition metal ions with H_8L from the aqueous phase, containing 0.1 M ethylenediamine $[C_2H_4(NH_2)_2]$ and 0.1 M trimethylenediamine $[(CH_2)_3(NH_2)_2]$ instead of ammonia, into the dichloromethane solution was also examined. It is worth noting that extraction of Co^{2+} and Ni^{2+} from the aqueous phase containing ethylene diamine was suppressed (Ex% = 0). On the other hand, Cu^{2+} and Cd^{2+} were extracted remarkably well (Cu^{2+} = 100% and Cd^{2+} = 90%) from the aqueous phase containing ethylene diamine. The interaction of calixarenes and amines in the dichloromethane solution likely involves the following two-step process: (i) proton transfers from the calixarene to the amine to form the amine cation and then (ii) the calixarene anion forms an endo-calix complex by association with the amine cation [35]. However, we think these reactions simultaneously take place and reach equilibrium at certain conditions.

To understand complexation in the aqueous phase, the distribution ratio of the $M(en)_2$ and $M(en)_3$ [$(en)_2$= ethylene diamine and $(en)_3$ = trimethylene diamine] species in the aqueous phase about each metal ion before extraction, using the formation constant with ethylene di and triamine, was calculated. Table 1 shows the distribution ratios of the $M(en)_2$ and $M(en)_3$ species in the aqueous phase before extraction and the extraction percentage for extraction with H_8L into dichloromethane. The distribution ratios for the complexations with H_8L are comparable and indicative of high efficiency (Table 1). The values of the distribution ratios of $M(en)_2$ and $M(en)_3$ species in the aqueous phase before extraction are integers. As mentioned above, the extraction

TABLE 1: Extraction percentage (Ex%) of the transition metal ions with H_8L from ethylene diamine into dichloromethane at 25°C and distribution ratio of $M(en)_2$ and $M(en)_3$ in the aqueous phase before extraction. Uncertainties are given in parentheses as standard errors of the mean (N = 3).

	%E	$M(en)_2$* (%)	$M(en)_3$** (%)
Co(II)	0	0	100
Ni(II)	0	0	100
Cu(II)	97.0 (2)	100	40
Zn(II)	46.0 (3)	3	97
Ag(I)	52.7 (5)	99	35
Cd(II)	90.1 (1)	15	85

*$(en)_2$ = ethylene diamine, **$(en)_3$ = ethylene triamine.

of Co^{2+} and Ni^{2+} metal ions was suppressed. Thus, most of the existing species in the aqueous phase are $M(en)_3$, while few $M(en)_2$ species were present, whereas in the case of Cu^{2+} most of the extracted species present in the aqueous phase were $M(en)_2$ species. These data suggest that the existence of $M(en)_2$ species diverts and/or controls the extraction with H_8L from the aqueous phase containing ethylene diamine to some extent. This can be explained by steric factors influencing the binding of the ligands to the metal ions. It means that probably the small molecule are encapsulated by larger molecular and steric barriers keep the guest from escaping the host. From the above results, masking effects of metal ions with amines were also observed, particularly metal ions showing high affinity with amines.

We additionally studied the composition of the Cu^{2+} and Cd^{2+} extracted species according to the molar ratio method. A plot of logD vs. $\log[H_8L]$ was constructed and a straight line was found with a slope of 1, indicating a molar ratio of H_8L:metal = 1:1, which is different from the composition of species in the case of the ammonia aqueous phase. The effect of pH on the extraction of Cu^{2+} was also verified. As logD increases linearly with an increase in pH, where a slope of 2 was obtained.

Thus, the extraction equilibrium can be expressed as (en= $H_2N(CH_2)_2NH_2$)

$$M_{(en)2}^{2+} + H_8L_{(O)} = M_{(en)2}H_6L_{(O)} + 2H^+ \qquad (6)$$

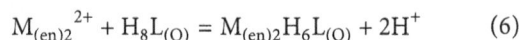

Table 2 shows the elemental analysis data of the Cu^{2+}-H_8L and Cd^{2+}-H_8L complexes. These data indicate that extracted species from the aqueous phase containing ethylene diamine was $M(en)_2$ not $M(en)_3$, in agreement with our initial assumption based on Table 1.

It has been reported that, in the extraction of metal ions using calix[n]arenes, the metal ion selectivity is related to the ring size of the calix[n]arene and to the radii of the metal ions [9, 10]. The present extraction study using H_8L accompanied by amines indicated that size is key to the selectivity, as the diameter of the H_8L ring (4-4.4 Å) is sufficient to fit the $M(en)_2$ complex (where the distance between the both ends of amino proton is ca. 4.0 Å). A possible structure of Cu^{2+}-H_8L is given in Scheme 1. An X-ray results is essential to

TABLE 2: Elemental analysis of Cu^{2+}-H_8L and Cd^{2+}-H_8L complexes and estimated chemical formula.

	Cu^{2+}: estimate chemical formula $Cu(en)_2H_6L.5H_2O$				Cd^{2+}: estimate chemical formula $Cd(en)_2 H_6L.3H_2O$		
	H(%)	C(%)	N(%)		H(%)	C(%)	N(%)
Obs.	8.37	70.89	3.56	Obs.	8.18	69.90	3.49
Calc.	8.73	70.40	3.57	Calc.	8.41	69.83	3.54

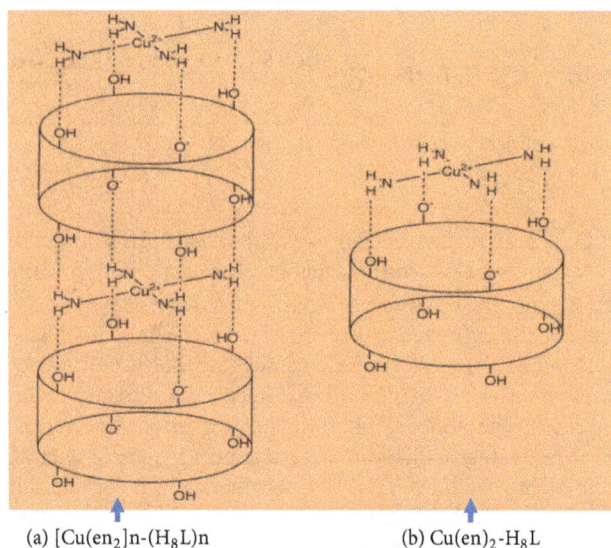

(a) $[Cu(en_2)n$-$(H_8L)n$ (b) $Cu(en)_2$-H_8L

SCHEME 1: Structure of $Cu(en)_2$-H_8L complex.

determine a compound structure. At present, our research is now progressing in this direction.

As mentioned above Cd^{2+} is very toxic and inhaling cadmium dust leads to kidney problems which can be fatal. Therefore, complexation studies of Cd^{2+}-H_8L complexes were also performed using FESEM, FTIR, and ^1H NMR spectroscopy for more information of Cd^{2+} extraction with H_8L. FESEM images of H_8L and Cd^{2+}-H8L showed very unique surface morphology. The H_8L had rod-like particles with almost 10 μm long and less than 1 μm wide [Figure 7(a)]. However, the Cd^{2+}-H_8L morphology was totally changed in compare with H_8L alone. The Cd^{2+}-H_8L sample had two types of particle sizes as shown in Figures 7(a) and 7(b), respectively. The nanosize spherical particles were Cd^{2+} ions and the square or bigger size particles were H_8L as evidence by EDX analysis. The EDX and elemental analysis showed almost the same results. It is interesting that the Cd^{2+} ions were homogeneously dispersed and/or distributed over the H_8L ligand. Figure 8(a) shows the EDX spectrum of big particles marked by red arrow in Figure 7(b). The atomic percentage of C was 69.26 for the big size particles. Very little amount of Cd^{2+} was also obtained during EDX analysis of big size particles. On the other hand, the spherical particles had mostly Cd^{2+} ion and atomic percentage of Cd was 33.78 [Figure 8(b), the EDX spectrum of spherical particle was based on Figure 7(c), marked by red arrow]. Almost the same

results were obtained for different location based on particles size. The EDX results shown in Figures 8(a) and 8(b) indicate the presence of Cd, C, Au, Cu, and O. The Au and Cu were found due to gold coating over the sample and Cu substrate was used.

The FTIR spectrum of H_8L and Cd^{2+}-H_8L is shown in Figures 9(a) and 9(b), respectively. In the FTIR spectra of the complexes, the intensities or wave numbers of the stretching vibration of the OH groups change drastically with complexation by breakage of the especially strong intramolecular hydrogen bonding existing in the free ligands [30]. The IR absorption bands at 3187 cm^{-1} due to vN-H: free ethylene diamine [36] shifted to 3361 cm^{-1} upon extraction of Cd^{2+} with H_8L. This indicates that hydrogen bonding was likely present in the $M(en)_2$-H_8L complex. The associated natures of $C(CH_3)_3$, $-CH_2-$, and hydroxyl group have been reported previously [30]. The band $C(CH_3)_3$ gives a sharp absorption band at ca. 2944 cm^{-1}, and this peak intensity was weakened after making a complex with Cd^{2+} cation. The ν(C=C) vibration bands shift by about 30 cm^{-1} (i.e., from 1605 to 1638 cm^{-1}) toward higher frequencies. It has been also recorded that the $-CH_2-$ vibrations almost disappeared and the very strong peak at 1489 cm^{-1} was disappeared. The differences in the infrared spectra may be caused by hydrogen bonding. As a whole, all the peaks of Cd^{2+}-H_8L composites were shifted to higher energy direction, indicating the strong interaction between Cd^{2+} and H_8L.

The solution behavior of the complexes was determined by ^1H-NMR spectroscopy in $CDCl_3$ at room temperature. In the spectra of H_8L [Figure 10(a)], and Cd^{2+}-H_8L [Figure 10(b)], each spectrum shows one singlet resonance for protons of -$C(CH_3)_3$ at δ = 1.25. For H_8L, singlet resonances of Ar-H are observed at δ = 7.12 and 7.14 ppm, respectively; for Ar-OH these resonances are at δ = 9.63, respectively. A new signal appeared at chemical shift of δ = 2.62 which was not observed in the H_8L. In our previous study, we also observed one singlet at around δ 3.00 for Ce^{3+}-H_8L and Ce^{3+}-H_6L probably for the bridging methane groups due to the different environments of the hydrogen atoms on the methylene group [30]. In the ^1H NMR spectra, peaks at 3.5 and 4.4 ppm were observed (due to the nonequivalent methylene protons (Ar-CH_AH_B-Ar) in the free H_8L which converge to 3.8 ppm in the $Cd(en)_2$-H_8L complex after extraction. This indicates that the cone conformation of H_8L converts into the 1,3,5,7-alternate conformation.

In this study, the H_8L-ester compounds were also synthesized according to published procedures [37]. The extraction percentage of any metal ion was poor throughout the entire

(a)

(b)

(c)

FIGURE 7: FESEM images of the samples: (a) H_8L, (b) Cd^{2+}-H_8L, and (c) sample (b) at high magnification.

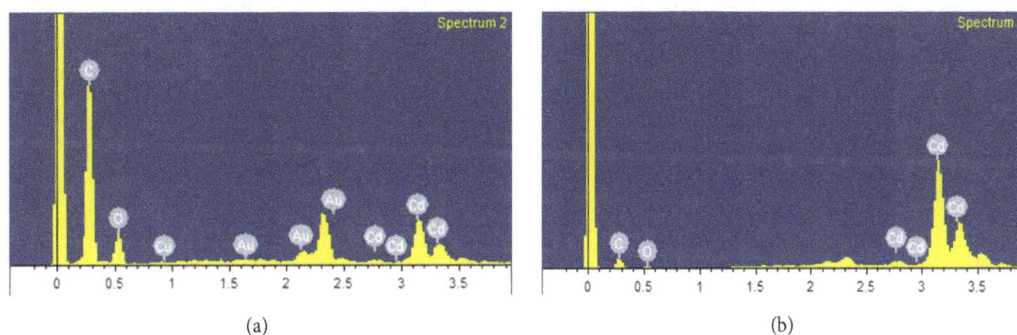

(a)

(b)

FIGURE 8: EDX spectra of (a) H_8L and (b) Cd^{2+}-H_8L. EDX spectra for the region marked by an arrow in (b), and (c) of Cd^{2+}-H_8L in Figure 7.

range of pH used. In fact, the liquid–liquid extraction ability of transition metal ions with the H_8L-ester/$CH_2COOC_2H_5$ derivative was low in comparison with H_8L alone. It is thought that the H_8L becomes an anion due to deprotonation. The H_8L-ester, which is modified from calix[8]arene, is more flexible than both the calix[4]arene and calix[6]arene derivatives; however, it did not show a good extraction capability for Co^{2+}, Ni^{2+}, Zn^{2+}, and Ag^+, as shown in Figure 11.

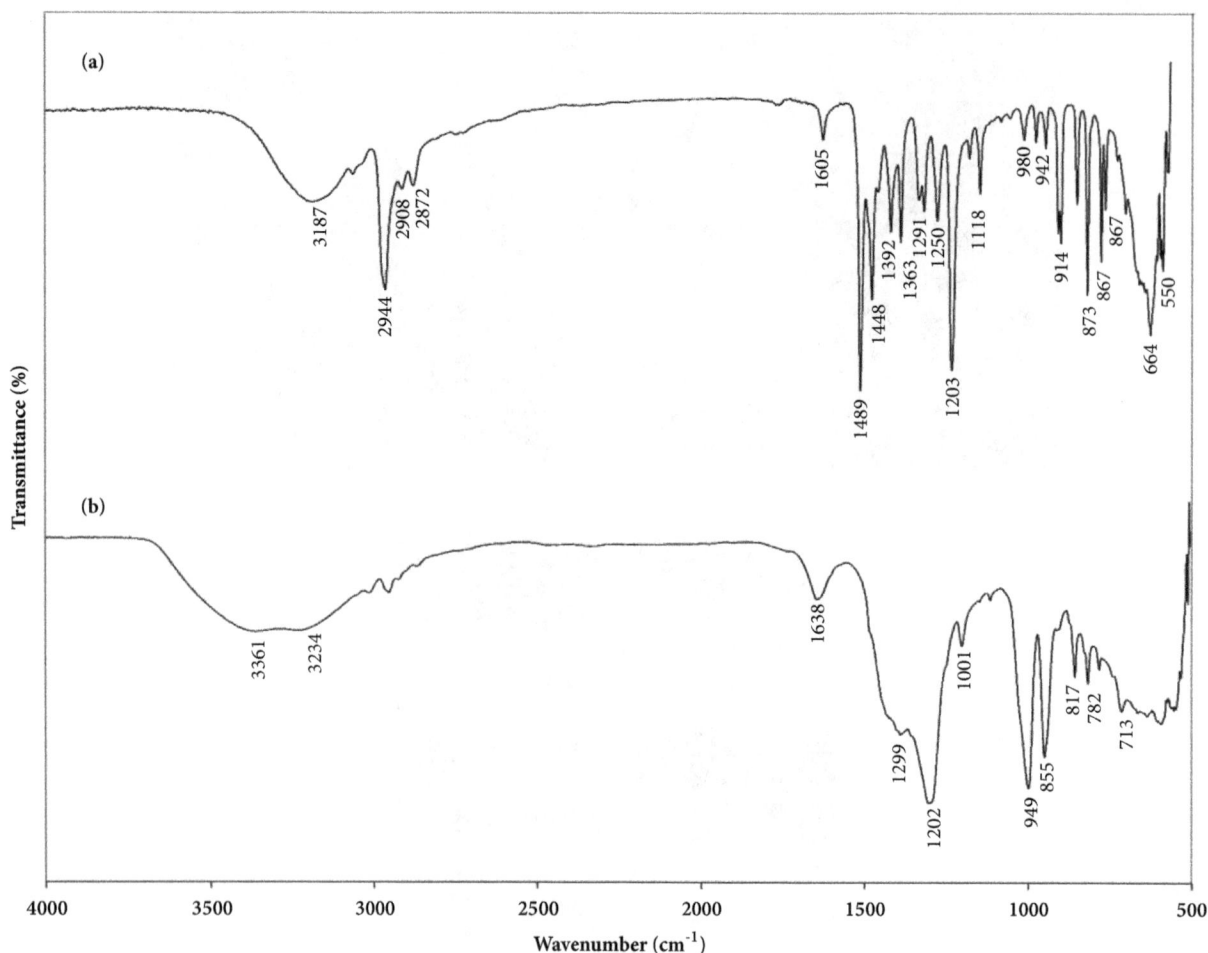

FIGURE 9: FTIR spectra of (a) H_8L and (b) Cd^{2+}-H_8L.

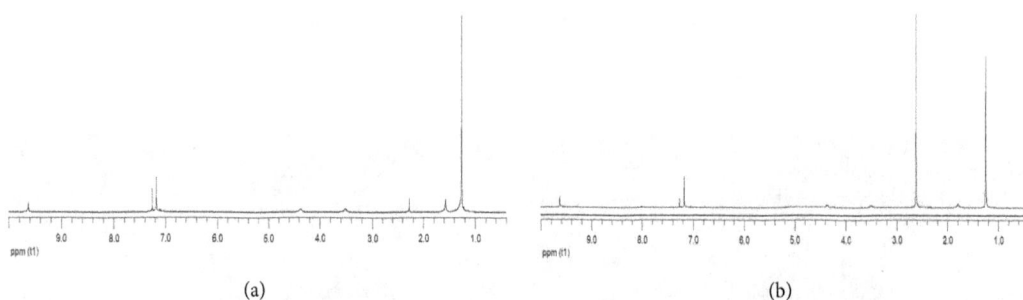

FIGURE 10: ^1H NMR spectra of (a) H_8L at 25°C in $CDCl_3$ and (b) Cd-H8L at 25°C in $CDCl_3$ at 400 MHz.

4. Conclusions

The host–guest extraction and/or complexation of divalent heavy metal cations and silver by H_8L and its ethyl ester host was investigated by changing various experimental parameters. The effect of the organic solvent on the extraction procedure was examined. In the solvents, the affinity of H_8L to bind transition metal cations was found to be much higher in the case of Cd^{2+} and Ni^{2+} when ammonia was used as the aqueous phase. The results from the extraction study

suggested that, in the case of ammonia, the ratio of the extracted species is 2:1 (H_8L:metal), whereas in the presence of ethylene diamine instead of ammonia, the composition of the extracted species is 1:1, indicating selectivity towards tetrahedral type metal ions. The ester-H_8L compounds were found to be not effective for the extraction of transition metal cations under the present reaction conditions. The extraction behavior of the metal ions was closely related to the pH at equilibrium and the matrix and/or medium (i.e., ammonia or amine) of the aqueous solution and organic phase.

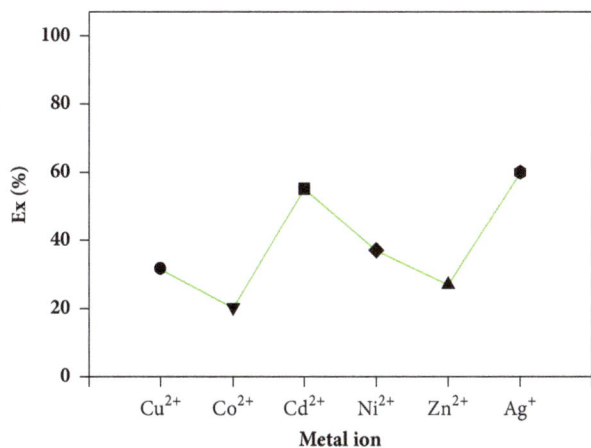

FIGURE 11: Metal ions percentage extraction (E%) at pH 11.5 by H_8L-ester. O/A = 1; T = 25°C.

References

[1] V. Böhmer and J. Vicens, "Special Calixarenes, Synthesis and Properties," in *Calixarenes: A Versatile Class of Macrocyclic Compounds*, vol. 3 of *Topics in Inclusion Science*, pp. 39–62, Springer Netherlands, Dordrecht, 1990.

[2] C. D. Gutsche, "Calixarenes," *The Royal Society of Chemistry, Cambridge*, p. 181, 1989.

[3] K. Ohto, M. Yano, K. Inoue et al., "Effect of coexisting alkaline metal ions on the extraction selectivity of lanthanide ions with calixarene carboxylate derivatives," *Polyhedron*, vol. 16, no. 10, pp. 1655–1661, 1997.

[4] N. N. M. Yusof, Y. Kikuchi, and T. Kobayashi, "Predominant hosting lead(II) in ternary mixtures of heavy metal ions by a novel of diethylaminomethyl-calix[4]resorcinarene," *International Journal of Environmental Science and Technology*, vol. 11, no. 4, pp. 1063–1072, 2014.

[5] N. N. M. Yusof, Y. Kikuchi, and T. Kobayashi, "Ionic imprinted calix[4]resorcinarene host for Pb(II) adsorbent using diallylaminomethyl-calix[4]resorcinarene copolymer," *Chemistry Letters*, vol. 42, no. 10, pp. 1119–1121, 2013.

[6] M. Martínez-Aragón, E. Goetheer, and A. de Haan, "Host–guest extraction of immunoglobulin G using calix[6]arenas," *Separation and Purification Technology*, vol. 65, no. 1, pp. 73–78, 2009.

[7] T. Oshima, M. Goto, and S. Furusaki, "Complex formation of cytochrome c with a calixarene carboxylic acid derivative: A novel solubilization method for biomolecules in organic media," *Biomacromolecules*, vol. 3, no. 3, pp. 438–444, 2002.

[8] R. Ludwig, "Calixarenes in analytical and separation chemistry," *Fresenius' Journal of Analytical Chemistry*, vol. 367, no. 2, pp. 103–128, 2000.

[9] C. D. Gutsche, B. Dhawan, K. H. No, and R. Muthukrishnan, "Calixarenes. 4. The Synthesis, Characterization, and Properties of the Calixarenes from p-tert-Butylphenol," *Journal of the American Chemical Society*, vol. 103, no. 13, pp. 3782–3792, 1981.

[10] C. D. Gutsche and K. C. Nam, "Calixarenes. 22. Synthesis, Properties, and Metal Complexation of Aminocalixarenes," *Journal of the American Chemical Society*, vol. 110, no. 18, pp. 6153–6162, 1988.

[11] X. Chen, R. A. Boulos, A. D. Slattery, J. L. Atwood, and C. L. Raston, "Unravelling the structure of the C60 and p-But-

[12] J. L. Atwood, L. J. Barbour, M. W. Heaven, and C. L. Raston, "Controlling van der Waals Contacts in Complexes of Fullerene C60," *Angewandte Chemie International Edition*, vol. 42, no. 28, pp. 3254–3257, 2003.

[13] C. Geraci, G. Chessari, M. Piattelli, and P. Neri, "Cation encapsulation within a ten-oxygen spheroidal cavity of conformationally preorganized 1,5-3,7-calix[8]bis-crown-3 derivatives," *Chemical Communications*, no. 10, pp. 921–922, 1997.

[14] C. Redshaw, "Coordination chemistry of the larger calixarenes," *Coordination Chemistry Reviews*, vol. 244, no. 1-2, pp. 45–70, 2003.

[15] T. Kajiwara, N. Iki, and M. Yamashita, "Transition metal and lanthanide cluster complexes constructed with thiacalix[n]arene and its derivatives," *Coordination Chemistry Reviews*, vol. 251, no. 13-14, pp. 1734–1746, 2007.

[16] R. Ludwig and N. T. K. Dzung, "Calixarene-based molecules for cation recognition," *Sensors*, vol. 2, no. 10, pp. 397–416, 2002.

[17] C. J. Liu, J. T. Lin, S. H. Wang, J. C. Jiang, and L. G. Lin, "Chromogenic calixarene sensors for amine detection," *Sensors and Actuators B: Chemical*, vol. 108, no. 1-2, pp. 521–527, 2005.

[18] R. M. Izatt, J. D. Lamb, R. T. Hawkins, P. R. Brown, S. R. Izatt, and J. J. Christensen, "Selective M+-H+ Coupled Transport of Cations through a Liquid Membrane by Macrocyclic Calixarene Ligands," *Journal of the American Chemical Society*, vol. 105, no. 7, pp. 1782–1785, 1983.

[19] S. R. Izatt, R. T. Hawkins, J. J. Christensen, and R. M. Izatt, "Cation Transport from Multiple Alkali Cation Mixtures Using a Liquid Membrane System Containing a Series of Calixarene Carriers," *Journal of the American Chemical Society*, vol. 107, no. 1, pp. 63–66, 1985.

[20] L. Mutihac and R. Mutihac, "Liquid-liquid extraction and transport through membrane of amino acid methylesters by calix[n]arene derivatives," *Journal of Inclusion Phenomena and Macrocyclic Chemistry*, vol. 59, no. 1-2, pp. 177–181, 2007.

[21] S. Ayata and M. Merdivan, "P-tert-Butylcalix[8]arene loaded silica gel for preconcentration of uranium(VI) via solid phase extraction," *Journal of Radioanalytical and Nuclear Chemistry*, vol. 283, no. 3, pp. 603–607, 2010.

[22] L. N. Puntus, A.-S. Chauvin, S. Varbanov, and J.-C. G. Bünzli, "Lanthanide complexes with a calix[8]arene bearing phosphinoyl pendant arms," *European Journal of Inorganic Chemistry*, no. 16, pp. 2315–2326, 2007.

[23] G. Suganthi, C. Meenakshi, and V. Ramakrishnan, "Molecular recognition of 1,5 diamino anthraquinone by p-tert-butyl-calix[8] arene," *Journal of Fluorescence*, vol. 20, no. 5, pp. 1017–1022, 2010.

[24] S. Erdemir, M. Bahadir, and M. Yilmaz, "Extraction of carcinogenic aromatic amines from aqueous solution using calix[n]arene derivatives as carrier," *Journal of Hazardous Materials*, vol. 168, no. 2-3, pp. 1170–1176, 2009.

[25] I. Yoshida, S. Fujii, K. Ueno, S. Shinkai, and T. Matsuda, " Solvent Extraction of Copper(II) Ion with ," *Chemistry Letters*, vol. 18, no. 9, pp. 1535–1538, 1989.

[26] E. Tashev, M. Atanassova, S. Varbanov et al., "Synthesis of octa(1,1,3,3-tetramethylbutyl)octakis(dimethylphosphinoyl-methyleneoxy) ca lix[8]arene and its application in the synergistic solvent extraction and separation of lanthanoids," *Separation and Purification Technology*, vol. 64, no. 2, pp. 170–175, 2008.

[27] E. Makrlík, P. Vaňura, and P. Selucký, "Solvent extrac-

tion of europium trifluoromethanesulfonate into nitrobenzene in the presence of p-tert-butylcalix[6]arene and p-tert-
butylcalix[8]arene," *Journal of Radioanalytical and Nuclear
Chemistry*, vol. 287, no. 1, pp. 277–280, 2011.

[28] F. Sansone, M. Fontanella, A. Casnati et al., "CMPO-substituted
calix[6]- and calix[8]arene extractants for the separation of
An3+/Ln3+ from radioactive waste," *Tetrahedron*, vol. 62, no.
29, pp. 6749–6753, 2006.

[29] R. Pathak and G. N. Rao, "Synthesis and metal sorption studies
of p-tert-butylcalix[8]arene chemically bound to polymeric
support," *Analytica Chimica Acta*, vol. 335, no. 3, pp. 283–290,
1996.

[30] Md. Hasan Zahir, "Synthesis and Characterization of Trivalent
Cerium Complexes of p-tert-Butylcalix[4,6,8]Arenes: Effect of
Organic Solvents," *Journal of Chemistry*, vol. 2013, pp. 1–9, 2013.

[31] Y. Masuda and Md. H. Zahir, "The host-guest extraction
chemistry of lighter lanthaniod (III) metal ions with p-t-
butylcalix[6]arene from the ammonia alkaline solution in
presence of succinic acid," *Analytical Sciences*, vol. 17, pp. a483–
a486, 2001.

[32] A. K. De, "Environmental Chemistry, New age international
Ltd," *Inorganic Chemistry and Analysis through Problems and
Exercises*, pp. 80-81, 2005.

[33] S. Petit, G. Pilet, D. Luneau, L. F. Chibotaru, and L. Ungur, "A
dinuclear cobalt(ii) complex of calix[8]arenes exibiting strong
magnetic anisotropy," *Journal of the Chemical Society, Dalton
Transactions*, no. 40, pp. 4582–4588, 2007.

[34] P. Thuéry and M. Nierlich, "The first metal complex of an acyclic
hexaphenol: structure of the binuclear complex of uranyl ions
with an analogue of p-tert-butylcalix[6]arene," *Journal of the
Chemical Society, Dalton Transactions*, no. 9, pp. 1481-1482.

[35] E. Akceylan, M. Bahadir, and M. Yilmaz, "Removal efficiency of
a calix[4]arene-based polymer for water-soluble carcinogenic
direct azo dyes and aromatic amines," *Journal of Hazardous
Materials*, vol. 162, no. 2-3, pp. 960–966, 2009.

[36] C. J. Pouchert, *The Aldrich library of IR spectra*, Aldrich
Chemical Company, Inc., 3rd edition.

[37] G. M. L. Consoli, F. Cunsolo, M. Piattelli, and P. Neri, "Study
on the esterification of p-tert-butylcalix[8]arene," *The Journal
of Organic Chemistry*, vol. 61, no. 6, pp. 2195–2198, 1996.

4

Two-Dimensional Thin Layer Chromatography-Bioautography Designed to Separate and Locate Metabolites with Antioxidant Activity Contained on *Spirulina platensis*

Margarita Cid-Hernández,[1] **Fernando Antonio López Dellamary-Toral,**[2]
Luis Javier González-Ortiz,[1] **María Judith Sánchez-Peña,**[1]
and Fermín Paul Pacheco-Moisés ⓘ[1]

[1]*Departamento de Química, Centro Universitario de Ciencias Exactas e Ingenierías, Universidad de Guadalajara, Blvd. Marcelino García Barragán 1421, 44430 Guadalajara, Jalisco, Mexico*
[2]*Departamento de Madera, Celulosa y Papel, Universidad de Guadalajara, km 15.5 de la Carretera Guadalajara-Nogales, 45220 Zapopan, Jalisco, Mexico*

Correspondence should be addressed to Fermín Paul Pacheco-Moisés; ferminpacheco@hotmail.com

Academic Editor: Bogusław Buszewski

Spirulina platensis contains several biologically active compounds, some of them with antioxidant activity. Nevertheless, not all of these compounds have been identified to date. As a first step to achieving such identification, a methodology to perform two-dimensional thin layer chromatography bioautographies on silica gel thin layer chromatography plates was proposed. Starting with a reference binary system, 5 other binary systems were tested, in which the relative polarity was systematically increased. To further improve the separation behavior, a phase modifier (NH_4OH) was used. The best separation results were obtained with the isopropyl alcohol/ethyl acetate/NH_4OH ternary system. This experimental system allowed four well-resolved spots showing antioxidant activity as well as two additional areas with mixtures containing antioxidant compounds. Although the proposed methodology was designed with a specific application, it would be predictable that its field of use could be considerably greater, making the convenient modifications on the solvent polarity and "masking level" produced by the ammonium derivatives.

1. Introduction

Spirulina platensis is a cyanobacterium that has been used in Mexico and other countries since ancient times [1, 2]. Recently, it has attracted worldwide attention due to its potential as a protein source [2, 3] and its therapeutic properties (e.g., immunomodulatory functions) [4]. *Spirulina* contains several known antioxidants, e.g., chlorophyll a, carotenoids, phycocyanins, glutathione, tocopherol, and others [5–7]. The relative amount of these compounds, as well as the possible presence of some additional ones, mainly depends on the environmental conditions during its culture [8–11]. The antioxidant activity of *Spirulina platensis* or its extracts has been reported for *in vitro* [9–12] and *in vivo* [13, 14] models.

However, it is clear that such activity has been produced by relatively complex mixtures that contain an unidentified number of components, most of which could possess low or negligible antioxidant activity. Due to the high number of components potentially present in *Spirulina* samples, it would be predictable that, to take advantage of beneficial properties of *Spirulina*, the isolation and characterization of its components as much as possible will be necessary.

Thin layer chromatography (TLC) is known as a very useful technique, utilized to separate complex mixtures, whose behavior depends on the balance of hydrophobic, hydrophilic, and steric interactions, as well as hydrogen bonding, occurring between the analytes and the mobile and stationary phases [15]. The most used stationary phase is the silica gel [16–18], which contains Si atoms bonded to none, one, or two hydroxyl groups [19]. Silanol groups (Si-OH) present on silica can be reversibly dehydrated, producing a siloxane group (Si-O-Si) from two Si-OH groups [16]. Therefore, silica surface must simultaneously contain both groups, being their relative amount dependent on the amount of the cationic groups which came from the phase modifier (e.g., ammonium hydroxide or salts), which are interacting with the silanol groups, decreasing the retention trend of the stationary phase toward the metabolites that are being separated by the TLC technique [20]. Since only the Si-OH groups are considered as strong adsorption sites [17, 21], the retention behavior can be externally modified (e.g., when water molecules are absorbed, preferentially on the Si-OH groups) [22]. Thus, depending on the stationary phase characteristics, as well as the mobile phase and sample compositions, the phase modifiers (e.g., ammonium salts or derivatives) may interact with the Si-OH groups and, therefore, modify the analytes retention behavior, especially with polar and basic compounds [18, 23].

In certain complex systems (e.g., those containing components covering a wide spectra of polarities), to improve the separation, it could be useful to simultaneously use two different chromatographic systems (e.g., different pairs of stationary phase/mobile phase). That is, a normal stationary phase (silica gel) developed with a nonaqueous mobile phase, which separates preferably nonpolar components and, in parallel, a reverse stationary phase (octadecyl silica) developed with an aqueous phase, used to separate preferably polar components [24, 25]. However, it is clear that this methodology duplicates the human and material cost of the process, which is an important factor to be considered when the objective is to obtain a considerable mass of the isolate components (e.g., to analyze them). An alternative methodology requires the use of a dual phase TLC plate, which have allowed complex mixtures to be separated, where great separation selectivity was reached [24, 25]. Dual phase TLC plates contain a narrow zone of SiO_2 and a wide zone of octadecyl silica (or vice versa). Unfortunately, at least in North America, these multi-K dual phase TLC plates are no longer available. Therefore, an alternative approach to effectively separate the compounds present in complex mixtures is to modify the interactions of the stationary phase, by means of the addition of a modifier to the mobile phase. Some amines have been used as phase modifiers, since they are able to mask silanol groups, reduce the silanophilic interactions with analytes, and, therefore, increase their retention factors (R_F) [20, 26, 27]. Furthermore, it has been reported that the chromatographic resolution is strongly affected by the pK_a of the amine used. In fact, the use of basic systems promotes the separation of closely related compounds with minor structural differences [28].

On the other hand, since component(s) that are practically incompatible(s) with the mobile phase remain very close to the bottom of the thin layer chromatograms, the presence of spots containing two or more compounds at R_F values near to zero could be possible. An equivalent statement can be expressed about components that are very compatible with the mobile phase, which migrate up to the top of the chromatogram (R_F=1.0). Therefore, to avoid the multicomponent spots presence, it is commonly preferred to consider it as well resolved, only those spots clearly separated (e.g., more distributed spots through the plate) that present retention factors not very near to zero or the unit [25, 29].

TLC bioautography (TLC bio) is an effective and inexpensive technique that combines the chromatographic separation with *in situ* localization of compounds with biological activity [30–32]. The reaction between the 1,1-diphenyl-2-picryl-hydrazyl (DPPH) radical and an active compound is a method commonly used to determine its antioxidant activity, which has been used to directly locate those types of compounds on TLC plates [33–37]. The characteristic reaction of this technique produces a pale yellow on the spots that contain compounds with antioxidant activity [34, 38].

Zarzycki *et al.* [33, 39] implemented a one-dimensional TLC technique (1D-TLC) to separate the chemical components present in four pharmaceutical formulations of *Spirulina platensis*. The cyanobacterium samples were extracted with methanol, acetone, or tetrahydrofuran, obtaining the best results with methanol [39, 40]. For all extracts, the spots on TLC plates were initially visualized using natural light and, to visualize additional spots, TLC plates were exposed to iodine vapors. Unfortunately, iodine vapors can react with some metabolites, interfering with their antioxidant activity, therefore impeding their evaluation with the DPPH technique.

The objective of this study was to provide a simple and cheap 2D-TLC biomethod for separation of compounds with antioxidant activity contained in *Spirulina platensis*. The ease of sample preparation, the quickness of this method, and the repeatability of the retention factors are the major novelties of this work.

2. Experimental

2.1. Materials and Reagents. Methanol (MeOH; Golden Bell, analytical grade), acetone (AcO; Karal, purity: 90%), isopropyl alcohol (IOH; Fermont, purity: 99.9%), ethyl acetate (EA; Karal, purity: 99.5%), and n-hexane (n-Hx; Fisher Chemical, purity: 99.9%) were distilled prior to use. Ammonium hydroxide aqueous solution (Fermont, concentration: 25-30%) and 1,1-diphenyl-2-picrylhydrazyl (DPPH; Sigma-Aldrich, purity: 97%) were used as received. TLC aluminum sheets were 20 cm × 20 cm (Merck; 1 mm thick, silica gel 60 F254), which were heated during 30 min at 100°C in an oven (Felisa, Model 292) and maintained on a glass desiccator until their use. *Spirulina platensis* used for the experiments was purchased from Natura Vitalis® GmbH in the form of tablets (Original spiruletten-1700 tablets; 400 mg of *Spirulina platensis*/tablet).

2.2. Extract. 25 g of *S. platensis* tablets was crushed to fine powder using mortar and pestle and transferred into an Erlenmeyer flask containing 500 mL of MeOH. This mixture was allowed to macerate for 48 hours under constant stirring. During maceration, the sample was protected from light and kept under a nitrogen atmosphere. The crude extract was filtered and concentrated with a rotary evaporator (Buchi R-3) to reduce the final volume to 125 mL.

TABLE 1: Relative volumetric proportion for the mobile phases used to perform the 1D-TLCs and 2D-TLCs.

Mobile phase	Solvent 1(δ_1*): Solvent 2 (δ_2*): modifier	Relative volumetric proportion
MP-1	AcO (20.3): n-Hx (14.9): non-used	3:7:0
MP-2	AcO (20.3): n-Hx (14.9): non-used	1:1:0
MP-3	AcO (20.3): EA (18.6): non-used	1:1:0
MP-4a	IOH (23.5): EA (18.6): non-used	0.16:1:0
MP-4b	IOH (23.5): EA (18.6): non-used	1:1:0
MP-4c	IOH (23.5): EA (18.6): non-used	1:0.16:0
MP-5a	IOH (23.5): EA (18.6): NH_4OH**	0.16:1:0.25
MP-5b	IOH (23.5): EA (18.6): NH_4OH**	1:1:0.25
MP-5c	IOH (23.5): EA (18.6): NH_4OH**	1:0.16:0.25

*Solubility parameter (δ_i; expressed in $(J/cm^3)^{1/2}$) [38]. **pK_b = 4.75.

2.3. TLC Chromatography.

TLC strips (2 cm × 10 cm) and TLC plates (20 cm × 20 cm) were used for 1D- and 2D-TLC, respectively. Nine different solvent systems (Table 1) were tested as mobile phases (MPs); the MP-1 was reported previously by Zarzycki et al., which was considered as the starting mobile phase [39, 40].

For 1D-TLC, a cylindrical glass chamber (10 cm x 11 cm; D x H) was used; its temperature was kept at 30± 1°C. Ten μL of the methanolic extract was spotted near the bottom of the TLC strip. Then, the solvent of the applied extract was evaporated completely at room conditions. 15 min before the TLC strip was developed, ten mL of the correspondent MP was poured inside the chamber. Immediately, the developed strips were dried at room temperature and were photographed under visible light (VL) or ultraviolet light at 366 nm (UVL_{366}). A Chromato-Vue CC-20 ultraviolet chromatography viewer, equipped with a UV filter from Ultra-Violet Products Inc., was used to allow the direct observation of the irradiated strips.

For 2D-TLCs, 100 μL of the methanolic extract was spotted near the bottom of the TLC plate; the solvent of the applied extract was evaporated completely at room temperature. A standard TLC glass chamber (rectangular TLC developing tank complete from Aldrich; 27 cm x 26.5 cm x 7.0 cm; L x H x W) was placed inside a recirculating water bath and kept at 30 ± 1°C; the temperature inside the TLC chamber was monitored continuously to avoid thermal variations. 100 mL of the corresponding mobile phase was added to this chamber. At the end of the first development, the plate was dried at room temperature and again placed in the chamber in a perpendicular direction from the original, to be developed in a second dimension. To finish the process, the plate was dried at room temperature and the plates were photographed under VL and UVL_{366} as above.

To corroborate the repeatability of the chromatographic procedure, the 2D-TLCs were performed by quintuplicate obtaining repetitive results, evaluated by their respective R_F values.

2.4. 2D-TLC Bioautographic Assay.

In order to locate spots with probable antioxidant activity, the 2D-TLC plates were carefully dipped for 3-5 seconds in a methanolic solution of DPPH (0.25 mM) and dried at room temperature for 30 seconds. Then, the first photographic record was taken (t_0); subsequently, more photographs were taken each hour, for 12 hours. Additional pictures were taken every 12 hours up to 48 hours and, for the MP-5c/MP-4a system (1st development/2nd development), some extra photographs were collected at longer times. Nevertheless, only the images taken at t_0 and when the spots showed maximum intensity (t_F; for the MP-1/MP-1 system, t_F =6 h and, for the MP-5c/MP-4a one, t_F =7 days) are shown.

3. Results and Discussion

3.1. 1D-TLCs.

In order to separate the compounds of the methanolic extract of *Spirulina platensis*, a "family" of 1D-TLCs was prepared. For this purpose, just as starting point, the mobile phase used by Zarzycki et al. [39] was used, performing a systematic modification of such mobile phase to increase its relative polarity. The relative polarity of mobile phases was qualitatively estimated considering the solubility parameters of the correspondent pure solvents (δ); such criteria are commonly used to design binary solvents with a gradual decreasing of its solvation capacity [41]. This methodology considers that when the relative amount of the solvents is kept constant, using a solvent with a higher solubility parameter instead of another with a lower δ value produces a binary solvent with a higher polarity. Similarly, the polarity of the binary system increases as the relative content of the solvent with higher δ value increases [41].

Representative photographs of the TLC strips developed with the indicated mobile phases are shown in Figure 1. Under visible light, eleven spots were obtained when the MP-1 was used; this result was equivalent to the reported previously [39], demonstrating that such experiment was successfully reproduced. Besides, six additional spots were observed under UVL_{366}. In contrast, only five additional spots were reported by Zarzycki when observed after iodine vapor exposure [39]. Nevertheless, since the well-resolved spots are especially important to this work, to establish the number of this type of spots is relevant. Thus, when analyzing the chromatograms obtained with the MP-1 (Figure 1) and their respective retention factors (Table 2), it was observed that only three well-resolved spots were obtained (since for spot #1, R_F= 0.97, there is uncertainty about whether that spot

TABLE 2: Retention factor (R_F) values of spots observed in the 1D-TLC plates, developed using the indicated mobile phases.

MP-1	VL		UVL$_{366}$	MP-2	VL		UVL$_{366}$	MP-3	VL		UVL$_{366}$
1	0.97	12	0.76	1	0.98	7	0.97	1	0.84	8	0.84
2	0.93*	13	0.66	2	0.93	8	0.79*	2	0.82	9	0.71*
3	0.90*	14	0.47	3	0.87*	9	0.70*	3	0.78	10	0.62*
4	0.84*	15	0.41	4	0.76*	10	0.26*	4	0.68*	11	0.36*
5	0.79	16	0.37	5	0.07	11	0.08	5	0.10*	12	0.13*
6	0.69	17	0.07	6	0.00	12	0.03	6	0.04	13	0.04
7	0.58					13	0	7	0.00	14	0
8	0.50										
9	0.11										
10	0.04										
11	0.00										

MP-4a	VL		UVL$_{366}$	MP-4b	VL		UVL$_{366}$	MP-4c	VL		UVL$_{366}$
1	0.92	7	0.92	1	0.83	8	0.83	1	0.89	8	0.81*
2	0.84	8	0.79*	2	0.77*	9	0.71*	2	0.87	9	0.59*
3	0.51*	9	0.23*	3	0.69*	10	0.53	3	0.77*	10	0.39*
4	0.43*	10	0.19*	4	0.32	11	0.33	4	0.70*	11	0.18*
5	0.07	11	0.04	5	0.11*	12	0.22*	5	0.49*	12	0.00
6	0.00	12	0.00	6	0.06	13	0.07	6	0.04		
				7	0.00	14	0	7	0		

MP-5a	VL		UVL$_{366}$	MP-5b	VL		UVL$_{366}$	MP-5c	VL		UVL$_{366}$
1	0.94	16	0.92	1	0.96	16	0.93	1	0.96	16	0.93
2	0.92	17	0.44*	2	0.93	17	0.70*	2	0.93	17	0.70*
3	0.88	18	0.37*	3	0.91	18	0.46	3	0.91	18	0.59*
4	0.80*	19	0.29*	4	0.58*	19	0.38*	4	0.77*	19	0.49*
5	0.41*	20	0.26*	5	0.54*	20	0.28	5	0.74*	20	0.43*
6	0.36*	21	0.16*	6	0.50*	21	0.16*	6	0.68*	21	0.39*
7	0.32*			7	0.42	22	0.09*	7	0.52*	22	0.32*
8	0.24*			8	0.32*	23	0.00	8	0.50*	23	0.27
9	0.20*			9	0.26			9	0.48*	24	0.17*
10	0.13			10	0.21*			10	0.42*	25	0.04*
11	0.11*			11	0.18*			11	0.38*		
12	0.10*			12	0.14*			12	0.30*		
13	0.09*			13	0.07			13	0.28		
14	0.04			14	0.03			14	0.11*		
15	0.00			15	0.00			15	0.00		

*Spots considered as well resolved.

is a pure component or a mixture [28, 29]; therefore, this spot was not considered well resolved). When the behavior shown by the other binary mobile phases is considered (Figure 1 and Table 2), it can be observed that a considerably higher number of well-resolved spots was obtained, the best results being with the MP-4c, where 7 well-resolved spots can be observed. Thus, although the highest number of total spots was obtained with the MP-1, the highest number of well-resolved spots (the goal of this work) was obtained with the MP-4c, followed by the mobile phases MP-4a and MP-4b.

Besides, to make a pseudo-reverse phase that allows compounds with a high polarity to be resolved more adequately, a small amount of phase modifier (NH_4OH) was added to the last three systems (MP-4a, b, and c), which promoted a considerable improvement in the chromatographic separation behavior (Figure 1). Thus, the number of total spots (T), as well as the number of well-resolved spots (W) obtained with phases MP-5a (T=21 and W=13), MP-5b (T=23 and W=11), and MP-5c (T=25 and W=18), was noticeably higher than the ones obtained with the equivalent phases without phase

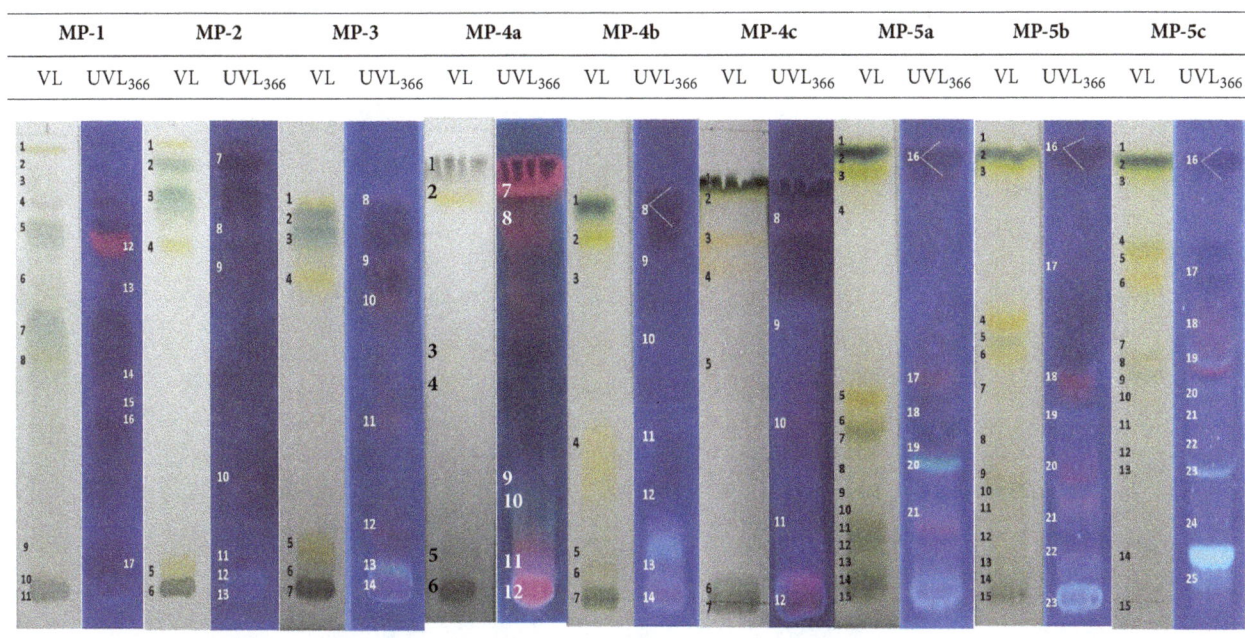

FIGURE 1: Representative photographs of 1D thin layer chromatograms for a methanolic extract of *Spirulina platensis*, developed with the indicated mobile phases and visualized under visible light (VL) or under ultraviolet light (UVL$_{366}$).

modifier (MP4a: T= 12 and W=5; MP-4b: T=14 and W=5; MP-4c: T=12 and W=7), which demonstrate the utility of the NH$_4$OH presence in the system (Table 2). A global analysis of the 1D-TLC results shows that, with the MP-5c, a suitable balance of the interactions among the solvent system, the stationary phase (partially masked by the NH$_4$OH), and the different metabolites was obtained, allowing for their gradual separation.

3.2. 2D-TLCs. To further improve the resolution of the components contained in the methanolic extract of *Spirulina platensis*, a wide experimental set of 2D-TLCs was performed, obtaining the best results with the system that used in the first development (1st) the mobile phase named MP-5c and, in the second one (2nd), the MP-4a. Thus, in Figure 2 representative photographs of chromatographic plates developed by 2D-TLC are presented, which used the following solvent systems: (a) 1st: MP-1 and 2nd: MP-1 (herein referred to as the starting system) and (b) 1st: MP-5c and 2nd: MP-4a (herein referred to as the "best tested system"). Further, in Table 3 the corresponding R$_F$ values for such experimental systems are presented.

Regarding the separating behavior of the starting system (a system of 2D-TLC using the same mobile phase in both developments), as it could be expected [42, 43], the separation quality was improved by the application of the second development. Thus, in the 1D-TLC that used the MP-1, seventeen spots were obtained, but only three well-resolved spots (Figure 1), whereas in the 2D-TLC, 20 total spots could be assessed and ten of them were considered well resolved (Figure 2).

In addition, a comparative analysis of the separating behavior of the last mentioned systems (Figure 2 and Table 3) shows that since the "best tested system" (MP-5c/MP-4a)

exhibits 28 spots, with fifteen of them being considered well-resolved spots, it can be affirmed that its separation behavior represents a considerable improvement on the chromatographic resolution, when it is compared to the starting system (MP-1/MP-1; 20 total spots and 10 well-resolved ones).

3.3. 2D-TLC Bioautographic Assays. Since the main goal of this work is to identify well-resolved spots with antioxidant activity, in a preliminary step, the yellowish characteristic produced by the DPPH technique was looked for only on the well-resolved spots (Table 3). Thus, in Figure 2 it is observed that only spots # 6 and # 8 are useful for providing material susceptible to being used in a subsequent identification procedure (e.g., well-resolved and containing compounds with antioxidant activity). In an equivalent analysis, but with the MP-5c/MP-4a system, a higher number of useful spots were identified, specifically, spots #12, #14, #16, and #29. It is important to mention that spot #29 could be observed neither with visible light nor with UV light, but it could be observed when its component(s) reacted as a consequence of the DPPH addition. Taking into account only the above-mentioned information, the improvement reached with the proposed 2D-TLC bioautographic assays is evident.

A more in-depth analysis of the MP-5c/MP-4a system showed in Figure 2 that although spots # 4 (green), #6 (more intense orange), and #9 (less intense orange) are qualitatively distinguishable by their respective colors, they exhibit a considerable overlapped area (spot #4 with #6 and #6 with #9); therefore, they were considered not well-resolved spots. Nevertheless, after DPPH application, spot #6 appeared very quickly (in the photo taken starting the process (t$_0$) a weak yellowish color could be assessed), hinting toward a strong antioxidant activity occurring in such spot. In the case of spot

Mobile phase	Before DPPH addition	After DPPH addition

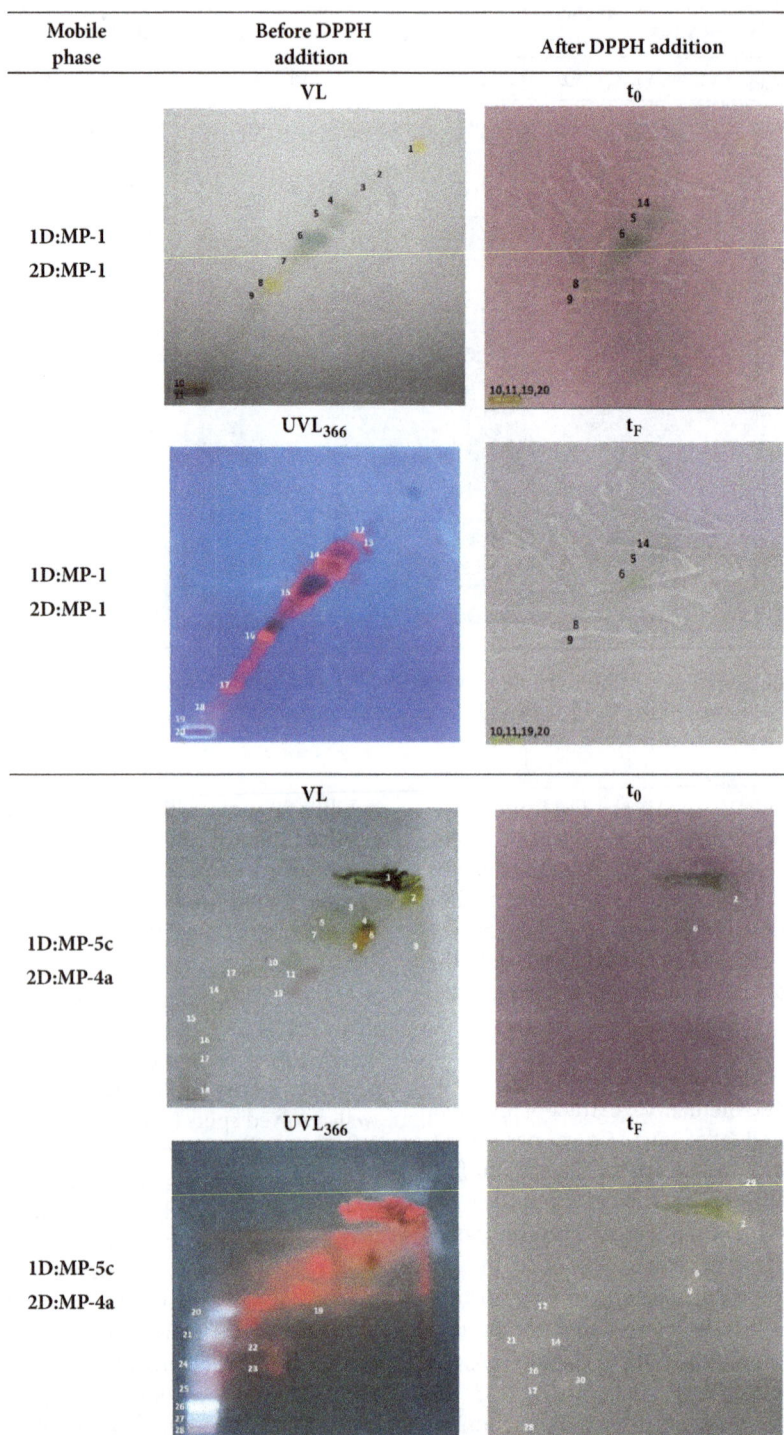

FIGURE 2: Representative photographs of 2D thin layer chromatograms for a methanolic extract of *Spirulina platensis*, developed with the indicated mobile phases and visualized under visible light (VL) and under ultraviolet light (UVL$_{366}$) (central column) and the correspondent 2D thin layer biochromatograms, obtained after DPPH treatment at t_0 and t_F (right column).

#9, its yellowish color could be assessed only after several hours, which can be interpreted as an antioxidant activity weaker than the one shown by component(s) present in spot #6. Finally, spot #4 never showed antioxidant activity. With such evidence, it could be useful to isolate the global area visualized in yellow in plate with DPPH to obtain a mixture

containing components with a noticeable global antioxidant activity. In addition, due to spot #28 remaining without displacement after both chromatographic developments, it was considered a non-well-resolved spot that possibly contained more than one component. Nevertheless, after the application of the DPPH, a very defined pale yellow spot appeared, which

TABLE 3: Retention factor (R_F) values observed in 2D-TLC plates, developed with the indicated mobile phases.

	VL			UVL$_{366}$	
Spot	MP-1	MP-1	Spot	MP-1	MP-1
1	0.74	0.76*	12	0.59	0.59*
2	0.66	0.66*	13	0.56	0.56*
3	0.62	0.62*	14	0.52	0.49
4	0.58	0.53*	15	0.41	0.38
5	0.55	0.49	16	0.30	0.26
6[+]	0.47	0.44*	17	0.15	0.15*
7	0.38	0.37	18	0.06	0.12*
8[+]	0.34	0.30*	19	0.04	0.02
9	0.31	0.26	20	0.00	0.00
10	0.03	0.02			
11	0.00	0.00			
	MP-5c	MP-4a		MP-5c	Mp-4a
1	0.79	0.79	19	0.46	0.59
2	0.74	0.86	20	0.46	0.09*
3	0.68	0.59*	21	0.38	0.06*
4	0.64	0.69	22	0.34	0.35
5	0.64	0.54*	23	0.26	0.35
6	0.60	0.71	24	0.26	0.03*
7	0.60	0.53*	25	0.16	0.02
8	0.56	0.88*	26	0.11	0.00*
9	0.56	0.68	27	0.05	0.00*
10	0.50	0.43*	28	0.00	0.00
11	0.45	0.48	29[+]	0.20	0.26
12[+]	0.45	0.29*			
13	0.38	0.43*			
14[+]	0.38	0.18*			
15	0.30	0.09*			
16[+]	0.22	0.00*			
17	0.15	0.00			
18	0.04	0.00			

*Spots considered as well resolved.
[+]Spots considered as well-resolved and containing compounds with antioxidant activity.

demonstrates that, regardless of the number of components, it is a mixture (or a pure component) potentially useful thanks to its antioxidant activity.

4. Conclusion

As a first step in the identification of the antioxidant compounds contained in methanolic extracts of *Spirulina platensis*, a simple and fast 2D-TLC biosystem was developed. The proposed experimental system allowed a suitable separation and localization of such type of components, whose dispersion on the plate was favored by the use of NH_4OH as a phase modifier. The last system was intentionally designed to be scaled to preparative TLC plates, which is a preliminary stage to the identification of components; at this time, we are working on that stage.

References

[1] G. Abdulqader, L. Barsanti, and M. R. Tredici, "Harvest of Arthrospira platensis from Lake Kossorom (Chad) and its household usage among the Kanembu," *Journal of Applied Phycology*, vol. 12, no. 3-5, pp. 493–498, 2000.

[2] O. Ciferri, "Spirulina, the edible microorganism," *Microbiology and Molecular Biology Reviews*, vol. 47, no. 4, pp. 551–578, 1983.

[3] E. W. Becker, "Micro-algae as a source of protein," *Biotechnology Advances*, vol. 25, no. 2, pp. 207–210, 2007.

[4] Z. Khan, P. Bhadouria, and P. S. Bisen, "Nutritional and therapeutic potential of *Spirulina*," *Current Pharmaceutical Biotechnology*, vol. 6, no. 5, pp. 373–379, 2005.

[5] S. K. Ali and A. M. Saleh, "Spirulina - An overview," *International Journal of Pharmacy and Pharmaceutical Sciences*, vol. 4, pp. 9–15, 2012.

[6] E. W. Becker and L. V. Venkataraman, "Production and utiliza-

tion of the blue-green alga Spirulina in India," *Biomass*, vol. 4, no. 2, pp. 105–125, 1984.

[7] E. Ross and W. Dominy, "The Nutritional Value of Dehydrated, Blue-Green Algae (Spirulina plantensis) for Poultry," *Poultry Science*, vol. 69, no. 5, pp. 794–800, 1990.

[8] J. C. M. Carvalho, F. R. Francisco, K. A. Almeida, S. Sato, and A. Converti, "Cultivation of Arthrospira (Spirulina) platensis (Cyanophyceae) by fed-batch addition of ammonium chloride at exponentially increasing feeding rates," *Journal of Phycology*, vol. 40, no. 3, pp. 589–597, 2004.

[9] L. M. Colla, E. B. Furlong, and J. A. V. Costa, "Antioxidant properties of Spirulina (Arthospira) platensis cultivated under different temperatures and nitrogen regimes," *Brazilian Archives of Biology and Technology*, vol. 50, no. 1, pp. 161–167, 2007.

[10] K. Kim, D. Hoh, Y. Ji, H. Do, B. Lee, and W. Holzapfel, "Impact of light intensity, CO2 concentration and bubble size on growth and fatty acid composition of Arthrospira (Spirulina) platensis KMMCC CY-007," *Biomass & Bioenergy*, vol. 49, pp. 181–187, 2013.

[11] S. Benelhadj, A. Gharsallaoui, P. Degraeve, H. Attia, and D. Ghorbel, "Effect of pH on the functional properties of Arthrospira (Spirulina) platensis protein isolate," *Food Chemistry*, vol. 194, Article ID 18081, pp. 1056–1063, 2016.

[12] P. Bermejo-Bescós, E. Piñero-Estrada, and Á. M. Villar del Fresno, "Neuroprotection by *Spirulina platensis* protean extract and phycocyanin against iron-induced toxicity in SH-SY5Y neuroblastoma cells," *Toxicology in Vitro*, vol. 22, no. 6, pp. 1496–1502, 2008.

[13] N. Kumar, S. Singh, N. Patro, and I. Patro, "Evaluation of protective efficacy of *Spirulina platensis* against collagen-induced arthritis in rats," *Inflammopharmacology*, vol. 17, no. 3, pp. 181–190, 2009.

[14] A. Karadeniz, M. Cemek, and N. Simsek, "The effects of Panax ginseng and Spirulina platensis on hepatotoxicity induced by cadmium in rats," *Ecotoxicology and Environmental Safety*, vol. 72, no. 1, pp. 231–235, 2009.

[15] A. Méndez, E. Bosch, M. Rosés, and U. D. Neue, "Comparison of the acidity of residual silanol groups in several liquid chromatography columns," *Journal of Chromatography A*, vol. 986, no. 1, pp. 33–44, 2003.

[16] G. B. Cox, "The influence of silica structure on reversed-phase retention," *Journal of Chromatography A*, vol. 656, no. 1-2, pp. 353–367, 1993.

[17] S. D. Rogers and J. G. Dorsey, "Chromatographic silanol activity test procedures: The quest for a universal test," *Journal of Chromatography A*, vol. 892, no. 1-2, pp. 57–65, 2000.

[18] S. Bocian and B. Buszewski, "Residual silanols at reversed-phase silica in HPLC - A contribution for a better understanding," *Journal of Separation Science*, vol. 35, no. 10-11, pp. 1191–1200, 2012.

[19] M. C. García-Alvarez-Coque, G. Ramis-Ramos, and J. J. Baeza-Baeza, "Reversed phase liquid chromatography," in *Analytical Separation Science*, vol. 1, pp. 159–198, Wiley, New York, NY, USA, 2015.

[20] S. Calabuig-Hernández, M. C. García-Alvarez-Coque, and M. J. Ruiz-Angel, "Performance of amines as silanol suppressors in reversed-phase liquid chromatography," *Journal of Chromatography A*, vol. 1465, pp. 98–106, 2016.

[21] R. P. W. Scott and P. Kucera, "Some aspects of the chromatographic properties of thermally modified silica gel," *Journal of Chromatographic Science (JCS)*, vol. 13, no. 7, pp. 337–346, 1975.

[22] M. L. Hair and W. Hertl, "Adsorption on hydroxylated silica surfaces," *The Journal of Physical Chemistry C*, vol. 73, no. 12, pp. 4269–4276, 1969.

[23] S. R. Cole and J. G. Dorsey, "Cyclohexylamine Additives for Enhanced Peptide Separations in Reversed Phase Liquid Chromatography," *Biomed. Chromatogr*, vol. 1, p. pp, 1997.

[24] T. Tuzimski and E. Soczewiński, "Correlation of retention parameters of pesticides in normal- and reversed-phase systems and their utilization for the separation of a mixture of 14 triazines and urea herbicides by means of two-dimensional thin-layer chromatography," *Journal of Chromatography A*, vol. 961, no. 2, pp. 277–283, 2002.

[25] T. Tuzimski and E. Soczewinski, "Use of a database of plots of pesticide retention (RF) against mobile-phase composition. Part I. Correlation of pesticide retention data in normal- and reversed-phase systems and their use to separate a mixture of ten pesticides by 2D TLC," *Chromatographia*, vol. 56, no. 3-4, pp. 219–223, 2002.

[26] A. Nahum and C. Horváth, "Surface silanols in silica-bonded hydrocarbonaceous stationary phases. I. Dual retention mechanism in reversed-phase chromatography," *Journal of Chromatography A*, vol. 203, no. C, pp. 53–63, 1981.

[27] K. E. Bij, C. Horváth, W. R. Melander, and A. Nahum, "Surface silanols in silica-bonded hydrocarbonaceous stationary phases. II. Irregular retetion behavior and efect of silanol masking," *Journal of Chromatography A*, vol. 203, pp. 65–84, 1981.

[28] A. Wehrli, J. C. Hildenbrand, H. P. Keller, R. Stampfli, and R. W. Frei, "Influence of organic bases on the stability and separation properties of reversed-phase chemically bonded silica gels," *Journal of Chromatography A*, vol. 149, no. C, pp. 199–210, 1978.

[29] C. Meyers and D. Meyers, *Thin-Layer Chromatography, in Current Protocols in Nucleic Acid Chemistry*, John Wiley & Sons, Inc, Hoboken, NJ, USA.

[30] L. Zhang, J. Shi, J. Tang et al., "Direct coupling of thin-layer chromatography-bioautography with electrostatic field induced spray ionization-mass spectrometry for separation and identification of lipase inhibitors in lotus leaves," *Analytica Chimica Acta*, vol. 967, pp. 52–58, 2017.

[31] M. Jamshidi-Aidji and G. E. Morlock, "Bioprofiling of unknown antibiotics in herbal extracts: Development of a streamlined direct bioautography using Bacillus subtilis linked to mass spectrometry," *Journal of Chromatography A*, vol. 1420, pp. 110–118, 2015.

[32] A. Marston, "Thin-layer chromatography with biological detection in phytochemistry," *Journal of Chromatography A*, vol. 1218, no. 19, pp. 2676–2683, 2011.

[33] B. K. Głód, P. M. Wantusiak, P. Piszcz, E. Lewczuk, and P. K. Zarzycki, "Application of micro-TLC to the total antioxidant potential (TAP) measurement," *Food Chemistry*, vol. 173, pp. 749–754, 2015.

[34] L. Gu, T. Wu, and Z. Wang, "TLC bioautography-guided isolation of antioxidants from fruit of *Perilla frutescens* var. *acuta*," *LWT - Food Science and Technology*, vol. 42, no. 1, pp. 131–136, 2009.

[35] Ł. Cieśla, J. Kryszeń, A. Stochmal, W. Oleszek, and M. Waksmundzka-Hajnos, "Approach to develop a standardized TLC-DPPH test for assessing free radical scavenging properties of selected phenolic compounds," *Journal of Pharmaceutical and Biomedical Analysis*, vol. 70, pp. 126–135, 2012.

[36] P. Molyneux, "The use of the stable free radical diphenylpicryl-

hydrazyl (DPPH) for estimating antioxidant activity Philip," *Songklanakarin Journal of Science and Technology*, vol. 26, pp. 211–219, 2004, http://pubsonline.informs.org/doi/abs/10.1287/isre.6.2.144.

[37] W. Brand-Williams, M. E. Cuvelier, and C. Berset, "Use of a free radical method to evaluate antioxidant activity," *LWT - Food Science and Technology*, vol. 28, no. 1, pp. 25–30, 1995.

[38] M. S. Blois, "Antioxidant determinations by the use of a stable free radical," *Nature*, vol. 181, no. 4617, pp. 1199-1200, 1958.

[39] P. K. Zarzycki, M. B. Zarzycka, V. L. Clifton, J. Adamski, and B. K. Głód, "Low-parachor solvents extraction and thermostated micro-thin-layer chromatography separation for fast screening and classification of spirulina from pharmaceutical formulations and food samples," *Journal of Chromatography A*, vol. 1218, no. 33, pp. 5694–5704, 2011.

[40] P. K. Zarzycki, M. M. Ślaogonekczka, M. B. Zarzycka, E. Włodarczyk, and M. J. Baran, "Application of micro-thin-layer chromatography as a simple fractionation tool for fast screening of raw extracts derived from complex biological, pharmaceutical and environmental samples," *Analytica Chimica Acta*, vol. 688, no. 2, pp. 168–174, 2011.

[41] F. Francuskiewicz, "Appendix," in *Polymer Fractionation*, pp. 189–191, Springer-Verlag, Berlin, Germany, 1994.

[42] B. Spangenberg, C. F. Poole, and C. Weins, "Theoretical Basis of Thin Layer Chromatography (TLC)," in *Quantitative Thin-Layer Chromatography*, pp. 13–52, Springer Berlin Heidelberg, Berlin, Germany, 2010.

[43] P. K. Zarzycki, "Simple horizontal chamber for thermostated micro-thin-layer chromatography," *Journal of Chromatography A*, vol. 1187, no. 1-2, pp. 250–259, 2008.

5

Rheological Behavior of Tomato Fiber Suspensions Produced by High Shear and High Pressure Homogenization and their Application in Tomato Products

Yong Wang ⓘ,[1,2] Ping Sun,[3] He Li ⓘ,[1] Benu P. Adhikari,[4] and Dong Li[5]

[1]Beijing Advanced Innovation Center for Food Nutrition and Human Health, Beijing Technology & Business University, 11 Fuchenglu, Beijing 100048, China
[2]Beijing Key Laboratory of Nutrition & Health and Food Safety, COFCO Nutrition & Health Research Institute, No. 4 Road, Future Science & Technology Park, Beijing 102209, China
[3]Processing Technology Research Center for Tomato, COFCO Tunhe, Changji, Xinjiang 831100, China
[4]School of Applied Sciences, RMIT University, City Campus, Melbourne, VIC 3001, Australia
[5]College of Engineering, China Agricultural University, P.O. Box 50, 17 Qinghua Donglu, Beijing 100083, China

Correspondence should be addressed to He Li; lihe@btbu.edu.cn

Academic Editor: Jiajia Rao

This study investigated the effects of high shear and high pressure homogenization on the rheological properties (steady shear viscosity, storage and loss modulus, and deformation) and homogeneity in tomato fiber suspensions. The tomato fiber suspensions at different concentrations (0.1%–1%, w/w) were subjected to high shear and high pressure homogenization and the morphology (distribution of fiber particles), rheological properties, and color parameters of the homogenized suspensions were measured. The homogenized suspensions were significantly more uniform compared to unhomogenized suspension. The homogenized suspensions were found to better resist the deformation caused by external stress (creep behavior). The apparent viscosity and storage and loss modulus of homogenized tomato fiber suspension are comparable with those of commercial tomato ketchup even at the fiber concentration as low as 0.5% (w/w), implying the possibility of using tomato fiber as thickener. The model tomato sauce produced using tomato fiber showed desirable consistency and color. These results indicate that the application of tomato fiber in tomato-based food products would be desirable and beneficial.

1. Introduction

Tomato (*Lycopersicon esculentum* Mill.) is one of the most popular fruits over the world because of its unique visual appeal, taste, and nutritional value as it contains ascorbic acid (vitamin C) and lycopene [1]. Processed tomato products such as purees and sauces are a primary source of tomatoes in contemporary diet. Considerable research has been undertaken in the past to quantify and elucidate the natural consistency and structure of tomato products [2].

From a structural point of view, most tomato products are aqueous dispersion containing aggregated or disintegrated cells and cell wall material dispersed in water soluble tomato components. The consistency of processed tomato products arises from the cell wall components such as cellulose, semicellulose, pectin, and interactions among these components [2]. Cellulose is major component of vegetable cell wall suspensions and it is also the main component that affects the rheology of processed tomato products. Pectins are embedded naturally within the cellulose backbone and they are also found in the serum phase. They are known to contribute to the structure of tomato products significantly depending on the processing conditions [3–6].

Homogenization is a key processing step in the production of ketchup, sauces, and other tomato products. The homogenization process decreases the mean particle size of the tomato suspensions and imparts smoother texture and higher viscosity. It also alters the nature of the suspensions network and increases the viscosity of the suspensions [7, 8]. During homogenization, tomato pulp is subjected to very high turbulence, shear, cavitation, and impact when it is forced through the homogenizer [9]. The homogenization process was found to alter particle size distribution, pulp sedimentation behavior, serum cloudiness, color, and microstructure of tomato juice, by disrupting the suspended pulp particles [10]. High pressure homogenization was reported to decrease the particle size due to the disruption of matrix and increase tomato product's Bostwick consistency, probably due to the formation of fiber network [11]. The large discrete cells and cell fragments of tomato suspensions were easily degraded by homogenization which resulted into higher water-holding capacity [6, 7, 11]. The high pressure homogenization reduced the mean particle size and narrowed the particle size distribution thereby increasing the total surface area and the interaction among the particles [12]. Bengtsson et al. reported that the nonhomogenized tomato suspensions had swollen cell structure with relatively few cell aggregates; however, the homogenized suspensions contained large number of degraded cell fragments [13].

Tomato peel is a by-product of tomato industry and fiber is extracted from tomato peel using chemical method [14]. Tomato peel fiber contains about 80% of total dietary fiber (mainly water insoluble fiber) much higher than other vegetable by-products [15]. Due to its unique chemical composition and functional properties, tomato peel fiber can be used as a food supplement to improve physical, chemical, and nutritional properties of food products. However, the color and flavor of tomato peel fiber must be considered carefully to avoid their negative impact on the sensorial characteristics of the final products [16]. To date the tomato fiber has received very little research attention despite its ability to contribute to desirable food texture and good mouth feel.

To the best of our knowledge, there is no study on the effect of high shear and high pressure homogenization on the tomato fiber. Thus, this study aimed to study the effects of high shear and high pressure homogenization on the morphological and rheological properties of tomato fiber suspensions. We also compared the morphological, rheological, and color parameters of homogenized tomato fiber suspensions with those of commercial tomato ketchup and a model tomato sauce formulated for comparison. We believe that the findings presented in this paper will provide better understanding of the functional properties of tomato fiber and help broaden its application as an important thickening ingredient in food industry.

2. Materials and Methods

2.1. Materials. The tomato fiber sample was kindly provided by COFCO Tunhe Co. Ltd., Beijing, China. The solid content of this fiber sample was determined and found to be 4.80% (w/w). This fiber sample contained 2.11% (w/w) insoluble dietary fiber as tested following the AOAC Official Method 991.43 [17] and 1.12% (w/w) protein as tested using China's national food safety standards [18]. The tomato fiber was produced by concentrating and separating the solid part out of the tomato paste (without tomato peels or seeds), by using high speech rotary mechanical instrument.

The food grade tomato paste (29.0° Brix cold break), tomato ketchup, sugar, soybean fiber, and salt used in this study were provided by COFCO Tunhe Co. Ltd., Beijing, China. Deionized water was used to prepare samples.

2.2. Mechanical Treatments. The tomato fiber suspensions were prepared in four concentrations (0.1%, 0.25%, 0.5%, and 1%, w/w) by mixing raw tomato fiber with adequate amount of deionized water as calculated based on the moisture content of tomato fiber.

The shearing treatments were carried out using a laboratory disperser (IKA Ultra-Turrax T25, Germany). The tomato fiber suspensions were subjected to 3400 rpm, 5000 rpm, 8000 rpm, 10000 rpm, 12000 rpm, and 14000 rpm for 12 minutes each.

The above-mentioned sheared samples were homogenized using a high pressure homogenizer (ATS AH100D, Shanghai, China), which is a lab-scale homogenizer equipped with valve. The maximum pressure of this homogenizer is 140 MPa. The homogenization was carried out for 2 passes at 0 MPa, 5 passes at 5 MPa, and then another 5 passes at 10 MPa.

2.3. Determination of Morphology. Twenty milliliter of untreated, sheared, and homogenized suspensions were separately placed in colorimetric tubes. Images were captured with a digital camera in order to compare the appearance of these suspensions. The microscope images of all the above-mentioned samples were acquired. Very small drop of each sample was placed on a microscope slide and the pictures were taken using a microscope (Olympus CX31, Japan) at 100x and 400x magnification.

2.4. Rheological Measurements. Rheological measurements were performed using AR2000ex rheometer (TA Instruments Ltd., Crawley, UK). This is a controlled stress, direct strain, and controlled rate rheometer coming with torque range from 0.0001 to 200 mN·m and high stability normal force from 0.01 to 50 N. The parallel plate was used for all the tests. The temperature was controlled by a water bath connected to the Peltier system in the bottom plate. A thin layer of silicone oil was applied on the edges of samples in order to prevent evaporation. The linear viscoelastic region was determined for each sample through strain sweeps at 1 Hz (data not shown). Viscoelastic properties [storage (G'), loss (G'') modulus, and loss tangent (δ)] of samples were determined within the linear viscoelastic region. The samples were allowed to equilibrate for 2 min before each measurement.

The steady shear tests were performed at 25°C over the shear rate range of 0.01–100 s^{-1} to measure the apparent viscosity. A steel cone geometry (60 mm diameter, 59 μm gap) was chosen for these measurements, since cone geometry is more preferable for viscosity measurement.

The frequency sweep tests were performed at 25°C over the angular frequency range of 0.1–10 rad/s. The strain amplitude of these frequency sweep measurements was selected to be 1% according to the strain sweep results (data not

TABLE 1: Composition of tomato sauce prepared by using tomato paste and tomato fiber or soybean fiber.

Sample	Tomato paste	2.5% homogenized tomato fiber	Soybean fiber
P101	75 g	0	0
P102	76 g	8.33 g	0
P103	77 g	18.75 g	0
P104	78 g	32.14 g	0
P105	79 g	0	0.74 g
P106	80 g	0	1.48 g
P107	81 g	0	2.96 g

TABLE 2: The composition of tomato sauce prepared by using tomato fiber or soybean fiber, tomato paste, sugar, and salt.

Sample	Tomato paste (%)	2.5% tomato fiber homogenized (%)	Soybean fiber (%)	Sugar (%)	Salt (%)	Water (%)	Total
P110	80	--	2.5	6.2	0.9	10.4	100
P111	80	13	--	6.1	0.9	--	100
P112	75	16	--	8.1	0.9	--	100
P113	70	19	--	10.1	0.9	--	100

shown) in order to confine these tests within linear viscoelastic region. An aluminum parallel plate geometry (40 mm diameter, 1 mm gap) was chosen for these measurements.

Creep experiments were carried out at a fixed shear stress of 7.958 mPa at 25°C. The variation in shear strain in response to the applied stress was measured over a period of 2 min. An aluminum parallel plate geometry (40 mm diameter, 1 mm gap) was chosen for these creep measurements.

2.5. Preparation of Tomato Sauce. The formulation of tomato sauce samples used in the first round of tests is provided in Table 1. The tomato paste and homogenized tomato fiber or soybean fiber were mixed according to this formulation. Required amount of water was added to make the mass of the sample to be 110 g. The homogenized tomato fiber with 2.5% concentration was prepared as described in Section 2.2.

The formulation of tomato sauce for second round of tests is shown in Table 2. Two hundred grams of sauce was prepared for each formulation by measuring and mixing ingredients listed in Table 2. The mixture was then heated at 95°C for 10 min in a water bath with continuous stirring. The sauce container was covered during heating to minimize the evaporation of water. The sauce was finally cooled down to ambient temperature.

2.6. Analysis of Physicochemical Properties. Bostwick consistency was determined using a standard 24 cm Bostwick Consistometer with 48 × 0.5 cm graduations (Endecotts ZXCON-CON1, London, UK). Seventy-five mL of sample was used to perform these tests. As the fluid flows down the instrument, the measurements were carried out after 30 seconds.

Colorimetric tests were performed using a spectrophotometer (Hunter Lab UltraScan VIS, Reston, US) in transmission mode. The samples were filled into a 10 mL quartz transmission cell with 10 mm path length. The *L*, *a*, and *b*

values were calculated by the averaging the data of triplicate runs. The suspensions were shaken to achieve uniformity in color immediately before measurement.

The pH and total acidity of samples were measured using an automatic acid analyzer (Metrohm 877 Titrino plus, Switzerland).

In order to measure the Bostwick consistency, color, pH, and total acidity of the tomato source samples, the total soluble solids content was adjusted to 12.5° Brix in order to keep the same test condition. A refractometer (Atogo RX-5000α, Japan) was used for this purpose.

2.7. Statistical Analysis. All of the above-mentioned tests were carried out in triplicate. The rheological data was obtained directly from the AR2000ex rheometer software (TA Instruments Ltd., Crawley, UK). The averaged value of triplicate runs was reported as the measured value along with the standard deviation.

3. Results and Discussion

3.1. Effect of Homogenization on Suspension Morphology. The effect of mechanical treatment on the appearance of tomato fiber suspensions at solid concentrations of 0.1–1.0% (w/w) is shown in Figure 1. The solid content was easily precipitated towards the bottom of the tube in all of the untreated samples irrespective of fiber concentration and the amount of sediment increased with increase in fiber concentration. The uniformity of suspensions greatly increased after shear homogenization or high pressure homogenization. The uniformity was relatively poor in shear homogenized samples at 0.1% and 0.25% (w/w) concentration compared with that of high pressure homogenized samples. The uniformity of suspensions produced by shear homogenization and high pressure homogenization was similar at 0.5% and 1.0% (w/w). It has been previously reported that the more stable

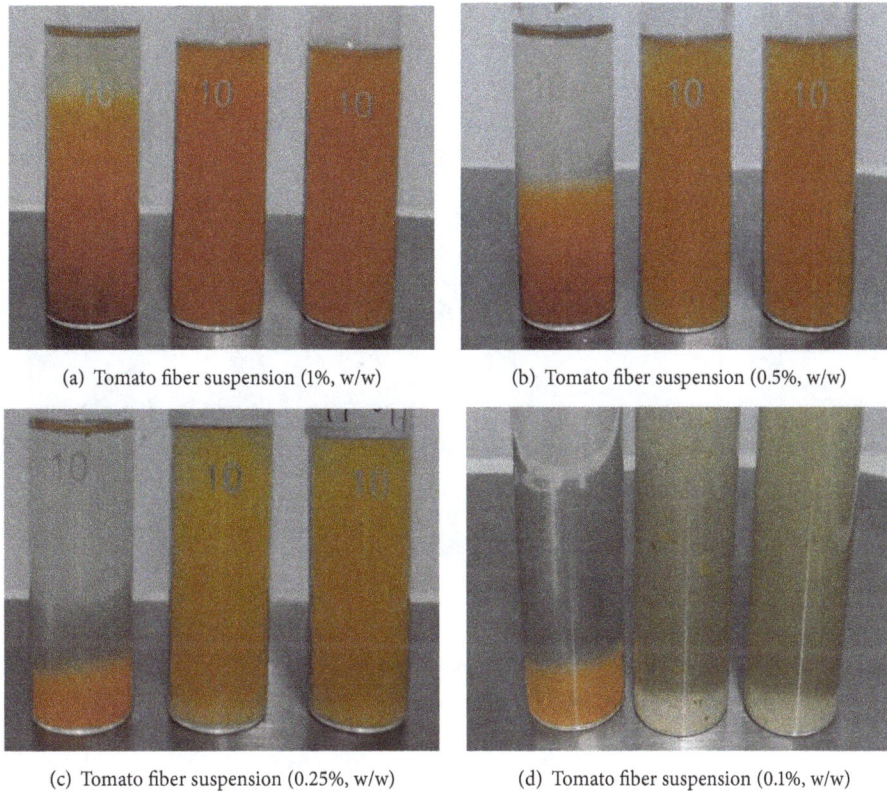

(a) Tomato fiber suspension (1%, w/w)

(b) Tomato fiber suspension (0.5%, w/w)

(c) Tomato fiber suspension (0.25%, w/w)

(d) Tomato fiber suspension (0.1%, w/w)

FIGURE 1: Photographs of tomato fiber suspensions at different concentration. In each photograph from left to right, untreated sample, high shear homogenized sample, and high pressure homogenized sample.

network structure can be formed in tomato fiber suspension when homogenized at 9 MPa [7]. It can be observed from photographs presented in Figure 1 that the shear homogenization affects only a part of the tomato fiber, most likely from tomato flesh. The fibers from tomato pericarp could only be fragmented under high pressure homogenization. The structural features of tomato fiber particles are drastically altered by the high pressure homogenization. It has been reported that the homogenized tomato fiber suspensions consisted of smashed cellular material which eventually formed fibrous-like network while the nonhomogenized suspensions consisted of a mixture of whole cells and dispersed cell wall materials [6].

The distribution of solids in tomato fiber suspensions is illustrated in Figure 2. Dark red discrete particles are observed in untreated and high shear homogenized samples at all concentrations, while the high pressure homogenized sample showed much better uniformity in solid distribution. The high pressure homogenized suspensions containing 0.5% or 1% (w/w) fiber began to exhibit water-holding properties, indicated by the increased height of tomato fiber sample on the glass (picture not shown). It was reported earlier that the homogenized tomato fiber suspensions showed higher water-holding capacity albeit at much higher solid concentrations (10% to 21.7%) [13]. This increased water-holding capacity would be a beneficial whenever the tomato fiber is used as an ingredient to impart desired texture in food products. The information presented in Figures 1 and 2 agree with

the findings in an earlier study [19] that the unhomogenized tomato juice showed whole cells with intact membranes and characteristic lycopene crystals while the homogenized samples showed large number of small particles composed of cell walls and internal constituents suspended in the juice serum.

The values of colorimetric parameters (L, a, and b) of unhomogenized tomato fiber suspensions at different concentration are presented in Table 3. The L and b values decreased with increase in fiber concentration while the a value showed substantial increase. The a/b value, which is of vital importance in the tomato processing industry, significantly ($p < 0.05$) increased with the increase in concentration. The a/b value of 2% (w/w) tomato fiber suspension suggested that this formulation has desirable color for potential application in tomato sauces. It has also been reported in an earlier study that the values for L^*, a^*, and b^* increased with the increase in homogenization pressure indicating that the tomato fiber suspensions became more saturated in red and yellow color [10].

The effects of high shear and high pressure homogenization on the 1% (w/w) tomato fiber suspension are shown in Figure 3. None of the L, a, or b parameters was significantly ($p > 0.05$) affected by the high shear homogenization or high pressure homogenization.

In order to illustrate the morphological changes caused by homogenization, the microscopic photographs of 1% (w/w) tomato fiber suspension are shown in Figure 4 before and

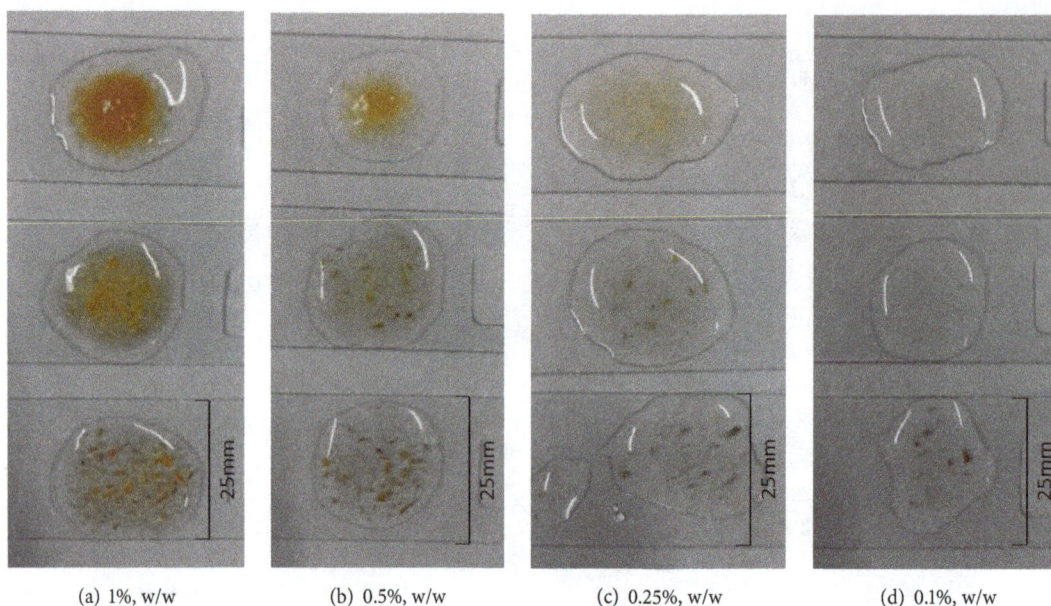

(a) 1%, w/w   (b) 0.5%, w/w   (c) 0.25%, w/w   (d) 0.1%, w/w

FIGURE 2: Microscopic photographs of tomato fiber suspensions showing distribution of fiber solids at different concentration. In each photograph from bottom to top, untreated sample, shear homogenized sample, and high pressure homogenized sample.

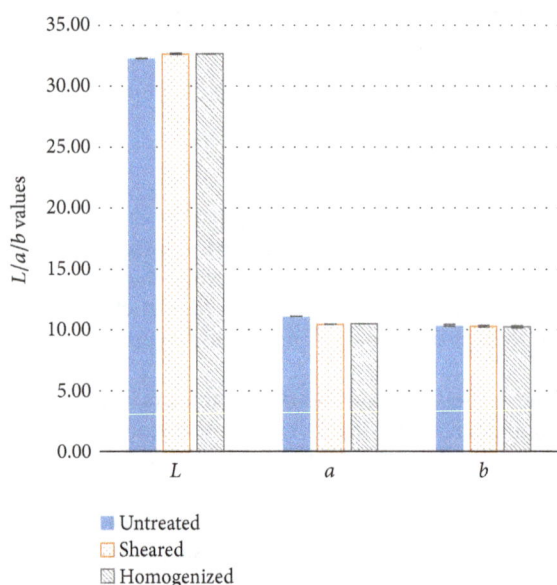

FIGURE 3: Effect of high shear and high pressure homogenization on the colorimetric parameters (L, a, and b) of 1% (w/w) tomato fiber suspension.

after homogenization. After high pressure homogenization, the solid tended to be evenly distributed at microscopic level (Figure 4(a)). The tomato fiber suspension showed a fibrous morphology with high degree of uniformity resembling a solution with negligibly very small amount of suspended solid after high pressure homogenization (shown in Figure 4(b)). The control samples showed unperturbed cells with intact membrane and the characteristic lycopene crystals. The homogenized samples showed a large number of small cell wall particles and internal cell constituents suspended in the

juice serum which agreed with Kubo et al. observation [10]. It has been reported that no intact cells were observed in tomato pulp subjected to high pressure (479 bar) homogenization and the internal cell constituents were found to be uniformly distributed in the homogenized pulp [9].

3.2. Effect of Homogenization on Rheological Properties. As shown in preceding section, the texture of tomato fiber suspensions could be significantly modified by homogenization. The effect of high shear and high pressure homogenization on the apparent viscosity is shown in Figure 5. All the tomato fiber suspensions showed shear-thinning behavior regardless of the concentration before and after homogenization. The apparent viscosity of all the samples increased with the increase in fiber concentration. The high shear homogenization significantly ($p < 0.05$) increased the apparent viscosity compared to the untreated sample. The application of high pressure homogenization increased the apparent viscosity the most (Figures 5(a)–5(d)). Augusto et al. reported that the viscosity of tomato juice (4.5° Brix) increased when the homogenization pressure increased from 50 MPa to 150 MPa [12]. Similar effect of high pressure homogenization which was on tomato suspensions was reported in various studies [6, 7, 20]. The cell wall of tomato cells could be broken even at moderate shear and this rupture is linked with the increase in viscosity.

The power law model (see (1)) was used to predict the variation of apparent viscosity with shear rate of tomato fiber suspensions.

$$\mu_a = K\dot{\gamma}^{n-1}, \tag{1}$$

where μ_a is the apparent viscosity (Pa·s), $\dot{\gamma}$ is the shear rate (s^{-1}), K is consistency coefficient (Pa·s^n), and n is the flow behavior index (dimensionless). The values of K and n for all the test samples were determined by fitting (1)

(a) High pressure homogenized (100x)

(b) High pressure homogenized (400x)

(c) Untreated (100x)

(d) Untreated (400x)

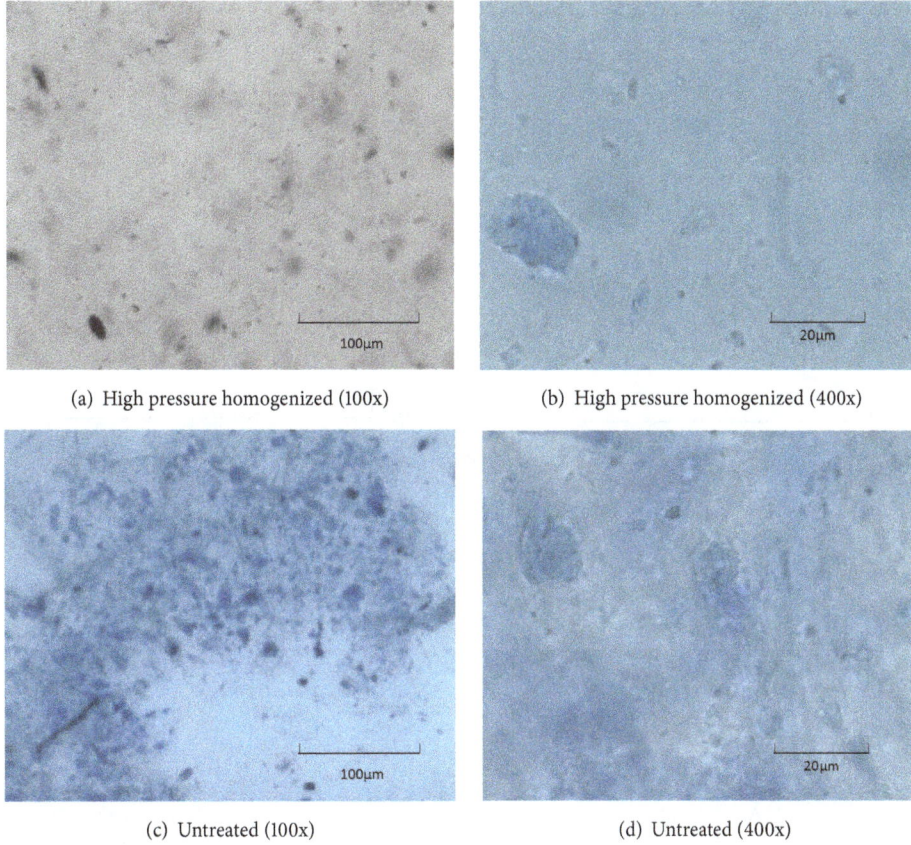

FIGURE 4: Microscopic photographs of 1% (w/w) tomato fiber suspension before and after mechanical treatment.

TABLE 3: Colorimetric parameters of unhomogenized tomato fiber suspensions at different concentration.

Conc.	L	a	b	a/b
0.25%	50.95 ± 1.79	4.63 ± 0.5	13.94 ± 0.27	0.33 ± 0.03
0.50%	40.35 ± 1.18	8.91 ± 0.25	14.80 ± 0.41	0.60 ± 0.033
1%	32.27 ± 0.02	11.08 ± 0.05	10.36 ± 0.01	1.07 ± 0.004
2%	32.23 ± 0.06	12.01 ± 0.02	10.39 ± 0.04	1.16 ± 0.004

to experimental apparent viscosity versus shear rate data presented in Figure 5 and are presented in Table 4. The flow behavior index (n) depends on the distribution of small and large particles and the rheology of the suspending fluid, while the consistency coefficient (K) depends on the maximum packing fraction (φ_m) and the distribution of small and large particles [21]. The K value increased very strongly with the increase of fiber concentration in all samples. The n value, which is indicator for shear-thinning behavior, was the lowest in pressure homogenized samples, the highest in the untreated samples, and intermediate in high shear homogenized samples at a given concentration. This means that the high pressure homogenized samples are most susceptible to shear thinning.

The values of storage modulus (G') of the homogenized and unhomogenized tomato fiber suspensions are shown in Figure 6. Both the homogenized and unhomogenized samples showed a slight increase of G' with the increase in angular frequency. At lower fiber concentrations (0.1%–1%),

the G' value of the high shear homogenized suspension increased more strongly compared to the unhomogenized sample. The increase of the G' value was the strongest in high pressure homogenized suspension which is similar to the variation of apparent viscosity with shear rate. This observation agrees with the earlier report that the homogenization process increases both storage and loss modulus of tomato suspension [7, 19].

The loss modulus (G'') of tomato fiber suspensions are presented in Figure 7. The G'' values increased with the increase in tomato fiber concentration. Both high shear and high pressure homogenization processes significantly ($p < 0.05$) increased the G'' values. The high pressure homogenization appears to be more effective in increasing G'' values as a function of angular frequency. All suspensions exhibited solid-like behavior with G' being higher than G''. Augusto et al. studied the effect of high pressure homogenization (up to 150 MPa) on the viscoelastic properties of tomato juice and found both G' and G'' when the juice was homogenized [22].

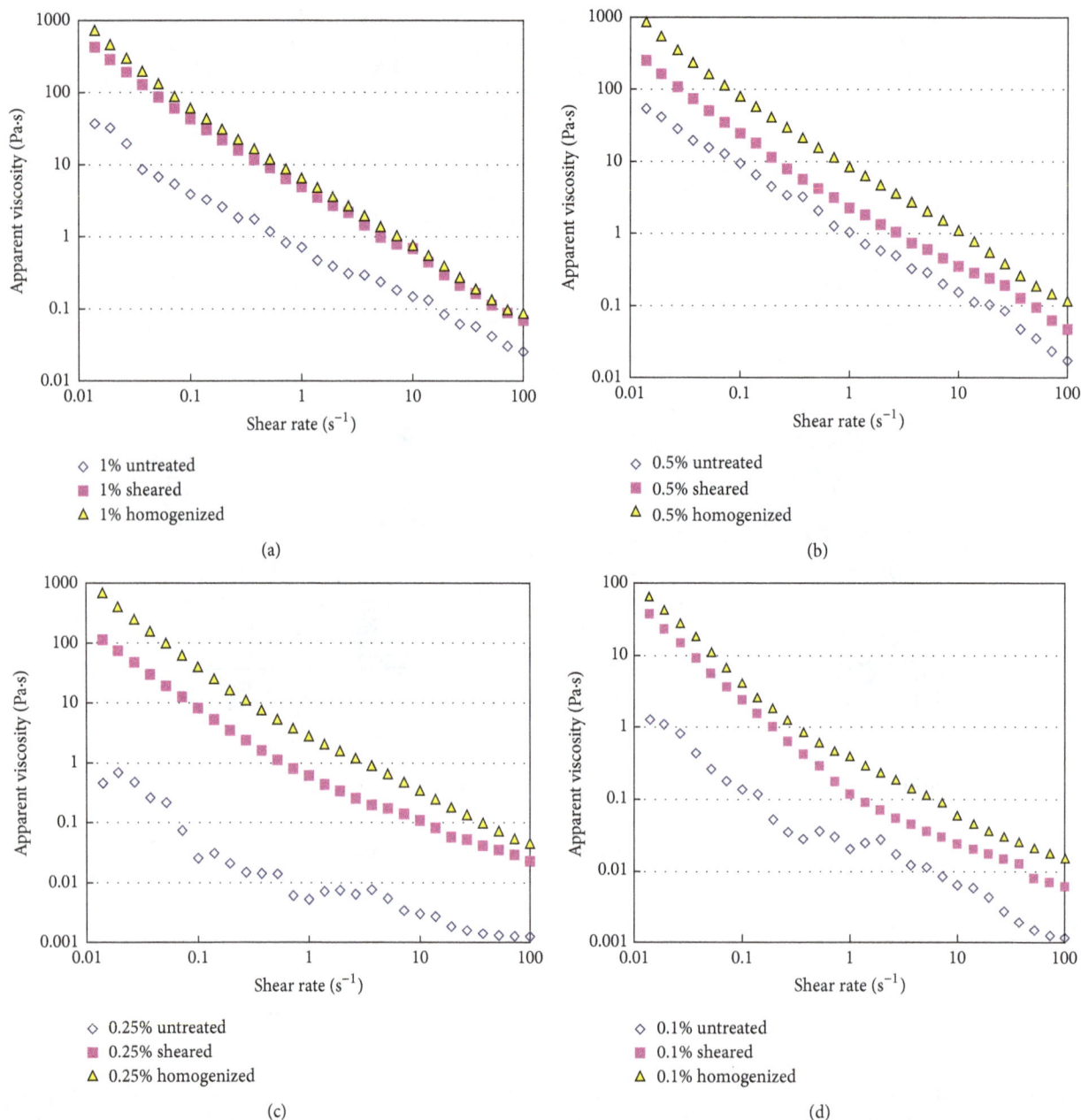

FIGURE 5: Flow behavior of tomato fiber suspensions before and after high shear or high pressure homogenization. Tomato fiber concentration: (a) 1%; (b) 0.5%; (c) 0.25%; (d) 0.1%; all in (w/w) basis.

The increase in homogenization pressure was also found to increase both G' (75.4 Pa to 212.2 Pa) and G'' (from 49.8 Pa to 80.9 Pa) in tomato suspensions [13].

The effect of homogenization on the creep behavior of tomato fiber suspensions is presented in Figure 8. At 1% (w/w) concentration, homogenized suspensions deformed less that the control sample under the same applied stress. The high pressure homogenized sample had the largest resistance to the applied stress among all the samples. This further indicates that homogenization helps build a stronger texture in the tomato fiber suspension, which could utilized to formulate food products with desirable texture. Figure 8 also shows that the slope of the creep curve is much smaller compared

to that of the control sample. This indicates that high shear and high pressure homogenized suspensions achieve an equilibrium state to maintain their solid-like structure sooner compared to the unhomogenized suspension. At the same stress, the unhomogenized suspension would continue to deform. This observation is consistent with earlier publication which reported that the homogenized tomato juice reduced the compliance of tomato juice due to stronger internal structure [19].

Based on all the rheological data presented above, it could be concluded that the rheological properties of tomato fiber could be significantly altered by the application of high shear or high pressure homogenization. The homogenized

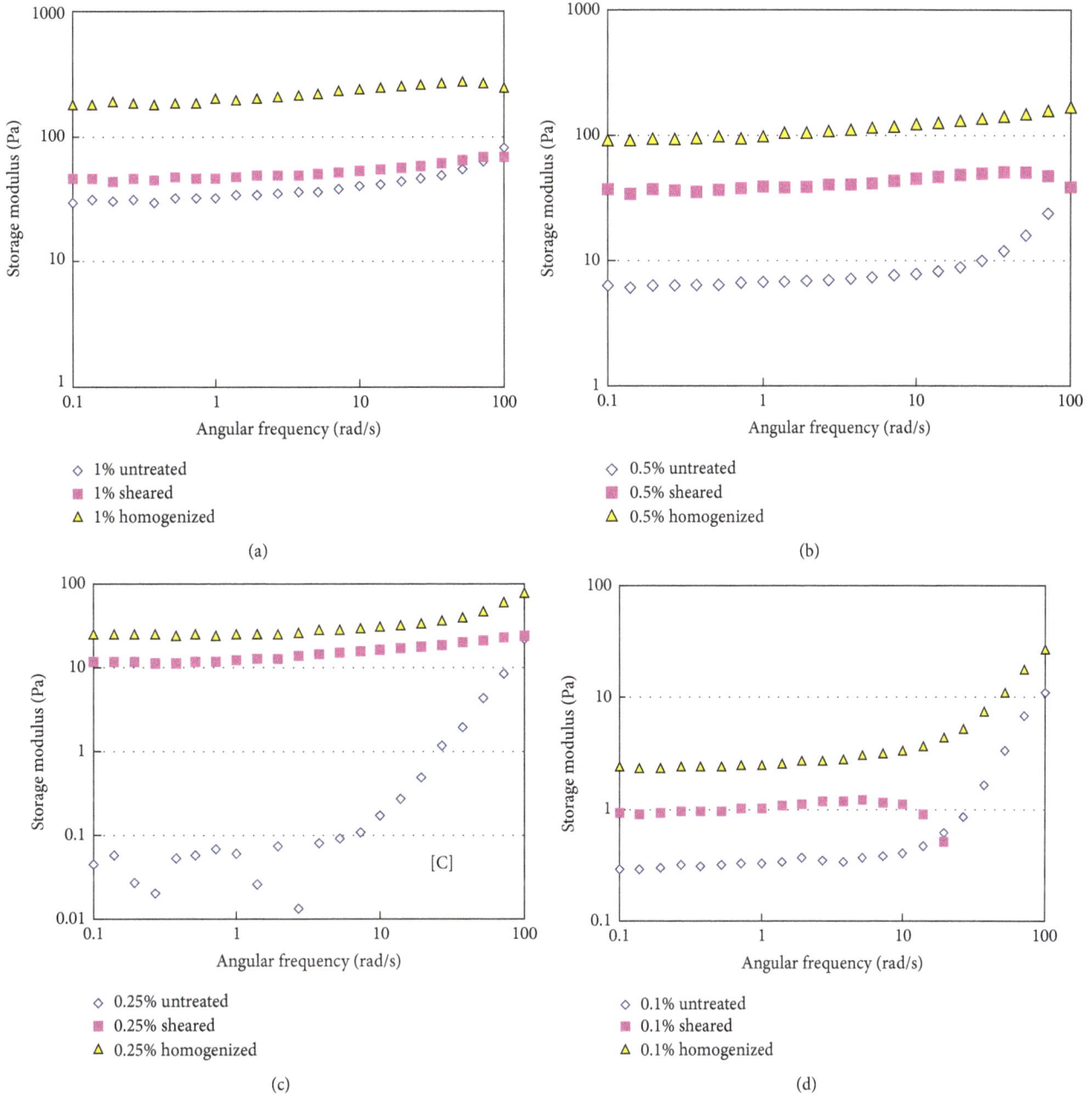

FIGURE 6: Storage modulus (G') of tomato fiber suspensions before and after homogenization. Tomato fiber concentration: (a) 1%; (b) 0.5%; (c) 0.25%; (d) 0.1% (w/w).

suspensions had higher apparent viscosity, higher G', and G'' and they could withstand larger external force and could maintain the solid-like structure better.

3.3. Comparison with Tomato Ketchup. Viscosity is a key indicator of quality of tomato paste and ketchup based on which consumers make their purchasing decision [23]. The apparent viscosity of high pressure homogenized tomato fiber suspension at 2.5% (w/w) fiber concentration was compared with that of tomato ketchup of 30° Brix (Figure 9). Despite the large difference in solid concentration between the two samples, they show similar shear-thinning behavior and

comparable apparent viscosity. Thus, the tomato fiber can replace other thickeners which might have been used in tomato ketchup, for example, pectin or xanthan gum.

The G' and G'' versus angular frequency curves of high pressure homogenized tomato fiber suspension (2.5%, w/w) and tomato ketchup (30° Brix) are presented in Figure 10. The curves of G'' versus angular frequency of these two samples were almost identical. The G' versus angular frequency curves of these samples bear similar trend. The storage modulus of the homogenized fiber suspension was higher than that of the tomato ketchup within the entire angular frequency range. This indicated that the fiber suspension had stronger

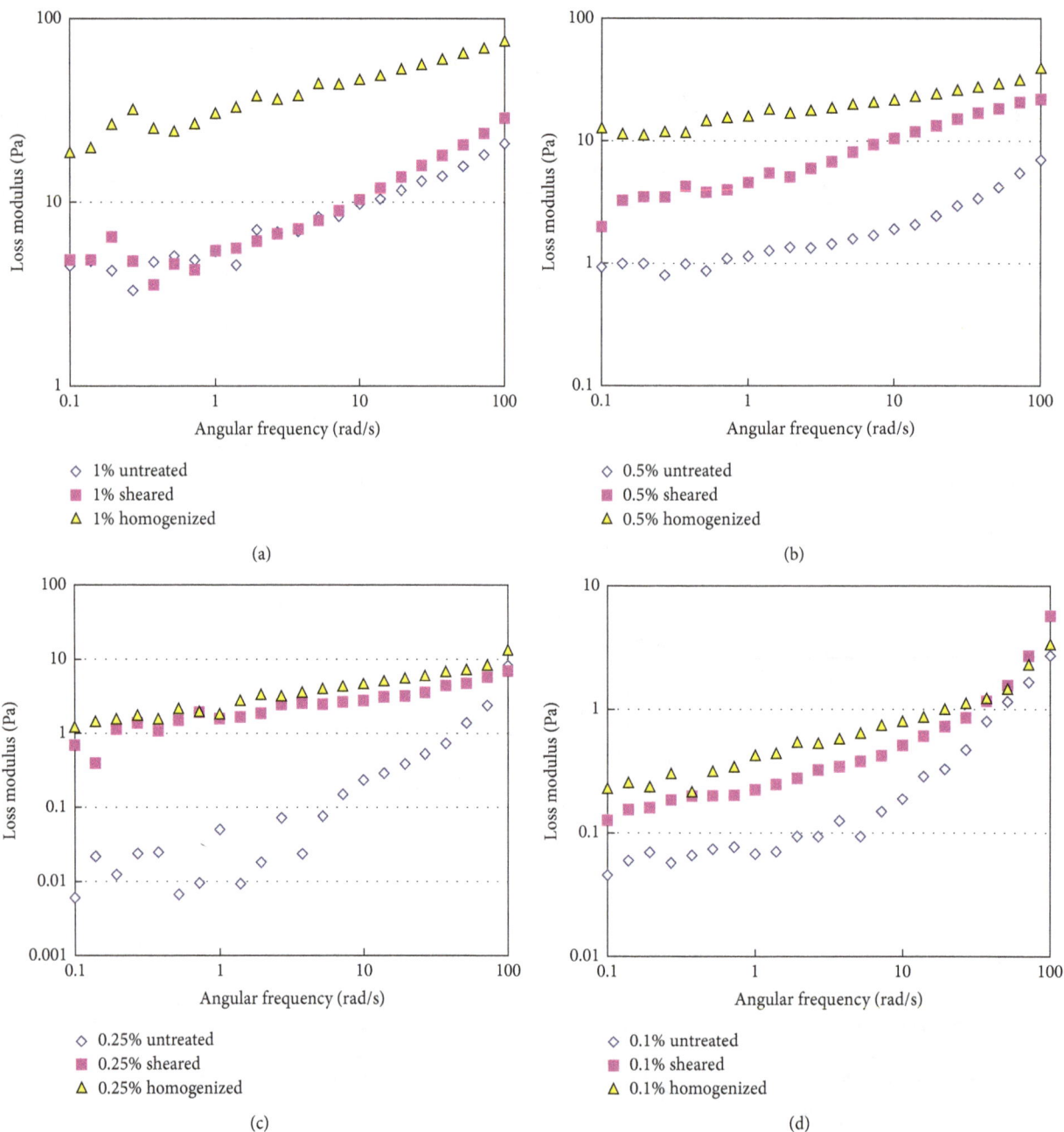

FIGURE 7: Loss modulus of tomato fiber suspension before and after homogenization. Tomato fiber concentration: (a) 1%; (b) 0.5%; (c) 0.25%; (d) 0.1% (w/w).

three-dimensional structure to resist external stress than the tomato ketchup. The viscoelastic characteristics of tomato sauce or ketchup are reported to depend on the diameter of the suspended particles water insoluble solids content [24]. The data presented in Figure 8 indicates that tomato fibers might be better choice if firmer or more solid-like texture is required.

The creep diagrams of high pressure homogenized tomato fiber suspension (2.5%, w/w) and tomato ketchup (30° Brix) are shown in Figure 11. The tomato fiber suspension deformed to a lesser extent than the tomato ketchup corroborating the fact that the tomato fibers provide firmer texture

than the ketchup, although the texture is also affected by concentration. According to a sensory evaluation data reported in earlier study the tomato suspension homogenized at 90 bar had significantly better thicker and smoother texture and significantly weaker graininess compared with the untreated sample [13].

3.4. Application of Tomato Fiber in the Formulation of Tomato Sauce. Dietary fibers such as soybean fiber are frequently added to produce tomato sauce. Thus, the effect of addition of homogenized tomato fiber or soybean fiber was measured and is presented in Figure 12. Bostwick consistency is

TABLE 4: Power law tomato fiber suspension before and after mechanical treatment.

	Sample	K (Pa·s^n)	n	R^2
Untreated	1%	0.73 ± 0.017	0.27 ± 0.0024	0.97
	0.5%	1.13 ± 0.031	0.12 ± 0.0013	0.82
	0.25%	0.01 ± 0.001	0.55 ± 0.0045	0.87
	0.1%	0.02 ± 0.001	0.36 ± 0.0028	0.85
High shear homogenized	1%	4.91 ± 0.038	0.07 ± 0.0006	0.85
	0.5%	2.65 ± 0.016	0.14 ± 0.0018	0.85
	0.25%	0.76 ± 0.008	0.18 ± 0.0014	0.75
	0.1%	0.18 ± 0.002	0.17 ± 0.0019	0.75
High pressure homogenized	1%	6.54 ± 0.045	0.04 ± 0.0004	0.75
	0.5%	8.82 ± 0.076	0.06 ± 0.0009	0.78
	0.25%	3.21 ± 0.032	0.04 ± 0.0006	0.74
	0.1%	0.43 ± 0.003	0.20 ± 0.0022	0.87

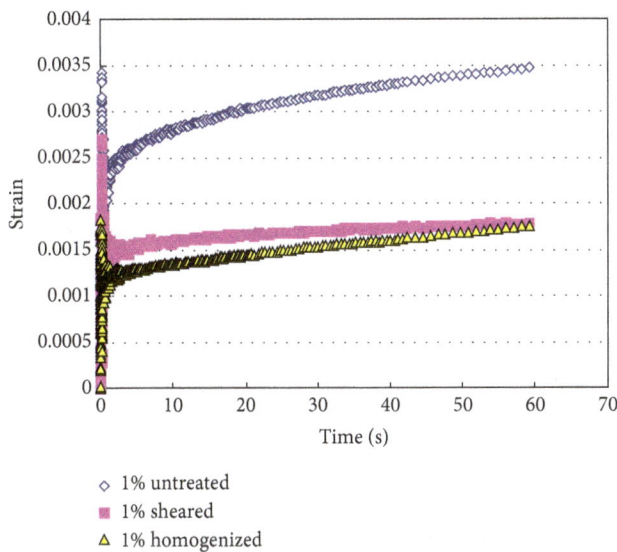

FIGURE 8: Creep diagrams of tomato fiber suspensions before and after homogenization. Tomato fiber concentration: 1% (w/w).

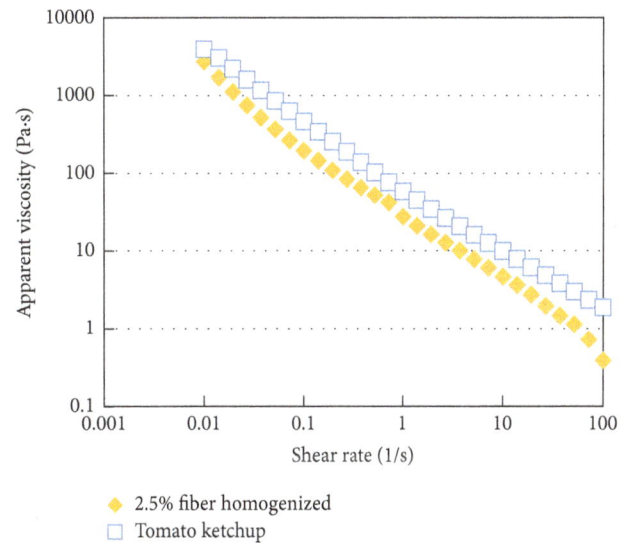

FIGURE 9: Apparent viscosity versus shear rate profiles of homogenized tomato fiber suspension and tomato ketchup (30° Brix).

employed in this section since it is more often used in the tomato industry than the rheological tests. A lower value of Bostwick consistency indicates a higher value of viscosity. As can be seen from this figure the addition of up to 0.5% (w/w) of tomato fiber could help the tomato sauce to achieve relatively high consistency. The amount of tomato fiber required would be one-third of the soybean fiber, to reach the same Bostwick consistency value. Typically, a tomato sauce with Bostwick consistency value of about 6–8 provides desirable texture or mouth feel of 0.2–0.5% dry fiber which is required.

A comparison of difference in color between the model tomato sauces prepared by using tomato fiber and soybean fiber is presented in Table 5. The Hunter color parameters (L, a, and b) and the ratio a/b are compared for these two formulations. A high value of a/b is desired in most tomato products. The a/b ratio containing tomato fiber is comparable but slightly higher compared to those containing soybean

fiber. A slight decrease in total acidity was also observed in sauce samples containing tomato fiber.

4. Conclusions

The effects of high shear and high pressure homogenization on the morphological and rheological properties of tomato fiber were investigated. Both the high shear and high pressure homogenization processes made these suspensions much more homogeneous which enabled even distribution of fiber particles. Both the high shear and high pressure homogenization significantly ($p < 0.05$) increased the apparent viscosity of the tomato fiber suspensions. The apparent viscosity of the high pressure homogenized suspension was 10 times higher than that of unhomogenized one. The storage and loss modulus of the homogenized suspensions were higher than those of the unhomogenized one within the angular frequency range

TABLE 5: The color parameters, pH, and total tomato sauce formulations prepared using high pressure homogenized tomato fiber or soybean fiber.

Sample	2.5% tomato fiber homogenized (%)	Soybean fiber (%)	a/b	L	a	b	pH	Total acid (%)
P110	--	2.5	2.17	24.38	30.05	13.82	4.09	1.9
P111	13	--	2.23	23.32	29.63	13.30	4.07	1.86
P112	16	--	2.27	23.61	29.12	12.84	4.07	1.79
P113	19	--	2.22	23.38	29.11	13.12	4.05	1.69

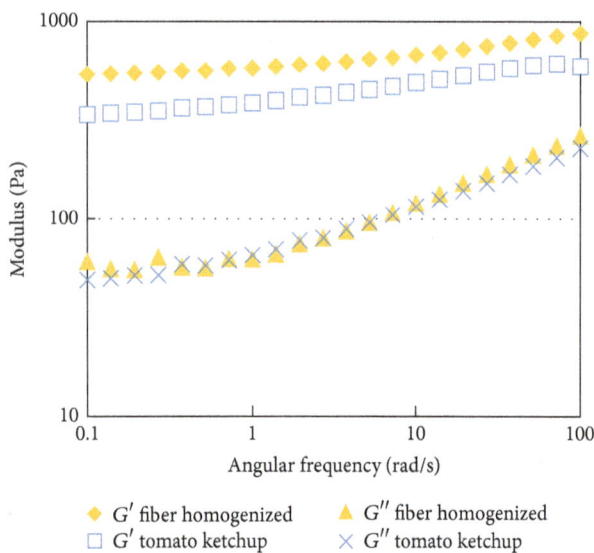

FIGURE 10: Storage and loss modulus of high pressure homogenized tomato fiber suspension (2.5%, w/w) and tomato ketchup (30° Brix).

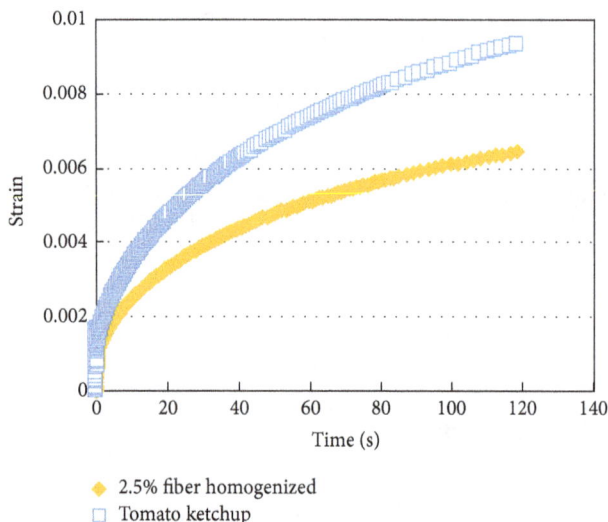

FIGURE 11: Creep diagram of high pressure homogenized tomato fiber suspension (2.5%, w/w) and tomato ketchup (30° Brix).

tested. The homogenized tomato fiber suspensions had more rigid structure compared to that of unhomogenized suspension and they resisted the deformation better (creep curve). The color and total acidity of model tomato sauce containing tomato fiber were more preferable than one containing soybean fiber at the same fiber content. The results presented

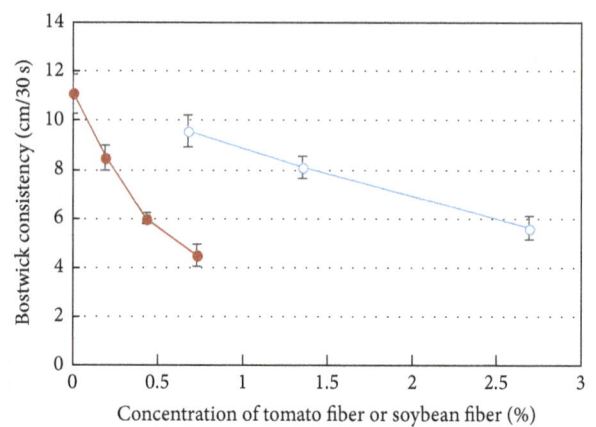

FIGURE 12: Bostwick consistency (cm/30 s) of tomato sauces prepared by using homogenized tomato fiber (0.19%–0.73%) or soybean fiber (0.67%–2.7%) using formula in Table 1. The red filled circles for tomato fiber; blue open circles for soybean fiber.

in this paper indicate that tomato fiber can be potentially used as food ingredient such as thickener or stablizer.

Acknowledgments

This research was supported by Beijing Advanced Innovation Center for Food Nutrition and Human Health Open Fund (20181043), China Postdoctoral Science Foundation Project (2012M520462), and Beijing Municipal Science and Technology Commission Project (D12110003112002). The authors wish to acknowledge COFCO Tunhe for kindly providing the test materials and access to the instruments. They also thank Mr. Jungang Zhao and Ms. Cheng Lv for their assistance in some of the experiments.

References

[1] J. L. Guil-Guerrero and M. M. Rebolloso-Fuentes, "Nutrient composition and antioxidant activity of eight tomato (Lycopersicon esculentum) varieties," *Journal of Food Composition and Analysis*, vol. 22, no. 2, pp. 123–129, 2009.

[2] A. K. Sankaran, J. Nijsse, L. Bialek, A. Shpigelman, M. E. Hendrickx, and A. M. Van Loey, "Enhanced electrostatic interactions in tomato cell suspensions," *Food Hydrocolloids*, vol. 43, pp. 442–450, 2015.

[3] S. K. Sharma, M. LeMaguer, A. Liptay, and V. Poysa, "Effect of composition on the rheological properties of tomato thin pulp," *Food Research International*, vol. 29, no. 2, pp. 175–179, 1996.

[4] N. Beresovsky, I. J. Kopelman, and S. Mizrahi, "The role of pulp interparticle interactions in determining tomato juice viscosity," *Journal of Food Processing and Preservation*, vol. 19, pp. 133–146, 1995.

[5] N. Takada and P. E. Nelson, "Pectineprotein interaction in tomato products," *Journal of Food Science*, vol. 48, pp. 1408–1411, 1983.

[6] E. Bayod, P. Månsson, F. Innings, B. Bergenståhl, and E. Tornberg, "Low shear rheology of concentrated tomato products. Effect of particle size and time," *Food Biophysics*, vol. 2, no. 4, pp. 146–157, 2007.

[7] E. Bayod and E. Tornberg, "Microstructure of highly concentrated tomato suspensions on homogenisation and subsequent shearing," *Food Research International*, vol. 44, no. 3, pp. 755–764, 2011.

[8] P. E. D. Augusto, A. Ibarz, and M. Cristianini, "Effect of high pressure homogenization (HPH) on the rheological properties of a fruit juice serum model," *Journal of Food Engineering*, vol. 111, no. 2, pp. 474–477, 2012.

[9] I. Colle, S. Van Buggenhout, A. Van Loey, and M. Hendrickx, "High pressure homogenization followed by thermal processing of tomato pulp: influence on microstructure and lycopene in vitro bioaccessibility," *Food Research International*, vol. 43, no. 8, pp. 2193–2200, 2010.

[10] M. T. K. Kubo, P. E. D. Augusto, and M. Cristianini, "Effect of high pressure homogenization (HPH) on the physical stability of tomato juice," *Food Research International*, vol. 51, no. 1, pp. 170–179, 2013.

[11] A. Panozzo, L. Lemmens, A. Van Loey, L. Manzocco, M. C. Nicoli, and M. Hendrickx, "Microstructure and bioaccessibility of different carotenoid species as affected by high pressure homogenisation: a case study on differently coloured tomatoes," *Food Chemistry*, vol. 141, no. 4, pp. 4094–4100, 2013.

[12] P. E. D. Augusto, A. Ibarz, and M. Cristianini, "Effect of high pressure homogenization (HPH) on the rheological properties of tomato juice: time-dependent and steady-state shear," *Journal of Food Engineering*, vol. 111, no. 4, pp. 570–579, 2012.

[13] H. Bengtsson, C. Hall, and E. Tornberg, "Effect of physicochemical properties on the sensory perception of the texture of homogenized fruit and vegetable fiber suspensions," *Journal of Texture Studies*, vol. 42, no. 4, pp. 291–299, 2011.

[14] F. Jiang and Y.-L. Hsieh, "Cellulose nanocrystal isolation from tomato peels and assembled nanofibers," *Carbohydrate Polymers*, vol. 122, pp. 60–68, 2015.

[15] P. García Herrera, M. C. Sánchez-Mata, and M. Cámara, "Nutritional characterization of tomato fiber as a useful ingredient for food industry," *Innovative Food Science and Emerging Technologies*, vol. 11, no. 4, pp. 707–711, 2010.

[16] I. Navarro-González, V. García-Valverde, J. García-Alonso, and M. J. Periago, "Chemical profile, functional and antioxidant properties of tomato peel fiber," *Food Research International*, vol. 44, no. 5, pp. 1528–1535, 2011.

[17] "AOAC Official Method 991.43 Total, Soluble, and Insoluble Dietary Fiber in Foods, AOAC International, Gaithersburg, Maryland, USA, 1994".

[18] "China National food safety standard GB 5009.5-2010 Determination of protein in foods, Ministry of Health of China, 2010".

[19] P. E. D. Augusto, A. Ibarz, and M. Cristianini, "Effect of high pressure homogenization (HPH) on the rheological properties of tomato juice: creep and recovery behaviours," *Food Research International*, vol. 54, no. 1, pp. 169–176, 2013.

[20] H. Bengtsson and E. Tornberg, "Physicochemical characterization of fruit and vegetable fiber suspensions. I: Effect of homogenization," *Journal of Texture Studies*, vol. 42, no. 4, pp. 268–280, 2011.

[21] C. Servais, R. Jones, and I. Roberts, "The influence of particle size distribution on the processing of food," *Journal of Food Engineering*, vol. 51, no. 3, pp. 201–208, 2002.

[22] P. E. D. Augusto, A. Ibarz, and M. Cristianini, "Effect of high pressure homogenization (HPH) on the rheological properties of tomato juice: viscoelastic properties and the Cox-Merz rule," *Journal of Food Engineering*, vol. 114, no. 1, pp. 57–63, 2013.

[23] E. Bayod, E. P. Willers, and E. Tornberg, "Rheological and structural characterization of tomato paste and its influence on the quality of ketchup," *LWT- Food Science and Technology*, vol. 41, no. 7, pp. 1289–1300, 2008.

[24] C. Valencia, M. C. Sánchez, A. Ciruelos, A. Latorre, J. M. Franco, and C. Gallegos, "Linear viscoelasticity of tomato sauce products: influence of previous tomato paste processing," *European Food Research and Technology*, vol. 214, no. 5, pp. 394–399, 2002.

6

Detection of Adulteration in Canola Oil by using GC-IMS and Chemometric Analysis

Tong Chen ⓘ**, Xinyu Chen, Daoli Lu, and Bin Chen** ⓘ

School of Food and Biological Engineering, Jiangsu University, Zhenjiang 212013, China

Correspondence should be addressed to Bin Chen; ncp@ujs.edu.cn

Academic Editor: David Touboul

The aim of the present study was to detect adulteration of canola oil with other vegetable oils such as sunflower, soybean, and peanut oils and to build models for predicting the content of adulterant oil in canola oil. In this work, 147 adulterated samples were detected by gas chromatography-ion mobility spectrometry (GC-IMS) and chemometric analysis, and two methods of feature extraction, histogram of oriented gradient (HOG) and multiway principal component analysis (MPCA), were combined to pretreat the data set. The results evaluated by canonical discriminant analysis (CDA) algorithm indicated that the HOG-MPCA-CDA model was feasible to discriminate the canola oil adulterated with other oils and to precisely classify different levels of each adulterant oil. Partial least square analysis (PLS) was used to build prediction models for adulterant oil level in canola oil. The model built by PLS was proven to be effective and precise for predicting adulteration with good regression ($R^2 > 0.95$) and low errors (RMSE ≤ 3.23).

1. Introduction

Canola oil is a vegetable oil derived from rapeseed which has low erucic acid content [1]. In the 1980s, in order to decrease the health concerns about erucic acid, rapeseed varieties free from erucic acid were developed by using selective breeding [2, 3]. And then, these varieties were called canola. Nowadays, China has become one of the major consumers of canola oil, and it is also popular with consumers in Canada, Europe, and South America. Canola oil can provide consumers with many health benefits that others cannot provide. For example, canola oil is low in saturated fat and high in polyunsaturated fats, with a good ratio of omega-6 to omega-3, which make it very suitable for cooking [4–6]. Since canola oil has its specific function to human body, it has become one of the most susceptible food materials adulterated with other vegetable oils of lower quality, which is a serious threat to the health of consumers. Therefore, it requires reliable tools and methods for analyzing the purity of edible vegetable oil.

Many techniques have been developed and used to detect adulteration in oil. These techniques include physical-chemical analysis, spectral analysis, gas chromatography (GC), gas chromatography-mass spectrometer (GC-MS), and electronic nose. Physical and chemical analysis includes sensory evaluation, colorimetry, centrifugation, and freezing. These traditional methods are simple and convenient and suitable for local monitoring. However, physical and chemical analysis methods are not accurate, require high degree technical expertise, and can only determine whether the sample is adulterated without finding out which specific component is adulterated. Spectral methods, e.g., Nuclear Magnetic Resonance Spectroscopy (NMR) [7], Raman [8], Fourier Transform Infrared (FTIR) [9], and Fluorescence [10], were shown to be useful for detection and quantification of adulteration in oil. However, their data analysis requires specialized software and complex algorithms which are difficult for common users to master. Chromatographic methods, such as GC-FID (flame ionization detector) [11], GC-MS [12], and high performance liquid chromatography (HPLC) [13], have been proven to be effective in detecting adulteration in oil. Nevertheless, the requirement for standard samples and high input of time and labor make them unsuitable for on-site analysis, thus limiting the wide use of them. Electronic nose [14] also has been used to evaluate the quality of oil. However, it needs electrode activation process during which sensor poisoning may occur depending on operation and ambient conditions.

Ion mobility spectrometry (IMS) is an analytical technique and was initially developed for the detection of explosives and chemical warfare agents [15]. At present, it has been widely used in novel application in the field of agricultural products and foods [16]. IMS is used to separate and identify ionized molecules in the gas phase based on their mobility in a carrier gas, which is considered as a screening technique due to its ability to identify the properties of samples at considerable low cost and short analysis time without pretreatment. On the other hand, IMS has limitations in detecting complex sample (e.g., food) for having low resolution and the risk of mutual interferences between analytes [17]. Yet, if combined with GC, the capability of IMS in separating various components is strengthened. Chromatographic elution of each target compound can be automatically analyzed and the obtained data information is richer because both retention time and drift time information are included [18]. GC-IMS has been shown to be able to characterize and discriminate adulteration in oil, wines, honey, and meat [19–23]. Successful applications have been reported on the determination of aldehydes in oil, adulteration in extra virgin olive oils by using UV-IMS [24], and determination of volatile compounds [25].

Most of the previous reports on canola oil analysis mainly focused on the adulteration detection and main component quantification of oil species [8, 26], with few studies performed on aroma differentiation. Odour is an important quality criterion for edible vegetable oil. The previous relevant studies often transformed the matrix to vector (like peaks selected manually as variables) for chemometric analysis by UV-IMS or GC-IMS, which may result in losing information of certain analytes. In addition, to the best of our knowledge, no recent work has been conducted using chemometrics for feature extraction of the two-dimensional data produced from GC-IMS instrument. Therefore, the potential use of GC-IMS for detection of canola oil adulteration was investigated. The aims of this study were (1) to investigate the use of GC-IMS combined with pattern recognition methods to detect the presence of adulterant in canola oil, (2) to apply a new method to extract information for two-dimensional data, (3) to build a model for content prediction of adulterated oil in canola oil, and (4) to develop a rapid method for adulteration detection in canola oil.

2. Materials and Methods

2.1. Preparation of Oil Blends. 3 canola oil samples were provided by Ningbo Entry-Exit Inspection and Quarantine Bureau (Zhejiang Province, China), while sunflower oil (2 samples), soybean oil (3 samples), and peanut oil (2 samples) were all purchased from Metro Supermarket at Zhenjiang, China. All the samples were stored at -5°C in the refrigerator before experimental process.

The adulterated samples were prepared by blending canola oil with sunflower oil, soybean oil, and peanut oil at levels of 0%, 5%, 10%, 20%, 30%, 40%, and 50% by volume, respectively. The mixed oil samples were brought to room temperature before detection. All the samples (147

samples) were homogenized with vortex for 60s and analyzed immediately after preparation.

2.2. Experimental Device. All prepared oil samples were analyzed with GC-IMS device (FlavourSpec®) from G.A.S. (Gesellschaft für Analytische Sensorysteme GmbH, Dortmund, Germany). The instrument was equipped with an incubating device intended to heat the sample and keep the headspace container at a constant temperature and a heated splitless injector for direct sampling of headspace volatile compounds from the oil samples into the GC-IMS. In addition, the device was coupled to an automatic sampler unit (CTC-PAL, CTC Analytics AG, Zwingen, Switzerland), which made the injection volume more accurate and repeatable without any human manipulation. Sample vials were transported into a heated incubator for preconditioning. After they reached equilibrium, a heated gas-tight syringe moved over the incubator and withdrew the headspace sample. After sample injection, the hot syringe was automatically cleaned by purging with inert gas. The experimental parameters used for this method were summarized in Table 1.

For analysis, 2 mL of oil sample was placed in a 10 mL vial which was sealed with a magnetic cap and heated at 90°C for 10 min in incubating box in order to generate volatile compounds from oil sample. 200 μL of headspace was automatically injected by heated syringe (95°C) into the heated injector (95°C) of the GC-IMS instrument. After that, the volatile organic compounds were pushed into the multicapillary column (MCC, 40°C) through a carrier gas (15 mL/min) for timely separation. And then, a drift gas (300 mL/min in counter flow) was generated to collide with the gaseous ions of separated components from the sample to produce the ion mobility spectra at room temperature and pressure, and finally gaseous ions were captured by a detector (a Faraday plate). Data of samples were obtained by IMS Control TFTP Server software and displayed by using LAV software (Version 1.3.1, from G.A.S).

2.3. Data Processing. The multidimensional signals of GC-IMS data required pretreatment before statistical analysis. The alignment is an important step before chemometric process of the data due to small deviation in temperature of MCC and flow velocity causing changes in retention time and small deviation in temperature of drift tube causing changes in drift time. Therefore, RIP (Reaction Ion Peak: reaction ions are generated by a cascade of reactions following the collision of a fast electron emitted from ionization source with the drift gas atmosphere; as a consequence, the so-called RIP representing the total of all ions available is formed) normalization was applied to align with a shift in x-axis by LAV software. Moreover, a Savitzky-Golay smoothing method (order 2 and window size 15) was used to standardize the measurement and improve the signal-to-noise rate.

Many studies [27, 28] have proposed various feature extraction methods, such as Gabor filters, local binary pattern, Haar, and histograms of oriented gradients (HOG). HOG is a feature descriptor used in computer vision and image processing for the purpose of object detection. The theory of HOG descriptor is that local object appearance

TABLE 1: Experimental conditions for GC-IMS analyses.

	Parameters	Value
	Headspace sampling volume	200 μL
	Incubation time	10 min
Automatic inject system	Sample volume	2 mL
	Incubation temperature	90°C
	Injector temperature	95°C
	MCC	OV-5 (nonpolar)
Column	Column temperature	40°C
	Length of column	30 cm
	Run time	15 min
	Ionization source	Tritium (6.5 KeV)
	Voltage	Positive drift
	Drift length	10 cm
	Carrier gas flow rate	15 mL min^{-1} (N$_2$ 5.0)
	Drift gas flow rate	300 mL min^{-1} (N$_2$ 5.0)
IMS	Equipment temperature	45°C
	Average	32
	Electric field strength	350 V cm^{-1}
	Grid pulse with	100 μs
	Trigger delay	100 ms
	Sampling frequency	150 kHz
	Repetition rate	21 ms

and shape within an image can be described by the distribution of intensity gradients or edge directions. The image is divided into small connected regions called cells, and for the pixels within each cell, a histogram of gradient directions is compiled by counting occurrences of gradient orientation in localized portions of an image. For the present, HOG is proven to offer better feature extraction that could significantly outperform existing feature sets for object detection.

Principle component analysis (PCA) is a classical feature extraction and data representation technique widely used in the area of pattern recognition and computer vision [29]. In PCA-based pattern recognition, the 2D (two-dimensional) matrices must be previously transformed into 1D (one-dimensional) vectors. However, the resulting vector usually leads to a high-dimensional vector space, which is difficult to evaluate the covariance matrix accurately due to its large size. Fortunately, a new method, called multiway principal component analysis (MPCA) [30], was developed for matrix feature extraction. Figure 1 summarizes the computation procedure of MPCA. As mentioned above, each sample was detected by GC-IMS and a 2D matrix was obtained. These sample data are arranged and merged to a three-dimensional matrix $X(I \times J \times K)$, where I is total samples, J is drift time, and K is retention time. First, the three-dimensional matrix X is split along the direction of the sample axis (see Figure 1) and formed a new matrix $X(I \times JK)$; then, the new matrix X is decomposed into the product of the score vector t_r with the load vector p_r, plus the residual matrix E (see (1)), which is similar to the two-dimensional principal component analysis method. R is the number of principal components; $\otimes$ is the product of Kronecker.

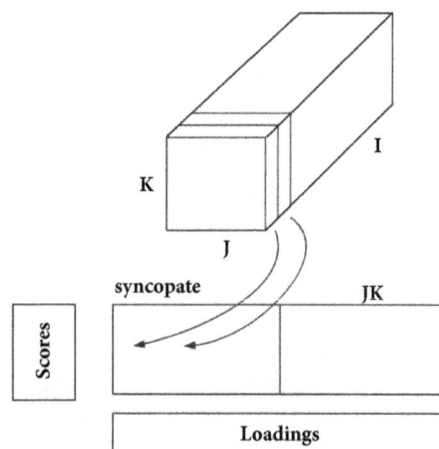

FIGURE 1: Computation procedure of MPCA method.

As opposed to conventional PCA, MPCA is based on 2D matrices rather than 1D vector. As a result, MPCA has two important advantages over PCA. (1) It is easier to evaluate the covariance matrix accurately. (2) Less time is required to determine the corresponding eigenvectors.

$$X = \sum_{r=1}^{R} t_r \otimes p_r^T + E \tag{1}$$

In our study, a preprocessing step to the statistical analysis of the GC-IMS data was performed to all oil samples which

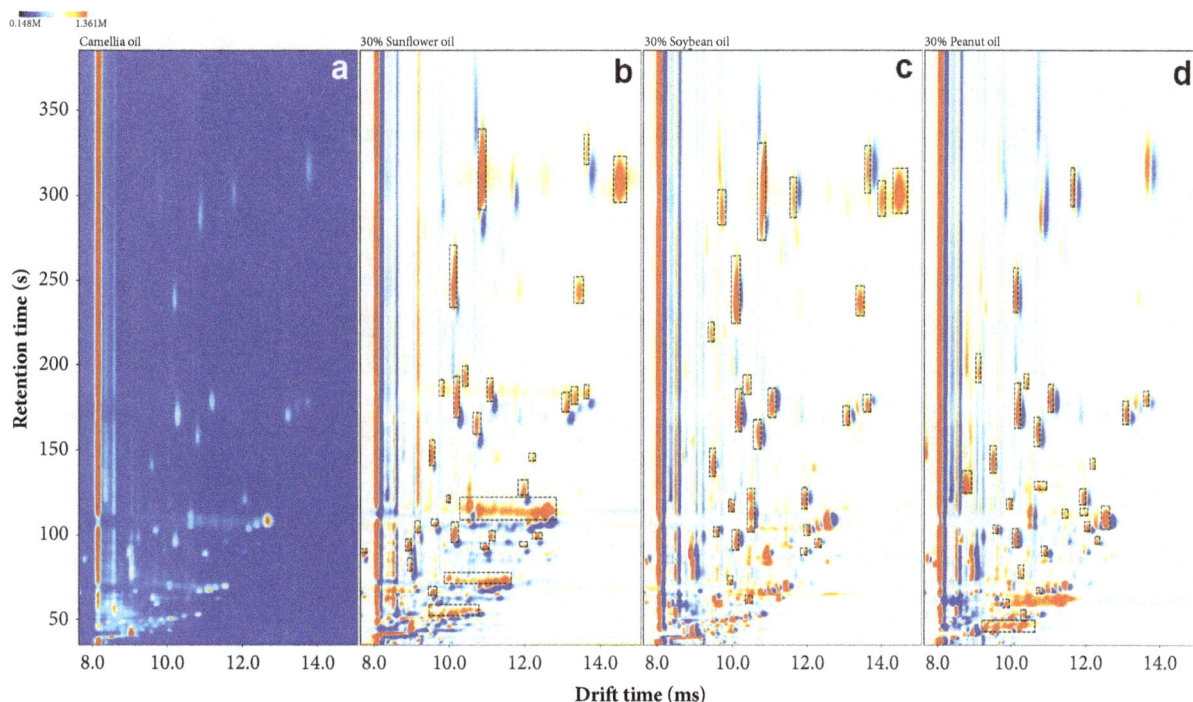

FIGURE 2: GC-IMS plot comparison of (a) pure canola oil, (b) 30% sunflower oil adulteration, (c) 30% soybean oil adulteration, and (d) 30% peanut oil adulteration.

consisted of a RIP normalization and a Savitzky-Golay smoothing filter. Firstly, a HOG method was used to extract texture and contour information. Later, as an unsupervised method, MPCA was applied to further extract the features and visualize the dataset retaining the maximum variability present in the original data and eliminating possible dependence between variables. Then, canonical discriminant analysis (CDA) algorithm was used to generate nonlinear boundaries between classes according to the content of adulterated oil. Lastly, partial least square analysis (PLS) was performed to study the predictive capacity of GC-IMS for the adulteration content of mixed oil.

For data analysis, the MATLAB R2009a (The Mathworks Inc, Natick, USA) and PLS Toolbox 5.5 (Eigenvector Research, Inc., Manson, WA, USA) were used.

3. Results and Discussion

3.1. Sample Analysis by GC-IMS. The size of raw data from each oil sample was huge, so the spectral area was cut (899×1114 dimension) and selected by limiting the retention time from 35.49 to 385.71 s and drift time from 7.666 to 15.086 ms on the basis of retaining the major information. As it is known, GC-IMS spectrum of a sample corresponded to a matrix and could be displayed as a pseudo color image for visualization [31]. A popular method for comparing two matrices is to form a difference image by subtracting the individual element values of one matrix from the corresponding element values of the other matrix. In this condition, a positive difference indicates that the analyzed matrix has a larger element value and a negative difference

indicates that the reference matrix has a larger element value. Therefore, taking canola oil sample as reference matrix, the other adulterated samples were regarded as analyzed matrix and colorized differences were formed by subtracting the canola oil sample from the adulterated sample. In this way, with the increment of sunflower oil content, the changes about volatile organic compounds from adulterated samples were clearly visualized. Figure 2 showed the GC-IMS pseudo color map for change visualization when a raw canola oil sample (Figure 2(a)) was adulterated with 30% sunflower oil (Figure 2(b)), 30% soybean oil (Figure 2(c)), or 30% peanut oil (Figure 2(d)), respectively. The red region indicated that the sample had more volatile components compared with reference sample. The deeper the color, the more components it had and the blue region was the opposite. As shown in Figure 2, there were obvious changes between pure canola oil sample and adulterated canola oil samples. Many new volatile compounds were produced, some peaks were marked by black dotted rectangle for effective observation, and those differences of red regions were reflected in retention time, drift time, and intensity of the corresponding peaks. On the other hand, levels of original volatile compounds in canola oil were weakened by different degree. Those peak differences (drift time, retention time, peak volume, etc.) from volatile organic compounds of each kind of vegetable oil are the key to qualitative analysis or quantitative detection. On the other hand, volatile organic compounds from samples adulterated with soybean oil appeared only in a short retention time, and the gas molecule materials could not be well separated, which may be caused by higher initial carrier gas flow-rate. However, only a general overall impression of the differences

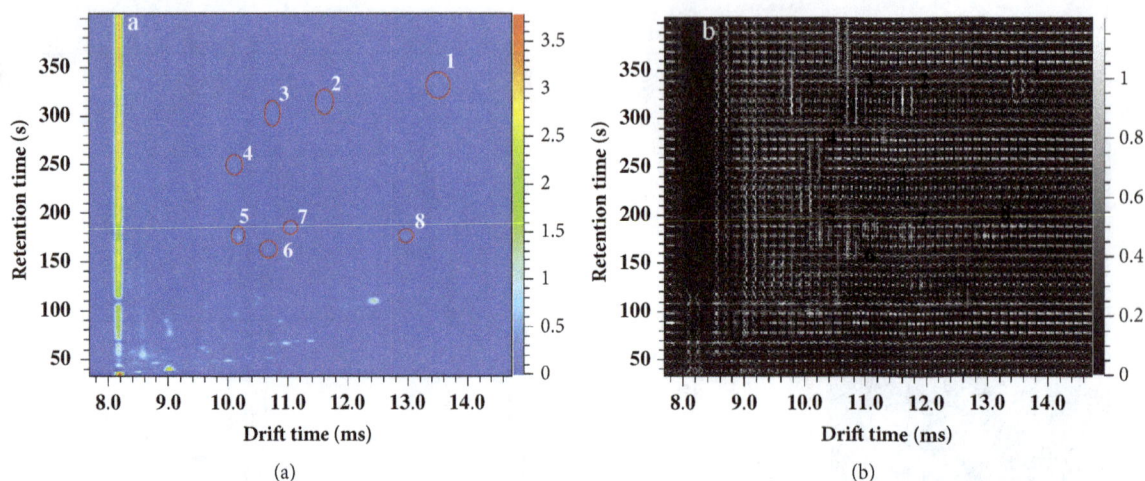

FIGURE 3: Visualization results of original data (a) and HOG feature extraction (b).

through the topographic plot was obtained. The content changes of volatile organic compounds were not regular and it was hard to realize digital characterization. Therefore, further analysis was necessary with the help of chemometric tools.

3.2. Adulteration Classification of Oil Samples.

Before MPCA analysis, the data set was processed by HOG algorithm (24×24 block size after optimization) in order to extract useful information and reduce the dimension of the original matrix. As shown in Figure 3, some peaks were selected and marked with red solid line ellipse and Arabic numerals for visual observation in Figure 3(a), which are corresponding to the marked area with the same Arabic numerals in Figure 3(b). It was clear that the visualization of HOG result (Figure 3(b)) could describe the characteristic information of the whole matrix with strong texture structure and contour information from the original data (Figure 3(a)), especially the peaks that were corresponding to volatile organic compounds. In addition, the dimension of the matrix from each sample was effectively reduced (777×966 dimension).

As mentioned earlier, each kind of adulterated samples after HOG pretreatment was arranged to a three-dimensional matrix and MPCA algorithm was used to process and analyze the matrix. Principal component scores obtained were sorted from high to low according to the cumulative contribution rate and the first 2 principal component score matrices were used to show the cluster of adulterated samples (see Figure 4). As shown in Figure 4, the data were mapped on two most important principal components PC1 and PC2, presenting the results of canola oil adulterated with sunflower oil (Figure 4(a)), with soybean oil (Figure 4(b)), and with peanut oil (Figure 4(c)). The axis heading in each figure was labeled with the respective contribution rates of PC1 and PC2 after MPCA process.

Figure 4(a) showed the PCA score plot of canola oil adulterated with sunflower oil, and PC1 and PC2 explained the 96.57% of the original information. It could be inferred that the first 2 PCs could give the most information of data

set. As shown, the pure canola oil samples were sufficiently well distinguished from adulterated samples with different levels of adulteration and each level of adulteration had its own cluster group. With the increase in the proportion of sunflower oil, the distribution of the samples moved from right to left between first two principal components. Figure 4(b) presented the result of MPCA analysis of the canola oil adulterated with soybean oil samples. Two principal components covered 86.19% of the original information. It was visible that the data points belonging to each class of adulteration rate were gathered in compact clusters and two principal components showed good separation in the direction of diagonal of axis except raw canola oil samples. Although two principal components contained large amount of original information, some clusters overlapped each other (for example, clusters 0% and 20% groups). The results of canola oil containing peanut oil samples were shown in Figure 4(c). The reconstructed information contained 95.77% of the variance. As can be observed, pure canola oil samples and adulterated samples were located in the opposite side of the axis, which indicated that the raw canola oils were well differentiated from adulterated samples. As a result, there were significant differences in aroma between canola oil and peanut oil. Combining with Figure 2(d), it could be inferred that volatile compounds from peanut oil could obviously cover up the original flavor of canola oils. On the other hand, different content of peanut oil also could be distinguished. However, 20% adulterant oil samples clustered in two-dimensional space overlapping with 30% and 40% groups, and its cluster region was a long and narrow strip. This phenomenon may be resulted from the origin differences between peanut oil samples purchased.

Finally, CDA was used as pattern recognition techniques for the authentication of canola oil. As seen in Figure 4, each kind of adulterated oil samples was all grouped into 7 distinct clusters. All canola oil samples adulterated with sunflower oil could be classified without any error (100 % of accuracy). Only 2 samples (one adulterated with soybean oil

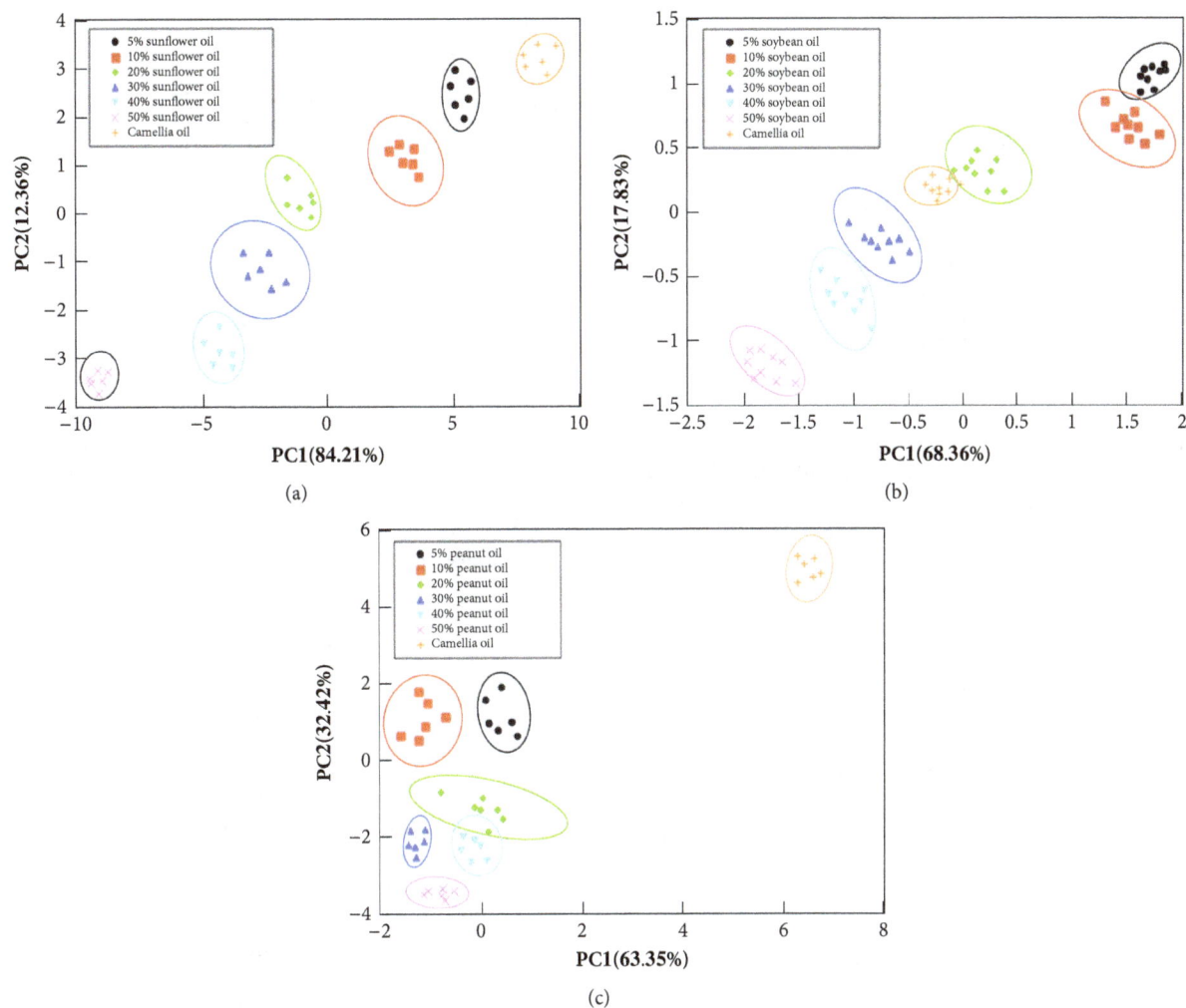

Figure 4: PC1 and PC2 scores of pure canola oil adulterated with (a) sunflower oil, (b) soybean oil, and (c) peanut oil.

and the other adulterated with peanut oil) were misclassified. The total accuracy of recognition was 98.64%, which showed excellent results of classification.

3.3. Rapid Characterization of Sunflower Oil Content in Canola Oil.

In order to establish relationship between the GC-IMS and the content of adulterant oil added to canola oil, an analytical method of partial least square (PLS) was used. Quantification of the percentage of adulterant added to canola oil was carried out by building separate calibration models for each kind of vegetable oil between 0 and 50% level.

The principal component scores of sunflower oil data set were selected as input variables and analyzed by PLS. Before building the model, the samples used for training and testing were randomly selected. Data set containing 70% canola samples was used for calibration and that containing the remaining 30% samples was used to predict the content of sunflower oil adulterated in canola oil. The other two kinds of adulterated samples were treated in the same way. The correlation coefficient (R^2) and root mean square error (RMSE) between predicted and experimental values were used to evaluate the performance of the model. The higher R^2 and lower RMSE mean better calibration model.

PLS is a multivariate projection method for building relationship between dependent variables and independent variables. Due to the limited amount of training samples, leave-one-out cross validation was applied and the first 3 principal components were determined. Correlation coefficient for calibration and prediction (R^2C and R^2P) and root mean square error of calibration and prediction (RMSEC and RMSEP) were tabulated in Table 2; good correlations of calibration were found between GC-IMS data and content of adulterant oil added to canola oil with high coefficient of determination ($R^2C>0.96$) and low errors (RMSEC $\leq$ 2.92). When PLS models were used to predict the testing data set, good prediction results for the content of adulterant oil added to canola oil were obtained with good coefficient ($R^2P>0.95$) and acceptable errors (RMSEC $\leq$ 3.23). The RMSEP value of sunflower oil adulterated in canola oil is lower than its RMSEC value and the possible reason may be that the separation of samples is uneven. The best results were obtained with the model carrying out with the canola oil adulterated

TABLE 2: Requisite parameters for adulteration level prediction in adulterant canola oil samples.

PLS Models	Calibration		Cross validation		Prediction	
	R^2C	RMSEC/(%)	R^2CV	RMSECV/(%)	R^2P	RMSEP/(%)
Adulterated with sunflower oil	0. 994	1.78	0.991	1.80	0.983	1.86
Adulterated with soybean oil	0.996	1.39	0.989	1.43	0.985	1.45
Adulterated with peanut oil	0.968	2.92	0.963	3.18	0.952	3.23

R^2C: coefficient of determination for calibration.
R^2CV: coefficient of determination for cross validation.
R^2P: coefficient of determination for prediction.
RMSEC: root mean square error of calibration.
RMSECV: root mean square error of cross validation.
RMSEP: root mean square error of prediction.

with soybean oils. All those models are acceptable and useful for detecting the adulteration in canola oil. Therefore, our present work verifies that adulteration in canola oil can be determined by PLS using the GC-IMS data.

4. Conclusions

In this paper, GC-IMS has been proposed to detect the adulteration of canola oil samples with low price vegetable oils. 147 samples were detected and the useful signals were extracted and analyzed by HOG and MPCA algorithms. A combination of these two methods was proved to be the most effective feature extraction method. PLS was used to predict the adulteration levels in canola oil precisely. The capacity in prediction showed that all those methods were satisfactory and applicable. In addition, the model of canola oil adulterated with soybean oil was proved to be the most effective.

The methodology developed has capability to detect adulteration in canola oil. The analysis time only needs about 15 min which is less than other techniques and no sample pre-treatment is required. Moreover, the device has been applied in industrial field broadly due to low cost and portability. Therefore, GC-IMS can be seen as a powerful authentication method with chemometrics for oil adulteration detection for its high efficiency and accuracy.

Acknowledgments

This work was funded by the National Natural Science Foundation of China (Project ID: 31772056).

References

[1] A. Abbadi and G. Leckband, "Rapeseed breeding for oil content, quality, and sustainability," European Journal of Lipid Science and Technology, vol. 113, no. 10, pp. 1198–1206, 2011.

[2] P. Roine, E. Uksila, H. Teir, and J. Rapola, "Histopathological changes in rats and pigs fed rapeseed oil," Zeitschrift für Ernährungswissenschaft, vol. 1, no. 2, pp. 118–124, 1960.

[3] B. Schiefer, F. M. Loew, V. Laxdal et al., "Morphologic effects of dietary plant and animal lipids rich in docosenoic acids on heart and skeletal muscle of cynomolgus monkeys," The American Journal of Pathology, vol. 90, no. 3, pp. 551–564, 1978.

[4] T. Issariyakul and A. K. Dalai, "Biodiesel from vegetable oils," Renewable & Sustainable Energy Reviews, vol. 31, pp. 446–471, 2014.

[5] H. Lei, L. L. Cai, L. L. Cao, and S. T. Jiang, "Effect of Erucic Acid Content in Rapeseed Oil on Food Intake Safety in Mice," Food Science, vol. 19, pp. 321–324, 2010.

[6] F. Hashempour-Baltork, M. Torbati, S. Azadmard-Damirchi, and G. P. Savage, "Vegetable oil blending: A review of physico-chemical, nutritional and health effects," Trends in Food Science & Technology, vol. 57, pp. 52–58, 2016.

[7] Z. Tan, E. Reyes-Suarez, W. Indrasena, and J. A. Kralovec, "Novel approach to study fish oil oxidation using 1H nuclear magnetic resonance spectroscopy," Journal of Functional Foods, vol. 36, pp. 310–316, 2017.

[8] S. Durakli Velioglu, H. T. Temiz, E. Ercioglu, H. M. Velioglu, A. Topcu, and I. H. Boyaci, "Use of Raman spectroscopy for determining erucic acid content in canola oil," Food Chemistry, vol. 221, pp. 87–90, 2017.

[9] E. Zahir, R. Saeed, M. A. Hameed, and A. Yousuf, "Study of physicochemical properties of edible oil and evaluation of frying oil quality by Fourier Transform-Infrared (FT-IR) Spectroscopy," Arabian Journal of Chemistry, vol. 10, pp. S3870–S3876, 2017.

[10] J. Tan, R. Li, Z.-T. Jiang et al., "Synchronous front-face fluorescence spectroscopy for authentication of the adulteration of edible vegetable oil with refined used frying oil," Food Chemistry, vol. 217, pp. 274–280, 2017.

[11] D. Peng, Y. Bi, X. Ren, G. Yang, S. Sun, and X. Wang, "Detection and quantification of adulteration of sesame oils with vegetable oils using gas chromatography and multivariate data analysis," Food Chemistry, vol. 188, pp. 415–421, 2015.

[12] X. Sun, L. Zhang, P. Li et al., "Fatty acid profiles based adulteration detection for flaxseed oil by gas chromatography mass spectrometry," LWT- Food Science and Technology, vol. 63, no. 1, pp. 430–436, 2015.

[13] R. Salghi, W. Armbruster, and W. Schwack, "Detection of argan oil adulteration with vegetable oils by high-performance liquid chromatography-evaporative light scattering detection," Food Chemistry, vol. 153, pp. 387–392, 2014.

[14] X. Peng, W. Chen, R. Luo, C. Liu, and H. Xu, "Detection of walnut oil adulterated with soybean oil, rapeseed oil/corn oil using electronic nose," Journal of the Chinese Cereals and Oils Association, vol. 30, no. 11, pp. 129–134, 2015.

[15] S. Armenta, M. Alcala, and M. Blanco, "A review of recent, unconventional applications of ion mobility spectrometry (IMS)," *Analytica Chimica Acta*, vol. 703, no. 2, pp. 114–123, 2011.

[16] Z. Karpas, "Applications of ion mobility spectrometry (IMS) in the field of foodomics," *Food Research International*, vol. 54, no. 1, pp. 1146–1151, 2013.

[17] G. W. Cook, P. T. LaPuma, G. L. Hook, and B. A. Eckenrode, "Using gas chromatography with ion mobility spectrometry to resolve explosive compounds in the presence of interferents," *Journal of Forensic Sciences*, vol. 55, no. 6, pp. 1582–1591, 2010.

[18] M. T. Jafari, M. Saraji, and H. Sherafatmand, "Design for gas chromatography-corona discharge-ion mobility spectrometry," *Analytical Chemistry*, vol. 84, no. 22, pp. 10077–10084, 2012.

[19] S. Liedtke, L. Seifert, N. Ahlmann, C. Hariharan, J. Franzke, and W. Vautz, "Coupling laser desorption with gas chromatography and ion mobility spectrometry for improved olive oil character-isation," *Food Chemistry*, vol. 255, pp. 323–331, 2018.

[20] Z. Karpas, A. V. Guamán, D. Calvo, A. Pardo, and S. Marco, "The potential of ion mobility spectrometry (IMS) for detection of 2,4,6-trichloroanisole (2,4,6-TCA) in wine," *Talanta*, vol. 93, pp. 200–205, 2012.

[21] M. T. Jafari, T. Khayamian, V. Shaer, and N. Zarei, "Determina-tion of veterinary drug residues in chicken meat using corona discharge ion mobility spectrometry," *Analytica Chimica Acta*, vol. 581, no. 1, pp. 147–153, 2007.

[22] N. Gerhardt, M. Birkenmeier, S. Schwolow, S. Rohn, and P. Weller, "Volatile-Compound Fingerprinting by Headspace-Gas-Chromatography Ion-Mobility Spectrometry (HS-GC-IMS) as a Benchtop Alternative to 1H NMR Profiling for Assessment of the Authenticity of Honey," *Analytical Chemistry*, vol. 90, no. 3, pp. 1777–1785, 2018.

[23] L. Zhang, Q. Shuai, P. Li et al., "Ion mobility spectrometry fingerprints: A rapid detection technology for adulteration of sesame oil," *Food Chemistry*, vol. 192, Article ID 17778, pp. 60–66, 2015.

[24] R. Garrido-Delgado, M. Eugenia Muñoz-Pérez, and L. Arce, "Detection of adulteration in extra virgin olive oils by using UV-IMS and chemometric analysis," *Food Control*, vol. 85, pp. 292–299, 2018.

[25] R. Garrido-Delgado, L. Arce, and M. Valcárcel, "Multi-capillary column-ion mobility spectrometry: A potential screening sys-tem to differentiate virgin olive oils," *Analytical and Bioanalyti-cal Chemistry*, vol. 402, no. 1, pp. 489–498, 2012.

[26] G. Ozulku, R. M. Yildirim, O. S. Toker, S. Karasu, and M. Z. Durak, "Rapid detection of adulteration of cold pressed sesame oil adultered with hazelnut, canola, and sunflower oils using ATR-FTIR spectroscopy combined with chemometric," *Food Control*, vol. 82, pp. 212–216, 2017.

[27] L. M. Surhone, M. T. Tennoe, S. F. Henssonow, and C. Vision, *Histogram of Oriented Gradients. Betascript Publishing*, vol. 12, no. 4, pp. 1368–1371, 2016.

[28] B. Chatterji, "Feature Extraction Methods for Character Recog-nition," *IETE Technical Review*, vol. 3, no. 1, pp. 9–22, 2015.

[29] H. Abdi and L. J. Williams, "Principal component analysis," *Wiley Interdisciplinary Reviews: Computational Statistics*, vol. 2, no. 4, pp. 433–459, 2010.

[30] S. Wold, P. Geladi, K. Esbensen, and J. Öhman, "Multi-way prin-cipal components-and PLS-analysis," *Journal of Chemometrics*, vol. 1, no. 1, pp. 41–56, 1987.

[31] S. Zhong, X. Jiang, and Z. Wei, "Pseudo-color coding with phase-modulated image density," *Advanced Materials Research*, vol. 403-408, pp. 1618–1621, 2012.

7

Development of a Robust UPLC Method for Simultaneous Determination of a Novel Combination of Sofosbuvir and Daclatasvir in Human Plasma: *Clinical Application to Therapeutic Drug Monitoring*

Naser F. Al-Tannak,[1] **Ahmed Hemdan,**[2,3] **and Maya S. Eissa** ⓘ[4]

[1]*Department of Pharmaceutical Chemistry, Faculty of Pharmacy, Kuwait University, AlJabriyah, Kuwait*
[2]*Department of Pharmaceutical Analytical Chemistry, Faculty of Pharmacy, Ahram Canadian University, Giza, Egypt*
[3]*Institute of Clinical Chemistry, University Medical Center Hamburg-Eppendorf (UKE), Martinistraße 52, 20246 Hamburg, Germany*
[4]*Department of Pharmaceutical Analytical Chemistry, Faculty of Pharmacy, Egyptian Russian University, Cairo, Egypt*

Correspondence should be addressed to Maya S. Eissa; maya-shaaban@hotmail.com

Academic Editor: Samuel Carda-Broch

A rapid and selective UPLC-DAD method was developed and validated for simultaneous analysis of the novel two-drug combination *Darvoni*® for the treatment of HCV: Sofosbuvir (SF)/Daclatasvir (DC) in human plasma using Ledipasvir as internal standard (IS) where the extraction process was conducted using automated SPE. Although the analysis of the combination after concomitant oral intake of two tablets of SF and DC individually was reported in literature, yet simultaneous analysis of this new combination in human plasma after a single oral dose was not previously reported. The adopted chromatographic separation was achieved on Waters® Acquity UPLC BEH C_{18} column (2.1×50 mm, 1.7 μm) as a stationary phase using isocratic elution using a mobile phase system of ammonium formate (pH 3.5; 5 mM) and acetonitrile (60:40 v/v) pumped at a flow rate of 0.2 mL.min^{-1}. The UV detection was carried out at 261 nm for SF and 318 nm for DC and IS. SF was eluted at 1.123 min while DC was eluted at 3.179 min. The proposed chromatographic method was validated in accordance with guidelines of FDA for bioanalytical method validation. A linear range was achieved in the range of 25-6400 and 50-12800 ng.mL^{-1} for SF and DC, respectively. The proposed UPLC-DAD method was found to be accurate with % bias ranging between -10.0-7.2 for SF and -6.9-8.0 for DC. Also it was proved to be precise with % CV for intraday precision ranging between 3.8-9.6 for SF and 2.8-9.2 for DC whereas interday precision ranged between 5.1-9.3 for SF and 3.7-9.1 for DC. Moreover, % extraction recovery ranged between 90.0-107.2 for SF and 93.1-108.0 for DC using the suggested method. The adopted chromatographic method was successfully applied to the therapeutic drug monitoring of SF and DC in healthy volunteers after the oral intake of one *Darvoni*® tablet.

1. Introduction

Infection due to hepatitis C virus (HCV) is a leading cause for severe chronic liver disease, which can result in progressive liver damage such as cirrhosis and hepatocellular carcinoma. Thus, it is considered to be a great worldwide health problem specifically in Egypt, which has the greatest prevalence of the epidemic problem of HCV in the world in accordance with the reported Egyptian Demographic Health Survey [EDHS] that had reached 14.7%. So prevention of HCV becomes a national priority [1].

The available treatment options for HCV infection until 2011 were restricted to ribavirin with pegylated interferon combination. This drug regimen has limited efficacy, especially in genotype 1 infected patients, and was also accompanied with dangerous side effects [2].

In 2014, the directly acting antivirals were introduced in the market as a new anti-HCV generation. The main goal of these new drug therapies is decreasing the incidence of possible side effects for HCV patients. These powerful drugs encompass Nonnucleoside Inhibitors (NNIs) and Nucleoside

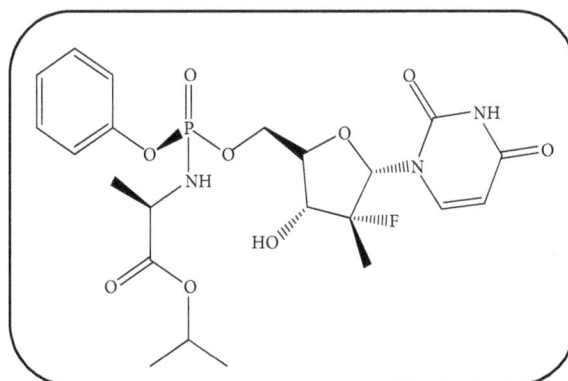

(a)

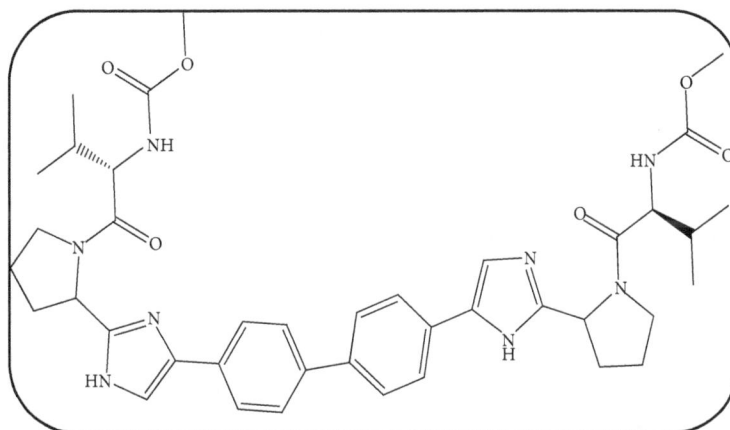

(b)

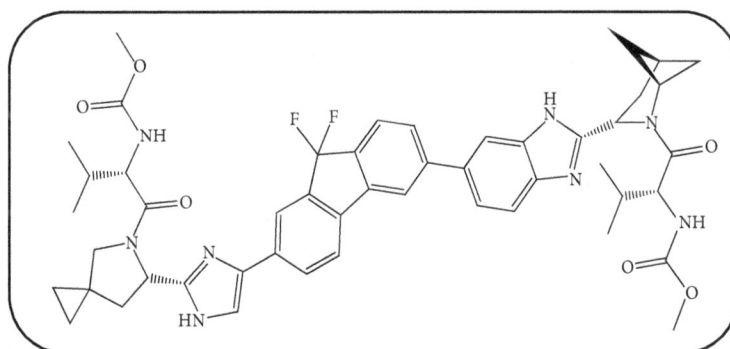

(c)

FIGURE 1: Chemical structure of SF (a), DC (b), and IS (c).

Inhibitors (NIs) of HCV RNA polymerase (NS5A/5B) and Protease Inhibitors (PIs). As shown in Figure 1(a), Sofosbuvir (SF) is the first released drug of the new generation and is an oral NI which is used alone or in combination with pegylated interferon ∝/Ribavirin or with PIs as well as with NNIs (such as Ledipasvir or Daclatasvir) [3]. SF is a white to off-white crystalline solid having a molecular weight of 529.458 g/mol, with a solubility ≥ 2 mg/mL across the pH range of 2-7.7 at 37°C, and is slightly soluble in water. It has Log P value=1.62 and has two pKa values; pKa1= 9.38 (amide); pKa2 = 10.30[4].

Daclatasvir (DC) is the first discovered drug as HCV RNA polymerase NS5A replication complex inhibitor [5].

Its chemical structure is presented in Figure 1(b). DC is effective against genotypes 1 and 3. DC (molecular weight = 738.89 g/mol) is a white to yellow crystalline nonhygroscopic powder. It is freely soluble in water, dimethyl sulfoxide, and methanol; soluble in ethanol (95%); practically insoluble in dichloromethane, tetrahydrofuran, acetonitrile, acetone, and ethyl acetate. It has Log P value =4.67 and a pKa value = 11.15 [6]. DC has illustrated a good safety profile upon concomitant intake with SF [7]. The once-daily oral dose of anti-HCV concomitant drug therapy of SF + DC has shown high rates of sustained virologic response in patients infected with HCV genotype 1 or 3 and also has been accompanied with better therapeutic outcomes [8]. Ledipasvir (IS) is used as internal standard; its chemical structure is presented in Figure 1(c). Ledipasvir is practically insoluble (<0.1 mg/mL) across the pH range of 3.0-7.5 and is slightly soluble below pH 2.3 (1.1 mg/mL). It has a molecular weight of 889.018 g/mol, Log p value = 3.8. It has two pKa values: pka1 = 4.0 and pka2 = 5.0. [9]

The increase in number of therapeutic options available for patients suffering HCV infections introduces a great challenge in the selection and management of HCV treatment. In regard to this, Therapeutic Drug Monitoring (TDM) is considered to be an important tool to assess the efficacy of drug regimen, help the clinicians to adjust the drug dosage, optimize the therapy or switch the treatment regimen, and overcome adverse events or therapeutic failure [10, 11].

Due to the extensive use of the mentioned drugs in combination therapy, the therapeutic drug monitoring of concentrations of SF and DC concentrations in patients undergoing HCV therapy is very important and critical, especially in determining the treatment optimization and the potential of drug interaction.

Therefore, there were several reported methods for SF and/or DC determination in human plasma using HPLC-UV detection [12–14], micellar LC method [15], and LC-MS/MS [16–23].

To the best of our knowledge, the adopted bioanalytical approach in this work will be the solely UPLC-PDA technique first developed for the simultaneous quantification of SF and DC in human plasma with a clinical application to their therapeutic drug monitoring after the oral intake of a coformulated tablet containing SF+DC (Darvoni® recently manufactured by Beacon Pharmaceuticals Limited) instead of the commonly two tablets dosage regimen of each one alone (Sovaldi® + Daklinza®). In accordance with US Food and Drug Administration (FDA) guidelines [24], the adopted UPLC-DAD method in this work was successfully developed and carefully validated. Also a successful clinical application to the adopted method was also conducted through therapeutic drug monitoring process after the intake of Darvoni® tablet to three healthy volunteers.

2. Materials and Methods

2.1. Chemicals and Reagents. Drug standards for SF, DC, and Ledipasvir (IS) were kindly supplied by Memphis Co. for Pharmaceutical and Chemical Industries, Cairo, Egypt, with certified purity of 99.98 ± 0.421 for SF, 99.93 ± 0.231 for DC, and 99.87 ± 0.642 for Ledipasvir. Darvoni film-coated tablets (coformualted with 400 mg SF and 60 mg DC) were purchased from Beacon Pharmaceuticals Limited, Bangladesh. Drug-free human plasma was obtained from Kuwait Blood Bank, Al Jabriyah, Kuwait. HPLC grade acetonitrile and other used chemicals in the adopted method were of analytical grade and obtained from Sigma Aldrich, Dor-set, UK. "In house" HPLC grade water was prepared with a MilliQfilter purchased from Millipore, Watford, UK. Syringe membrane filters (13mm) were purchased from kinesis scientific expert, Cambridgeshire, UK. Nylon solvent filters (0.45 um) used for

TABLE 1: Quality control samples for SF and DC.

Prepared QC samples (ng.mL^{-1})	SF	DC
LLOQ	25	50
QL	50	100
QM	400	1600
QH	3200	6400
ULOQ	6400	12800

solvent filtration and Water 20-positions Extraction Manifold with SPE cartridges (Sep-Pak® Vac C18) used for sample preparation were purchased from Water Corporation, Milford, USA. SPE eluates were dried using DRI-BLOCK DB-3 evaporator which was purchased from Techne, Stone, UK.

2.2. Instrumentation and Chromatographic Conditions. Chromatographic separation was achieved using Waters® Acquity UPLC separation module with quaternary Solvent Manager (H-Class), online degasser with autosampler injector, and photodiode array detector coupled with Empower® software for data acquisition (Waters®, Milford, USA). Waters® Acquity UPLC BEH C$_{18}$ column (2.1 mm × 50 mm, 1.7 μm particle size) was used as the stationary phase for the development of the chromatographic separation, optimization, and method validation. An isocratic elution was conducted using a mobile phase system of 5 mM of ammonium formate in water and acetonitrile (60:40 v/v). The flow rate was set at 0.2 mL.min^{-1}. Column temperature was adjusted at 45°C and samples were injected at 2 μL injection volume and determined at a wavelength of 261 nm for SF and 318 nm for DC and IS.

2.3. Solutions. Due to the possible degradation of DC in solution at high-intensity UV and visible light as previously reported by [25], all the solutions and samples containing DC must be protected from daylight during its preparation, storage, and analysis. Two stock solutions for each of SF and DC were prepared in methanol as 1.0 mg.mL^{-1}. Each stock standard solution was then further diluted with the same solvent to produce working standard solutions at the following concentration levels: 1.25, 2.5, 5.0, 10.0, 20.0, 80.0, 160.0, and 320.0 μg.mL^{-1} for SF and 2.5, 5.0, 10.0, 20.0, 80.0, 160.0, 320.0, and 640.0 μg.mL^{-1} for DC. Ledipasvir (IS) working standard solution was prepared as 50 μg.mL^{-1} in methanol. Plasma calibration standards were freshly prepared at the following concentration levels: 25, 50, 100, 200, 400, 1600, 3200, and 6400 ng.mL^{-1} for SF and 50, 100, 200, 400, 1600, 3200, 6400, and 12800 ng.mL^{-1} for DC by dilution from their respective working solutions in control human plasma and subjected to subsequent analysis on the same day. The quality control (QC) samples used were subsequently prepared by further dilution of the working standard solutions in human plasma as presented in Table 1.

2.4. Preparation of Plasma Samples. Sample clean-up was conducted using the following SPE procedure: The cartridges were first preconditioned with acetonitrile (1 mL) then equilibrated by water (1 mL), in a positive pressure manifold. A volume of 450 μL of acetonitrile and a volume of 50 μL of IS were added to 500 μL of the spiked human plasma samples, vortexed for one min approximately, and then centrifuged at room temperature for 10 min at 15,000 rpm to permit protein precipitation. Supernatants were then loaded onto SPE cartridges on a 20 position extraction manifold operated under positive vacuum. Then, the cartridges were washed with water (1 mL) and the analytes were successively eluted with acetonitrile. After that, the eluates collected from the previous step were evaporated under a N_2 stream to dryness at 40°C. Then the dried residues were reconstituted with 125 μL of mobile phase and centrifuged at 15,000 rpm for 7 min at room temperature and then filtered using syringe membrane filters (13mm) kinesis to be ready for injection and analysis into the UPLC system.

2.5. Method Validation. The assay validation was carried out in accordance with guidance for bioanalytical method validation recommended by the FDA [24]. Method development and validation for the adopted bioanalytical chromatographic method in plasma samples include the demonstration of (1) selectivity; (2) calibration curve, linearity, and sensitivity; (3) accuracy and precision; (4) recovery; and (5) stability of the analytes in the spiked plasma samples.

2.5.1. Selectivity. The potential interferences from endogenous matrix components were investigated by evaluating ten different lots of human plasma as blank and at the LLOQ level of the spiked SF and DC. Drug-free plasma samples chromatograms were compared with those of the spiked plasma to ensure the absence of analytical interferences from endogenous substances present in plasma samples.

2.5.2. Calibration Curve, Linearity, and Sensitivity. Eight-point calibration standard curves of SF and DC in plasma, ranging from 25 to 6400 ng.mL^{-1} and 50 to 12800 ng.mL^{-1}, respectively, were prepared in triplicate for each run. The LLOQ is the lowest concentration in the standard calibration curve that back-calculates with good precision that does not exceed 20% of the CV and satisfactory accuracy which does not exceed 20% of the nominal concentration.

2.5.3. Accuracy and Precision. Accuracy, intraday precision, and interday precision values were determined by the analysis of spiked human plasma samples with five different concentrations for each of SF and DC, corresponding to the LLOQ, low, medium, high QC samples, and ULOQ three times on the same day and on three separate days. Accuracy was expressed as the % of deviation between the nominal and measured concentration (% error). Precision was calculated as coefficient of variation % (CV).

2.5.4. Recovery. The overall recovery of SF and DC from spiked human plasma was determined at the LLOQ, low,

medium, high QC samples, and ULOQ. The ratio of the peak area response of extracted QC samples for SF or DC to that of the IS was compared to that of unextracted standards obtained by injecting the corresponding concentration of DC and IS in the mobile phase and analyzed in triplicate. The extraction recovery was computed as previously published [12, 26], using the ratio of the response and the assay concentration factor (500:125, since during SPE procedure 500 μL spiked plasma samples was extracted, evaporated to dryness under a N_2 stream at 40°C, and then reconstituted with 125 μL of the mobile phase), and was expressed as % of the response of the calculated amount of SF or DC diluted in mobile phase and directly injected into the UPLC system, which corresponded to 100% recovery.

2.5.5. Stability of the Analytes in the Spiked Plasma Samples. Stability tests were carried out under various conditions simulating those that a real human plasma samples may be subjected to during routine analysis. Stability studies of SF and DC in human plasma included the following:

(a) Freeze and thaw stability after three freeze-thaw cycles of stored plasma samples; (b) bench-top stability of plasma samples after storage at RT for 48 h; (c) long-term stability of plasma samples after storage at−80ȖçC for 6 months; stability of SF and DC stock solutions (d) kept in the refrigerator and (e) kept on bench; and processed sample stability (f) after the heating process for the frozen plasma samples (60°C for 60 min) in addition to (g) the stability of the dried extracts (after SPE procedure) at −20°C for 6 days and (h) the stability of the reconstituted extracts in the mobile phase after being kept at 4°C for 4 days in the autosampler.

For each of the previously mentioned conditions, three series of LLOQ, LQ, MQ, HQ, and ULOQ spiked plasma samples were analyzed. The SF and DC concentrations in the analyzed plasma samples were compared to freshly made QC samples. For all of the previously mentioned stability studies, SF and DC in plasma samples were considered as stable if the stability sample results were within 15% of nominal concentrations according to the FDA guidelines for bioanalytical method validation [24].

3. Results and Discussion

3.1. Chromatographic Separation Conditions and Sample Extraction Procedure. The chromatographic separation of SF and DC was carried out using different HPLC columns and various mobile phases. The proper chromatographic separation was achieved on Waters® Acquity UPLC BEH C_{18} column (2.1 × 50 mm, 1.7 μm) using isocratic elution with a mobile phase of ammonium formate (pH 3.5; 5 mM) and acetonitrile (60:40 v/v) pumped at a flow rate 0.2 mL.min^{-1}. According to these chromatographic conditions, SF was eluted at 1.123 min while DC was eluted at 3.179 min with a perfect separation from plasma endogenous peaks as presented in Figure 2. The total run time is considered to be short which allows faster analysis of multiple samples during routine work using simple isocratic elution. Figure 2 represents the chromatograms of (a) drug-free human plasma;

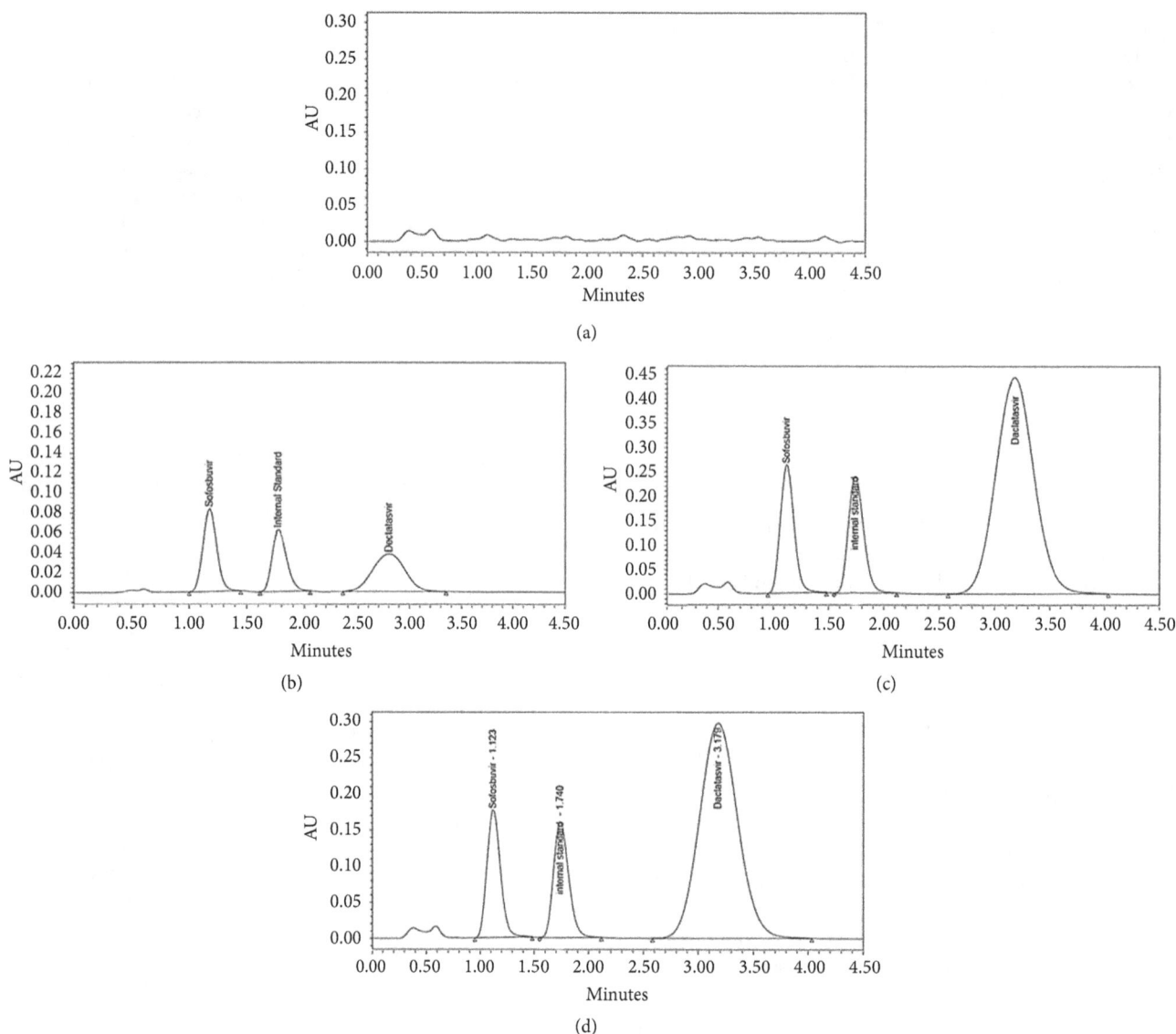

FIGURE 2: Typical UPLC chromatograms of (a) drug-free human plasma; (b) blank plasma spiked with SF and DC at LLOQ, (c) blank plasma spiked with SF and DC at ULOQ, and (d) plasma sample from a volunteer 0.5 h after administration of Darvoni® (400 mg of SF/60 mg of DC).

(b) blank plasma spiked with SF and DC at LLOQ; (c) blank plasma spiked with SF and DC at ULOQ; and (d) plasma sample from a volunteer 0.5 h after administration of Darvoni® (400 mg of SF/60 mg of DC).

A satisfactory drug recovery in sample extraction is crucial for the simultaneous estimation of SF and DC in human plasma at low concentration levels. Various extraction procedures were tried such as liquid–liquid extraction with different solvents and protein precipitation techniques, but they gave very low drug recovery which lead to significant interference from plasma peaks background. Deproteinization using acetonitrile was carried out due to the fact that SF and DC are ~ 61-65% [27] and ~99% [28] bound to human plasma proteins, respectively, followed by a neat SPE procedure employing Water 20-positions Extraction Manifold with

SPE cartridges (Sep-Pak® Vac C_{18}) which was considered as a reliable and fast technique for determination of SF and DC in human plasma samples. This extraction strategy allowed the improvement of sample clean-up without lowering drug recovery leading to high and constant recovery of SF and DC from human plasma.

3.2. Method Validation

3.2.1. Selectivity. Plasma sample extraction and chromatographic separation procedures were carried out to obtain a selective simultaneous determination for SF and DC. Various lots of blank human plasma from different sources were carefully evaluated for interference from endogenous matrix components. A typical chromatogram of blank human plasma is illustrated in Figure 2(a) which does not present any significant interfering endogenous peaks from human plasma at the retention times of SF, DC, or IS. The absence of any analytical interference was also assured using the peak purity checker system and matching with the library of the Empower® software.

3.2.2. Calibration Curve, Linearity, and Sensitivity.

Two calibration standard curves of SF and DC in plasma, ranging from 25 to 6400 ng.mL^{-1} and 50 to 12800 ng.mL^{-1}, respectively, were prepared in triplicate for each run. The calibration plots showed linearity over the previously mentioned concentration ranges and the determination coefficient (r^2) was not less than 0.998. The LLOQ is the lowest concentration in the standard calibration curve that back-calculates with good precision that does not exceed 20% of the CV and good accuracy within 20% of the nominal concentration. The LLOQ of SF and DC was 25 and 50 ng.mL^{-1}, respectively. A typical chromatogram for a spiked plasma sample containing the LLOQ for both SF and DC is presented in Figure 2(b). This figure presents satisfactory sensitivity for the routine analysis of human plasma samples in clinical application of therapeutic drug monitoring.

3.2.3. Accuracy and Precision.

Data results for accuracy (expressed as % bias = [measured concentration – nominal concentration] / nominal concentration × 100) and precision (expressed as % CV) presented in Table 2 were less than 15% from the nominal concentrations and < 20% for LLOQ, according to FDA guidelines [24]. These results show that the adopted chromatographic method provides a high degree of accuracy and reproducibility.

3.2.4. Extraction Recovery.

The average of % extraction recovery for SF and DC in their respective concentration ranges varied from 90.0 to 107.2 % and 93.1 to 108.0 % for SF and DC, respectively, with % CV values ranging from 1.2 to 3.5 % and 1.1 to 3.6 % for SF and DC, respectively, as presented in Table 2. These data results demonstrate that the extraction procedure for human plasma samples produces clean extracts and good recovery to reach to satisfactory sensitivity for the analysis.

3.2.5. Stability Studies.

The stability of SF and DC in human plasma samples at five levels (LLOQ, LQ, MQ, HQ, and ULOQ) was investigated under the following conditions that the common and routine clinical samples are usually subjected to [29–31]: (a) after three freeze-thaw cycles (24 hours at −80°C to room temperature) on plasma samples spiked with SF and DC, the recovered concentrations of SF and DC were very similar to those of freshly spiked plasma samples. In addition, no decomposition products of SF or DC were detected during (b) short-term storage (48 h) of plasma samples at RT and (c) long-term storage at −80°C for 6 months as illustrated in Table 3 which confirms the high stability of both SF and DC in human plasma samples both at RT and at −80°C. Stock solution stability for SF and DC was

also assessed. The stability of the standard solutions kept in refrigerator and those kept on bench was compared to freshly prepared standard solutions. It was found that solutions kept in (d) refrigerator are stable up to 10 days while those kept on (e) bench are stable for only 5 days. Moreover, the processed sample stability for SF and DC after the (f) heating process for the frozen plasma samples (60°C for 60 min) in addition to the stability of the (g) dried extracts (after SPE procedure) at −20°C for 6 days and in the processed plasma samples (h) reconstituted in the mobile phase after being kept at 4°C for 4 days in the autosampler was also tested. After the heating process for the frozen plasma samples (60°C for 60 min) and after 6 days storage of the dried extracts at −20°C or 4 days storage for the reconstituted extracts in the autosampler at 4°C, concentrations of the processed plasma samples were within ±15% of their nominal concentrations and < 20 % for LLOQ as presented in Table 3, indicating that both drugs are stable under the stated conditions to which routine plasma samples are subjected.

3.3. Application of the Adopted Method to the Plasma Samples of Healthy Volunteers.

Concentrations of SF and DC were estimated using the adopted UPLC-PDA method in the plasma samples obtained from three healthy male volunteers after the oral intake of one Darvoni® tablet coformulated with 400 mg of SF and 60 mg of DC. To confirm the clinical application of the suggested method, a typical UPLC chromatogram of a plasma sample obtained from one of the volunteers after 0.5 h from the oral intake of one Darvoni® tablet is presented in Figure 2(d). The shown chromatogram assures satisfactory and good clinical application of the adopted chromatographic method during routine therapeutic drug monitoring. The good recovery of SF produced by the application of the proposed UPLC-PDA method makes no need to quantify the main metabolite of SF (GS-331007).

4. Conclusion

A rapid, accurate, precise, sensitive, and selective UPLC-PDA method was developed for simultaneous quantification of SF and DC in plasma samples has been adopted in this work for the first time. The developed method had been validated according to FDA guidelines for bioanalytical method validation. The method applicability was confirmed by the analysis of plasma samples of three healthy volunteers after the oral intake of coformulated Darvoni® tablet. So far, the previously published analytical methods developed for pharmacokinetic investigations of SF and DC were all based on costly LC-MS/MS equipment after the oral intake of two tablets for each drug, which decreased the feasibility of the routine analysis of SF and DC. The adopted UPLC-UV method is considered to be greatly applicable for the routine TDM of SF and DC in plasma at conventional clinical laboratories where LC-MS/MS equipment is not present. The suggested method will be suitable for standard clinical laboratories that do not possess LC-MS/MS equipment. With respect to large-scale pharmacokinetic analysis, the proposed

TABLE 2

(a) Accuracy, intraday, and interday precision and % extraction recovery for SF and DC in their QC samples in human plasma (n=6)

SF

Nominal Concentration (ng.mL^{-1})	Average of the measured concentration (ng.ml^{-1})	Accuracy (% Bias)	Intra-day precision (% CV)	Inter-day precision (% CV)	% Extraction recovery ± %CV
25	24	-4.0	9.6	8.6	96.0 ± 1.2
50	45	-10.0	4.6	5.1	90.0 ± 3.5
400	425	6.3	5.2	9.3	106.3 ± 1.3
3200	3430	7.2	3.9	7.4	107.2 ± 2.8
6400	6123	-4.3	3.8	6.1	95.7 ± 1.5

DC

Nominal Concentration (ng.mL^{-1})	Average of the measured concentration (ng.ml^{-1})	Accuracy	Intra-day precision	Inter-day precision	% Extraction recovery ± %CV
50	54	8.0	6.5	3.7	108.0 ± 2.3
100	95	-5.0	9.2	9.1	95.0 ± 1.7
1600	1490	-6.9	4.7	7.5	93.1 ± 1.1
6400	6690	4.5	5.4	6.4	104.5 ± 2.0
12800	12357	-3.5	2.8	8.2	96.5 ± 3.6

(b) Accuracy, intraday, and interday precision and % extraction recovery for SF and DC in their QC samples in human plasma (n=6)

SF

Nominal Concentration (ng.mL^{-1})	Average of the measured concentration (ng.ml^{-1})	Accuracy (% Bias)	Intra-day precision (% CV)	Inter-day precision (% CV)	% Extraction recovery ± %CV
25	24	-4.0	9.6	8.6	96.0 ± 1.2
50	45	-10.0	4.6	5.1	90.0 ± 3.5
400	425	6.3	5.2	9.3	106.3 ± 1.3
3200	3430	7.2	3.9	7.4	107.2 ± 2.8
6400	6123	-4.3	3.8	6.1	95.7 ± 1.5

DC

Nominal Concentration (ng.mL^{-1})	Average of the measured concentration (ng.ml^{-1})	Accuracy	Intra-day precision	Inter-day precision	% Extraction recovery ± %CV
50	54	8.0	6.5	3.7	108.0 ± 2.3
100	95	-5.0	9.2	9.1	95.0 ± 1.7
1600	1490	-6.9	4.7	7.5	93.1 ± 1.1
6400	6690	4.5	5.4	6.4	104.5 ± 2.0
12800	12357	-3.5	2.8	8.2	96.5 ± 3.6

method will be a satisfactory, simple, cheap, fast, and easier to set up alternative UPLC chromatographic method coupled with DAD. Stability studies of SF and DC under various common conditions to which both drugs may be subjected to during sample handling and analysis through routine TDM process illustrated that both drug concentrations remained approximately unchanged in plasma, in their stock solutions, and in processed plasma samples under different storage conditions. All in all, the adopted UPLC-DAD technique

and the data results of the stability studies analyses could be valuable for dosing both drugs and appropriately dealing with their plasma samples both in routine clinical application and in TDM. This clinical application gave the opportunity to the optimization of the drug treatment which can improve the life quality and increase the therapy efficacy itself. It can lead also to a cost saving outcome, reducing side effects, and consequently a better clinical cost for patient's care.

TABLE 3: Stability of SF and DC in plasma samples under several storage conditions.

| | | Nominal Concentration (ng.ml^{-1}) | | | | | | | | | |
| | | SF | | | | | DC | | | | |
		25	50	400	3200	6400	50	100	1600	6400	12800
(a) After three freeze/thaw cycles	% Bias	-2.9	5.2	-5.4	4.2	6.1	8.5	-3.5	-5.8	-3.1	-9.4
	% CV	5.4	6.5	7.2	5.3	9.4	5.8	1.5	8.5	3.2	5.4
(b) short-term storage (48 h) at RT	% Bias	4.1	4.1	7.3	-5.7	-4.8	5.1	6.4	9.2	8.4	-5.0
	% CV	1.1	6.5	5.5	3.2	5.2	6.6	7.3	1.5	2.4	5.1
(c) long-term storage at −80°C for 6 months	% Bias	3.5	-3.7	3.4	5.1	7.3	8.4	9.4	-2.0	-5.7	-6.3
	% CV	2.1	4.4	3.0	5.9	8.6	3.5	1.3	7.0	6.9	3.5
(d) Standard solutions kept in refrigerator up to 10 days	% Bias	-2.5	6.5	3.8	2.8	4.1	7.3	-5.7	-3.9	-2.8	1.7
	% CV	1.0	3.2	4.3	6.5	7.6	8.6	5.7	9.3	1.0	1.3
(e) Solutions kept on bench are stable for 5 days	% Bias	3.9	6.3	8.2	9.0	1.8	4.7	3.9	2.8	-3.7	-4.5
	% CV	6.4	8.3	9.5	4.6	2.7	3.6	6.5	1.6	4.9	1.7
(f) Frozen plasma samples which undergo heating process (60°C for 60 min)	% Bias	-6.7	-5.7	3.7	4.1	5.9	8.6	4.7	8.1	-9.4	-7.2
	% CV	5.4	7.2	6.5	2.7	3.8	1.7	5.2	7.4	6.9	1.5
(g) Dried extracts (after SPE procedure) kept at −20°C for 6 days	% Bias	9.5	4.7	5.8	-9.3	-6.3	-7.8	9.5	-5.2	-6.0	-3.7
	% CV	5.3	6.2	8.3	6.4	9.3	6.2	5.3	8.1	4.3	3.8
(h) Reconstituted extracts in the mobile phase kept at 4°C for 4 days in the auto sampler	% Bias	-5.7	-4.0	-5.2	4.9	8.1	7.2	7.6	6.1	8.5	5.3
	% CV	4.6	1.5	4.6	4.1	5.3	9.4	8.3	1.7	3.2	8.4

References

[1] Y. A. Mohamoud, G. R. Mumtaz, S. Riome, D. Miller, and L. J. Abu-Raddad, "The epidemiology of hepatitis C virus in Egypt: a systematic review and data synthesis," *BMC Infectious Diseases*, vol. 13, no. 1, article 288, 2013.

[2] M. G. Ghany, D. B. Strader, D. L. Thomas, and L. B. Seeff, "Diagnosis, management, and treatment of hepatitis C: an update," *Hepatology*, vol. 49, no. 4, pp. 1335–1374, 2009.

[3] B. J. Kirby, W. T. Symonds, B. P. Kearney, and A. A. Mathias, "Pharmacokinetic, Pharmacodynamic, and Drug-Interaction Profile of the Hepatitis C Virus NS5B Polymerase Inhibitor Sofosbuvir," *Clinical Pharmacokinetics*, vol. 54, no. 7, pp. 677–690, 2015.

[4] https://pubchem.ncbi.nlm.nih.gov/compound/sofosbuvir.

[5] G. Bertino, A. Ardiri, M. Proiti et al., "Chronic hepatitis C: this and the new era of treatment," *World Journal of Hepatology*, vol. 8, no. 2, pp. 92–106, 2016.

[6] https://pubchem.ncbi.nlm.nih.gov/compound/Daclatasvir.

[7] F. Poordad, E. R. Schiff, J. M. Vierling et al., "Daclatasvir with sofosbuvir and ribavirin for hepatitis C virus infection with advanced cirrhosis or post-liver transplantation recurrence," *Hepatology*, vol. 63, no. 5, pp. 1493–1505, 2016.

[8] A. F. Luetkemeyer, C. McDonald, M. Ramgopal, S. Noviello, R. Bhore, and P. Ackerman, "12 Weeks of Daclatasvir in Combination with Sofosbuvir for HIV-HCV Coinfection (ALLY-2 Study): Efficacy and Safety by HIV Combination Antiretroviral Regimens," *Clinical Infectious Diseases*, vol. 62, no. 12, pp. 1489–1496, 2016.

[9] https://pubchem.ncbi.nlm.nih.gov/compound/Ledipasvir.

[10] J. Vionnet, A.-C. Saouli, M. Pascual et al., "Therapeutic drug monitoring for sofosbuvir and daclatasvir in transplant recipients with chronic hepatitis C and advanced renal disease," *Journal of Hepatology*, vol. 65, no. 5, pp. 1063–1065, 2016.

[11] S. Burgess, N. Partovi, E. M. Yoshida, S. R. Erb, V. M. Azalgara, and T. Hussaini, "Drug Interactions With Direct-Acting Antivirals for Hepatitis C: Implications for HIV and Transplant Patients," *Annals of Pharmacotherapy*, vol. 49, no. 6, pp. 674–687, 2015.

[12] G. Nannetti, L. Messa, M. Celegato et al., "Development and validation of a simple and robust HPLC method with UV detection for quantification of the hepatitis C virus inhibitor daclatasvir in human plasma," *Journal of Pharmaceutical and Biomedical Analysis*, vol. 134, pp. 275–281, 2017.

[13] S. Miraghaei, B. Mohammadi, A. Babaei, S. Keshavarz, and G. Bahrami, "Development and validation of a new HPLC-DAD method for quantification of sofosbuvir in human serum and its comparison with LC–MS/MS technique: Application to a bioequivalence study," *Journal of Chromatography B*, vol. 1063, pp. 118–122, 2017.

[14] S. Madhavi and A. P. Rani, "Bioanalytical method development and validation for the determination of sofosbuvir from human plasma," *International Journal of Pharmacy and Pharmaceutical Sciences*, vol. 9, no. 3, pp. 35–41, 2017.

[15] D. W. Zidan, W. S. Hassan, M. S. Elmasry, and A. A. Shalaby, "Investigation of anti-Hepatitis C virus, sofosbuvir and daclatasvir, in pure form, human plasma and human urine using micellar monolithic HPLC-UV method and application to pharmacokinetic study," *Journal of Chromatography B*, vol. 1086, pp. 73–81, 2018.

[16] S. Notari, M. Tempestilli, G. Fabbri et al., "UPLC–MS/MS method for the simultaneous quantification of sofosbuvir, sofosbuvir metabolite (GS-331007) and daclatasvir in plasma of HIV/HCV co-infected patients," *Journal of Chromatography B*, vol. 1073, pp. 183–190, 2018.

[17] O. M. Abdallah, A. M. Abdel-Megied, and A. S. Gouda, "Development and validation of LC-MS/MS method for simultaneous determination of sofosbuvir and daclatasvir in human Plasma: Application to pharmacokinetic study," *Biomedical Chromatography*, no. 1, pp. 1–9, 2018.

[18] A. Ariaudo, F. Favata, A. De Nicolò et al., "A UHPLC-MS/MS method for the quantification of direct antiviral agents simeprevir, daclatasvir, ledipasvir, sofosbuvir/GS-331007, dasabuvir, ombitasvir and paritaprevir, together with ritonavir, in human plasma," *Journal of Pharmaceutical and Biomedical Analysis*, vol. 125, pp. 369–375, 2016.

[19] M. R. Rezk, E. B. Basalious, and M. E. Amin, "Novel and sensitive UPLC–MS/MS method for quantification of sofosbuvir in human plasma: application to a bioequivalence study," *Biomedical Chromatography*, vol. 30, no. 9, pp. 1354–1362, 2016.

[20] M. R. Rezk, E. R. Bendas, E. B. Basalious, and I. A. Karim, "Development and validation of sensitive and rapid UPLC-MS/MS method for quantitative determination of daclatasvir in human plasma: Application to a bioequivalence study," *Journal of Pharmaceutical and Biomedical Analysis*, vol. 128, pp. 61–66, 2016.

[21] H. Jiang, H. Kandoussi, J. Zeng et al., "Multiplexed LC-MS/MS method for the simultaneous quantitation of three novel hepatitis C antivirals, daclatasvir, asunaprevir, and beclabuvir in human plasma," *Journal of Pharmaceutical and Biomedical Analysis*, vol. 107, pp. 409–418, 2015.

[22] L. Qu, W. Wang, D. Zeng, Y. Lu, and Z. Yin, "Quantitative performance of online SPE-LC coupled to Q-Exactive for the analysis of sofosbuvir in human plasma," *RSC Advances*, vol. 5, no. 119, pp. 98269–98277, 2015.

[23] M. R. Rezk, E. B. Basalious, and I. A. Karim, "Development of a sensitive UPLC-ESI-MS/MS method for quantification of sofosbuvir and its metabolite, GS-331007, in human plasma: Application to a bioequivalence study," *Journal of Pharmaceutical and Biomedical Analysis*, vol. 114, pp. 97–104, 2015.

[24] FDA, "Guidance for Industry: Bio analytical Method Validation," 2013, https://www.fda.gov/downloads/drugs/guidances/ucm070107.Pdf.

[25] "European Medicines Agency Daklinza: Assessment Report, 2014," http://www.ema.europa.eu/docs/en_GB/document_library/EPAR_Public_assessment_report/human/003768/WC500172849.pdf.

[26] A. Loregian, S. Pagni, E. Ballarin, E. Sinigalia, S. G. Parisi, and G. Palù, "Simple determination of the HIV protease inhibitor atazanavir in human plasma by high-performance liquid chromatography with UV detection," *Journal of Pharmaceutical and Biomedical Analysis*, vol. 42, no. 4, pp. 500–505, 2006.

[27] E. Cholongitas and G. V. Papatheodoridis, "Sofosbuvir: a novel oral agent for chronic hepatitis C," *Annals of Gastroenterology*, vol. 27, pp. 331–337, 2014.

[28] R. E. Nettles, M. Gao, M. Bifano et al., "Multiple ascending dose study of BMS-790052, a nonstructural protein 5A replication complex inhibitor, in patients infected with hepatitis C virus genotype 1," *Hepatology*, vol. 54, no. 6, pp. 1956–1965, 2011.

[29] N. F. Al-Tannak, "UHPLC-UV method for simultaneous determination of perindopril arginine and indapamide hemihydrate in combined dosage form: a stability-indicating assay method," *Scientia Pharmaceutica*, vol. 86, p. 7, 2018.

[30] N. F. Al-Tannak and A. Hemdan, "UHPLC-ESI-QToF analysis of trandolapril and verapamil hydrochloride in dosage form and spiked human plasma using solid phase extraction: stability indicating assay method," *Current Pharmaceutical Analysis*, vol. 14, no. 6, pp. 595–603, 2018.

[31] S. A. Mohammed, M. S. Eissa, and H. M. Ahmed, "Simple protein precipitation extraction technique followed by validated chromatographic method for linezolid analysis in real human plasma samples to study its pharmacokinetics," *Journal of Chromatography B*, vol. 1043, pp. 235–240, 2017.

8

Insights into Geographic and Temporal Variation in Fatty Acid Composition of Croton Nuts using ATR-FTIR

Nathan W. Bower ⓘ,[1] **Murphy G. Brasuel** ⓘ,[1] **Eli Fahrenkrug** ⓘ,[1] and **Matthew D. Cooney**[2]

[1]*Chemistry and Biochemistry Department, Colorado College, Colorado Springs, CO 80903, USA*
[2]*GIS Technical Director, Colorado College, Colorado Springs, CO 80903, USA*

Correspondence should be addressed to Nathan W. Bower; nbower@coloradocollege.edu

Academic Editor: Adil Denizli

Croton megalocarpus seedcake oils from 30 different locations in south central Kenya were analyzed for their fatty acid composition using ATR-FTIR to determine the efficacy of a simple procedure for measuring initial geographic and subsequent temporal variation during five months of seed storage. To our knowledge, this is the first report showing variation in how oils in untreated nuts from different locations change during storage, and how these differences are correlated with local environments. These variations are important to forensic authentication efforts and they provide insights into ways to optimize Croton oil composition.

1. Background

Croton megalocarpus Hutch. (Euphorbiaceae) is a fast-growing tree indigenous to East Africa with a range extending from moist montane forests to dry savannas [1]. It has seen some use in traditional medicine [2], but recent use has focused on its potential as a local source of biodiesel fuel [3]. In order to optimize performance and minimize maintenance of engines that use biodiesel, a uniform and reproducible oil is desirable [4]. Because of the range of ecological habitats where the trees are found, it seemed there might be differences in the oil composition due to different climatic variables and/or to different varieties/ecotypes of *Croton megalocarpus*. This study was undertaken to determine whether there are differences in the nut oil composition from different locations, due to either varietal differences or habitat variation. It was also undertaken to test the efficacy of ATR-FTIR (attenuated total reflectance Fourier transform infrared) spectroscopy as a faster and less costly method than the wet-chemical and instrument-intensive methods typically used for tracking changes in the oil composition during storage.

Geographic differences of olive oil varietals have been observed and used to help with the authentication of these high-value, registered commodities. These determinations usually require time-consuming sample preparation and analysis with instruments such as gas chromatography–mass spectrometry (GC-MS) or liquid chromatography–mass spectrometry (LC-MS) [5]. Recent research has explored faster and simpler preparation methods such as NIR (near infrared) [6] and ATR-FTIR spectroscopies coupled with chemometric methods such as PLS (partial least squares) regression after centering and scaling each spectrum [7]. Because of the large number of variables involved when using an entire spectrum, spectral processing techniques such as jackknife resampling, cross-validation, and principle components are commonly employed [8]. Subsequent chemometric analysis such as pattern recognition using soft independent modeling of class analogy algorithm (SIMCA) with partial least squares regression (PLSR) has been used to monitor oxidation products and to detect adulterants [9].

While these IR methods require less preparation and analysis time than chromatographic procedures, complex spectral processing and chemometric analyses can often camouflage artifacts and make understanding of the data more difficult for an operator inexperienced with either the software or its interpretation. To minimize these issues, we used standard ATR-FTIR spectral analyses from a portable instrument coupled with known oils to calibrate the spectra and the fatty acid methyl ester (FAME) analyses obtained using GC-MS. Finally, we optimized the choice of wavelengths to simplify the data analyses, making this process commensurate with the simplicity of the ATR-FTIR instrumentation.

2. Methods

Croton megalocarpus nuts from 30 different locations ranging from approximately 0.4901 N and 35.7434 E to -2.91895 N and 37.5058 E were obtained from Eco Fuels Kenya, Ltd. in the fall of 2017. (For a complete listing, see Supplementary Material, Table 1.) Upon receipt, these were dried at 22°C and 15% RH in paper bags over a period of 10–14 days until the average mass loss per day (attributed to water) fell below 0.5%. At that time the first analyses using cold-pressed oils and ATR-FTIR (see below) were conducted and the nuts were transferred to covered polypropylene containers and stored in a 0 to 1°C refrigerator (RH = 0 to 1%) for subsequent analyses conducted over a 4.5 month period.

Cold (22°C) pressing of individual weighed seeds from the Croton nuts (usually holding three seeds) was accomplished after manually removing and weighing the calyx, shell (ovary), and seed coat (testa). An in-house apparatus consisting of a close-fitting 5 mm diameter piston placed inside of an Al cylinder-cup with a small hole drilled in the side allowed access to the expressed oil. The apparatus was placed in a 10 cm bench vice to provide the necessary pressure. A Pasteur pipet was used to transfer a drop of oil to the diamond window of a Bruker ALPHA Platinum ATR-FTIR. To assure a consistent maximum peak absorbance of 0.295 ± 0.001 (1s) at 1743 cm^{-1}, a drop thickness 0.5 – 1 mm was maintained for all analyses.

The ATR-IR spectra were collected using the instrument's default conditions (24 spectra averaged together with data collected every 2 cm^{-1}), giving a resolution of 4 cm^{-1}. The ATR window was cleaned with isopropanol, allowed to dry, and a background collected every 10–15 min. Spectra were baseline subtracted using the instrument's default settings, and both the raw spectra and the baseline-corrected wavenumber versus absorbance files were transferred to Excel for subsequent data processing. (This was done by clicking on the "history" tab and the "AB" tab in the upper left of the spectrum window. The two columns were copied and pasted into Excel.)

Sixteen known vegetable oils (see Supplementary Material, Table 2) containing 8–24% saturated fatty acids (SFA), 10–85% monounsaturated fatty acids (MUFA), and 2–72% polyunsaturated acids (PUFA) were also analyzed by ATR-FTIR to create linear calibrations for these three components. Although PLS regression of all of the spectra may be used, for simplicity, the percentages of these three components were obtained by plotting the frequency of the cis C=C-H stretch near 3009 cm^{-1} versus the log [% PUFA/% MUFA] to obtain the ratio of these two components. The % PUFA was obtained by plotting the ratio of the absorbances measured at 2922 cm^{-1} and 3009 cm^{-1} versus % PUFA. The % SFA was obtained by subtracting the % PUFA and % MUFA from 100.

The fatty acids fractions were also obtained by solvent extraction of unpressed seeds using dichloromethane. This solvent was chosen for its low water solubility, low vapor pressure, and dipole character to replace the more toxic chloroform-methanol solvent traditionally used to extract lipids. Reagent grade (Fisher) CH_2Cl_2 is a single, uniform solvent that does not leave residues for the GC-MS and ATR-FTIR analyses. Grinding pistachios in CH_2Cl_2 has been shown to give fatty acid profiles nearly identical to those obtained using Soxhlet extraction [12].

The CH_2Cl_2 extraction was accomplished by removing the seed coat and grinding individual seeds in 5 mL of CH_2Cl_2 with a glass pestle in a porcelain evaporating dish. The extract was filtered through glass microfiber paper (Whatman 934AH) with three washes of 5 mL CH_2Cl_2. The CH_2Cl_2 was evaporated in a fume hood using a stream of dry air and the crude oil and protein contents were obtained by weighing the products. The oil was analyzed directly using ATR-FTIR or by GC-MS after conversion to the fatty acid methyl esters (FAMEs). If the oil could not be analyzed immediately, it was stored at -10°C.

Conversion of the fatty acids to FAMES for the GC-MS analyses was accomplished following the European Union's Regulation (EU) 2015/1833 for olive oils [13]. This involved dissolving 0.01 – 0.02 g of oil in 2 mL of heptane, followed by addition of 0.2 mL of methanolic KOH (0.11 g of KOH per mL of dry methanol), shaking for 30 sec, and letting the mixture settle for 2–5 min. A milliliter of the supernatant was then transferred to a capped vial containing anhydrous Na_2SO_4. This was shaken and allowed to settle for 5 min to remove any residual water and KOH.

The Na_2SO_4-dried heptane-FAME mixture was subsequently diluted (0.025 mL/1.5 mL CH_2Cl_2) and analyzed using an Agilent GC-MS (7890A GC and 597C MS) with a 30 m x 250 μm x 0.25 μm HP-5MS column using a temperature program (100°C for 1 min, ramped to 270°C at 3°C/min). This program gave near baseline resolution for all detected FAMEs (C14 to C22) except for C18:3 and cis C18:1, which coeluted (see Supplementary Material Table 4 and Supplementary Material Figure 1). These two were quantified using the areas of their 292.3 and 296.3 peaks obtained from selective ion monitoring. The relative amounts of the other FAMEs were determined by integrating the areas of the peaks in the total ion chromatograms (TICs) after calibrating with known oils prepared and analyzed using the same EU 2015/1833 procedure.

Statistical analysis of the data and plots of the spatial and temporal variation were conducted using Microsoft Excel 2016, Minitab ver. 18.1 (Minitab, Inc., State College PA 16801-3210, USA), and ArcGIS ver. 10.4.1 with Spatial Analyst Extension. Interpolation between points used an inverse square of the distance for the weighting of points.

3. Results and Discussion

Table 1 gives the average mass of a complete nut after the initial week of drying and corrected to a mass that includes a full calyx. (Many nuts had only a partial or no calyx.) The

TABLE 1: Composition of the nuts.

Component	Mean	1s
Nut (g) [a]	4.16	1.21
Calyx (%)	15.23	2.75
Ovary (%)	48.68	6.83
Testa (%)	14.64	3.57
Nut meat (%)	21.44	5.57
% of Nut meat:		
Meal (%)	56.32	14.19
Oil (%)	42.64	5.27
Ash (%)	1.04	0.52

[a] As received, corrected for any missing calyx.

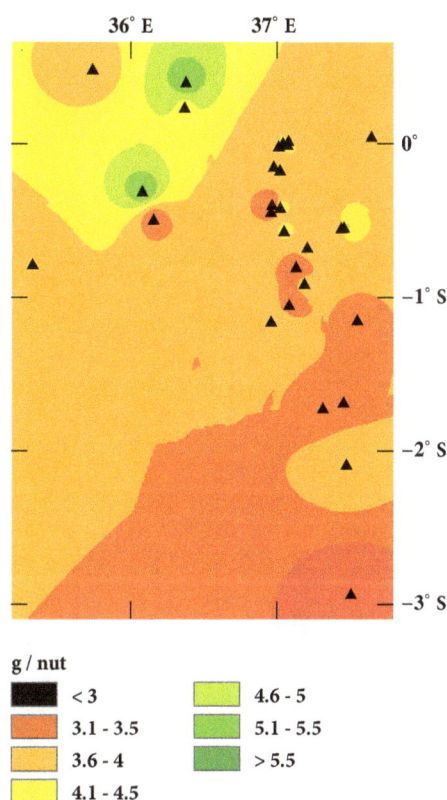

FIGURE 1: Interpolated map of the average "as received" nut mass at each location.

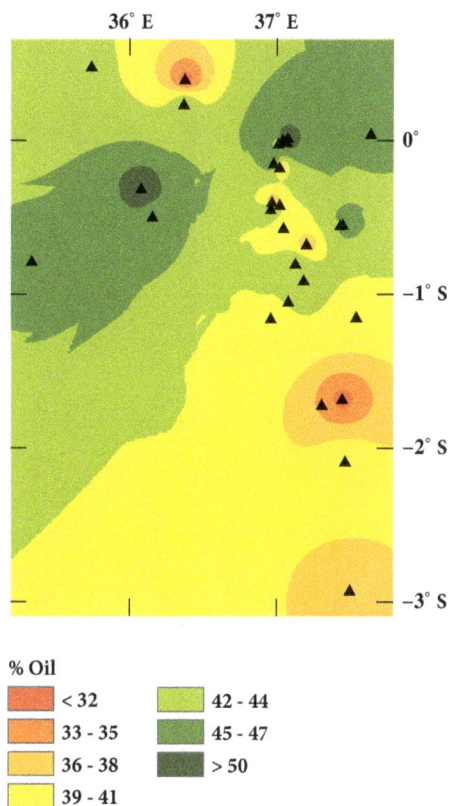

FIGURE 2: Interpolated map of the mass % oil in the deshelled seed after four months of storage.

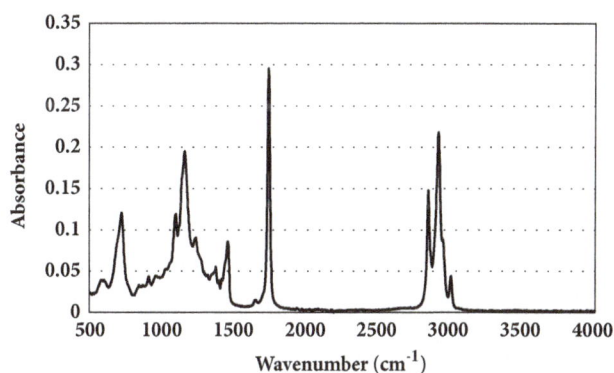

FIGURE 3: ATR-FTIR spectrum of Croton oil from Sipili, Kenya, before baseline subtraction and normalization to the C=O stretch at 1743 cm^{-1} attributed to the ester carbonyl of the triglycerides.

ovary and any associated calyx are removed and repurposed, while the seeds (testa + nut meat) are pressed to obtain the oil and a seedcake that may be used as a feed supplement, as the nut meat contains 55.7% crude protein (Carlo Erba C/N analyzer).

Figures 1 and 2 show the variation in the mass and mass percentage of oil extracted from the nutmeat for the various locations. These variables are often associated with different varieties of a plant species, but they are also impacted by the growing conditions, such as temperature, precipitation, and nutrient availability. For example, preliminary analyses of elemental content suggest % Mn and % P (PANalytical XRF) are higher in seeds with higher oil percentages.

Figure 3 presents a typical spectrum obtained from Croton nut oils. All spectra were similar enough that they appear identical at this scale. However, there are subtle differences that are related to the geography and to the time in storage. Figures 4(a) and 4(b) present the spatial variation observed within the first and last months of the study based on a plot of the absorbance measured at 3009 cm^{-1} after normalizing it to the 1743 cm^{-1} ester peak (set to unity using the baseline-corrected spectra). Sample locations are indicated by black triangles.

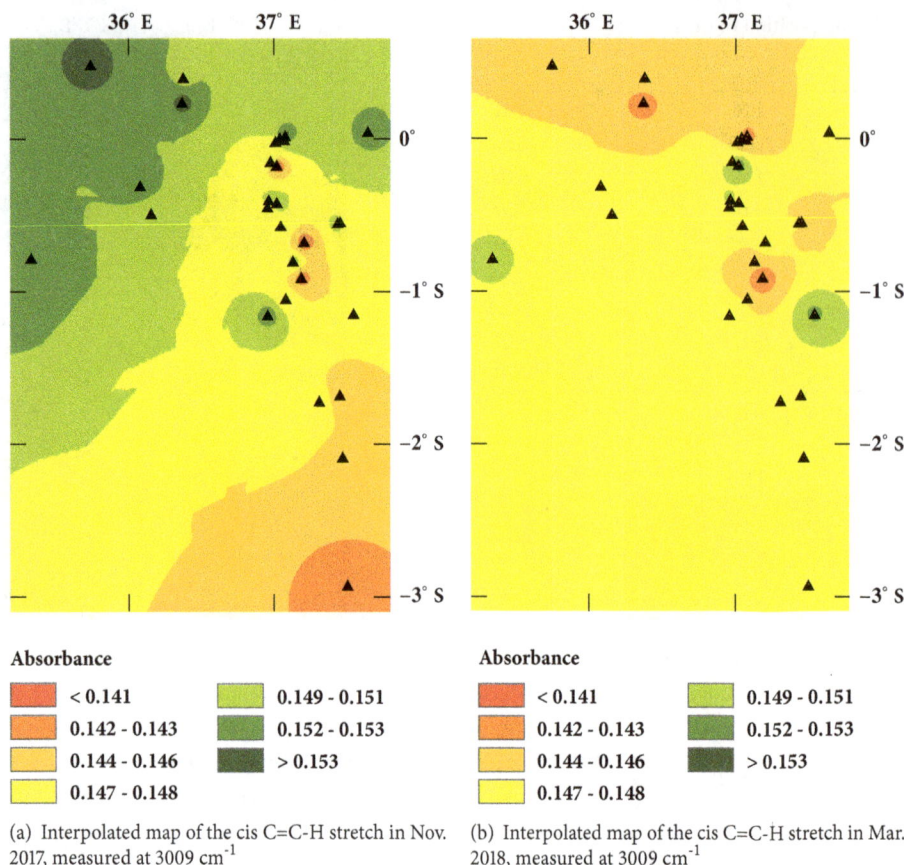

(a) Interpolated map of the cis C=C-H stretch in Nov. 2017, measured at 3009 cm^{-1}.

(b) Interpolated map of the cis C=C-H stretch in Mar. 2018, measured at 3009 cm^{-1}.

FIGURE 4

Agroclimatic zone maps for the region (see Supplementary Material, Table 5 and Figure 2) indicate the red to orange areas experience mean annual temperatures of 20–22°C, while the green to yellow areas are cooler with mean annual temperatures from 15–20°C [14]. (Some of the interpolated regions without dots contain high montane or arid regions where Croton trees do not grow.) In some instances, the higher absorbances at 3009 cm^{-1} correspond to cool, dry regions which suggests humidity and water availability also play a role in the fatty acid profiles.

We expected the geographic variation would reflect varietal/ecotypic differences in the Croton nuts based on similar studies of olive oils [6–8]. However, the correspondence of the variation in Figure 4(a) with the average nut mass (Figure 1), percent oil (Figure 2), and morphology (see Supplementary Material, Figure 3) suggests that the distribution of unsaturation represented by the absorbance in Figure 4(a) may be due to local climate differences. Cooler temperatures and drier environments favor cells producing lower melting oils in order to maintain plasma membrane flexibility [15]. This tendency is illustrated in Figure 5, where the mean annual temperature and precipitation have been standardized (using their z-scores) and are plotted versus the standardized absorbance. Lower temperatures have higher absorbances at 3009 cm^{-1}. (Higher absorbances correlate with higher unsaturation; $R^2 = 0.83$, $P < 0.001$, $DF = 9$).

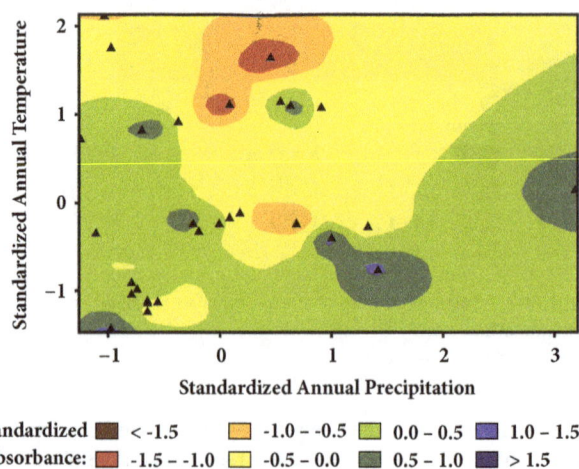

FIGURE 5: Interpolated map of the March ATR-FTIR standardized absorbance data, which correlates with higher unsaturation, versus the standardized temperature and precipitation.

Although less pronounced, average temperatures and average precipitation in the middle of the plot also favor lower levels of saturation. It is the extremes that push the plant cells to make more of the unsaturated fatty acids that give the plasma membranes their flexibility.

The subsequent changes in unsaturation observed during storage at $0°C$ support this explanation for the variation in Figure 4(a). The average unsaturation increased by a few percent relative to the initial profile, but in a manner that leveled geographic differences. The largest increases in unsaturation (Figure 4(b)) occurred in nuts from sites with higher average annual temperatures (for an interpolated map of temperature and the corresponding changes in the unsaturation, see Supplementary Material, Figure 4). Thus, enzymes that catalyze dehydrogenation reactions at lower temperatures may be involved. If so, then the apparent geographic variation may simply be a plastic response to mean temperature and precipitation differences rather than varietal/ecotype differences [16].

Based on these observations, we attribute the variation in the different levels of unsaturation (saturated, mono- and polyunsaturated) over time shown in Figure 6 which appeared when Croton nuts were placed in a $1°C$ refrigerator to the plasticity of living seeds as they respond to a temperature change. This increase only lasted about a month before a slower process that we attribute to chemical oxidation took over that decreased the overall unsaturation. This decrease is similar to what happens with drying of oils like linseed oil [17] or with cold storage of pine nuts [18] and almonds [19]. Based on GC-MS analyses of the extracted oils, over the three-month period from the highest level of unsaturation in Figure 6 to the lowest level at 4.5 months, the saturated fatty acids (primarily C18:0 and C16:0) and the triply unsaturated fatty acid (C18:3) decreased by about 2, 1, and 3%, respectively. At the same time the mono- and di-unsaturated fatty acids (C18:1 and C18:2) increased by about 2.5% each. Figures 6 and 7 summarize the changes in the unsaturation and fatty acid composition as measured by ATR-FTIR in the cold-pressed oils during the five-month study.

Although a frequency shift in the 3009 cm^{-1} peak to lower values (3008 cm^{-1} and below) is expected as the double bonds are broken, only a small shift in the frequency of the peak was observed despite the declining percentage of the polyunsaturated fatty acid, C18:3. This may be because C18:2 was increasing over the same time interval, or the refrigeration slowed the reaction sufficiently that more time is needed to see a noticeable effect. However, extracted oil samples kept in stoppered flasks at room temperature ($22°C$) for 3 months did exhibit a significant frequency shift to smaller wavenumbers that was largest for those which were the most oxidized ($R^2 = 0.57$, $P = 0.018$, $DF = 7$). Other components that may reflect geographic differences or biological and chemical reactions over time, such as the iodine value and saponification number [20], the peroxide value [21], the free fatty acid (FFA) [22] content, or the trans-fat content [23], may also be measured. However, the components that relate to degradation of the oils were all below 0.1% even at the end of this study. Nuts contaminated by mold did show a jump in the FFA content to around 0.2% of the total oil over the 4.5 months.

Table 2 presents the fatty acid composition from CH$_2$Cl$_2$-extracted and cold-pressed oils from nuts that were dried and stored long enough to achieve stable values. Differences may

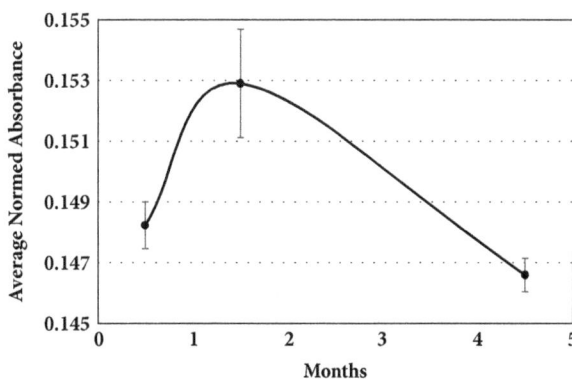

FIGURE 6: Variation (± 1 SE) in the unsaturation proxy (the normed absorbance at 3009 cm^{-1} due to the cis C=C-H stretch) as measured by ATR-FTIR during the 5 month study.

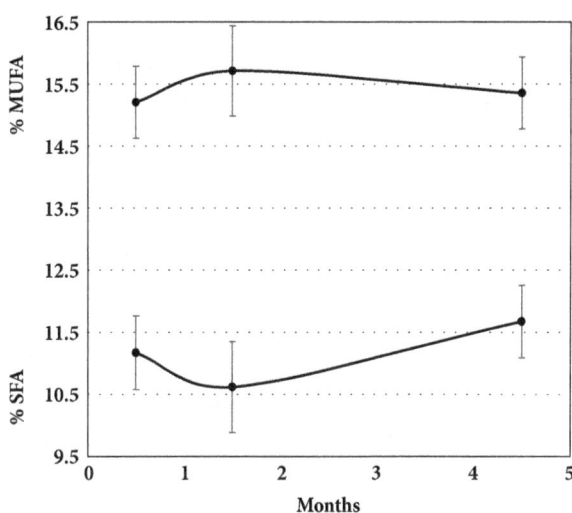

FIGURE 7: Average mass percentages (± 1 SE) of monounsaturated and saturated fatty acids, calculated from the ATR-FTIR spectra. (Polyunsaturated fatty acids comprise the remainder.)

be expected between the two methods used for cell disruption and oil extraction (cold-pressing and grinding in CH$_2$Cl$_2$), as cold-pressed oils are expected to have higher levels of the more fluid components, notably the low molecular weight saturated and the polyunsaturated fatty acids [24, 25]. Thus, solvent extraction data could appear to have come from regions with lower annual temperatures than data obtained using a cold press if the method is not taken into account.

Despite these expectations, there are no significant differences between the fatty acid profiles of the samples that had stabilized in the refrigerator using these two extraction methods. The results from this study are in agreement with literature values for commercial Croton oil from Tanzania (see Table 2), especially given the decrease in C16:0 and increase in C18:2 during cold storage. However, the higher levels of C18:3 in the Kenyan nuts compared to the Tanzanian values are probably real. Although ATR-FTIR is not the best way to obtain accurate oil composition data, the average Croton nut PUFA, MUFA, and SFA values after 4.5 months

TABLE 2: GC-MS relative percent fatty acid values.

	CH$_2$Cl$_2$ Extract[a]		Cold Press[a]		Cold Press	Not stated
FA	Mean	± 1s	Mean	± 1s	Wu [10]	Kafuku [11]
C14:0	0.03	0.02	0.02	0.01	0.04	0.10
C16:0	3.49	0.78	3.56	0.73	6.23	6.50
C18:0	4.34	0.84	4.72	0.95	4.37	3.80
C18:1 [b]	9.11	2.01	10.06	2.54	9.95	11.60
C18:2 [b]	76.57	2.61	75.46	3.39	74.31	72.70
C18:3	5.74	1.72	5.39	1.63	3.62	3.50
C20:0	0.17	0.12	0.18	0.11	0.92	
C20:1	0.48	0.29	0.55	0.32		0.90
C20:2	0.07	0.07	0.05	0.04		0.20

[a]Trace levels (< 0.02%) of C16:1, C16:2, C17:0, C17:1, C17:2 and C22:0 were also detected.
[b]Multiple isomers were observed in each chromatogram and these were added together.

were 73.5 ± 5% (1s), 15.6 ± 1.3%, and 10.9 ± 5% compared to the GC-MS values of 80.9%, 10.6%, and 8.5% for these composites of fatty acids obtained using a hand press.

4. Conclusions

ATR-FTIR provides a faster and simpler method (operators can be easily trained) than GC-MS for monitoring changes in Croton oil composition over time and space. The accuracy and precision of oil component analyses are somewhat lower than what can be provided by a chromatographic analysis. However, the ability to obtain results on-site in a matter of minutes instead of hours, the lack of expensive reagents or toxic waste, and the ability to store spectra that can be revisited and reprocessed at a later time, such as for peroxide or iodine values or with more sophisticated chemometric data analyses [26], are important advantages.

Kenyan *Croton megalocarpus* nut oils were very homogeneous across the geographic range studied despite differences in size and morphology even though these are not plantation trees. Although only a single season was sampled, we think this study is representative of what will occur in other years.

Croton megalocarpus oil has high levels of polyunsaturated fats comparable to the levels found in safflower, hemp, and walnut oils. Many of these have been found to offer significant health benefits compared to saturated fats. This study did not find any significant evidence for species or varietal differences that are related to geography, though freshly collected nuts did exhibit subtle differences in unsaturation that could be discerned using ATR-FTIR analyses of cold-pressed oils obtained with a simple hand press.

Because the geographic pattern observed early in the study faded over time during cold storage, the geographic variation may simply indicate an underlying factor such as higher levels of residual water in larger nuts, or it may reflect the plasticity found within this plant species. The portable ATR-FTIR allows analyses to be conducted in the field before these variations have been leveled by the uniformity of storage conditions. These geographic variations are of great interest as scientists try to understand how climate change will affect both natural plant distributions and the yields that may be expected from agricultural crops. In any case, these observations provide a cautionary tale for applications that rely on oil composition obtained via ATR-FTIR to identify the origins and/or adulteration of geographically registered food commodities.

This study also suggests that the relative percentage of unsaturated fatty acids can be increased simply by storing the nuts at a low temperature for a few weeks, potentially increasing the value of the oil for uses beyond the production of biodiesel fuels. ATR-FTIR may also prove useful at a production facility where blending or processing of oils for improvements in engine performance is being conducted.

Finally, we would be remiss if we did not point out that this study was based on only a single season, so results may differ in years with different weather patterns.

Acknowledgments

We thank the 2017-18 Colorado College Instrumental Analysis class for their interest and enthusiasm and for their contributions to the analyses. We also thank Professor Shane Heschel (Organismal Biology and Ecology Department, Colorado College) for his discussion of the plant biology. Finally, we thank Eco Fuels Kenya for their donation of the Croton nuts analyzed in this study. This research was funded by faculty research grants from the Dean of Colorado College and by the Department of Chemistry and Biochemistry at Colorado College.

Supplementary Materials

The Supplementary Material provides the following additional Tables. (1) Table 1 is a listing of the 30 different sample locations with their coordinates, average nut masses, % oil in the deshelled seed, and length to width ratio. (2) Table 2 is a listing of the known oils used to calibrate the ATR-FTIR analyses. (3) Table 3 is a listing of the baseline-corrected raw and normalized ATR-FTIR absorbance data used to construct Figures 6 and 7. (4) Table 4 contains GC-MS data with a worked example used to construct Table 2. (5) Table 5 contains the GIS interpolated data for the mean annual precipitation and temperature at each sample site and the z-scores of the ATR-FTIR absorbance data. Additional figures are also supplied, including Figure 1, a typical GC-MS TIC; Figure 2, GIS interpolated maps for the temperature and precipitation specific to the sampled tree sites; Figure 3, interpolated map of the nuts' length to width ratios; and Figure 4, interpolated map of the change between December and March in the unsaturation proxy at 3009 cm^{-1}. *(Supplementary Materials)*

References

[1] R. Kindt, J.-P. B. Lillesø, and P. Van Breugel, "Comparisons between original and current composition of indigenous tree species around Mount Kenya," *African Journal of Ecology*, vol. 45, no. 4, pp. 633–644, 2007.

[2] E. N. Matu and J. van Staden, "Antibacterial and anti-inflammatory activities of some plants used for medicinal purposes in Kenya," *Journal of Ethnopharmacology*, vol. 87, no. 1, pp. 35–41, 2003.

[3] B. Aliyu, B. Agnew, and S. Douglas, "Croton megalocarpus (Musine) seeds as a potential source of bio-diesel," *Biomass & Bioenergy*, vol. 34, no. 10, pp. 1495–1499, 2010.

[4] G. Kafuku and M. Mbarawa, "Biodiesel production from Croton megalocarpus oil and its process optimization," *Fuel*, vol. 89, no. 9, pp. 2556–2560, 2010.

[5] D. Ollivier, J. Artaud, C. Pinatel, J. P. Durbec, and M. Guérère, "Triacylglycerol and fatty acid compositions of French virgin olive oils. Characterization by chemometrics," *Journal of Agricultural and Food Chemistry*, vol. 51, no. 19, pp. 5723–5731, 2003.

[6] O. Galtier, N. Dupuy, Y. Le Dréau et al., "Geographic origins and compositions of virgin olive oils determined by chemometric analysis of NIR spectra," *Analytica Chimica Acta*, vol. 595, no. 1-2, pp. 136–144, 2007.

[7] O. Galtier, Y. Le Dréau, D. Ollivier, J. Kister, J. Artaud, and N. Dupuy, "Lipid compositions and French registered designations of origins of virgin olive oils predicted by chemometric analysis of mid-infrared spectra," *Applied Spectroscopy*, vol. 62, no. 5, pp. 583–590, 2008.

[8] V. Alba, V. Bisignano, A. Rotundo, G. B. Polignano, and E. Alba, "Characterization of olive germplasm by chemical oil components and morphological descriptors in Basilicata region (Italy)," *Plant Genetic Resources*, vol. 10, no. 2, pp. 145–151, 2012.

[9] D. P. Aykas and L. E. Rodriguez-Saona, "Assessing potato chip oil quality using a portable infrared spectrometer combined with pattern recognition analysis," *Analytical Methods*, vol. 8, no. 4, pp. 731–741, 2016.

[10] D. Wu, A. P. Roskilly, and H. Yu, "Croton megalocarpus oil-fired micro-trigeneration prototype for remote and self-contained applications: experimental assessment of its performance and gaseous and particulate emissions," *Interface Focus*, vol. 3, no. 1, pp. 20120041-20120041, 2012.

[11] G. Kafuku, M. K. Lam, J. Kansedo, K. T. Lee, and M. Mbarawa, "Croton megalocarpus oil: A feasible non-edible oil source for biodiesel production," *Bioresource Technology*, vol. 101, no. 18, pp. 7000–7004, 2010.

[12] A. Abdolshahi, M. H. Majd, J. S. Rad, M. Taheri, A. Shabani, and J. A. Teixeira da Silva, "Choice of solvent extraction technique affects fatty acid composition of pistachio (Pistacia vera L.) oil," *Journal of Food Science and Technology*, vol. 52, no. 4, pp. 2422–2427, 2015.

[13] "EUR-Lex, Commission implementing regulation (EU) 2015/1833 of 12 October 2015 amending regulation (EEC) No 2568/91 on the characteristics of olive oil and olive-residue oil and on the relevant methods of analysis," https://eur-lex.europa.eu/eli/reg_impl/2015/1833/oj.

[14] H. M. H. Braun, *Agro-Climatic Zone Map of Kenya*, Republic of Kenya, Ministry of Agriculture Kenya Soil Survey, Nairobi, 1980, https://esdac.jrc.ec.europa.eu/content/agro-climatic-zone-map-kenya-appendix-2-report-no-e1.

[15] S. L. Neidleman, "Effects of temperature on lipid unsaturation," *Biotechnology & Genetic Engineering Reviews*, vol. 5, no. 1, pp. 245–268, 1987.

[16] C. R. Linder, "Adaptive evolution of seed oils in plants: Accounting for the biogeographic distribution of saturated and unsaturated fatty acids in seed oils," *The American Naturalist*, vol. 156, no. 4, pp. 442–458, 2000.

[17] M. Lazzari and O. Chiantore, "Drying and oxidative degradation of linseed oil," *Polymer Degradation and Stability*, vol. 65, no. 2, pp. 303–313, 1999.

[18] L. Cai, C. Liu, and T. Ying, " Changes in quality of low-moisture conditioned pine nut (," *CyTA - Journal of Food*, vol. 11, no. 3, pp. 216–222, 2013.

[19] I. Kazantzis, G. D. Nanos, and G. G. Stavroulakis, "Effect of harvest time and storage conditions on almond kernel oil and sugar composition," *Journal of the Science of Food and Agriculture*, vol. 83, no. 4, pp. 354–359, 2003.

[20] F. R. van de Voort, J. Sedman, G. Emo, and A. A. Ismail, "Rapid and direct Iodine value and saponification number determination of fats and oils by attenuated total reflectance/fourier transform infrared spectroscopy," *Journal of the American Oil Chemists' Society*, vol. 69, no. 11, pp. 1118–1123, 1992.

[21] F. R. van de Voort, A. A. Ismail, J. Sedman, J. Dubois, and T. Nicodemo, "The determination of peroxide value by fourier transform infrared spectroscopy," *Journal of the American Oil Chemists' Society*, vol. 71, no. 9, pp. 921–926, 1994.

[22] A. A. Ismail, F. R. van de Voort, G. Emo, and J. Sedman, "Rapid quantitative determination of free fatty acids in fats and oils by fourier transform infrared spectroscopy," *Journal of the American Oil Chemists' Society*, vol. 70, no. 4, pp. 335–341, 1993.

[23] M. M. Mossoba, A. Seiler, J. K. G. Kramer et al., "Nutrition labeling: Rapid determination of total trans fats by using internal reflection infrared spectroscopy and a second derivative procedure," *Journal of the American Oil Chemists' Society*, vol. 86, no. 11, pp. 1037–1045, 2009.

[24] S. M. Ghazani, G. García-Llatas, and A. G. Marangoni, "Micronutrient content of cold-pressed, hot-pressed, solvent extracted and RBD canola oil: Implications for nutrition and quality," *European Journal of Lipid Science and Technology*, vol. 116, no. 4, pp. 380–387, 2014.

[25] S. Gharby, H. Harhar, D. Guillaume et al., "Chemical investigation of Nigella sativa L. seed oil produced in Morocco," *Journal of the Saudi Society of Agricultural Sciences*, vol. 14, no. 2, pp. 172–177, 2015.

9

Comprehensive Assessment of Degradation Behavior of Simvastatin by UHPLC/MS Method, Employing Experimental Design Methodology

Maja Hadzieva Gigovska [ID],[1] Ana Petkovska,[1] Jelena Acevska,[2] Natalija Nakov,[2] Packa Antovska,[1] Sonja Ugarkovic,[1] and Aneta Dimitrovska[2]

[1]Research & Development, Alkaloid AD, Blvd. Aleksandar Makedonski 12, 1000 Skopje, Macedonia
[2]Faculty of Pharmacy, University "Ss Cyril and Methodius", Mother Theresa 47, 1000 Skopje, Macedonia

Correspondence should be addressed to Maja Hadzieva Gigovska; hadzievam@yahoo.com

Academic Editor: David Touboul

This manuscript describes comprehensive approach for assessment of degradation behavior of simvastatin employing experimental design methodology as scientific multifactorial strategy. Experimental design methodology was used for sample preparation and UHPLC method development and optimization. Simvastatin was subjected to stress conditions of oxidative, acid, base, hydrolytic, thermal, and photolytic degradation. Using 2^n full factorial design degradation conditions were optimized to obtain targeted level of degradation. Screening for optimal chromatographic condition was made by Plackett–Burman design and optimization chromatographic experiments were conducted according to Box-Behnken design. Successful separation of simvastatin from the impurities and degradation products was achieved on Poroshell 120 EC C18 50 × 3.0 mm 2.7 μm, using solutions of 20 mM ammonium formate pH 4.0 and acetonitrile as the mobile phase in gradient mode. The proposed method was validated according to International Conference on Harmonization (ICH) guidelines. Validation results have shown that the proposed method is selective, linear, sensitive, accurate, and robust and it is suitable for quantitative determination of simvastatin and its impurities. Afterwards, the degradation products were confirmed by a direct hyphenation of liquid chromatograph to ion-trap mass spectrometer with heated electrospray ionization interface. This study highlights the multiple benefits of implementing experimental design, which provides a better understanding of significant factors responsible for degradation and ensures successful way to achieve degradation and can replace the trial and error approach used in conventional forced degradation studies.

1. Introduction

To meet the demands of modern pharmaceutical analysis, the employment of chemometrical tools in every possible way during analysis is necessary, since many variables can be simultaneously controlled to achieve the desired results through limited experimental trials. The use of an experimental design (DoE) approach by which multivariate data can be handled and fitted to an empirical function is justifiable because it offers a better choice over one factor at time (OFAT) for identification and control of critical factors [1, 2].

In this direction, use of such systematic approach would be a necessity for any extensive study, such as forced degradation studies (FDS) for stability assessment of the active

pharmaceutical ingredients (APIs) and finished dosage forms (FDFs). However, despite availability of diverse literature reports that defines the concept of forced degradation, detailed information about a forced degradation strategy is not provided and the experimental conditions to conduct forced degradation are described in a general way without description of the exact stress conditions to be applied [3–10]. Generally, a trial and error approach is adopted to select the strength, temperature, and time of exposure to achieve loss of active substance from 5 to 20% [5, 11, 12]. Till date, far from our knowledge, none of the reported analytical procedures describe simple and satisfactory sample preparation methodology where the influences of the stressor strength, time of expose, and temperature are evaluated in detail. Due to the considerable cost, time consumption, scientific expertise, and high incidence of random results, the need of more systematic approach is recognized.

The objective of this study was to present optimization of forced degradation condition using DoE.

Simvastatin (SIM) was chosen as model API for couple of reasons: (1) because of the great scientific community interest for its potential use in brain diseases and different types of cancers besides its well-known antihyperlipidemic activity and (2) due to its proven instability [13].

This study focuses on systematic evaluation of SIM instability in hydrolytic, oxidative, or photolytic condition, as currently available data mainly indicate the instability of simvastatin as a part of selectivity of the stability-indicating methods, while the influences of the stressor straight, time of expose, and temperature were not evaluated in detail [14–16].

Therefore, the goals of the present study are to explore the degradation behavior of SIM under different stress conditions (acidic, alkaline, oxidative, thermal, and photolytic) using simplified FDS by adopting DoE concept.

The research aimed to employ DoE approach for development and optimization of UHPLC/MS method to resolve all the possible degradation products, followed by method validation studies for ensuring the robust performance.

The proposed methodology would enable studying the combination of conditions where optimal degradation is obtained, as well as evaluation of the effect of each factor with the change in level of the other factors. This approach was chosen because it is efficient and easily accomplishable and allows interactions to be detected. Furthermore, this approach reduces the number of experiments, time, and cost and obtains good prediction of desired degradation.

2. Materials and Methods

2.1. Chemicals and Standards. Simvastatin CRS (purity 99.7% "as is"), Simvastatin impurity E (Lovastatin), and Simvastatin for peak identification were provided by European Directorate for the Quality of Medicines and Health Care Council of Europe (EDQM-Strasbourg, France). Simvastatin impurity B (Simvastatin acetate ester), Simvastatin impurity C (Anhydro simvastatin), and Simvastatin impurity D (Simvastatin dimer) were purchased by LGC GmbH, Im Biotechnologiepark, TGZ II, Germany. Also, in house standard of methyl simvastatin was used.

Simvastatin API samples with certificate of suitability to the monographs of the European Pharmacopeia (CEP) were kindly supplied by Teva Pharmaceutical Industries Ltd., Israel.

All the reagents used (acetonitrile, ammonium formate, formic acid, sodium hydroxide, hydrochloric acid, and hydrogen peroxide) were analytical grade, purchased from Merck (Darmstadt, Germany). Water was purified by a Werner water purification system.

Regenerated cellulose membrane syringe filters with pore size 0.2 μm, purchased from Phenomenex (Torrance, CA. USA), were used.

2.2. Experimental Conditions

2.2.1. Ultra High Performance Liquid Chromatography (UHPLC). Chromatographic analysis was performed on Waters Acquity Ultra Performance (Waters Corporation, Milford, USA) equipped with a Binary Solvent Manager, UHPLC Column Compartment, Sample Manager Heater/Cooler and autosampler, and photo-diode array detector. Instrument control, data acquisition, and processing were done by using Empower 2 build 2154 software.

The separation was performed on Poroshell 120 EC C18 50 × 3.0 mm, 2.7 μm (Agilent Technologies, USA), using buffer (20 mM ammonium formate, pH 4.0) and acetonitrile (ACN) as a mobile phase in a gradient mode as follows: T (min)/ACN (%) 0/40; 10/40; 20/85; 25/85; 30/40; 35/40. The column temperature was 35°C. Flow rate was 0.7 mL/min. Injection volume was 10 μL. The UV detection was performed at 248 nm.

2.2.2. Liquid Chromatography-Tandem Mass Spectrometry (LC-MS). The LC/MS analyses were conducted on Dionex UltiMate™ 3000 UHPLC-UV-DAD (Thermo Fisher Scientific, Waltham, MA, USA), interfaced with linear iontrap mass spectrometer (LTQ XL) equipped with heated electrospray-ionization source operated in the positive ionization mode. Instrument control and results processing was done using Dionex Chromeleon 7.2 (for UHPLC-DAD analyses) and Thermo Xcalibur v2.2 SP1 (for UHPLC-DAD/MS analyses). Structural confirmation and fragment elucidation were performed using Mass Frontier v7.0.

Optimized mass parameters were as follows: ion source heater temperature was set at 280°C and capillary temperature at 200°C; capillary voltage was 20 V with collision energy 35 eV.

Nitrogen was used as nebulizing gas at pressure of 50 psi and the flow was adjusted to 10 L/min. MS data were acquired in the positive ionization mode. The full scan covered the mass range at m/z 100-1200. Collision-induced fragmentation experiments were performed in the ion trap using helium as collision gas, with voltage ramping cycle from 0.3 up to 2 V. Maximum accumulation time of ion trap and the number of MS repetitions to obtain the MS average spectra were set at 500 ms and 3, respectively.

2.2.3. Screening and Optimization of
the UHPLC Method Using DoE

(1) Application of Plackett–Burman Design for Screening Significant Parameters. The Plackett–Burman design (PB design) was used to study the effects of nine independent factors, i.e., molarity of ammonium formate (x_1), flow rate (x_2), wavelength (x_3), volume of injection (x_4), detector acquisition rate (x_5), column temperature (x_6), percent of organic modifier in the initial mobile phase composition (x_7), different column lots (x_8), and injector temperature (x_9).

The range and the levels of experimental investigated variables are presented in Table 1. For each of the 12 experiments 3 solutions (diluent, system suitability solution, and standard solution) were injected. The experiments were performed in

TABLE 1: Plan of Plackett–Burman design and experimentally obtained results.

Number of experiments	Factor levels									Responses		
	x_1	x_2	x_3	x_4	x_5	x_6	x_7	x_8	x_9	y_1	y_2	y_3
1	20	0.7	228	7	25	35	60	lot 2	25	0.72	2.50	0.92
2	20	0.3	228	3	10	35	40	lot 2	5	0.92	4.25	1.25
3	20	0.3	228	7	10	45	40	lot 1	25	1.20	3.28	6.23
4	20	0.3	240	3	25	45	60	lot 1	5	1.02	3.72	1.22
5	20	0.7	240	7	10	35	60	lot 1	5	1.03	2.02	2.21
6	20	0.7	240	3	25	45	40	lot 1	25	1.56	3.44	7.78
7	5	0.7	228	7	25	45	40	lot 2	5	1.25	3.10	1.59
8	5	0.3	228	3	25	35	60	lot 1	25	0.88	3.72	2.16
9	5	0.7	228	3	10	45	60	lot 2	5	0.71	3.56	1.26
10	5	0.3	240	7	25	35	40	lot 2	5	1.08	3.89	6.41
11	5	0.7	240	3	10	35	40	lot 2	25	1.59	4.56	4.52
12	5	0.3	240	7	10	45	60	lot 1	25	0.80	3.35	2.53

x_1: molarity of ammonium formate (mM); x_2: flow rate (mL/min); x_3: wavelength (nm); x_4: volume of injection (μL); x_5: detector acquisition rate (Hz); x_6: column temperature (°C); x_7: percent of organic modifier in the initial mobile phase composition (%); x_8: column with the same composition but different lots (lot 1: USCFZ13194/B13243; lot 2: USCFZ13193/B13243); x_9: injector temperature (°C); y_1: Rs E/F; y_2: Rs G/SIM; and y_3: Rs B/C.

TABLE 2: Box-Behnken experimental design matrixes of the selected independent variables and studied responses.

Number of experiments	Factors levels			Responses		
	x_1	x_2	x_3	y_1	y_2	y_3
1	3.8	0.6	40	3.06	1.27	1.72
2	4.2	0.6	40	2.79	1.33	1.51
3	3.8	0.8	40	3.66	1.40	1.60
4	4.2	0.8	40	4.02	1.04	1.41
5	3.8	0.7	35	3.12	1.51	1.58
6	4.2	0.7	35	4.25	1.07	1.43
7	3.8	0.7	45	3.91	1.37	1.38
8	4.2	0.7	45	2.58	1.52	1.32
9	4.0	0.6	35	3.59	1.39	1.59
10	4.0	0.8	35	3.85	1.30	1.40
11	4.0	0.6	45	3.05	1.44	1.56
12	4.0	0.8	45	3.25	1.35	1.78
13	4.0	0.7	40	4.58	1.06	2.06
14	4.0	0.7	40	4.48	1.12	2.15
15	4.0	0.7	40	4.28	1.03	2.12

x_1: pH of mobile phase; x_2: flow rate (mL/min); x_3: content of acetonitrile (%); y_1: Rs between impurity G and SIM; y_2: tailing factor SIM (T); and y_3: Rs between impurities B and C.

randomized order to minimize the effects of uncontrolled variables that may influence the results. Three responses were measured for each experiment: resolution (Rs) between impurity E and F (y_1), Rs between Impurity G and SIM (y_2), and Rs between impurities B and C (y_3).

(2) Application of Box-Behnken Design for Optimization of the Chromatographic Conditions. Next in this study, the important chromatographic factors selected based on the obtained results from PB design were optimized by a Box-Behnken (BB) design.

The independent variables were pH of ammonium formate (x_1), flow rate (x_2) contents of ACN in the initial mobile phase composition (x_3). The low, centre, and high levels of

each variable are given in Table 2. Three responses were measured for each experiment: Rs between impurity G and SIM (y_1), tailing of SIM (y_2), and Rs between impurities B and C (y_3).

A total of 15 tests (including 3 replicates of the centre point) were carried out in random order, in accordance with the BB design (Table 2).

2.3. Standard and Sample Preparation

2.3.1. Standard Preparation. Standard solution of SIM in final concentration of 1 μg/mL was used for quantitative determination. Standard solutions of all impurities were prepared individually and in a mixture to final concentration

of 4 μg/mL each, using ACN and water in ratio 50 : 50 as solvent.

7.5 mg of "simvastatin for peak identification", corresponding to a mixture of SIM spiked with its related impurities (A, B, C, D, E, and F), was dissolved in 5.0 mL solvent and used as system suitability solution.

2.3.2. Sample Preparation. In all experiments the concentration of SIM in the sample solution was 1000 μg/mL. Simvastatin was subjected to stress under acidic, alkaline, oxidative, thermal, and photolytic conditions.

In the preliminary experiments, SIM was subjected to 0.1M HCl for 3 hours and 0.1M NaOH for one hour at ambient temperature. The oxidation stress was done with 30% H_2O_2 solution for 3 h, at ambient temperature. For thermal degradation SIM was exposed at 60°C for 24 hours. Photo degradation study was performed by exposing the drug powder, spread as a thin film in a transparent quartz Petrii plates covered with a transparent quartz cover and exposed to direct sunlight for one and two days. Additionally, control study in dark was run simultaneously. All stress studies were performed in amber color glassware to protect the solutions from light degradation.

(1) Sample Preparation according to Full Factorial Design for Acid, Alkali, Oxidative, and Thermal Degradation. The forced degradation experiments set-up based on 2^n full factorial design was performed. The experiments were designed considering variables including time of exposure, temperature, and stressor strength at two levels.

Acid and alkali degradation was performed using 2^3 factorial design (three variables considered at two levels: 0.01 M and 0.1 M HCl/ NaOH heated at 25°C and 40°C for 15 and 45 min). Set-up of eight experiments for each stressor was conducted as described in Table 3.

At the end of exposure, the samples were neutralized with appropriate amount of NaOH or HCl respectively (0.01 M or 0.1 M) and diluted to final concentration of 1000 μg/mL with solvent.

For oxidative degradation also three variables were considered at two levels (the high level for H_2O_2, temperature and time of exposure were 30%, 40°C and 60 minutes, and the low levels were 3%, 25°C and 15 minutes, respectively).

2^2 factorial designs were conducted to set up thermal degradation where the high-level values were 105°C and 5 h and the low levels were 80°C and 3 h, respectively.

2.4. Statistical Evaluation. The experimental design and statistical analysis of data for the optimization and robustness testing along with forced degradation sample preparation were performed with Design-Expert software, Version 7.0.0 (Stat-Ease, Inc., Minneapolis, MN, USA).

3. Results and Discussion

3.1. Optimization of Chromatographic Conditions. The method conditions were evaluated to obtain good quality of separation and ideal peak shape and maintain resolution between all impurities in minimum analysis time. Decisions

concerning the type of stationary phase, solvent type, and water phase nature were made based on prior knowledge from literature. Although some methods have been developed for the determination of SIM and its impurities [14–18] including the two official methods reported in European Pharmacopeia and United State pharmacopoeia utilizing HPLC gradient elution, to the best of our knowledge, there are no references in the literature concerning chemometric approach to the development and validation of the UHPLC method, intended for the quantitative analysis of SIM and its impurities. These methods employed a time-consuming trial and error approach for giving potential information concerning the sensitivity of the factors on the separation and it did not provide the information concerning interaction between factors.

Therefore, within this study, science-based approach was employed to develop, optimize, and validate sensitive, robust, and cost-effective UHPLC method for determination of SIM and its impurities.

SIM and its impurities have very similar physical-chemical properties. The log$P$ values are 4.39, 3.85, 4.94, 5.55, 3.68, 4.12, and 3.68 for SIM, impurity A, impurity B, impurity C and impurity E, impurity G, and impurity F, respectively (Table S1, supplementary material). The C_{18} packing columns were shown to be the most suitable according to the lipophilic nature of the compounds. Initially, four columns were examined (Zorbax Extend C_{18}, Zorbax Eclipse C_{18}, Zorbax XDB C_{18}, and Poroshell 120 EC C18) and it was decided to continue the investigation on Poroshell 120 EC C18 50 × 3.0 mm, 2.7 μm. This decision was based on the properties of this column, which is packed with specific, spherical core shell particles, allowing high efficiency of separation. The use of this column enabled tight, symmetrical peak with good resolution between impurities E and F. Among organic modifiers used in HPLC, it was decided to use ACN based on Rs between impurities E and F, critical Rs between B and C, and shorter analysis runtime. The addition of buffer was inevitable, and several buffers suitable for LC-MS analysis were examined (ammonium formate, ammonium acetate, and trifluoroacetic acid). Concentrations of these buffers were varied in the range from 5 to 20 mM (Table 1). Next, in order to obtain complete information about method behavior, nine factors were assessed in twelve experiments according to PB design (Table 1). PB design was chosen because of its high efficiency with respect to the number of runs required. The model was validated by the analysis of variance (ANOVA). The statistical analysis showed (data presented in Table S2, supplementary material) that the model represents the phenomenon quite well and the variation of the response was correctly, thus useful in predicting the effects of the factors on the selected responses.

Next, the qualitative contribution of each factor and, respectively, responses were analysed for all various conditions of degradation. Each response coefficient was studied for its statistical significance by half-normal plot and Pareto charts as shown in Figure 1. Pareto charts establish t value of the effect by two limit lines, namely, the Bonferroni limit line and t limit line. Coefficients with t value of effect above the Bonferroni line are designated as certainly significant.

TABLE 3: Experimental conditions and results from 2^n full factorial design for acid, alkali, oxidative, and thermal degradation.

Exp. No	Experimental conditions Factor levels			Acid degradation			Alkali degradation Responses (%)			Oxidative degradation			Experimental conditions Factor levels		Thermal degradation Responses (%)		
	x_1	x_2	x_3	y_1	y_2	y_3	y_1	y_2	y_3	y_1	y_2	y_3	x_4	x_5	y_1	y_2	y_3
1.	-	-	-	7.32	0.15	6.83	15.28	0.17	11.23	1.09	0.03	0.12	-	-	3.98	0.04	3.38
2.	+	-	-	14.43	0.12	13.93	16.23	0.10	12.53	1.16	0.04	0.13	-	+	6.33	0.09	5.38
3.	-	+	-	5.79	0.25	5.04	26.65	0.12	15.23	1.45	0.06	0.14	+	-	20.12	0.26	17.10
4.	+	+	-	22.04	0.17	21.74	39.43	0.08	23.50	1.55	0.08	0.16	+	+	25.79	0.28	21.92
5.	-	-	+	16.71	0.05	16.39	37.25	0.09	35.89	2.19	0.10	2.04					
6.	+	-	+	41.05	0.09	40.73	48.32	0.15	40.23	3.29	0.17	2.16					
7.	-	+	+	18.76	0.23	18.47	58.23	0.11	45.62	4.79	0.15	2.56					
8.	+	+	+	39.89	0.29	38.32	68.89	0.19	58.26	6.28	0.19	2.89					

x_1: stressor strength 0.01 M and 0.1 M HCl/NaOH or 3% and 30% H_2O_2 for hydrolysis and oxidative degradation respectable; x_2: temperature 25°C and 40°C; x_3: time 15 and 45 minutes; x_4: temperature: 80°C and 105°C; x_5: time 180 and 300 minutes; y_1: amount of total impurities (%); y_2: amount of impurity with RRT 1.16 (%); and y_3: amount of impurity A (%). The high level of each factor was considered as "+" and low level as "-".

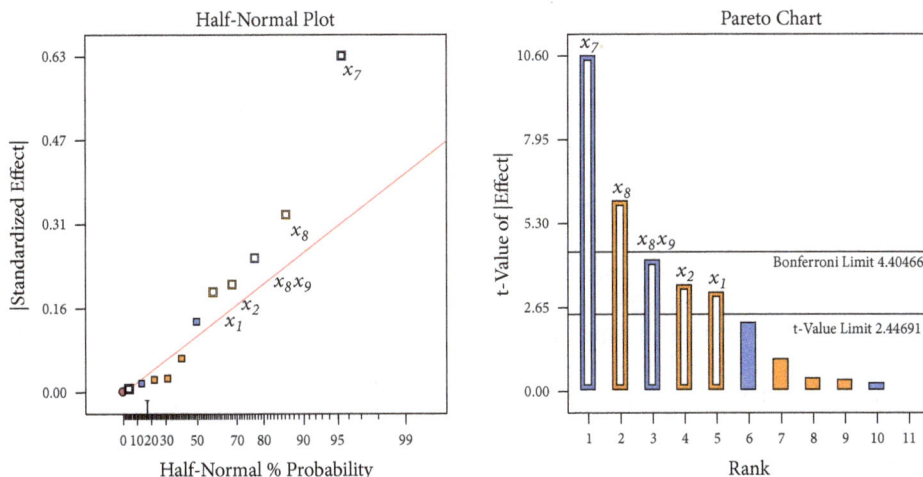

(a)

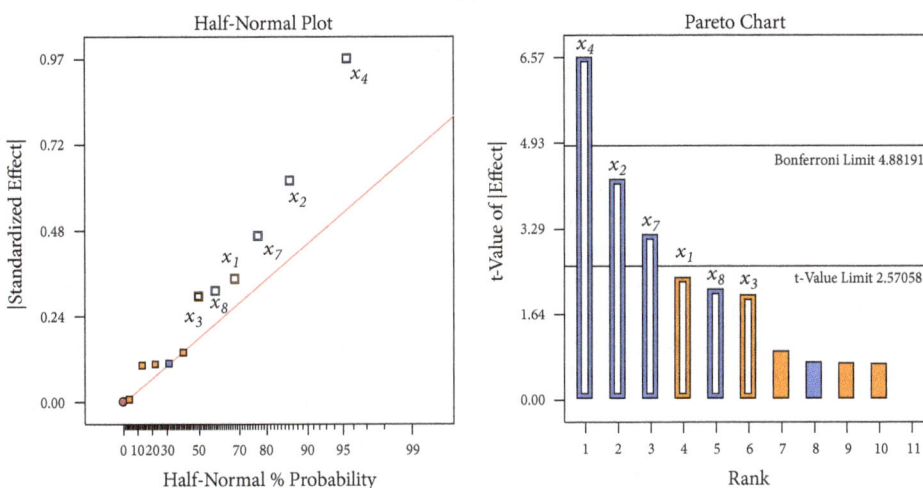

(b)

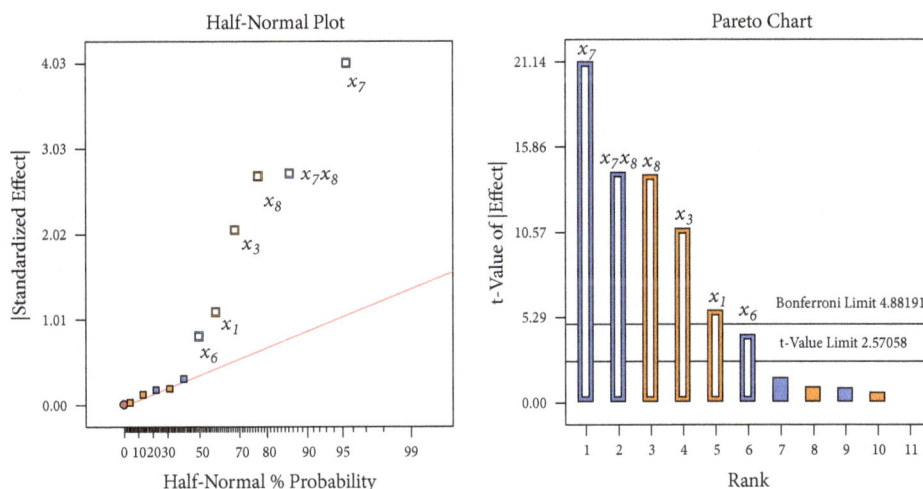

(c)

FIGURE 1: Half-normal plot and Pareto chart showing the significant effects based on the observation of Plackett–Burman design for the investigated responses (a) Rs between impurity E and F (y_1); (b) Rs between SIM and impurity G (y_2); (c) Rs between impurity B and C (y_3) where x_1 is molarity of ammonium formate (mM); x_2 is flow rate (mL/min); x_3 is wavelength (nm); x_4 is volume of injection (μL); x_5 is detector acquisition rate (Hz); x_6 is column temperature (˚C); x_7 is percent of organic modifier in the initial mobile phase composition (%); x_8 is column of the same composition but of different lot; and x_9 is injector temperature (˚C). Positive effects are marked with white bar box with orange frame and orange bar box for significant and insignificant factor, respectable. Negative effects are marked with white bar box with blue frame and blue bar box for significant and insignificant, respectable.

Coefficients with t value of effect between Bonferroni line and t limit line are termed as coefficients likely to be significant, while t value of effect below the t limit line is statistically insignificant.

Pareto chart analyses revealed that the molarity of ammonium formate, percent of organic modifier in the initial mobile phase composition (%), and column type have significantly affected all investigated responses.

Molarity of ammonium formate have been shown to have positive effect on all responses meaning that higher molarity revealed increased resolution and therefore all experiments in the optimization phase are performed using 20 mM ammonium formate. As a critical parameter pH value in range ±2 units were evaluated.

The statistical evaluation shows that the ratio of the mobile phase composition has a large effect on peak separation, which decreases with increasing organic portion. A negative relationship between flow rate varied from 0.3 to 0.7 mL/min and resolution between SIM and impurity G, and a positive relationship with resolution between impurities E and F was observed.

Additionally, an important factor affecting all responses was the column batch. It is very likely that during the ongoing investigation, column aging occurs and the related change in chromatographic separation might cause an inconclusive statistical evaluation. The other parameters investigated showed only little influence on resolution as displayed in Figure 1.

Hence, systematic scouting resulted in selection of three key critical parameters (flow rate, pH of buffer, and ACN content), which were optimized using BB experimental design (Table 2). Among the various experimental designs, BB design, as response surface design was preferred for the prediction of nonlinear response, due to its flexibility, in terms of experimental runs and information related to the factor's main and interaction effects. The model was validated by analysis of variance (ANOVA) using Design-Expert software (data presented in Table S3, supplementary material). Decision concerning the evaluated responses was made taking into consideration problems in the method described in European Pharmacopeia, where unsatisfactory separation between E/F and B/C is seen.

Comparison of different proposed models from experimental trials for all responses favored quadratic model as best fitted model. The data revealed in Table S3 indicates that the chosen quadratic models fit the data well and have high predictive powers for new observations. To get more realistic model insignificant terms with corresponding value higher than 0.05 were eliminated through backward elimination process. The coefficients of the second-order polynomial model were estimated by the least square regression analysis, and the function of responses related to the three selected factors was obtained. The obtained results are presented in Table S3.

From the obtained results it could be concluded that Rs between impurity G and SIM is negatively influenced by the content of ACN and positive influenced by the flow rate.

On the other hand, Rs between peak of impurities B and C is negatively influenced by pH of the water phase.

All the selected variables significantly influence peak symmetry, but the values obtained for the peak tailing ranged from 1.02 to 1.59 being therefore within the acceptable limits (≤2) for all determinations.

Also, the results suggested that two-factor interactions of investigated factors were significant as well, indicating the necessity of their simultaneous influence examination rather than isolated single factor at the time evaluation. Furthermore, the quadratic term indicates a nonlinear curvilinear trend.

In order to facilitate the visualization of factor interactions 3D response surface plots were created and presented in Figure 2. Response surface plots for all the evaluated responses were created keeping one factor constant (flow rate, content of acetonitrile and pH for Rs between impurity G and SIM; tailing and Rs between peak of impurities B and C as chosen responses, respectably)

Analyzing Figures 2(a) and 2(c), it can be concluded that simultaneous increase of pH of the water phase and decrease of ACN content in the mobile phase tend to increase Rs between impurity G and SIM, and simultaneous increase of ACN content in the mobile phase and flow rate enhances the Rs between peak of impurities B and C.

Direct determination of optimal factor setting was very difficult regarding the number and the antagonist influence of interaction and quadratic terms implicated in the model; therefore, optimal chromatographic conditions were chosen using desirability function, where Rs between peaks pair as critical parameter was considered at maximal.

No specific limitations were imposed to the tailing factor, as its value falls within the acceptable range in all cases in the experimental model (Figure 2(b)). From the desirability plot presented in Figure 2(d), it can be concluded that a set of coordinates producing high desirability value (D = 0.943) were pH value of 4, flow rate of 0.7 mL/min, and 40% ACN. These conditions were selected for further validation.

The representative chromatogram of the peak identification obtained under optimized conditions is presented in Figure 3. Under the proposed chromatographic conditions, satisfactory separation (Rs 4.4) of SIM and impurity G was achieved, and the method can separate all known impurities with resolution more than 1.5, which is much better than obtained with existing monograph methods.

As it can be seen, the proposed methodology represents an efficient and easily accomplishable approach to resolving the problem of searching for optimum RP-UHPLC conditions.

Optimization of chromatographic method using experimental design methodology allow improvement of the accuracy and precision of the method by achieving better chromatographic separation of simvastatin from interfering chromatographic peak of impurity G. DoE approach is a systematic, scientifically approach that can reveal information that can be easily overseen when applying one factor at a time approach for method development and optimization. Additionally, it saves time investing the possible interaction between variables.

Employing such an approach, we have obtained the maximum amount of information with the smallest possible

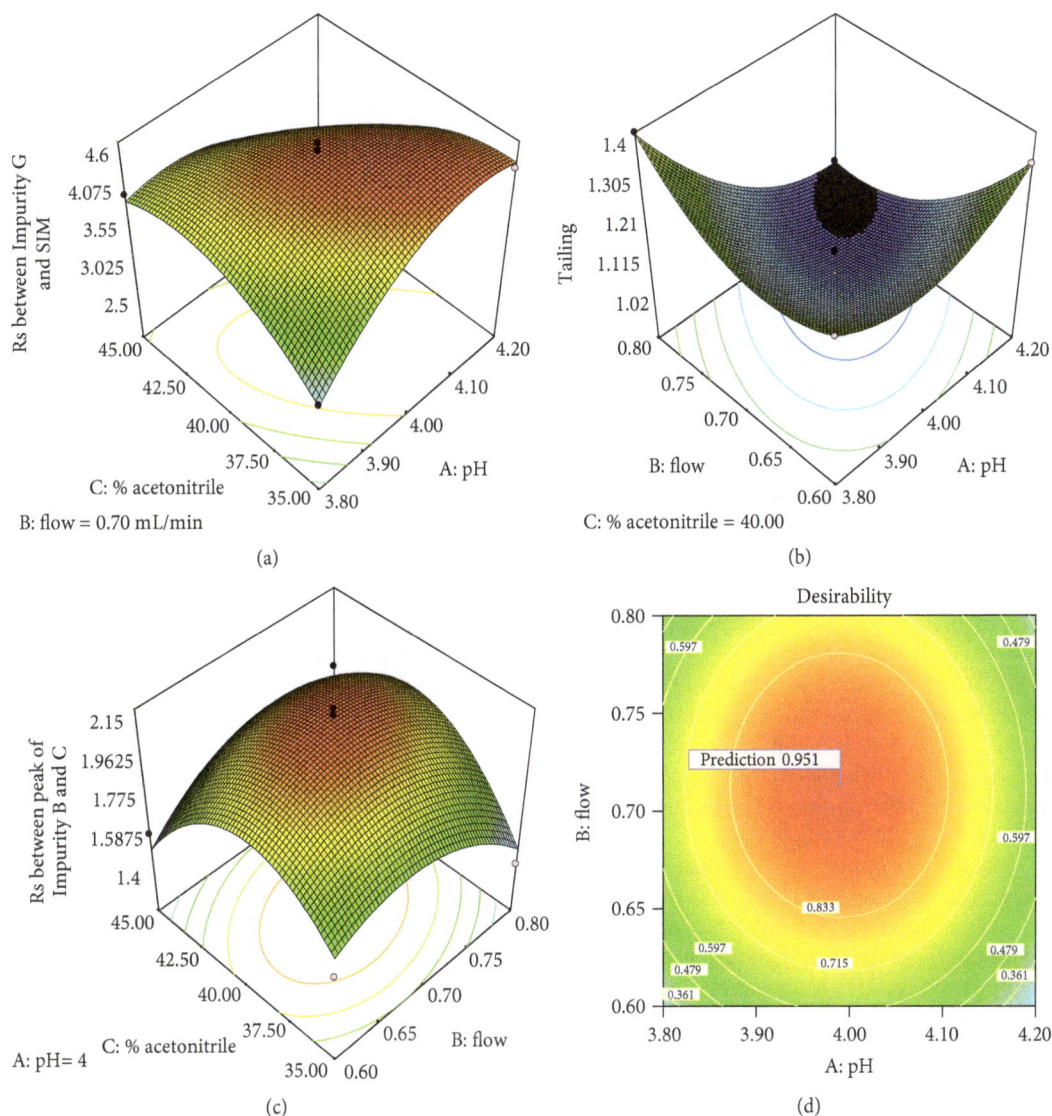

FIGURE 2: 3D surface plot representing the (a) *Rs* between impurity G and SIM, (b) tailing factor, and (c) *Rs* between peaks of impurities B and C, as a function of % organic modifier, pH, and flow rate. (d) Desirability plot for optimization of the selected responses where the red area corresponds to the optimum chromatographic conditions while ACN content maintained constant at 40%. Color change from blue to red represents increasing response values (min ▬▬▬ max).

number of experiments. It provides an improved perspective and knowledge of the analytical procedure. The main advantage of this approach is evident: all the factors that potentially influence the separation can be studied simultaneously.

3.2. Method Validation.

The developed and optimized method was validated as per ICH guidelines [19]. The validation results indicate than the method is specific, linear (0.4 to 6.0 μg/mL for all impurities and 0.4 to 1.5 μg/mL for SIM), accurate, and precise (Table 4).

Stability of the standard solution and sample solution was evaluated by analyzing the same sample immediately after preparation and after time interval (0, 18, and 38 h) by keeping the solution at room temperature. From the stability study, it was concluded that both standard and samples are stable for 38 hours in room temperature. The difference in % between

centrifuged and filtered solutions was found to be within the limits (≤2) when samples are filtered through regenerate cellulose (RC-0.2 μm) membrane filter, so this is suitable for filtration.

To test the capacity of this newly developed analytical procedure to withstand small deliberate changes in the method, various factors within the robustness testing were deliberately changed, like: column temperature (±5°C), organic content of mobile phase (±5%), flow rate (±0.1 units), and wavelength (±2 nm).

Robustness study confirmed that method could be considered robust because changes of factors in defined ranges do not influence the responses (Table 5).

3.3. LC-MS/MS Study on Forced Degradation Samples.

The proposed method was transferred to a UHPLC/MS system

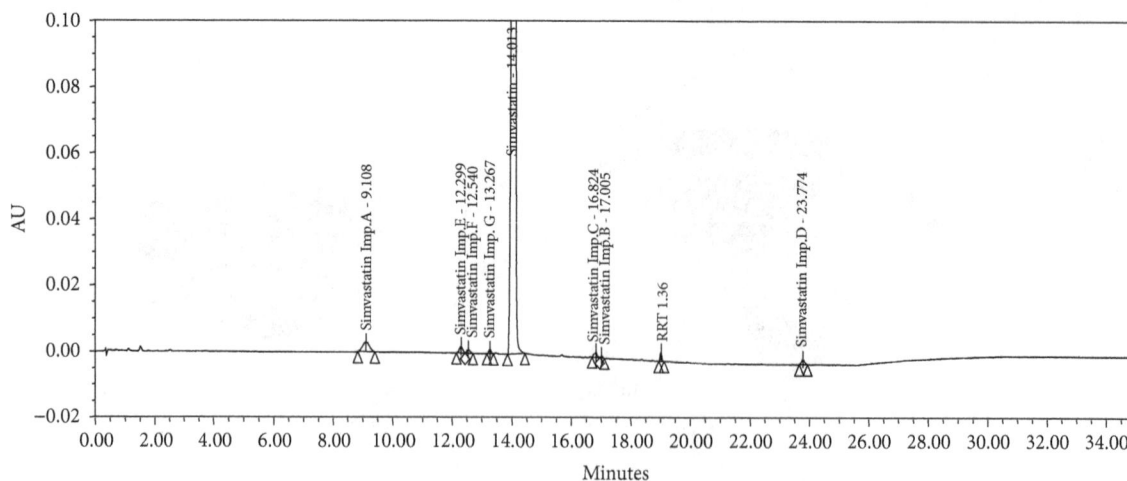

FIGURE 3: Representative chromatogram of peak identification solution.

to carry out deeper analysis of the behavior of SIM. Mass spectra and fragmentation patterns of all impurities were recorded, analyzed with Mass Frontier 7.0 fragmentation software, and confirmed by data published in literature [17, 18]. The obtained results are presented in Table S4, supplementary material. All the specified impurities [20], as well as unspecified impurity methyl simvastatin, were confirmed using commercially available standards. Additionally, other unknown impurities seen in a sample obtained by 'worst-case forced degradation' samples were evaluated. Among these, the highest peak was unspecified impurity with relative retention time (RRT) 1.16 related to SIM retention time. As can be seen from the results presented in Table S4, the MS spectra of the impurity with RRT 1.16 revealed modifications of SIM molecule on the lactones ring. The finding that the molecular mass of this impurity has mass 46 amu higher compared to SIM implies that it is a derivative of SIM (presented in Table S1). The fragmentation patterns of this impurity follow distinct fragmentation as SIM, which are in good agreement with literature [18]. The proposed method was found suitable for detection of both specified and main unspecified degradation products in the samples obtained by force degradation.

3.4. Optimization of the Sample Preparation and Experimental Condition for Forced Degradation

3.4.1. Stability of SIM under Different Forced Degradation Conditions. SIM according to the results obtained from our preliminary experiments, described in Section 2.3.2, was found to be susceptible to degradation under acid and alkali hydrolysis and it is slightly degraded under photo, oxidative, and thermal degradation. According to the requirements stated in European Pharmacopeia, two specified impurities (impurity E and impurity F) should be evaluated [20].

In the preliminary experiments, it was found that impurities E and F remained unaffected by all stress conditions applied. Simvastatin impurity B and impurity G were not detected at all, whereas impurity C only slightly increased in heated sample solution. Impurity D is proven to be formed in

lactonization and was found in all stress conditions applied, with maximal obtained degradation of 0.08%. Simvastatin impurity A was found to be sensitive to all stress conditions applied, especially after acid and alkali hydrolysis. Therefore, beside the percentage of total impurities, amount of formed impurity A was chosen to be evaluated with the DoE approach, together with the major unspecified impurity with RRT 1.16.

3.4.2. Optimization of Various Forced Degradation Conditions by DoE Approach

(1) Selection of Independent Variables, Dependent Variables, and Model. The *independent variables* evaluated by the full factorial DoE enclosed in Section 2.3.2. (1) were selected based on preliminary experiments and detailed literature survey which provided valuable information about the experimental region and definition of factors intervals. Stressor strength, time of exposure, and temperature were identified as factors which should be analysed.

Selection of the type and concentrations of stressor for oxidative degradation, acid, or base was made considering the results from our previous conducted preliminary experiments. HCl and NaOH at range of 0.01 M and 0.1 M were evaluated as suitable reagents for hydrolysis. H_2O_2 with concentration of 3% and 30% was used for oxidative forced degradation. The effect of temperature on acid, alkali, and oxidative degradation of SIM was studied at two levels 25°C and 40°C, respectively. Usually, hydrolytic degradation usually is performed at room temperature, and if no degradation is observed, then the temperature can be increased. However, implementing the DoE approach enables simultaneous evaluation of the effect of the temperature in just few experiments.

In order to gain information about degradation in short time, the time of exposure was chosen based on the minimum length. The low was set at 15 minutes and the high level was set at 45 minutes.

Following the general recommendation stated in ICH guideline [21], thermal degradation studies were performed

TABLE 4: Obtained results from validation of the proposed method.

	Imp.E	Imp.G	SIM	MSIM	Imp.B	Imp.C	Imp.D
System suitability							
Resolution	/	4.44	4.27	6.83	11.20	1.56	38.26
NTP	79895	110461	169356	268485	253612	184417	292527
T	1.01	1.02	1.04	1.05	1.07	1.02	1.02
Linearity [1]							
Regression coefficient	0.9969	0.9994	0.9920	0.9980	0.9990	0.9970	0.9997
Slope	33806	15647	31722	10155	27192	21212	10765
Intercept	6077	388	6375	368	253	2171	401
Response factor	1.50	0.55	1.0	0.36	0.96	0.75	0.38
Precision [2]							
Method precision	7.26	NA	NA	9.04	6.12	NA	NA
Intermediate precision (F –test)	1.86	1.02	NA	1.85	1.05	1.02	1.03
Accuracy given as recovery (%) [3]							
50	98.8 ± 0.5	100.4 ± 0.6	100.5 ± 0.5	98.8 ± 0.6	99.5 ± 1.5	101.0 ± 0.2	100.7 ± 1.0
100	98.3 ± 0.3	99.1 ± 1.4	100.2 ± 0.1	99.1 ± 0.7	99.7 ± 0.8	98.3 ± 0.9	100.3 ± 0.9
150	98.8 ± 0.2	99.1 ± 0.1	99.9 ± 0.6	99.1 ± 0.1	99.2 ± 0.2	101.5 ± 0.2	99.6 ± 1.0
Sensitivity [4]							
LOD (µg/mL)	0.12	0.06	0.03	0.12	0.12	0.12	0.12
RSD	11.37	30.18	6.92	9.12	13.60	10.77	2.58
LOQ (µg/mL)	0.4	0.4	0.1	0.4	0.4	0.4	0.4
RSD	3.98	1.30	9.29	4.79	1.58	3.30	6.64

(1) Nine solutions of SIM in the concentration ranging from 0.1 µg/mL to 1.5 µg/mL and nine solutions of all impurities in the concentration ranging from 0.4 µg/mL to 6 µg/mL were analyzed.
(2) The repeatability was shown by 6 replicate injections of the standard solution in concentration of 1 µg/mL and the intermediate precision was performed on 6 samples in the two following days using the same equipment.
(3) Determined in triplicate at three concentration levels of 50%, 100%, and 150% by spiking the prequantified samples with a known amount standard of impurities.
(4) The LOD and LOQ were estimated at a signal-to-noise ratio of 3:1 and 10:1, respectively, for each impurity by injecting a series of dilute solutions with known concentration.
NA: not applicable, MSIM: methyl simvastatin, and RSD: relative standard deviation.

TABLE 5: Obtained results from validation of the proposed method: robustness.

	Simvastatin			Res E/F	Res G/SIM	Res B/C
	Rt (min)	T	NTP			
Flow (mL/min)						
0.6 mL/min	14.05	0.99	109188	1.52	4.48	1.66
0.8 mL/min	13.78	1.08	128239	1.50	4.46	1.53
Content of acetonitrile in initial phase						
35%	14.77	1.05	234052	1.53	4.05	1.61
45%	14.74	1.02	236896	1.58	4.17	1.66
Column temperature						
30°C	13.42	1.01	93614	1.51	4.68	1.89
40°C	13.83	1.02	109994	1.53	4.43	1.78
Another column	15.07	1.05	169635	1.59	4.47	2.15

at 80°C and it was considered as low level and 105°C was considered as high level. Time of exposure was chosen to be 3 and 5 hours.

The regulatory guidance does not specify the initial concentration of a compound for FDS [21] and several studies recommended range from 100 to 1000 μg/mL [5, 11, 12]. In order to get even minor decomposition products in the range of detection, initial concentration of 1000 μg/mL was chosen for this study.

Amount of total impurities (%), Simvastatin impurity A, and unknown impurity RRT 1.16 were chosen as *dependent variables*.

(2) Statistical Verification of the Proposed Model. The adequacy of the proposed design was statistically assessed by several statistical criteria, such as coefficient of determination (R^2), adjusted R^2, predicted R^2, and adequate precision. As can be seen in Table 6 the calculated values of the R-squared ($R^2 > 0.9$ in all cases) and adjusted R^2 indicate that the model reasonably fits the experimental data.

The predicted R-square value was in acceptable concordance with the adjusted R-square value for all responses. The differences between the predicted R values and the adjusted R values are small, and thus, they are in reasonable agreement.

Adequate precision defined as a signal-to-noise ratio greater than 4 is desirable, and the obtained ratio for all the responses indicated an adequate signal (Table 6).

(3) SIM Degradation Behavior Evaluated by the Proposed Model. The significance of the effects of each variable was evaluated by ANOVA and the obtained results are presented in Table 6. Values of coefficients b_1 for y_1 and y_3, and especially the values of coefficients b_3 for all responses, demonstrate that SIM under acidic condition is most affected by strength of HCl and time of exposure. Values of the coefficients for the two-factor interaction, b_{13} for all investigated responses, confirmed the main effects of these factors. Both factors have positive sign, meaning that increase of the concentration of HCl and longer exposition is followed by an increase of the amount of total formed impurities, especially amount of impurity A.

As can be seen from the analysis of the percent of formed impurity A, it follows the same pattern as the total impurities discussed above.

The amount of formed impurity with RRT 1.16 follows different pattern, and in this case its formation is mainly affected by the temperature.

The assessment of the simultaneous influence of the time of exposure and strength of HCL on the formation of impurity with RRT 1.16 was based on interaction coefficient b_{13} given in Table 6. The absolute value of the coefficient characterizes the magnitude of the effect, whereas the sign of the coefficient shows whether the increase of the factor value increases ("+" sign) or decreases ("−" sign). Values of the coefficients for acid degradation for amount of formed impurity with RRT 1.16 showed that the interaction between time of exposure and strength of HCl has a significant effect on the degradation process. Individually, time of exposure and strength of HCL were not significant within the range evaluated, at the 95% confidence interval. This highlights the advantage of the proposed methodology, because this synergistic interaction might be overseen by traditional approach for conducting the degradation studies. Also, significant interaction between temperature and time of exposure was observed (Table 6). For alkali hydrolysis all the investigated factors have effect on the degradation of SIM. It was found that the degradation rate of SIM (and formation of impurity A) is strongly dependent on the time of exposition. Evaluating the obtained results can be seen that SIM is very sensitive in alkaline conditions and in some of the experiments degradation of more than 50% was observed. Performing the experiments using DoE allows an overview of the degradation behavior in wider region, so the risk of obtaining irrelevant results from secondary degradation is minimized, thus pointing out the additional advantage of the proposed design.

In addition, same as observed in acid degradation, this approach provides important information on interactions between time of exposure and strength of NaOH and their effect of the formation of impurity with RRT 1.16. Individually, these variables were not significant within the range evaluated, at the 95% confidence interval, but their synergistic

TABLE 6: Statistical parameters of ANOVA and obtained regression coefficients for different degradation.

	R^2	R^2 Predicted	R^2 Adjusted	Adequate precision	Regression coefficients							
					b_0	b_1	b_2	b_3	b_{12}	b_{23}	b_{13}	b_{123}
Acid degradation												
y_1	0.9736	0.8943	0.9537	16.724	20.75	8.60	/	8.35	/	/	2.76	/
y_2	0.9823	0.9293	0.9691	23.526	0.17	/	0.066	/	/	0.029	0.026	/
y_3	0.9701	0.8806	0.9477	15.756	20.21	8.53	/	8.33	/	/	2.53	/
Alkali degradation												
y_1	0.9810	0.9241	0.9668	22.878	38.79	4.43	9.51	14.39	/	/	/	/
y_2	0.9590	0.8362	0.9283	15.000	0.13	/	/	$8.7E^{-3}$	/	0.016	0.031	/
y_3	0.9730	0.8920	0.9527	17.544	30.31	3.32	5.34	14.69	/	/	/	/
Oxidative degradation												
y_1	0.9328	0.7312	0.8824	9.504	2.73	/	0.79	1.41	/	0.61	/	/
y_2	0.9595	0.8382	0.9292	14.839	0.10	0.017	0.017	0.050	/	/	/	/
y_3	0.9943	0.9771	0.9900	29.558	1.28	/	0.16	1.14	/	0.15	/	/
Thermal degradation												
y_1	0.9439	0.7755	0.9158	8.203	14.05	8.90	/	/	/	/	/	/
y_2	0.9918	0.9672	0.9877	22.000	0.16	0.11	/	/	/	/	/	/
y_3	0.9383	0.7531	0.9074	7.797	12.06	7.45	/	/	/	/	/	/

Linear mathematical model of the measured response $y = b_0 + b_1 x_1 + b_2 x_2 + b_3 x_3 + b_{12} x_1 x_2 + b_{13} x_1 x_3 + b_{23} x_2 x_3 + b_{123} x_1 x_2 x_3$, where y is the response [y_1: amount of total impurities (%); y_2: amount of impurity with RRT 1.16 (%); y_3: amount of Simvastatin impurity A (%)], x_i is investigated factors [for acid, alkali, and oxidative degradation; x_1 represents stressor strength (0.01 M and 0.1 M HCl/NaOH or 3% and 30% H_2O_2 for hydrolysis and oxidative degradation respectable); x_2 represents temperature and x_3 represents time of exposure. For thermal degradation x_1 is temperature and x_2 is time of exposure]; b_0 is the intercept b_1, b_2 and b_3, b_{12}, b_{23}, b_{12} and b_{123} as regression coefficients for the variables and interaction between the variables.

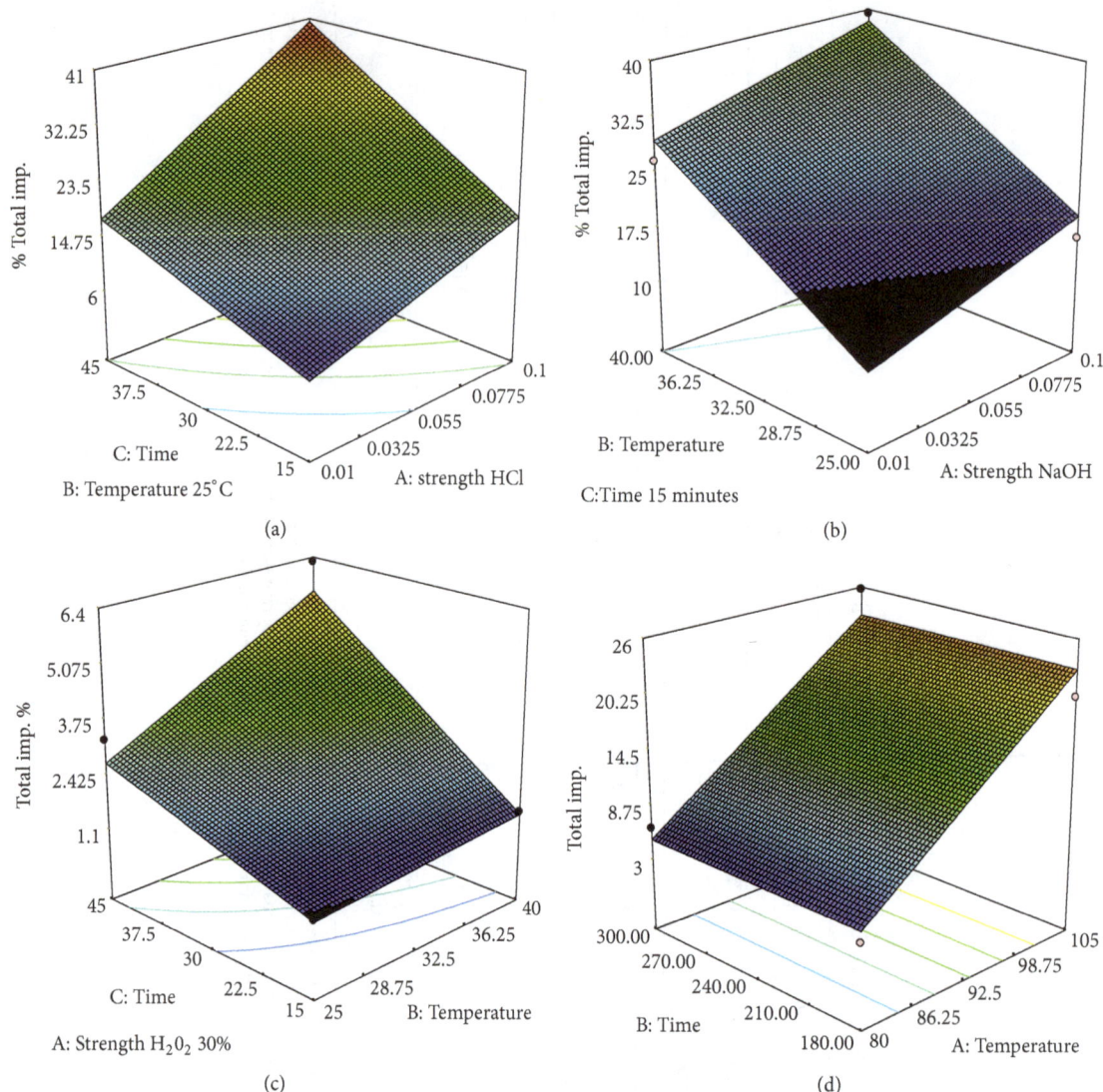

FIGURE 4: 3D response surface plots showing the desired degradation under various conditions: (a) acid degradation; (b) alkali degradation; (c) oxidative degradation; and (d) thermal degradation. Color change from blue to red represents increasing degradation (min ▬▬▬ max).

effect on the degradation process favored formation of impurity with RRT 1.16.

These analyses indicated that, for oxidative degradation, temperature and the time of exposure were most significant factors for all the evaluated responses. Additionally, the strength of H_2O_2 was important factor for formation of impurity with RRT 1.16.

As it could be expected in thermal degradation, the temperature has biggest impact. This kind of effect was expected because it is known that kinetic constants have exponential dependency with reaction temperature (Arrhenius law) and this has also been reported by other authors [22].

(4) Prediction Possibilities of the Proposed Model: Response Surface Methodology. Next in the evaluation phase, response surface plots were generated for the most significant factors for each of the various degradation conditions, providing prediction of the conditions for optimum degradation (Figure 4). Each response surface plot represents a number of combinations of two test variables with all other variables at low levels.

The response surface plot for acid degradation was generated by keeping the temperature at minimum value 25°C (Figure 4(a)). As discussed before, increase of the concentration of HCl and longer exposition is followed by an increase of the amount of total impurities. In some of the experiments, the amount of total impurities (Table 4) was greater than 20%. Although there are references in the literature that mention wider recommended range, the more extreme conditions often provide data that are confounded with secondary degradation products; therefore, need for systematic approach for optimization is recognized [23–25].

The response surface plot for oxidative degradation was generated by keeping the H_2O_2 at maximal value (Figure 4(c)). This plot demonstrated that combination of temperature and time of exposure has positive effect meaning that longer exposure at temperature of 40°C will result with higher degradation, but maximum obtained degradation was

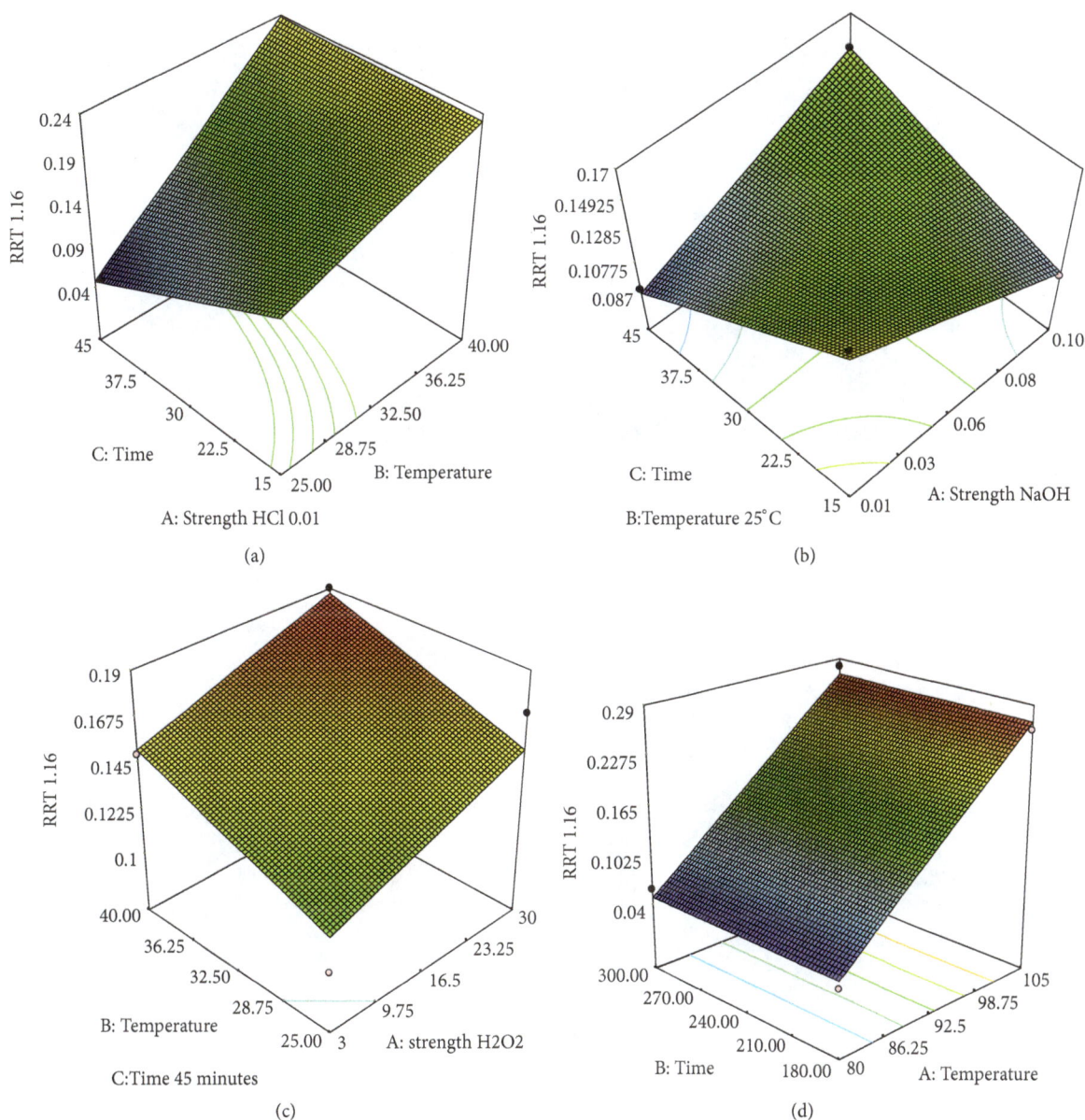

FIGURE 5: 3D response surface plots showing the formation of impurity RRT 1.16 under various conditions: (a) acid degradation; (b) alkali degradation; (c) oxidative degradation; and (d) thermal degradation. Color change from blue to red represents increasing degradation (min ▬▬ max).

about 7%. For oxidative degradation, it has been observed that 5% degradation would be achieved by treating with 30% H_2O_2 at 40°C for 45 min. When these conditions were adopted in practice, the resulting degradation was 5.68%.

The response surface plots obtained from thermal degradation study demonstrated that an increase in temperature from 80°C to 105°C favored the degradation significantly. The targeted drug degradation (Figure 4(d)) was obtained by heating the solution at 80°C for 5 h (5.15%), where minimum formation of impurity RRT 1.16 (0.05%) is achieved.

Photolytic studies were performed on classical manner and about 6.63% degradation has been obtained after exposure of UV light for 2 days.

Similarly, as for the total degradation product, response surface methodology can explain the formation of unspecified impurity with RRT 1.16. As can be seen in Table 3 and graphically presented in Figure 5, elevated temperatures favor formation of the impurity RRT 1.16. In acid degradation it was observed in maximal values, which is consistent with the evaluated literature data [18]. Some literature data suggests that formation of impurity with RRT 1.16 is not affected by applied stress condition (using 0.1 M HCl, 0.1 M NaOH, and 3% H_2O_2) [18]. However, the proposed experimental design indicated that with simultaneous evaluation of the time of the exposure and temperature it is possible to follow the degradation behavior of this impurity.

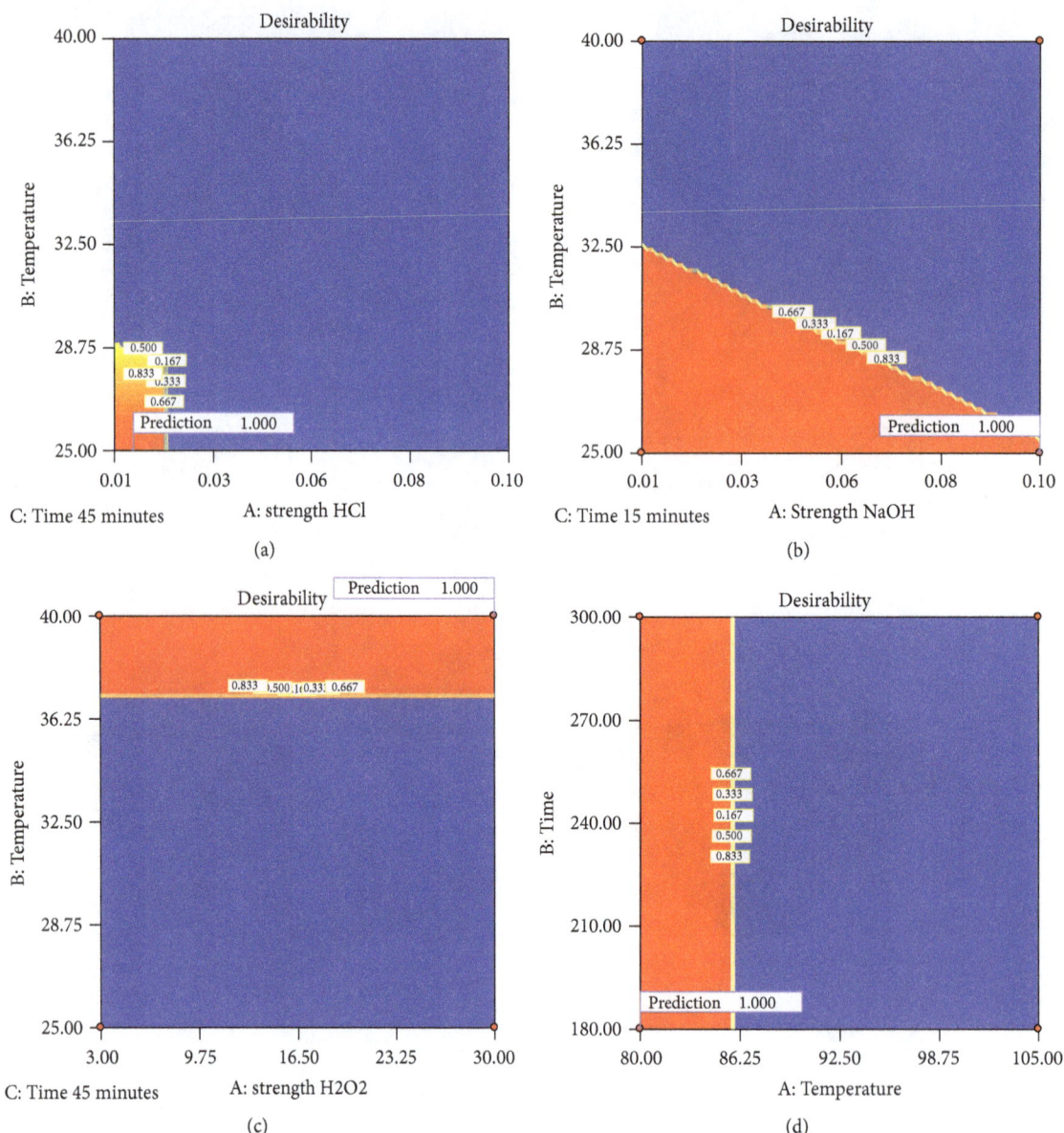

FIGURE 6: Optimization of the selected responses by means of the desirability function. The red area corresponds to the optimum conditions while time maintained constant: (a) acid degradation; (b) alkali degradation; (c) oxidative degradation; and (d) thermal degradation.

Furthermore, it is possible to generate an explanatory model with possibility of relating these types of data in a single experiment.

(5) Prediction Possibilities of the Proposed Model: Desirability Plot. Optimal conditions were chosen using desirability function, where % of impurity (RRT 1.16) was targeted to 0.01 %, and targeted degradation was set in the range 5-20%, following the general literature recommendations [5, 11, 12].

No specific limitations were imposed to the % impurity A, as its value falls within the same range in all case in the experimental model as total impurities.

From the desirability plot presented in Figure 6(a), it can be concluded that a set of coordinates producing high desirability value (D = 1.00) are 0.01 M HCl, 25°C, and 43.5

minutes, which were selected as experimental condition in the verification study.

The predicted response values corresponding to the above optimum condition are given in Table 7. Comparison between obtained and predicted results was made and noticeable difference was not clearly observed (Table 7). The results of the experiments confirmed that the chosen model was adequate for reflecting the expected optimization. Good predictability of the desirability plot provides valuable information about proposed methodology, saving considerable amounts of chemicals and experimental time.

In alkali degradation response surface gradually increased with increasing the concentration from 0.01 M to 0.1 M NaOH and with increasing the temperature. However, as discussed above, this model could be used for prediction and

TABLE 7: Comparison of experimental and predictive values of different responses under optimal conditions.

Parameters	Predicted (%)		Obtained (%)		Predicted Error	
	Total imp.	RRT 1.16	Total imp.	RRT 1.16	Total imp.	RRT 1.16
Acid degradation	18.25	0.06	18.47	0.065	1.21	8.63
Alkali degradation	14.65	0.16	14.24	0.19	2.88	-15.78
Oxidative degradation	5.36	0.19	5.68	0.19	2.53	NA
Thermal degradation	5.18	0.05	4.88	BDL	-5.24	NA

Predicted error = (obtained values – predicted)/predicted * 100 BDL (below disregard limit) (0.05%); NA: not applicable.

optimal conditions were chosen using desirability function, where time of exposure was kept at minimal value.

From the desirability plot presented in Figure 6(b), it can be concluded that a set of coordinates producing high desirability value (D = 1.00) are 0.01 M NaOH at 25°C, followed by immediately neutralization with 0.01 M HCl.

3.4.3. Advantages of the Proposed Model for Forced Degradation Studies. The proposed model is applicable for evaluation of degradation behavior of simvastatin. Generally, implementation of DoE in optimization of experimental conditions in forced degradation study give better data quality with less laboratory work and lead to a decrease in a cost of analysis. The use of DoE to identify theoretical values of variables for optimum degradation was successful, because when the proposed parameters were put in practice, the obtained results matched the predicted degradation.

In fact, the main significant advantage of the present methodology is the simplicity of the sample preparation since measurements are made directly on the liquid samples and optimum degradation was achieved in minimal experimental trials.

As explained with the degradation behavior of SIM through evaluation of total impurities and formation of unspecified impurity with RRT 1.16 under optimized degradation conditions, simultaneous evaluation of the time of the exposure and temperature gives information that can easily been overseen with traditional approach. The results enclosed in this study shows that it is possible to generate an explanatory model with possibility of relating these types of data in a single design methodology.

It is therefore hoped that the results reported here will provide useful guidelines for conducting forced degradation study.

4. Conclusions

The proposed methodology represents an efficient and easily accomplishable approach for searching optimum degradation conditions for conducting the stability studies on APIs. This study showed that DoE is an excellent tool and could successfully be used to develop empirical equation for the prediction and understanding of the degradation process. The obtained results showed sufficiently good correlation between the experimental data and predictive value throughout the studied parameters. This suggests that proposed full factorial design approach can replace the trial and error

approach used to achieve optimum degradation in forced degradation studies.

The investigation also showed that chromatographic techniques coupled with chemometric tools provide useful information of separation and elution time, making this combined methodology a powerful analytical tool. The proposed optimized method for determination of simvastatin and its impurities gives rapid and efficient separation and represents an improvement over the existing reported methods especially in the terms of sensitivity and low cost per sample. The validation study supported the selection of the chromatographic conditions by confirming that the method was specific, accurate, linear, precise, and robust.

References

[1] S. B. Ganorkar and A. A. Shirkhedkar, "Design of experiments in liquid chromatography (HPLC) analysis of pharmaceuticals: Analytics, applications, implications and future prospects," *Reviews in Analytical Chemistry*, vol. 36, no. 3, 2017.

[2] D. B. Hibbert, "Experimental design in chromatography: a tutorial review," *Journal of Chromatography B*, vol. 910, no. 1, pp. 2–13, 2012.

[3] S. Klick, P. G. Muijselaar, J. Waterval et al., "Toward a generic approach for: Stress testing of drug substances and drug products," *Pharmaceutical Technology*, vol. 29, no. 2, pp. 48–66, 2005.

[4] G. Ngwa, "Forced degradation as an integral part of HPLC stability-indicating method development," *Drug Delivery Technology*, vol. 10, no. 5, pp. 56–59, 2010.

[5] M. Blessy, R. D. Patel, P. N. Prajapati, and Y. K. Agrawal, "Development of forced degradation and stability indicating studies of drugs - A review," *Journal of Pharmaceutical Analysis*, vol. 4, no. 3, pp. 159–165, 2014.

[6] M. K. Sharma and M. Murugesan, "Forced Degradation Study an Essential Approach to Develop Stability Indicating Method," *Journal of Chromatography & Separation Techniques*, vol. 08, no. 01, 2017.

[7] K. M. Alsante, L. Martin, and S. W. Baertschi, "A stress testing benchmarking study," *Pharmaceutical Technology*, vol. 27, no. 2, pp. 60–72, 2003.

[8] M. Bakshi and S. Singh, "Development of validated stability-indicating assay methods—critical review," *Journal of Pharmaceutical and Biomedical Analysis*, vol. 28, no. 6, pp. 1011–1040, 2002.

[9] S. Shubhangi, D. Chaitali, and S. Joshi, "Force Degradation Study to Stability Indicating Method," *World Journal of Pharmacy and Pharmaceutical Science*, vol. 3, no. 8, pp. 863-73, 2014.

[10] T. P. Aneesh and A. Rajasekaran, "Forced Degradation Studies- a Tool for Determination of Stability in Pharmaceutical Dosage Forms," *International Journal of Biological & Pharmaceutical Research*, vol. 3, no. 5, pp. 699–702, 2012.

[11] S. Ranjit and Z. Rehman, "Current Trends in Forced Degradation Study for Pharmaceutical Product Development," *J Pharm Educ Res*, vol. 3, no. 1, pp. 54–64, 2012.

[12] S. James, N. Carolina, L. L. Augsburger, H. G. Brittain, A. J. Hickey, and C. Hill, "Pharmaceutical Stress Testing © 2005," *Drugs and the Pharmaceutical Sciences*, 2005.

[13] M. Bifulco and A. Endo, "Statin: New life for an old drug," *Pharmacological Research*, vol. 88, pp. 1-2, 2014.

[14] G. D. Sankar, R. Kondaveni, T. V. Raghava Raju, and M. Vamsi Krishna, "Gradient stability indicating RP-HPLC method for impurity profiling of simvastatin in tablet dosage forms," *Asian Journal of Chemistry*, vol. 21, no. 6, pp. 4294–4300, 2009.

[15] Krishna, S. Radha, G. R. Deshpande, B. M. Rao, and N. S. Rao, "A Stability-Indicating RP-LC Method for the Determination of Related Substances," *in Simvastatin*, vol. 2, no. 1, pp. 91–99, 2010.

[16] R. S. Plumb, M. D. Jones, P. Rainville, and J. M. Castro-Perez, "The rapid detection and identification of the impurities of simvastatin using high resolution sub 2 μm particle LC coupled to hybrid quadrupole time of flight MS operating with alternating high-low collision energy," *Journal of Separation Science*, vol. 30, no. 16, pp. 2666–2675, 2007.

[17] A. Álvarez-Lueje, C. Valenzuela, J. A. Squella, and L. J. Núñez-Vergara, "Stability study of simvastatin under hydrolytic conditions assessed by liquid chromatography," *Journal of AOAC International*, vol. 88, no. 6, pp. 1631–1636, 2005.

[18] M. Vuletić, M. Cindrić, and J. D. Koružnjak, "Identification of unknown impurities in simvastatin substance and tablets by liquid chromatography/tandem mass spectrometry," *Journal of Pharmaceutical and Biomedical Analysis*, vol. 37, no. 4, pp. 715–721, 2005.

[19] Ich, "ICH Topic Q2 (R1) Validation of Analytical Procedures: Text and Methodology," in *International Conference on Harmonization*, 1994.

[20] 9th Pharmacopoeia, European, "Simvastatin monograph No 1563, 01/2017.

[21] ICH, "Stability Testing of New Drug Substances and Products Q1A(R2)," in *Proceedings of the International Conference on Harmonization*, 2003.

[22] R. G. Simões, H. P. Diogo, A. Dias et al., "Thermal stability of simvastatin under different atmospheres," *Journal of Pharmaceutical Sciences*, vol. 103, no. 1, pp. 241–248, 2014.

[23] T. Rawat and I. P. Pandey, "Forced degradation studies for drug substances and drug products- scientific and regulatory considerations," *Journal of Pharmaceutical Sciences and Research*, vol. 7, no. 5, pp. 238–241, 2015.

[24] N. G. Shinde et al., "Pharmaceutical Forced Degradation Studies with Regulatory Consideration," *Asian J. Res. Pharm. Sci*, vol. 3, no. 4, pp. 178–188, 2013.

[25] WHO Technical Report Series, No. 929, 2005, Annex 5, Guidelines for Registration of Fixed-Dose Combination Medicinal Products, Appendix 3, Pharmaceutical Development (or Preformulation) Studies.

10

Simultaneous Determination of Nitroimidazoles and Quinolones in Honey by Modified QuEChERS and LC-MS/MS Analysis

Haiyan Lei,[1] Jianbo Guo,[2] Zhuo Lv,[2] Xiaohong Zhu,[2] Xiaofeng Xue,[3] Liming Wu,[3] and Wei Cao ⓘ[1]

[1]Department of Food Science and Engineering, School of Chemical Engineering, Northwest University, Xi'an, Shaanxi 710069, China
[2]Shaanxi Institute for Food and Drug Control, Xi'an, Shaanxi 710069, China
[3]Institute of Apiculture Research, Chinese Academy of Agricultural Sciences, Beijing 100093, China

Correspondence should be addressed to Wei Cao; caowei@nwu.edu.cn

Academic Editor: Troy D. Wood

This study reports an analytical method for the determination of nitroimidazole and quinolones in honey using liquid chromatography-tandem mass spectrometry (LC-MS/MS). A modified QuEChERS methodology was used to extract the analytes and determine veterinary drugs in honey by LC-MS/MS. The linear regression was excellent at the concentration levels of 1–100 ng/mL in the solution standard curve and the matrix standard curve. The recovery rates of nitroimidazole and quinolones were 4.4% to 59.1% and 9.8% to 46.2% with relative standard deviations (RSDs) below 5.2% and the recovery rates of nitroimidazole and quinolones by the matrix standard curve ranged from 82.0% to 117.8% and 79% to 115.9% with relative standard deviations (RSDs) lower than 6.3% in acacia and jujube honey. The acacia and jujube honeys have stronger matrix inhibition effect to nitroimidazole and quinolones residue; the matrix inhibition effect of jujube honey is stronger than acacia honey. The matrix standard curve can calibrate matrix effect effectively. In this study, the detection method of antibiotics in honey can be applied to the actual sample. The results demonstrated that the modified QuEChERS method combined with LC-MS/MS is a rapid, high, sensitive method for the analysis of nitroimidazoles and quinolones residues in honey.

1. Introduction

Nitroimidazoles and quinolones (Figure 1) are a group of antibacterial compounds that have been widely used in medical domain. There are many antibiotics left in honey because of the illegal addition of beekeepers [1–4], which directly threatens the health and safety of consumers. Nosemosis of bees is one of the protozoa infections in adult honeybee, which is very destructive to honeybee colonies and is an infectious disease caused by *Cryptosporidium parvum*. Nitroimidazoles, for example, metronidazole, can be used to prevent and treat nosemosis of bees. Quinolones, for instance, ofloxacin, can potentially be used to prevent and treat honeybees piroplasmosis. However, the misuse and illegal use of nitroimidazoles drugs may cause potential hazards of cell mutagenicity and carcinogenic radionuclide, quinolones drugs can lead to the reaction of certain degree

and hepatotoxicity or even death [5, 6]. Therefore, nitroimidazoles and quinolones have been banned in honey. Hence, the control of nitroimidazoles and quinolones is highly significant for the agricultural environment and food industry.

High-performance liquid chromatography (HPLC) with diode array detector (DAD) [7], liquid chromatography with fluorescence detection (LC-FD) [8], liquid chromatographic-mass spectrometric (LC-MS) [9, 10], and liquid chromatography-tandem mass spectrometry (LC-MS/MS) have been used to analyze nitroimidazoles and quinolones in food industry (e.g., milk powder, bovine milk, butter, fish tissue, eggs, chicken meat, pig plasma, bovine meat, swine tissues, honey, feed hair, and water) [11–19]. LC-MS/MS is one of the most promising techniques for the analysis of antibiotics in foodstuff because of its sensitivity and accuracy. Sample preparation generally use QuEChERS approach, SPE clean-up or liquid-liquid extraction. Compared with SPE clean-up

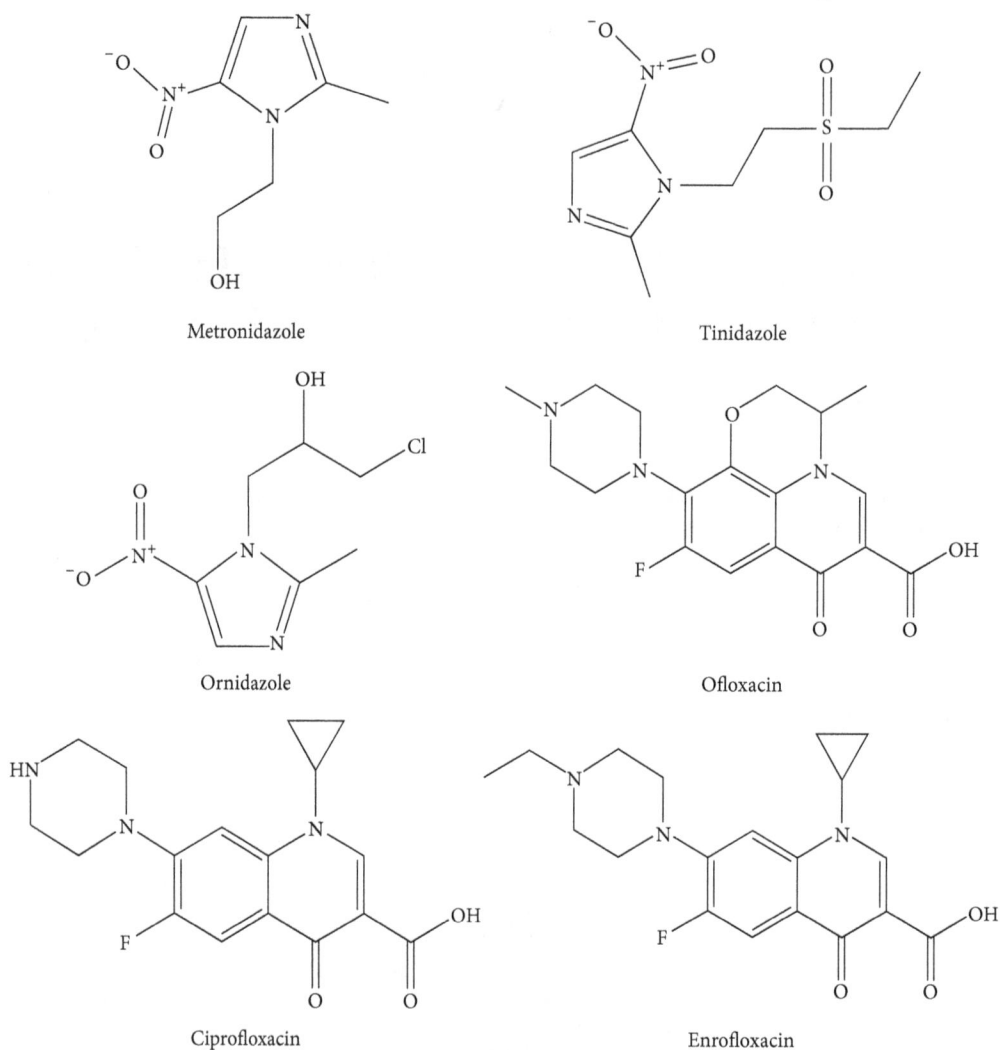

FIGURE 1: Chemical structures of six nitroimidazoles and quinolones.

and liquid-liquid extraction, QuEChERS approach has almost the same purifying effect, but it requires minimum operational steps and solvent and has higher accuracy and wider application [16, 20–24]. It was initially developed for the analysis of pesticide residues in fruits and vegetables and was extended to the analysis of veterinary drugs and environmental pollutants residues [25].

In this study, we developed a multiresidue test method based on the application of LC-MS/MS combined with modified QuEChERS sample preparation methodology for rapid determination of nitroimidazoles and quinolones residues. The improvement of the method is shown in Figure 2.

The innovation of this method is to eliminate the matrix effect in honey via matrix effect standard curve; the matrix effect of acacia and jujube honey has stronger inhibitory action to nitroimidazoles and quinolones residue with the matrix inhibition effect of jujube honey being stronger than acacia honey. The matrix standard curve can effectively correct the matrix effect in honey. This method is more accurate than the previous published method.

2. Experimental

2.1. Materials and Reagents. Acacia and jujube honey samples were purchased from consumer stores and provided by beekeepers. The samples were stored at ambient temperature (25°C) before analysis.

Analytical standard substances, including metronidazole (purity = 100.0%) CAS: 443-48-1, batch lot: 100191-201507; ornidazole (purity = 100.0%) CAS: 16773-42-5, batch lot: 100608-201102; tinidazole (purity > 99.9%) CAS: 19387-91-8, batch lot: 100336-200703; ofloxacin (purity > 99.5%) CAS: 82419-36-1, batch lot: 130454-201206; ciprofloxacin (purity > 99.5%) CAS: 85721-33-1, batch lot: 130451-201203, were obtained from Institute of Pharmaceutical and Biological Products (Beijing, China); enrofloxacin (purity > 99%) CAS: 93106-60-6, batch lot: 107071, was purchased from Dr. Ehrenstorfer GmbH (Augsburg, Germany).

These reagents of sodium chloride, sodium hydroxide, anhydrous sodium sulfate, anhydrous magnesium sulfate, citric acid, disodium hydrogen phosphate, and glacial acetic acid are of analytical purity (Sinopharm, Beijing, China).

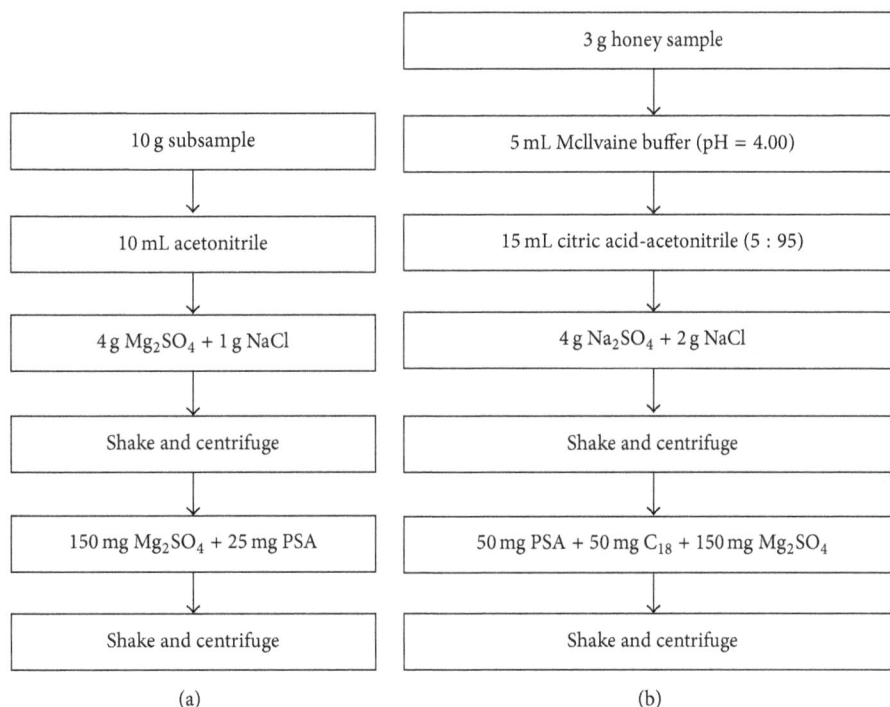

FIGURE 2: Original QuEChERS methodology (a) and modified QuEChERS methodology (b).

Ammonium formate (Fluka, Tianjin, China), formic acid (Fluka, Tianjin, China), acetonitrile (Fisher, Fairlawn, USA), and methanol (Tedia, Fairfield, USA) are of HPLC grade. PSA and C_{18} (40 μm) are of also analytical purity (Agela, Beijing, China). A Milli-Q ultrapure water system (Millipore, Bedford, MA, USA) was used to obtain the HPLC-grade water.

Anhydrous sodium sulfate was baked at 600°C for 3 h and moved the sealed container for conservation.

Mcilvaine buffer (pH = 4.00) was prepared by dissolving 19.2 g disodium hydrogen phosphate and 8.9 g citric acid in 1.625 L of Milli-Q water and the pH was adjusted with 4 mol·L^{-1} sodium hydroxide solution.

2.2. Standard Solutions. The individual stock standard solutions of metronidazole, ornidazole, tinidazole, ofloxacin, ciprofloxacin, and enrofloxacin were prepared in methanol at the concentration of 1 mg/mL. The mixed working standard solutions (1 μg/mL) were prepared by diluting stock solutions with methanol. All standard solutions were stored at −20°C in dark bottles.

Preparation of standard solutions: mixture working solutions (1 μg/mL) were diluted with methanol/water (50/50, v/v) at the concentration of 1, 5, 10, 20, 40, and 100 ng/mL.

The matrix-matched working standard solutions: 3.0 g homogenized negative acacia and jujube honey samples were weighed and placed into 50 mL polypropylene centrifuge tube, respectively; then various amounts mixture working solutions were added.

2.3. Sample Preparation. The modified QuEChERS methodology was used. An aliquot of 3.0 g homogenized samples was weighed and placed into 50 mL polypropylene centrifuge tube, and 5 mL Mcilvaine buffer (pH = 4.00) was added. The mixture was shaken in a vortex mixer for 30 s; then 15 mL citric acid- acetonitrile (5 : 95) was added. Subsequently, 2.0 g sodium chloride and 4.0 g anhydrous sodium sulfate were added to this mixture and vigorously shaken in a vortex for 2 min; afterwards, the tube was centrifuged at 10000 rpm for 10 min. Next, 10 mL supernatant solution was transferred into a 15 mL centrifuge tube and evaporated to 2 mL under nitrogen at 40°C. Then 50 mg PSA, 50 mg C_{18}, and 100 mg anhydrous Mg_2SO_4 were added to the tube in a vortex for 2 min and then centrifuged at 10000 rpm for 10 min. Next, 1 mL supernatant solution was transferred into a 15 mL centrifuge tube and evaporated to dryness under nitrogen at 40°C. The residue was reconstituted in 1 mL methanol/water (50/50, v/v) and filtered through a 0.45 μm filter before LC-MS/MS analysis.

2.4. LC-MS/MS Conditions

2.4.1. LC Conditions. Chromatographic analyses were performed by Waters 2695 series HPLC System (Milford, MA, USA); chromatographic separation was achieved by Waters XTerra RP18 (2.1 mm × 150 mm, 5 μm) analytical column. The injection volume was 10 μL, and the temperature of the column was maintained at 35°C. The mobile phases were acetonitrile (mobile phase A) and 10 mM ammonium formate + 0.5% formic acid in Milli-Q water (mobile phase B) at a flow rate of 0.3 ml/min. The total chromatographic runtime was 15 min. Mobile phase gradient flow program was shown in Table 1.

TABLE 1: Mobile phase gradient flow program.

Time/min	A/%	B/%
0.00	5	95
3.00	20	80
8.00	25	75
8.10	50	50
10.00	50	50
10.10	5	95
15.00	5	95

A: acetonitrile; B: 10 mM ammonium formate + 0.5% formic acid in Milli-Q water.

TABLE 2: Retention time and MS/MS conditions for the target compounds.

Name	Qualification ion (m/z)	Quantification ion (m/z)	Cone voltage (V)	Collision energy (eV)	ion ratio	Retention time (min)
Metronidazole	172.0/128.0 172.0/81.9	172.0/128.0	25	15 25	3.69	2.70
Tinidazole	248.0/92.8 248.0/128.0	248.0/92.8	30	20 15	6.63	5.23
Ornidazole	220.0/127.9 220.0/81.9	220.0/127.9	30	15 25	6.45	6.47
Ofloxacin	361.9/261.0 361.9/221.0	361.9/261.0	35	25 35	5.15	6.72
Ciprofloxacin	331.9/314.0 331.9/231.0	331.9/314.0	40	20 35	3.17	6.93
Enrofloxacin	360.0/245.0 360.0/203.0	360.0/245.0	35	25 36	5.37	7.35

2.4.2. MS/MS Conditions. Waters Quattro Micro API triple quadruple tandem MS coupled to electrospray ionization (ESI) interface and Waters Jet Stream Ion Focusing (Waters, USA) was used for mass analysis and quantification of target analytes. The MS was operated in the positive ion mode and utilized multiple reaction monitoring (MRM). The tuning parameters were optimized for the target analytes: the gas temperature was set at 350°C with a flow rate of 800 L/h and the gas was high purity nitrogen. Capillary voltage was 2.5 kV, ion source temperature was 120°C, cone gas flow was 50 L/h, and the gas was high purity nitrogen. The system operation, data acquisition, and analysis are controlled and processed by the MassHunter software.

3. Results and Discussion

3.1. Optimization of Mass Spectrometry Conditions. Both positive and negative ionization modes in ESI were used to evaluate the signal responses of target analytes [26]. After evaluation, the target compounds (metronidazole, tinidazole, ornidazole, ofloxacin, ciprofloxacin, and enrofloxacin) were analysed in positive ESI mode ([M + H]+) due to higher response. The size of collision energy and cone voltage influence sensitivity and fragmentation [22]. In the full-scan mass analysis, the parent ion is obtained; then the optimum cone voltage was optimized according to higher response. The fragmentation ion of every target compound was obtained via product scan in optimum cone voltage. The fragment ion of metronidazole is 128.0 m/z and 81.9 m/z, the fragment ion of tinidazole is 128.0 m/z and 92.8 m/z, the fragment ion of ornidazole is 127.9 m/z and 81.9 m/z, the fragment ion of ofloxacin is 261.0 m/z and 221.0 m/z, the fragment ion of metronidazole is 314.0 m/z and 231 m/z, and the fragment ion of metronidazole is 245.0 m/z and 203 m/z. The collision energies were optimized for each individual analyte to give the best response. The most intense and stable fragmentation ions were selected for quantification ion, and the second most abundant ions were used for qualification ion [27]. Eventually, the best collision energies of quantification ion and qualification ion were optimized by the maximum intensity response. All the MS/MS parameters are presented in Table 2.

3.2. Matrix Effects. Matrix is components of exclusive tested matter, which has a significant interference for linearity, accuracy, precision, limits of detection, and quantification. The interference was said to be matrix effect. Recently, they have been discussed in several review articles [27, 28]. Matrix effects (MEs) are a major problem affecting the quantitative accuracy of liquid chromatography-electrospray ionization mass spectrometry (LC-ESI-MS) when analyzing complicated samples [1]. The influence of sample substrate on target compounds determination is derived from endogenous components of sample, which are organic and inorganic components and exist in extracting solution after sample preparation; the components included ion particle

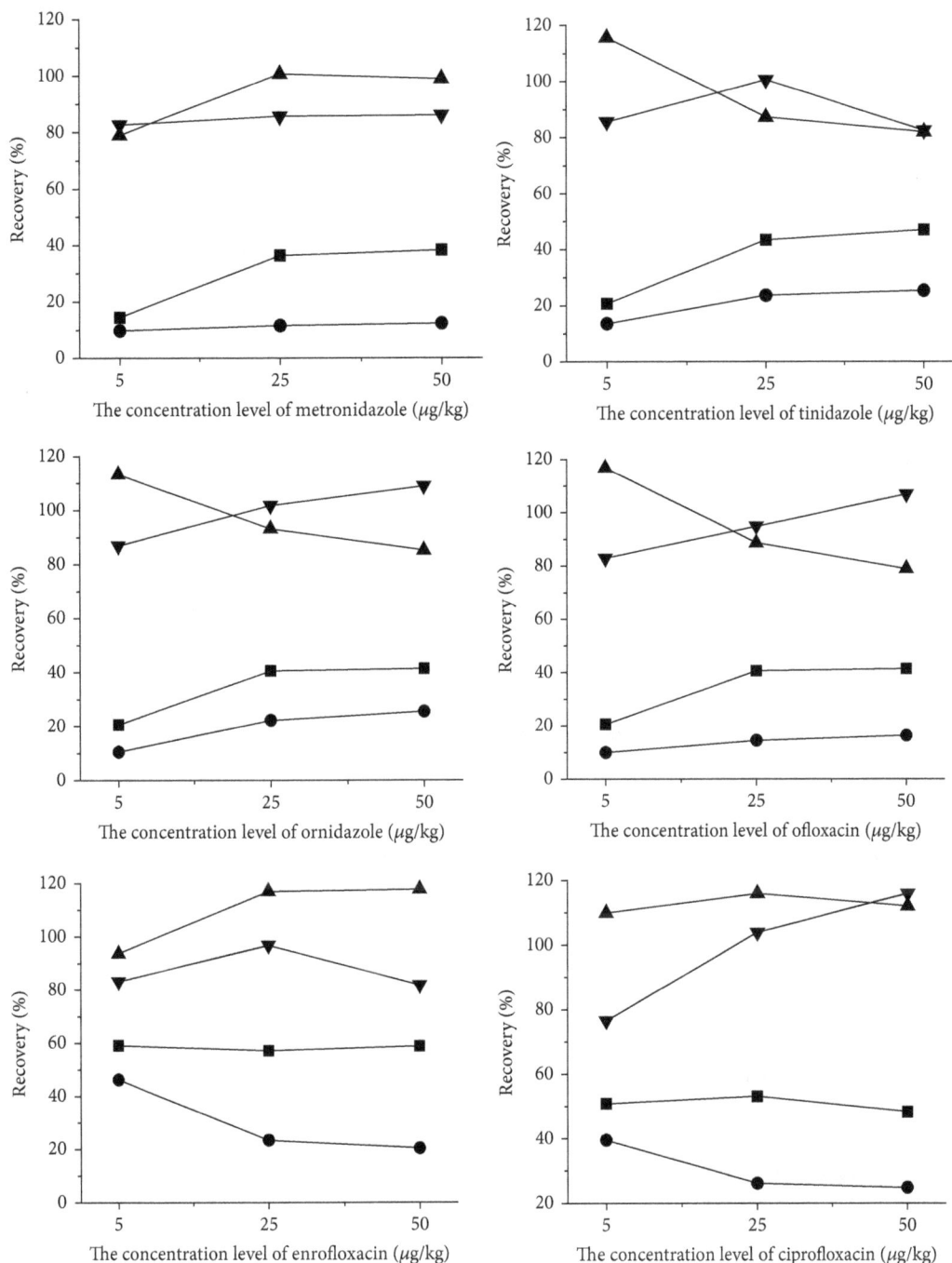

FIGURE 3: Absolute and relative recovery of target compounds. ■, ●: the recovery rates of six kinds of drugs by the solution standard curve in acacia and jujube honey, respectively; ▼, ▲: the recovery rates of six kinds of drugs by the matrix standard curve in acacia and jujube honey, respectively.

composition (electrolytes and salts), strong polar compounds (phenols and pigment), and organic components (sugars, amines, carbamide, lipoid, congener, and metabolism of target object); when these substances and target compounds fly into the ion source at the same time after honey samples preparation, they will affect the ionization process of target compounds [29]. In the study, the matrix effects in these samples were evaluated by the spiked honey samples at 5, 25, and 50 μg/kg. According to Figure 3, compared with

matrix-matched calibration curve, the recovery of the neat standard calibration curve was lower, and ME indicated matrix suppression. ME phenomena were obvious in honey sample, which may be impressed by sugars, phenols, pigment, protein, and flavonoids. Moreover, jujube honey contain high contents of pigment and flavonoids that make it dark; therefore, recovery of jujube honey is lower than acacia honey and the matrix effect of jujube honey was stronger than acacia honey. Hence, matrix-matched standard curve

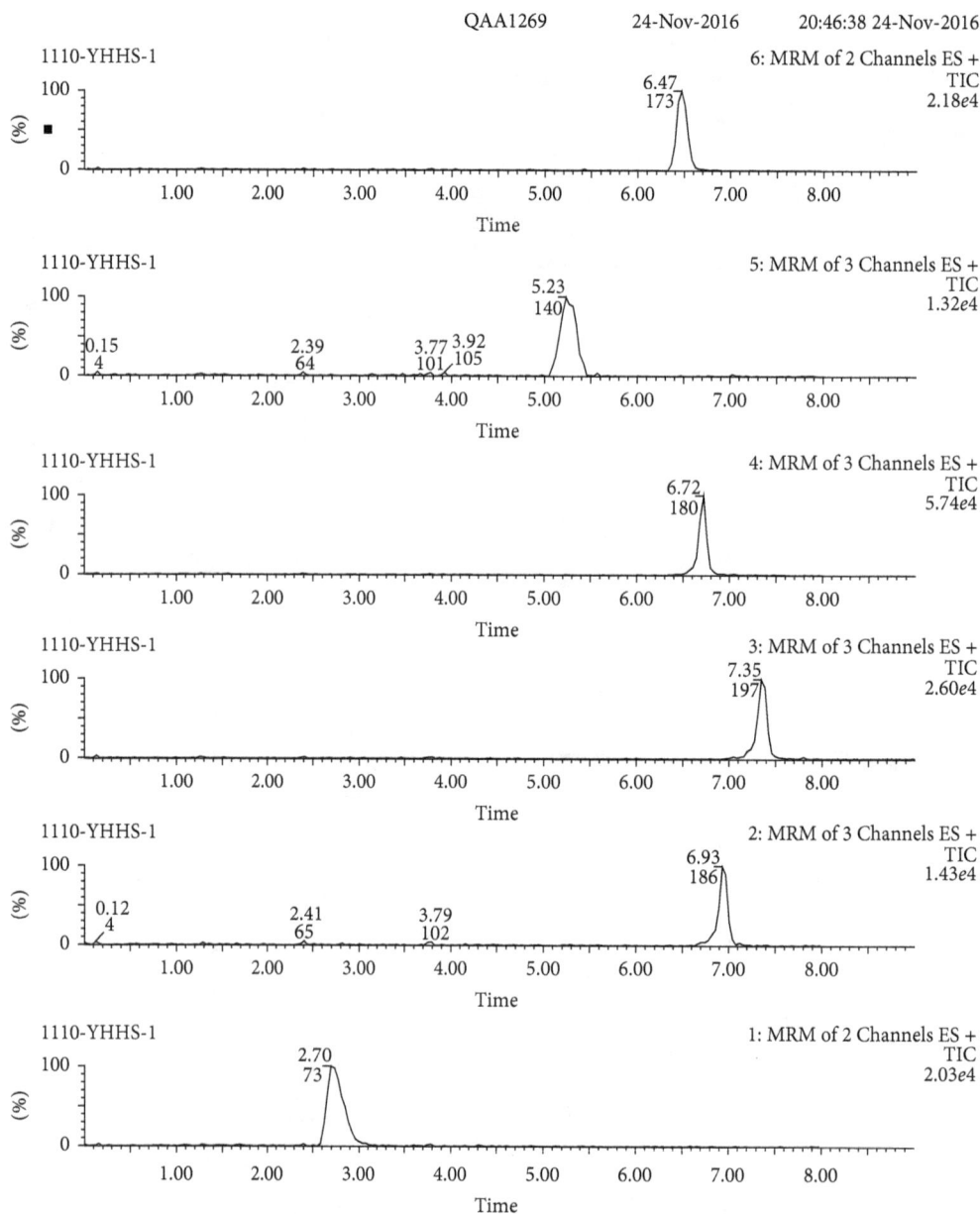

FIGURE 4: Typical chromatograms of MRM transitions for metronidazole, tinidazole, ornidazole, ofloxacin, ciprofloxacin, and enrofloxacin at the concentration of 25 μg/kg in acacia honey.

has been used to overcome the matrix effect and signal irreproducibility, matrix interference, and loss of recovery. Figure 3 showed that the matrix-matched working standard curves for each compound can meet detection requirements and effectively correct matrix effects in acacia and jujube honey.

3.3. Linearity. A serial dilution of the standard mixture was prepared (1–100 ng/mL) and analyzed using the optimized assay conditions. And the correlation coefficients (R^2) ranged within 0.994–0.999 using this method. The negative acacia and jujube honeys were extracted and tested according to Sections 2.3 and 2.4. The matrix-matched calibration curve

was constructed by determining the peak area of metronidazole, ornidazole, tinidazole, ofloxacin, ciprofloxacin, and enrofloxacin at six concentration levels in the range of 10–100 ng/mL curve in acacia and jujube honey and the correlation coefficients (R^2) ranged within 0.990–0.999 and 0.992–0.996, respectively. In general, the linearity of the matrix-matched working standard curves and the solution standard curve was excellent. The data were treated by using the Waters QuanLynx module of the MassLynx software. The result is shown in Table 3.

3.4. Limit of Detection and Quantification. The limits of detection (LOD) and limits of quantification (LOQ) were

TABLE 3: Linear range, regression equation, and correlation coefficients.

Compound	Linear range (ng/mL)	Calibration equation			Correlation coefficients, R^2		
		Standard solutions	Acacia honey	Jujube honey	Standard solutions	Acacia honey	Jujube honey
Metronidazole	10–100	$y = 149.667x + 228.847$	$y = 84.2987x + 17.3078$	$y = 36.6167x + 33.0998$	0.998	0.997	0.996
Tinidazole	10–100	$y = 69.3581x + 46.8099$	$y = 47.1843x - 59.0018$	$y = 33.5304x + 8.3281$	0.997	0.997	0.993
Ornidazole	10–100	$y = 86.7363x + 162.9$	$y = 57.8037x + 8.9570$	$y = 38.1225x + 13.5625$	0.999	0.999	0.992
Ofloxacin	10–100	$y = 159.216x - 487.671$	$y = 62.2512x + 56.6425$	$y = 51.2493x + 5.2237$	0.994	0.990	0.993
Ciprofloxacin	10–100	$y = 79.9202x - 207.609$	$y = 40.8103x + 4.2569$	$y = 33.0771x + 25.5369$	0.996	0.999	0.996
Enrofloxacin	10–100	$y = 88.4074x - 347.422$	$y = 69.1707x + 7.48835$	$y = 42.9161x + 12.0416$	0.995	0.998	0.992

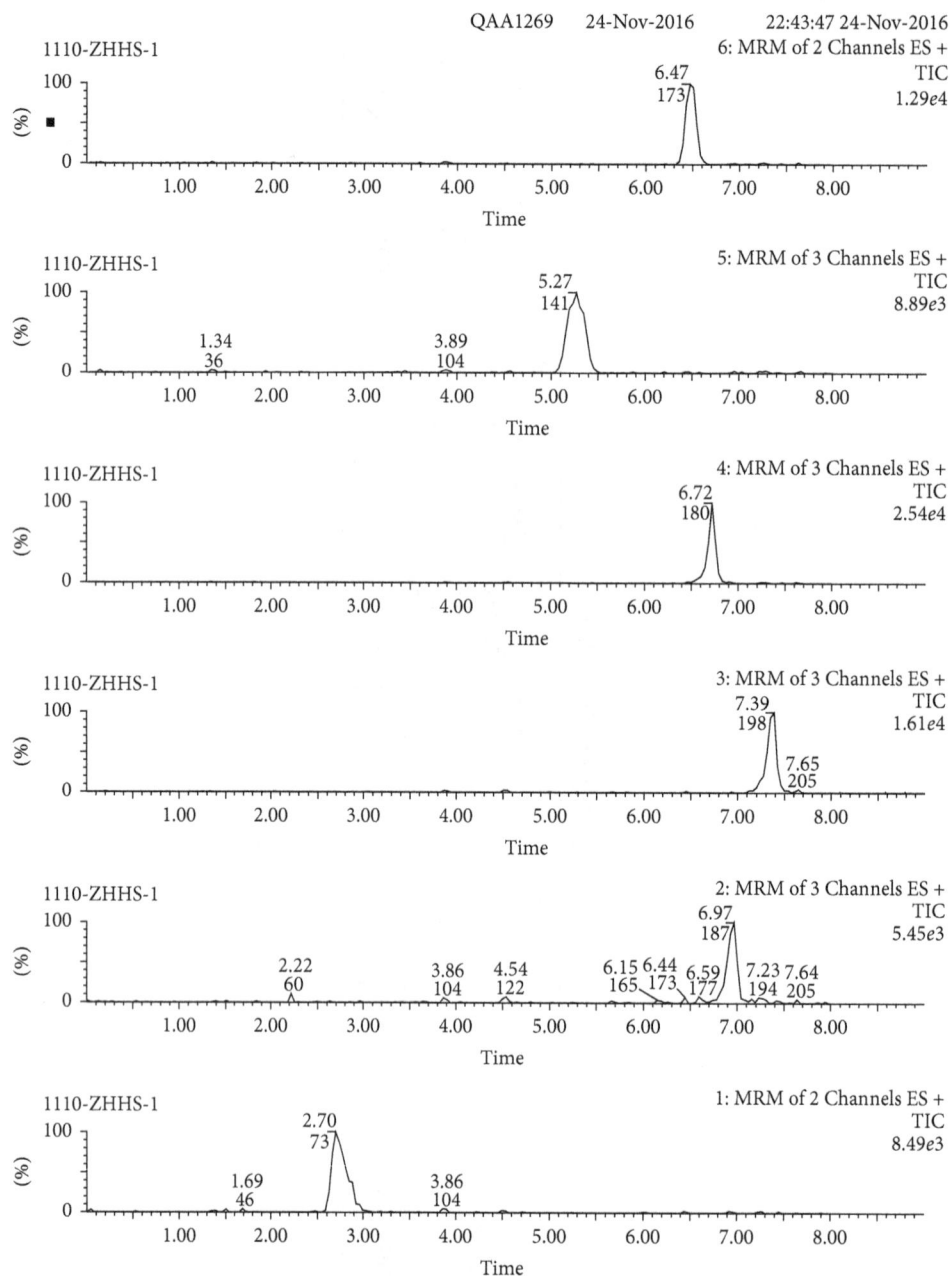

FIGURE 5: Typical chromatograms of MRM transitions for metronidazole, tinidazole, ornidazole, ofloxacin, ciprofloxacin, and enrofloxacin at the concentration of 25 μg/kg in jujube honey.

determined (n = 3) via the signal-to-noise ratio (SNR) of the analyte; LOD and LOQ were estimated as 3 × S/N and 10 × S/N, respectively. The LODs of the target compounds were achieved in the range of 0.64–1.41 μg/kg and 0.83–1.58 μg/kg for acacia and jujube honey, respectively. The LOQs of the target compounds were achieved in the range of 2.13–4.70 μg/kg and 2.77–5.27 μg/kg for acacia and jujube honey, respectively. The result is shown in Table 4.

3.5. Recovery and Precision. Recovery and precision were determined at three concentration levels (5, 25, and 50 μg/kg)

for acacia and jujube honey. Recoveries of the analytes ranged within 81.9%–116.8% and 81.0%–115.9% in acacia and jujube honey, respectively. In detail, recoveries range within 81.0%–100.7% (metronidazole), 82.5%–115.6% (tinidazole), 85.2%–113.3% (ornidazole), 81.9%–116.7% (ofloxacin), 81.8%–116.8% (ciprofloxacin), and 86.5%–115.9% (enrofloxacin). Recoveries for target compounds were higher than 80% and RSDs were lower than 6.3% in Table 5. These are highly acceptable values for the honey samples. Typical chromatograms of MRM transitions about six drugs are shown in Figures 4 and 5 at the concentration of 25 μg/kg in acacia and jujube honey.

TABLE 4: LODs and LOQs for the target compounds.

Compound	Acacia honey		Jujube honey	
	LOD/(μg/kg)	LOQ/(μg/kg)	LOD/(μg/kg)	LOQ/(μg/kg)
Metronidazole	0.64	2.13	0.83	2.77
Tinidazole	1.25	4.17	1.51	5.03
Ornidazole	1.41	4.70	1.58	5.27
Ofloxacin	0.77	2.56	0.99	3.30
Ciprofloxacin	0.92	3.07	1.17	3.91
Enrofloxacin	1.13	3.76	1.24	4.13

TABLE 5: Recovery rates and RSDs of target compounds from honey samples using by LC-MS/MS.

Compound	Concentration level (μg/kg)	Recovery% (RSD%)	
		Acacia honey	Jujube honey
Metronidazole	5	82.7 (0.044)	81.0 (0.058)
	25	85.7 (0.031)	100.7 (0.050)
	50	86.1 (0.020)	99.0 (0.024)
Tinidazole	5	115.6 (0.050)	85.60 (0.027)
	25	87.2 (0.031)	100.5 (0.051)
	50	82.6 (0.012)	82.5 (0.039)
Ornidazole	5	113.3 (0.044)	86.8 (0.017)
	25	93.2 (0.049)	101.8 (0.036)
	50	85.2 (0.043)	108.9 (0.007)
Ofloxacin	5	116.7 (0.046)	82.9 (0.057)
	25	88.6 (0.054)	94.8 (0.039)
	50	81.9 (0.019)	106.5 (0.023)
Ciprofloxacin	5	83.1 (0.027)	83.1 (0.027)
	25	96.8 (0.040)	96.8 (0.040)
	50	116.8 (0.050)	81.8 (0.033)
Enrofloxacin	5	109.9 (0.053)	86.5 (0.063)
	25	115.9 (0.044)	103.9 (0.062)
	50	112.0 (0.031)	115.9 (0.020)

The repeatability and intermediate precision results were expressed as relative standard deviation (RSD); number of determinations (n) = 3.

3.6. Application to Real Samples. The proposed modified QuEChERS method with LC-MS/MS was applied to 46 actual honey samples from honey producers and cooperatives located in the cities of Shaanxi, Hebei, Gansu, Chongqing, Hubei, and Shanxi in China. The samples were extracted and analysed according to the protocols described in Sections 2.3 and 2.4. Quantifications of metronidazole, tinidazole, ornidazole, ofloxacin, ciprofloxacin, and enrofloxacin were performed using the external standard. The results are shown in Table 6.

selectivity, recovery, and accuracy. Therefore, it has been successfully applied to the determination of nitroimidazoles and quinolones residues. Matrix standard calibration curves were successfully employed in correction for matrix effect. It can be seen that some honeys have the residues of nitroimidazole and quinolones in Table 6. Hence, the control of antibiotics in food is highly significant for the agricultural environment and food industry.

4. Conclusion

A simple and rapid method of modified QuEChERS combined with liquid chromatography-tandem mass spectrometry for the determination of multiresidue in honey samples was established. The method was fast, efficient, reliable and could be used in the monitoring of antibiotic in honey, which is fit for the purpose and satisfactory in terms of

Acknowledgments

This work is financially supported by the Agricultural Science and Technology Innovation Program of Shaanxi Province (2012NKC01-21), the Agricultural Science and Technology Project of Shaanxi Province (2013K01-47-01), and Social

TABLE 6: Concentrations of antibiotics detected in acacia and jujube honey.

Number	Concentration of target compounds (μg/kg) ($n = 3$)					
	Metronidazole	Tinidazole	Ornidazole	Ofloxacin	Ciprofloxacin	Enrofloxacin
(1)	ND	ND	ND	ND	ND	ND
(2)	ND	ND	ND	ND	ND	ND
(3)	66.95	ND	ND	ND	ND	ND
(4)	ND	ND	ND	ND	ND	ND
(5)	ND	ND	ND	ND	ND	ND
(6)	ND	ND	ND	ND	ND	ND
(7)	ND	ND	ND	ND	ND	ND
(8)	ND	ND	ND	ND	ND	ND
(9)	ND	ND	ND	ND	ND	ND
(10)	ND	ND	ND	ND	ND	ND
(11)	ND	ND	ND	ND	89.43	ND
(12)	ND	ND	ND	ND	ND	ND
(13)	ND	ND	ND	ND	ND	ND
(14)	ND	ND	ND	ND	ND	ND
(15)	ND	ND	ND	ND	ND	ND
(16)	ND	ND	ND	ND	ND	ND
(17)	ND	ND	ND	ND	ND	ND
(18)	ND	ND	ND	ND	ND	ND
(19)	ND	ND	ND	ND	ND	ND
(20)	ND	ND	ND	ND	ND	ND
(21)	ND	ND	ND	ND	ND	ND
(22)	ND	ND	ND	ND	ND	ND
(23)	ND	ND	ND	ND	ND	ND
(24)	ND	ND	ND	ND	ND	ND
(25)	ND	ND	ND	ND	ND	ND
(26)	ND	ND	ND	ND	ND	ND
(27)	ND	ND	ND	ND	ND	ND
(28)	5.87	ND	ND	ND	ND	ND
(29)	ND	ND	ND	ND	ND	ND
(30)	ND	ND	ND	ND	ND	ND
(31)	ND	ND	ND	ND	ND	ND
(32)	ND	ND	ND	ND	ND	ND
(33)	ND	ND	ND	ND	ND	ND
(34)	ND	ND	ND	ND	ND	ND
(35)	29.35	ND	ND	ND	52.91	ND
(36)	ND	ND	ND	ND	ND	ND
(37)	ND	ND	ND	ND	ND	ND
(38)	ND	ND	ND	ND	ND	ND
(39)	ND	ND	ND	ND	ND	ND
(40)	ND	ND	ND	ND	ND	ND
(41)	ND	ND	ND	ND	ND	ND
(42)	ND	ND	ND	ND	ND	ND
(43)	6.12	ND	ND	ND	ND	ND
(44)	ND	ND	ND	ND	ND	ND
(45)	ND	ND	ND	ND	ND	ND

ND: not detected; numbers (1)–(33): acacia honey; numbers (34)–(45): jujube honey.

Development of Science and Technology Project (2016SF-425).

References

[1] A. H. Shendy, M. A. Al-Ghobashy, S. A. Gad Alla, and H. M. Lotfy, "Development and validation of a modified QuEChERS protocol coupled to LC-MS/MS for simultaneous determination of multi-class antibiotic residues in honey," *Food Chemistry*, vol. 190, pp. 982–989, 2016.

[2] J. M. Płotka, M. Biziuk, and C. Morrison, "Determination of antibiotic residues in honey," *Trends in Analytical Chemistry*, vol. 30, pp. 1035–1041, 2011.

[3] P. Butaye, L. A. Devriese, and F. Haesebrouck, "Differences in antibiotic resistance patterns of Enterococcus faecalis and Enterococcus faecium strains isolated from farm and pet animals," *Antimicrobial Agents and Chemotherapy*, vol. 45, no. 5, pp. 1374–1378, 2001.

[4] R. Venable, C. Haynes, and J. M. Cook, "Reported prevalence and quantitative LC-MS methods for the analysis of veterinary drug residues in honey: A review," *Food Additives & Contaminants. Part A, Chemistry, Analysis, Control, Exposure & Risk Assessment*, vol. 31, no. 4, pp. 621–640, 2014.

[5] Y. Z. Chen and B. Hu, "The harm of veterinary drug residues in animal foods and its reason analysis," *Journal of Food and Biological Technology*, vol. 28, pp. 162–166, 2009.

[6] M. Y. Chen, T. Sun, and X. M. Wang, "Veterinary drug residue and its risks," *Progress in Animal Medicine*, vol. 26, pp. 111–113, 2005.

[7] Y.-J. Yu, H.-L. Wu, S.-Z. Shao et al., "Using second-order calibration method based on trilinear decomposition algorithms coupled with high performance liquid chromatography with diode array detector for determination of quinolones in honey samples," *Talanta*, vol. 85, no. 3, pp. 1549–1559, 2011.

[8] D. Terrado-Campos, K. Tayeb-Cherif, J. Peris-Vicente, S. Carda-Broch, and J. Esteve-Romero, "Determination of oxolinic acid, danofloxacin, ciprofloxacin, and enrofloxacin in porcine and bovine meat by micellar liquid chromatography with fluorescence detection," *Food Chemistry*, vol. 221, pp. 1277–1284, 2017.

[9] A. Di Corcia and M. Nazzari, "Liquid chromatographic-mass spectrometric methods for analyzing antibiotic and antibacterial agents in animal food products," *Journal of Chromatography A*, vol. 974, no. 1-2, pp. 53–89, 2002.

[10] W.-X. Zhu, J.-Z. Yang, Z.-X. Wang, C.-J. Wang, Y.-F. Liu, and L. Zhang, "Rapid determination of 88 veterinary drug residues in milk using automated TurborFlow online clean-up mode coupled to liquid chromatography-tandem mass spectrometry," *Talanta*, vol. 148, pp. 401–411, 2016.

[11] M. E. Dasenaki and N. S. Thomaidis, "Multi-residue determination of 115 veterinary drugs and pharmaceutical residues in milk powder, butter, fish tissue and eggs using liquid chromatography-tandem mass spectrometry," *Analytica Chimica Acta*, vol. 880, pp. 103–121, 2015.

[12] X. Xia, Y. Wang, X. Wang et al., "Validation of a method for simultaneous determination of nitroimidazoles, benzimidazoles and chloramphenicols in swine tissues by ultra-high performance liquid chromatography-tandem mass spectrometry," *Journal of Chromatography A*, vol. 1292, pp. 96–103, 2013.

[13] R. Galarini, G. Saluti, D. Giusepponi, R. Rossi, and S. Moretti, "Multiclass determination of 27 antibiotics in honey," *Food Control*, vol. 48, pp. 12–24, 2015.

[14] Á. Tölgyesi, V. K. Sharma, S. Fekete, J. Fekete, A. Simon, and S. Farkas, "Development of a rapid method for the determination and confirmation of nitroimidazoles in six matrices by fast liquid chromatography-tandem mass spectrometry," *Journal of Pharmaceutical and Biomedical Analysis*, vol. 64-65, pp. 40–48, 2012.

[15] A. Gadaj, V. Di Lullo, H. Cantwell, M. McCormack, A. Furey, and M. Danaher, "Determination of nitroimidazole residues in aquaculture tissue using ultra high performance liquid chromatography coupled to tandem mass spectrometry," *Journal of Chromatography B*, vol. 960, pp. 105–115, 2014.

[16] G. Stubbings and T. Bigwood, "The development and validation of a multiclass liquid chromatography tandem mass spectrometry (LC-MS/MS) procedure for the determination of veterinary drug residues in animal tissue using a quechers (Quick, Easy, Cheap, Effective, Rugged and Safe) approach," *Analytica Chimica Acta*, vol. 637, no. 1-2, pp. 68–78, 2009.

[17] K. Bousovaa, H. Senyuva, and K. Mittendorf, "Quantitative multi-residue method for determination antibiotics in chicken meat using turbulent flow chromatography coupled to liquid chromatography-tandem mass spectrometry," *Journal of Chromatography A*, vol. 1274, pp. 19–27, 2013.

[18] L. R. Guidi, F. A. Santos, A. C. S. R. Ribeiro, C. Fernandes, L. H. M. Silva, and M. B. A. Gloria, "A simple, fast and sensitive screening LC-ESI-MS/MS method for antibiotics in fish," *Talanta*, vol. 163, pp. 85–93, 2017.

[19] M. T. Martins, F. Barreto, R. B. Hoff et al., "Multiclass and multi-residue determination of antibiotics in bovine milk by liquid chromatography-tandem mass spectrometry: Combining efficiency of milk control and simplicity of routine analysis," *International Dairy Journal*, vol. 59, pp. 44–51, 2016.

[20] B. K. Matuszewski, M. L. Constanzer, and C. M. Chavez-Eng, "Strategies for the assessment of matrix effect in quantitative bioanalytical methods based on HPLC-MS/MS," *Analytical Chemistry*, vol. 75, no. 13, pp. 3019–3030, 2003.

[21] A. Posyniak, J. Zmudzki, and K. Mitrowska, "Dispersive solid-phase extraction for the determination of sulfonamides in chicken muscle by liquid chromatography," *Journal of Chromatography A*, vol. 1087, no. 1-2, pp. 259–264, 2005.

[22] G. Chen, P. Cao, and R. Liu, "A multi-residue method for fast determination of pesticides in tea by ultra performance liquid chromatography-electrospray tandem mass spectrometry combined with modified QuEChERS sample preparation procedure," *Food Chemistry*, vol. 125, no. 4, pp. 1406–1411, 2011.

[23] M. Anastassiades, S. J. Lehotay, D. Stajnbaher, and F. J. Schenck, "Fast and easy multiresidue method employing acetonitrile extraction/partitioning and "dispersive solid-phase extraction" for the determination of pesticide residues in produce," *Journal of AOAC International*, vol. 86, no. 2, pp. 412–431, 2003.

[24] K. H. Park, J.-H. Choi, A. M. Abd El-Aty et al., "Quantifying fenobucarb residue levels in beef muscles using liquid chromatography-tandem mass spectrometry and QuEChERS sample preparation," *Food Chemistry*, vol. 138, no. 4, pp. 2306–2311, 2013.

[25] M. Tenon, N. Feuillère, M. Roller, and S. Birtić, "Development and validation of a multiclass method for the quantification of veterinary drug residues in honey and royal jelly by liquid chromatography-tandem mass spectrometry," *Food Chemistry*, vol. 221, pp. 1298–1307, 2017.

[26] H. C. Liu, T. Lin, and X. L. Cheng, "Simultaneous determination of anabolic steroids and β-agonists in milk by QuEChERS and ultra high performance liquid chromatography tandem mass

spectrometry," *Journal of Chromatography B*, vol. 1043, pp. 176–186, 2017.

[27] V. Gresslera, A. R. L. Franzenb, G. J. M. M. de Lima, F. C. Tavernari, O. A. D. Costa, and V. Feddern, "Development of a readily applied method to quantify ractopamine residue in meat and bone meal by quechers-lc-ms/ms," *Journal of Chromatography B*, vol. 1015-1016, pp. 192–200, 2016.

[28] H.-W. Liao, G.-Y. Chen, I.-L. Tsai, and C.-H. Kuo, "Using a postcolumn-infused internal standard for correcting the matrix effects of urine specimens in liquid chromatography-electrospray ionization mass spectrometry," *Journal of Chromatography A*, vol. 1327, pp. 97–104, 2014.

[29] L. Q. Wang, L. M. He, Z. L. Zeng, and J. X. Chen, "Study of matrix effects for liquid chromatography tandem mass spectrometry analysis of veterinary drug residues," *Journal of Mass Spectrometry*, vol. 32, pp. 321–331, 2011.

11

Gas Chromatography Coupled to High Resolution Time-of-Flight Mass Spectrometry as a High-Throughput Tool for Characterizing Geochemical Biomarkers in Sediments

Hector Henrique Ferreira Koolen (iD),[1,2] **Clécio Fernando Klitzke,**[3] **Joe Binkley,**[3] **Jeffrey Patrick,**[3] **Ana Cecília Rizatti de Albergaria-Barbosa,**[4,5] **Rolf Roland Weber,**[5] **Márcia Caruso Bícego,**[5] **Marcos Nogueira Eberlin,**[1] and **Giovana Anceski Bataglion**[1,6]

[1]*ThoMSon Mass Spectrometry Laboratory, Institute of Chemistry, University of Campinas (UNICAMP), 13083-970, Campinas, SP, Brazil*
[2]*Metabolomics and Mass Spectrometry Research Group, Amazonas State University (UEA), 69065-001, Manaus, AM, Brazil*
[3]*LECO Corporation, 49085, St. Joseph, MI, USA*
[4]*Laboratory of Marine Geochemistry, Geoscience Institute, Federal University of Bahia (UFBA), 40170-020, Salvador, BA, Brazil*
[5]*Marine Organic Chemistry Laboratory, Oceanographic Institute, University of São Paulo (USP), 05508-120, São Paulo, SP, Brazil*
[6]*Department of Chemistry, Federal University of Amazonas (UFAM), 69077-000, Manaus, AM, Brazil*

Correspondence should be addressed to Hector Henrique Ferreira Koolen; hectorkoolen@gmail.com

Academic Editor: Eladia M. Pena-Mendez

The performance of gas chromatography coupled to high-resolution time-of-flight mass spectrometry (GC-HRTofMS) for characterizing geochemical biomarkers from sediment samples was evaluated. Two approaches to obtain the geochemical biomarkers were tested: (1) extraction with organic solvent and subsequent derivatization and (2) in-situ derivatization thermal desorption. Results demonstrated that both approaches can be conveniently applied for simultaneous characterization of many geochemical biomarkers (alkanes, alkanols, sterols, and fatty acids), avoiding conventional time-consuming purification procedures. GC-HRTofMS reduces both sample preparation time and the number of chromatographic runs compared to traditional methodologies used in organic geochemistry. Particularly, the approach based on in-situ derivatization thermal desorption represents a very simple method that can be performed in-line employing few milligrams of sediment, eliminating the need for any sample preparation and solvent use. The high resolving power ($m/\Delta m_{50\%}$ 25,000) and high mass accuracy (error ≤ 1 ppm) offered by the "zig-zag" time-of-flight analyzer were indispensable to resolve the complexity of the total ion chromatograms, representing a high-throughput tool. Extracted ion chromatograms using exact m/z were useful to eliminate many isobaric interferences and to increase significantly the signal to noise ratio. Characteristic fragment ions allowed the identification of homologous series, such as alkanes, alkanols, fatty acids, and sterols. Polycyclic aromatic hydrocarbons were also identified in the samples by their molecular ions. The characterization of geochemical biomarkers along a sedimentary core collected in the area of Valo Grande Channel (Cananéia-Iguape Estuarine-Lagunar System (São Paulo, Brazil)) provided evidences of environmental changes. Sediments deposited before opening of channel showed dominance of biomarkers from mangrove vegetation, whereas sediments of the pos-opening period showed an increase of biomarkers from aquatic macrophyte (an invasive vegetation).

1. Introduction

Organic matter (OM) is ubiquitously present in all natural waters and sediments, playing a central role in many environmental processes such as global carbon and nutrient cycles [1]. It can encompass many classes of compounds, such as alkanes, alkanols, sterols, and fatty acids, which are widely considered as geochemical biomarkers [2–4].

The organic geochemistry field comprehends diversified studies about OM in sediments; thus the biomarkers of interest may vary depending on the focus. Compounds classes, such as *n*-alkanes, alkanols, sterols, and fatty acids, are valuable to identify changes in the relative contributions of autochthonous and allochthonous OM sources to the sedimentary record over historical time-scale [5, 6]. They have also been used to assess changes in biogeochemical processes related to aquatic productivity [7], diagenetic alterations [8], and climate changes [9]. Hydrocarbons, such as *n*-alkanes, hopanes, and steranes, are also important for investigating contamination by crude oil and its derivatives in recent sediments [10, 11]. Sterols are of special interest to evaluate the contamination levels from domestic sewage discharge [12,13]. Additionally, several studies have employed a comprehensive approach based on the assessing of diverse classes of geochemical biomarkers to increase the reliability of the interpretations [14, 15].

A classic methodology based on the traditional gas chromatography coupled to mass spectrometry (GC-MS) technique is widely employed for characterizing geochemical biomarkers in sediment samples. Briefly, an extraction step is followed by purification of the raw organic extract on silica/alumina columns to yield fractions with fewer compounds [11, 14, 15]. The need of a purification step is due to the chemical complexity of organic matter that would complicate the separation from raw extract by the traditional GC-MS technique, resulting in unresolved complex mixture (UCM) [16]. This term refers to a raised "baseline hump" in chromatograms composed by a mixture of compounds unresolved by capillary columns [17]. The problem related to UCM regions can be partly resolved by the previous purification step, although this procedure requires large solvent quantity and is time-consuming.

High-resolution mass spectrometry (HRMS) coupled to GC is rarely used in environmental analysis, even though its benefits are well known [18, 19]. For instance, the advantages of GC-HRMS for qualitative and quantitative environmental analyses have been demonstrated for persistent organic compounds in sediments [18, 20]. Such studies have reported the elimination of matrix problems using accurate mass measurements [21, 22]. In this sense, gas chromatography coupled to high-resolution time-of-flight mass spectrometry (GC-HRTofMS) seems attractive to simultaneously characterize diverse classes of geochemical biomarkers from the raw extract without any time-consuming purification step on alumina/silica columns. There is also a possibility of chemical characterization directly from sediment samples using thermal desorption interfaced to GC-HRTofMS. That technique is based on the thermal desorption of compounds from a small amount (mg) of sediment, and then they are transferred to the GC inlet by a carrier gas [23, 24]. A similar procedure is based on pyrolysis, in which occurs thermal degradation of macromolecular OM, such as biopolymer, generating low-molecular weight products [25]. Thermal desorption, however, allows the characterization of volatile and semivolatile compounds in their intact form using lower temperature than those commonly used in pyrolysis [23, 24]. Despite the advantages of both

GC-HRTofMS and thermal desorption, these approaches are rarely used together in organic geochemistry of recent sediments.

The main objective of the present study was to investigate the performance of GC-HRTofMS in characterizing simultaneously diverse classes of geochemical biomarkers. For that, we tested and compared two approaches to obtain geochemical biomarkers: (1) extraction with organic solvent and subsequent derivatization and (2) in-situ derivatization thermal desorption. Then, as a field-testing application, geochemical biomarkers were characterized along a sedimentary core from the Cananéia-Iguape estuarine-lagoonal system (Brazil) to investigate changes in OM sources through the depositional period. The study area is in the World Heritage List under natural criteria of the United Nations Educational Scientific and Cultural Organization (UNESCO) because of its extensive areas of mangrove and Atlantic forest. However, an opening of an artificial channel (Valo Grande Channel) may had disturbed organic matter cycle.

2. Experimental

2.1. Study Area and Samples. A sedimentary core (198 cm) was collected in the Cananéia-Iguape Estuarine-Lagunar System (São Paulo, Brazil), more specifically in front of the Valo Grande Channel area, at latitude and longitude of $24°45'2.45''$ S and $47°37'3.70''$ W, respectively, with the aid of a Rossfelder VT-1 vibracorer. The study area and sampling station are shown in Figure S1 in Supplementary Material. The sedimentary core was sampled continuously at intervals of 2.0 cm immediately after collection. The samples were frozen at -18°C and freeze-dried at -50°C and pressure of 270 mbar with a Thermo Savant Modulyo D freeze dryer (Thermo Electron Corporation, Madison, USA). That environmental system is located in the southeastern Brazilian coast and consists of a complex of lagoonal channels with a broad and relatively well-preserved Atlantic forest and mangrove vegetation. The Cananéia-Iguape Estuarine-Lagunar System is considered by the UNESCO as a Biosphere Reserve. However, environmental changes have occurred in that system in the last two centuries because of the artificial opening of a channel (Valo Grande Channel) connecting the Ribeira de Iguape River to the lagoonal system. The original dimensions (4.4 m wide and 2.0 m deep) of the channel rapidly increased and currently are about 250 m of wide and 7 m of deep [26]. The average percentage of clay, silt, and sand of the sediment samples were 5.4, 43.4, and 51.1, respectively, whereas average percentage of total organic carbon (TOC) was 1.92%. The content of fine particles (silt and clay) as well as TOC presented higher values in the first 138 cm of the sedimentary core.

2.2. Extraction with Organic Solvent and Subsequent Derivatization. The freeze-dried sediment samples were macerated, and 20 g was extracted using the microwave system MarsX (CEM Corporation, Mathews, USA) operating at 1600 W. The extractions were performed with 50 mL of *n*-hexane and dichloromethane (1:1, v/v) with pressure and temperature gradients reaching 200 psi and 85°C in 5 min. The system was kept isothermal for 15 min, returning to the initial pressure and temperature conditions. The organic extracts were evaporated to 1.0 mL, and then an aliquot of 50.0 μL was derivatized using bis(trimethylsilyl) trifluoroacetamide (BSTFA) with 1% trimethylchlorosilane (TMCS) for 120 min at 70°C. After reaction, the residual reagents were evaporated to dryness with a gentle flow of nitrogen stream and the extract was redissolved to exactly 200 μL of n-hexane. All solvents used were of HPLC grade, purchased from Mallinckrodt Chemicals, and used as received.

2.3. In-Situ Derivatization Thermal Desorption. An amount of 20 mg of the homogenized and freeze-dried sediment sample was added into a small quartz tube, which is heated up with a pyroprobe using a platinum filament coil. Quartz wool was properly packed at the top and bottom to avoid possible leakage, or the sample being blown out. An aliquot of 10 μL of a 25% methanolic tetramethylammonium hydroxide (TMAH) solution was added to the sample inside the quartz tube to provide deeper insights in alkanol, sterol, and fatty acid compositions. Using TMAH, hydroxyl groups of carboxylic acids and alcohols were in-situ derivatized in the same step of thermal desorption and converted to the corresponding methyl esters and methyl ethers, respectively. The compounds present in the sediment samples were thermally desorbed at 400°C during 20 s, using a CDS Analytical Pyroprobe® 5200 (Chemical Data Systems, Oxford, USA) coupled to a Leco Pegasus® GC-HRT MS. Pyroprobe interface, valve oven and transfer line were at 300°C. Helium was used as the carrier gas (1.0 mL min^{-1}) to transfer the analytes during the course of thermal desorption directly to the GC inlet during 4 min at 300°C.

2.4. GC-HRTofMS Analysis. The equipment used, Leco Pegasus® GC-HRT MS, consists of an Agilent 7890 GC (Agilent Technologies, Palo Alto, USA) and a "zig-zag" TOF analyzer (LECO Corporation, St. Joseph, USA). A Restek Rxi-5MS column (30 m, 0.25 mm I.D., 0.25 μm film thickness) was used at initial temperature of 50°C for 2 min, raised to 320°C with the rate 6°C min^{-1}, and maintained for 20 min. A split ratio of 50:1 was used to inject liquid samples and splitless mode was used for the thermal desorption experiments. Helium was used as the carrier gas with constant flow rate of 1.0 mL min^{-1}. GC inlet and transfer line to the ion source were at 300 and 320°C, respectively. Electron ionization (EI) source was at 70 eV, 3 mA, and 250°C. Data were recorded in full scan mode from 50-550 m/z with resolving power ($m/\Delta m_{50\%}$) > 25,000 and mass error < 1.0 ppm using acquisition rate of 6 spectra s^{-1} and extraction frequency of 1.75 kHz. Data were acquired and processed using Leco ChromaTOF HRT software (LECO Corporation, St. Joseph, USA) and chromatographic peaks were identified based on spectra from NIST library, retention time of standards, and elution order.

3. Results and Discussion

A representative total ion chromatogram (TIC) obtained by using the approach based on in-situ derivatization thermal desorption is shown in Figure 1.

Note the difficulty in resolving the entire wide range of compounds present in sediment samples. Although the complexity of the chromatogram, chemical characterization of important geochemical biomarkers can be achieved because of the high resolution offered by the "zig-zag" TOF analyzer. The extracted ion chromatograms (XIC) for characteristic exact m/z and the elution patterns allowed the identification of aliphatic compounds classes, such as n-alkanes, n-alkanols, and n-fatty acids, as shown in Figure 1.

The class of n-alkanes was identified by extracting chromatograms for the m/z 57.06983 (exact mass: 57.06988, error: -0.87 ppm), which corresponds to the common fragment $CH_3CH_2CH_2CH_2^+$. Note that a series of n-alkanes obtained via thermal desorption approach is not present as doublets of n-alk-1-ene/n-alkane. Such doublets are formed through the homolytic cleavage of long alkyl chains of aliphatic biopolymers upon pyrolysis [27]. The absence of n-alk-1-ene/n-alkane doublets shows that thermal desorption is able to characterize the aliphatic composition of sedimentary OM without degrading biopolymers, which would complicate interpretation of results. The TMAH reagent converts fatty acids and alkanols in their methyl ester and methyl ether derivatives, respectively, which are more easily analyzed by GC. The classes of n-alkanols and n-fatty acids were identified by extracting chromatograms for the m/z 97.10110 (exact mass: 97.10118, error: -0.82 ppm) and m/z 74.03617 (exact mass: 74.03623, error: -0.81 ppm), respectively. The ion at m/z 97.10118 is due to the fragment $CH_3(CH_2)_4CHCH^+$, which allows extracting only n-alkanols without interference of n-fatty acids that are coeluted. The ion at m/z 74.03623 is due to the fragment $CH_3OCOHCH_2^+$ that was generated by McLafferty rearrangement ion for methylated fatty acids.

For the approach using extraction with organic solvent and subsequent derivatization, a representative TIC is shown in Figure 2. The class of n-alkanes was also identified by extracting chromatograms for the m/z 57.06983 (exact mass: 57.06988, error: -0.87 ppm). Considering that the conventional reagent BSTFA was used for derivatization, the classes of n-alkanols and n-fatty acids were identified by extracting chromatograms for the m/z 103.05729 (exact mass: 103.05737, error: -0.77 ppm) and m/z 117.03663 (exact mass: 117.03653, error: -0.85 ppm), respectively. These ions correspond to the fragments $(CH_3)_3SiOCH_2^+$ and $(CH_3)_3SiOCO^+$ generated by α-cleavage from the molecular ions of n-alkanols and n-fatty acids, respectively.

Usually, chromatograms for n-alkanes, n-alkanols, and n-fatty acids classes are obtained by individual chromatographic runs of purified fractions from raw extract. In this study, the results obtained for both approaches show that GC-HRTofMS allows characterizing simultaneously n-alkanes, n-alkanols, and n-fatty acids, reducing both sample preparation time and number of chromatographic runs, which is a novel aspect in the organic geochemistry field. Particularly, the approach based on in-situ derivatization thermal desorption

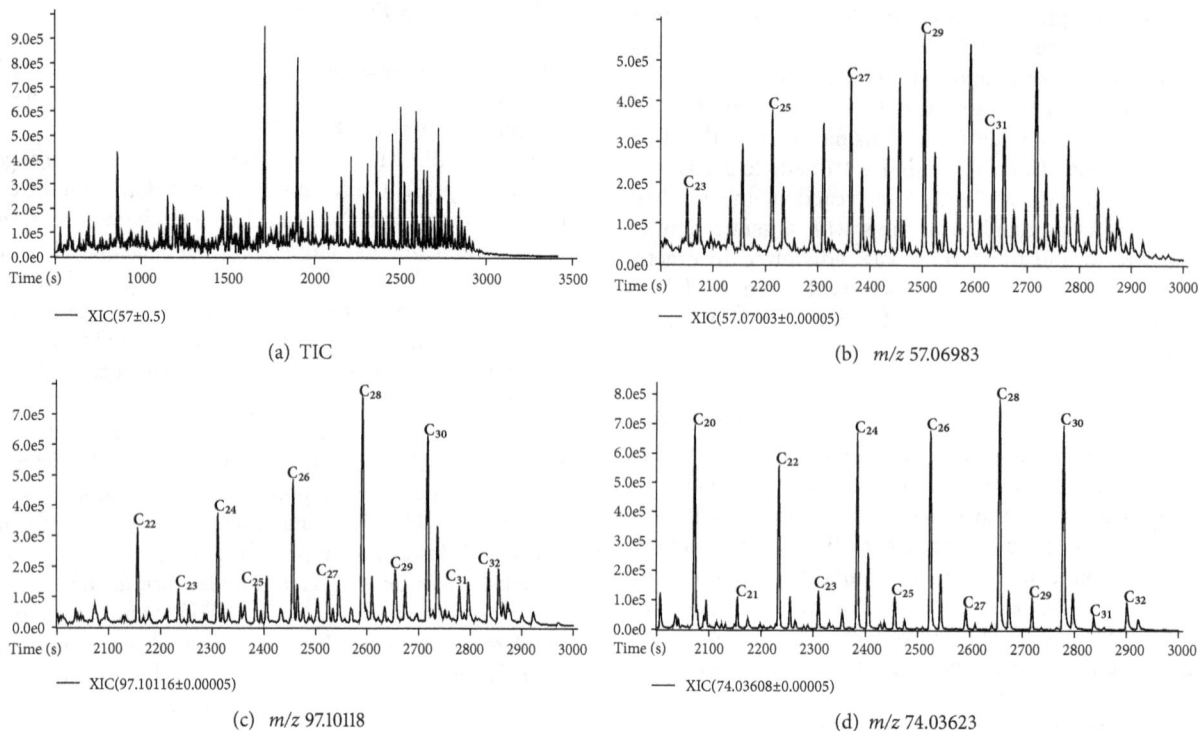

FIGURE 1: (a) TIC obtained by using the approach based on in-situ derivatization thermal desorption for a sediment sample; (b) expansion (2000-3000 s) of XIC for m/z 57.06983 relative to the series of n-alkanes, (c) expansion (2000-3000 s) of XIC for m/z 97.10118 relative to the series n-alkanols, and (d) expansion (2000-3000 s) of XIC for m/z 74.03623 relative to the series of n-fatty acids. n in Cn is the C number.

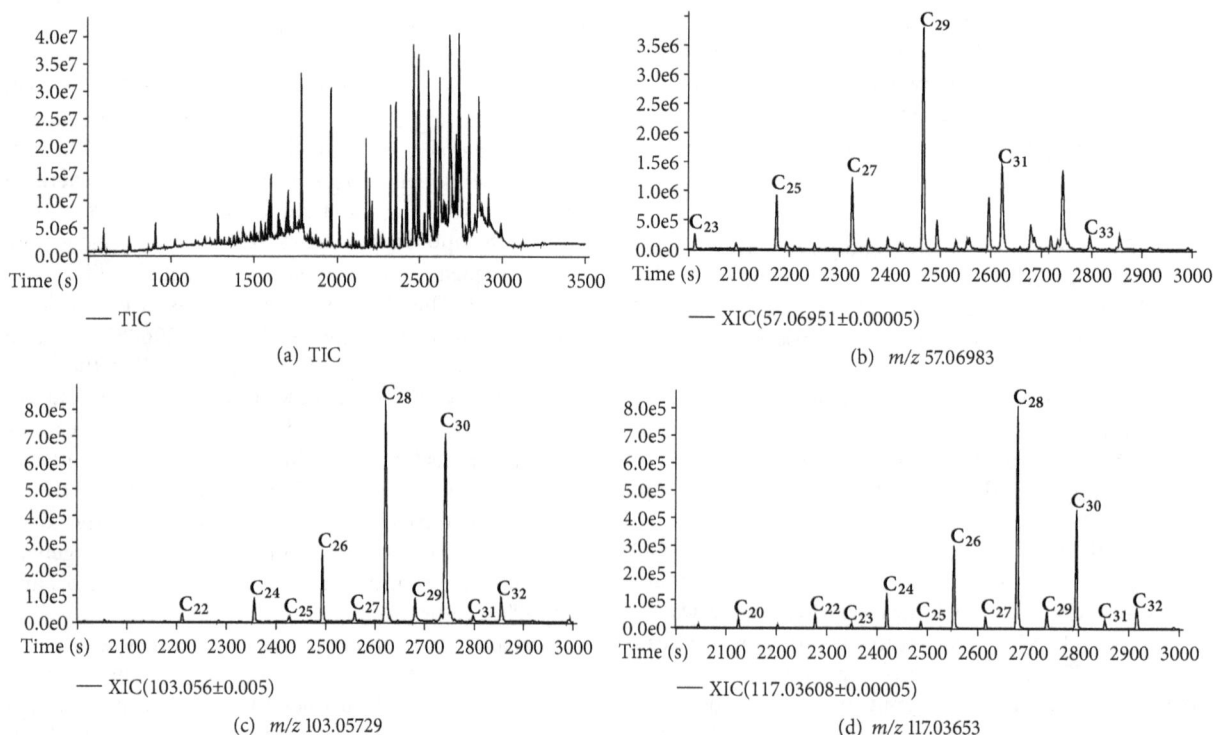

FIGURE 2: (a) TIC obtained by using the approach based on extraction with organic solvent and subsequent derivatization for a sediment sample; (b) expansion (2000-3000 s) of XIC for m/z 57.06983 relative to the series of n-alkanes, (c) expansion (2000-3000 s) of XIC for m/z 103.05729 relative to the series n-alkanols, and (d) expansion (2000-3000 s) of XIC for m/z 117.03653 relative to the series of n-fatty acids. n in Cn is the C number.

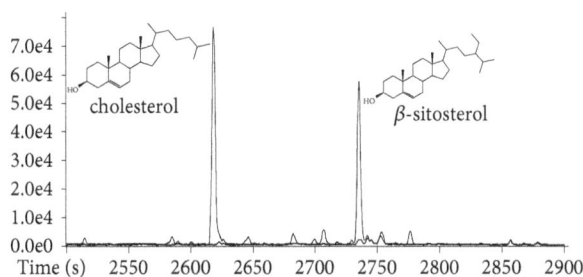

FIGURE 3: Expansion (2500-2900 s) of XIC for m/z 368.34354 and m/z 396.37481, which corresponds to cholesterol (molecular mass: 386.35432 u) and β-sitosterol (molecular mass: 414.38562 u), respectively. Mass spectra of cholesterol and β-sitosterol are shown in Figure S2 in Supplementary Material.

allows characterizing directly the free lipids fraction from milligrams of sediment, requiring no organic solvents and leading to elimination of all issues related to their use and disposal, which is also a novel aspect in the organic geochemistry field. However, this approach may not be adequate for the analysis of sediments with very low content of organic carbon; thus the approach based on extraction with organic solvent and subsequent derivatization may be an option in these cases. We highlight that the high resolving power ($m/\Delta m_{50\%}$ 25,000) and high mass accuracy (error ≤ 1 ppm) offered by the "zig-zag" time-of-flight analyzer were indispensable to resolve the complexity of TIC. By using GC-HRTofMS, many isobaric interferences were eliminated and the signal to noise ratios (S/N) were significantly increased in the XIC from exact m/z in comparison to those from unitary m/z for all geochemical biomarker classes identified in both approaches. It is also noteworthy to mention that, for a same sediment sample, results found by in-situ derivatization thermal desorption are equivalent to those obtained by extraction with organic solvent and subsequent derivatization. Therefore, thermal degradation used in this study can desorb free lipids while compounds bonded to insoluble polymeric material remain preserved in the sediment.

Sterols are also geochemical biomarkers commonly used in the assessment of OM sources in aquatic environments and can be detected by extracting chromatograms for some key fragment ions, as shown for cholesterol and β-sitosterol in Figure 3.

For the approach based on in-situ derivatization thermal desorption, sterols are identified by extracting fragment ions correspondent to water loss from their intact molecules or methanol from derivatized ones, as shown in Figure 3 for cholesterol (exact mass: 368.34375, error: -0.57 ppm) and for β-sitosterol (exact mass: 396.37505, error: -0.60 ppm). For the approach based on the extraction with organic solvent and subsequent derivatization using the conventional reagent BSTFA, those ions correspond to the fragments formed after the loss of the group $(CH_3)_3SiOH$. Beyond cholesterol and β-sitosterol, cholestanol, brassicasterol, campesterol, stigmasterol, and stigmastanol were also identified based on exact m/z of such fragment ions and elution order. Triterpenols,

such as lupeol, α-amyrin, and β-amyrin, presented fragmentation behavior similar to sterols.

Polycyclic aromatic hydrocarbons (PAH) are another class of compounds that may be present in sediments, especially those from industrial areas. They are emitted mainly during incomplete burning of coal, oil, gas, coke, wood, garbage, or other organic materials. These sources are called pyrogenic, whereas crude oil, refined petroleum, coal, coal tar, pitch, and asphalts represent petrogenic sources. Natural sources, including volcanism and diagenetic processes, are also possible [3, 28, 29]. Several PAH can be identified by extracting chromatograms for their molecular ions, such as m/z 142.07760 (methyl naphthalene, exact mass: 142.07770, error: -0.70 ppm), m/z 178.07759 (phenanthrene/anthracene, exact mass: 178.07770, error: -0.62 ppm), m/z 202.07757 (fluoranthene/pyrene, exact mass: 202.07770, error: -0.64 ppm), and m/z 252.09311 (benzo[a]pyrene, mass exact: 252.09335, error: -0.95 ppm), as shown in Figure 4. Also, the effectiveness of high resolving power and mass accuracy offered by the "zig-zag" time-of-flight analyzer for PAH is shown in Figure 5 with the comparison between ion chromatograms extracted in high and low resolution as measured by typical quadrupole analyzer.

Figure 5 shows that the high resolving power and mass accuracy were also important to eliminate many isobaric interferences and increase S/N for PAH.

As a proof of concept of the ability of GC-HRTofMS as a high-throughput tool in organic geochemistry, OM characterization was done along a sedimentary core to assess the historical record in the area of the Valo Grande Channel, artificially created in the Cananéia-Iguape Estuarine-Lagunar System in 1852 [26]. n-Alkanes, n-alkanols, and n-fatty acids are straight chain hydrocarbons produced by many organisms and the dominant chain lengths and carbon number distributions vary depending on the source organism [25]. The long-chain n-alkanes (C_{27} to C_{31}) and n-alkanols and n-fatty acids (C_{26} to C_{32}) are the main components of epicuticular waxes of higher plants, whereas algae are dominated by shorter-chain n-alkanes (C_{17} to C_{21}) and n-alkanols and n-fatty acids (C_{14} to C_{18}). The mid-chain n-alkanes (C_{21} to C_{25}) and n-alkanols and n-fatty acids (C_{20} to C_{24}) are dominant in aquatic plants, such as macrophytes [30, 31]. Profiles of n-alkanes, n-alkanols, and n-fatty acids for samples from the sedimentary core were obtained by extracting chromatograms for the m/z 57.06983, 103.05729, and 117.03663, respectively, as shown is Figure 6.

Figure 6 shows that changes in the sources of OM delivered to the sedimentary core of the Valo Grande Channel can be detected by the distribution of n-alkanes, n-alkanols, and n-fatty acids. n-Alkanes series for recent sediments usually exhibits strong odd over even carbon number predominance, whereas n-alkanols and n-fatty acids exhibit strong even over odd carbon number predominance [30]. The n-alkanes series in the sedimentary core showed unimodal distribution ranging from C_{19} to C_{33}, with strong odd over even carbon number predominance (Figure 6). The n-alkanols and n-fatty acids series ranged from C_{14} to C_{34} and C_{14} to C_{32}, respectively, with strong even over odd carbon number predominance (Figure 6). Such distribution profiles are typical of

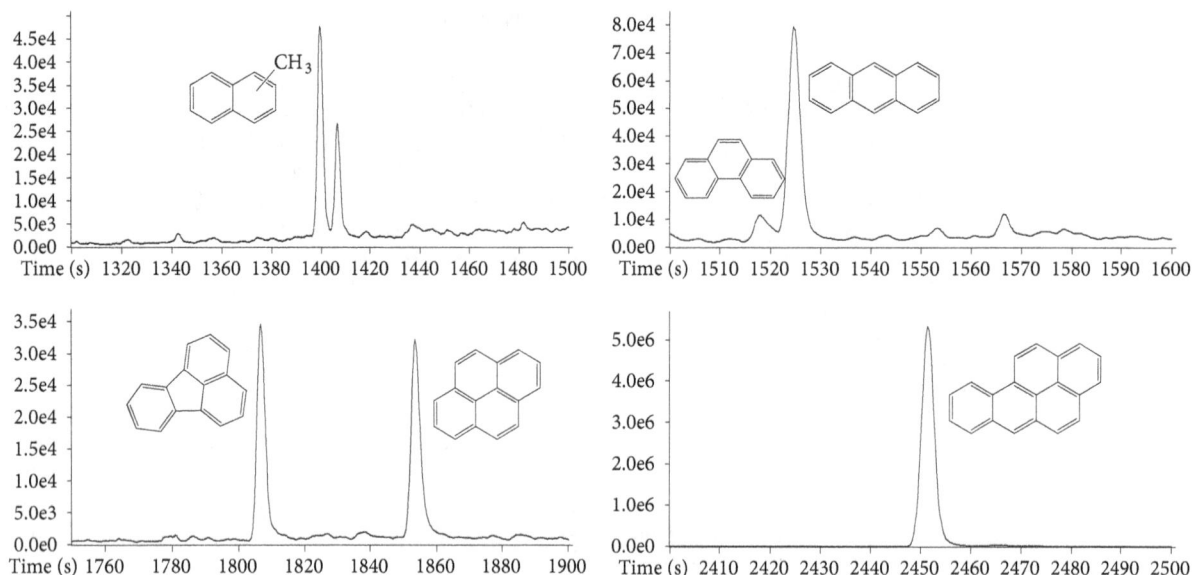

FIGURE 4: Expansion of XIC for *m/z* 142.07760 (methyl naphthalene), *m/z* 178.07759 (phenanthrene/anthracene), *m/z* 202.07757 (fluoranthene/pyrene), and *m/z* 252.09335 (benzo[a]pyrene).

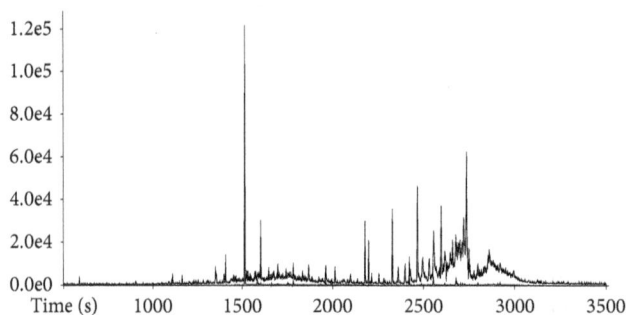

FIGURE 5: XIC for C_2 naphthalene in low resolution (*m/z* 156, dark chromatogram) and high resolution (*m/z* 156.0559, red chromatogram).

areas with predominance of natural sources of organic matter [30], indicating that anthropogenic sources are negligible in the studied area.

The *n*-alkanes series in the sediments correspondent to the pre-opening channel period showed predominance of C_{29} (Figure 6), with similar distribution to that observed for the species Laguncularia racemosa (white mangle) and Rhizophora mangle (red mangle) [32]. In fact, these mangle species are the most abundant in the Cananéia-Iguape Estuarine-Lagunar System that present extensive areas of mangrove vegetation [13]. The *n*-alkanes distribution observed for the sediments correspondent to the pos-opening period, however, presented higher relative contribution of the mid-chain *n*-alkanes, C_{21}, C_{23}, and C_{25}, and also similar contribution of C_{27}, C_{29}, and C_{31} (Figure 6). The *n*-alkanes distribution observed for sediments of the pos-opening period was similar to that observed for leaves of the macrophyte Spartina alterniflora and for sediments subject to that

vegetation [32]. Other submerged and floating macrophytes also present profiles of *n*-alkanes with similar contribution of the homologous from C_{23} to C_{29} [31, 33, 34]. An increase of aquatic macrophyte has been observed in the Cananéia-Iguape Estuarine-Lagunar System lately and associated with the opening of the Valo Grande Channel [13].

The *n*-alkanols and *n*-fatty acids series also showed changes in their distribution profiles between the two deposition periods. The *n*-alkanols and *n*-fatty acids series in the sediments correspondent to the pre-opening channel period showed predominance of C_{26}, C_{28} and C_{30} with a Gaussian-like distribution (Figure 6). Similar profile for *n*-alkanols was observed for sediments from the Pichavaram mangrove complex, located between two estuaries in India [35]. Similar profiles of both *n*-alkanols and *n*-fatty acids were also observed for sediments from estuaries where the mangrove vegetation, Rhizophora mangle (red mangle) and Avicennia germinans (black mangle), is dominant [36]. In

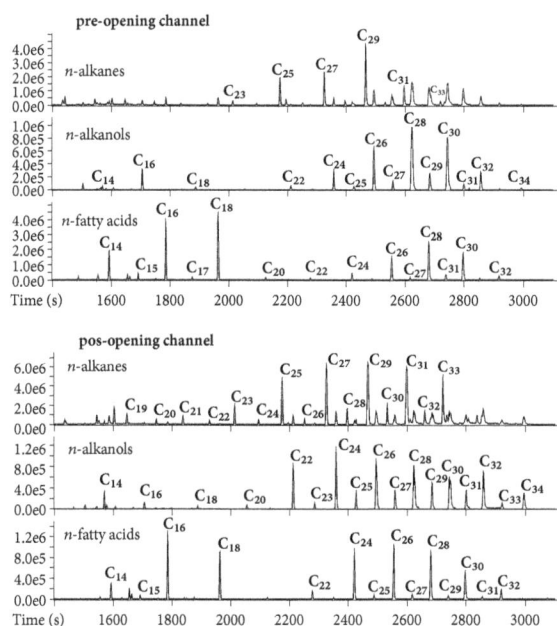

FIGURE 6: Expansion (1500–3000 s) of XIC for *n*-alkanes, *n*-alkanols, and *n*-fatty acids for representative samples of both pre-opening and pos-opening periods of the Valo Grande Channel in the Cananéia-Iguape Estuarine-Lagunar System. *n* in C*n* is the carbon number. *n*-alkanes, *n*-alkanols, and *n*-fatty acids series were obtained by extracting chromatograms for the *m/z* 57.06983, 103.05729, and 117.03663, respectively.

contrast, in the sediments correspondent to the pos-opening channel period, the profiles of *n*-alkanols and *n*-fatty acids showed higher relative contribution of the mid-chain homologous, C_{22}, C_{24}, and C_{26}. Profiles of distribution of both *n*-alkanols and *n*-fatty acids presenting high contribution of these mid-chain homologous have been observed for a variety of macrophytes aquatic [31].

The geochemical biomarker variations as function of depth can be better visualized by choosing a homologous of each class to represent terrestrial vegetation input and another to represent aquatic vegetation input, as shown in Figure 7.

Results presented in Figures 6 and 7 indicate that the reduction of the long-chain homologous dominant in higher plants is accompanied by the increase of the mid-chain ones, which are dominant in floating and submerged aquatic plants. Thus, the results reveal that the sediments deposited in the pre-opening channel period are characterized by the strong predominance of biomarkers of higher plants (mangrove vegetation), whereas the sediments deposited in the pos-opening channel period present an increase of biomarkers of aquatic macrophytes. In addition, Figure 7 indicates that the more abrupt changes in the relative abundance of biomarkers of different types of vegetation occur around 110-112 cm.

A previous study concerned to anthropogenic metal input in this estuarine-lagoonal system observed an abrupt increase of the concentration of Cr, Cu, Zn, and Sc in the same depth [26]. The authors also determined the sedimentation rates

via ^{210}Pb gamma spectrometry. Based on this analysis, the sedimentary core covers the period between 1723±15 and 2008. The sediment portion 110-112 cm, which corresponds to approximately 35±8 years after the conclusion of construction of the Valo Grande Channel (1827-1852), readily evidences the increase of biomarkers from aquatic vegetation. The channel construction resulted in an increase of freshwater flow from the Ribeira de Iguape River to the estuarine-lagoonal system, creating proper conditions for the reproduction of aquatic macrophytes [35]. Thus, the first period represents an environment subject to OM input from terrestrial plants, Atlantic Forest and mangrove vegetation. In a second period, a similar contribution of OM from terrestrial and aquatic vegetation is due to the invasive species of macrophytes. Equivalent discussion can be raised by the variations in the relative abundances of some sterols and β-amyrin as function of depth along the sedimentary core collected in Valo Grande Channel, as shown in Figure 8.

The compounds stigmasterol, β-sitosterol, and stigmastanol showed higher relative contribution in the pos-opening channel period, whereas the triterpenol β-amyrin was strongly dominant in the pre-opening channel period. Phytosterols, such as stigmasterol and β-sitosterol, are produced by plants in general, including aquatic macrophytes [14]. The triterpenol β-amyrin, however, is considered a highly specific biomarker for higher plants and its abundant presence in sediments confirms the widespread input of terrestrial OM [4, 14]. High abundance of β-amyrin was observed for sediments from estuaries dominated by mangrove vegetation [36]. The change in the relative contribution of such sterols and β-amyrin corroborates those observed for the classes of *n*-alkanes, *n*-alkanols, and *n*-fatty acids, indicating that MO input from aquatic vegetation has increased significantly after opening of the Valo Grande Channel.

Alkylated homologous PAH, such as methylnaphthalene (C_1), dimethylnaphthalene (C_2), and trimethylnaphthalene (C_3), phenanthrene, anthracene, fluoranthene, pyrene, and benzo[a]pyrene were detected in all samples along the sedimentary core. These compounds have been classified as priority pollutants by the United States Environmental Protection Agency (USEPA) because of their toxic, carcinogenic, and mutagenic characteristics [30]. Source appointment of PAH in aquatic environments is difficult because of the multiples possible sources. Petrogenic PAH are predominantly of low molecular weight and alkylated, whereas pyrolitic HPA are dominated by the high molecular weight compounds [37]. The sediment samples deposited in the pre-opening channel period presented a mixture of low and high molecular weight and alkylated PAH with similar proportion, whereas those deposited in the pos-opening period presented an abrupt increase of the contribution of benzo[a]pyrene that is a five-member PAH well-known by the International Agency for Research on Cancer (IARC) as carcinogenic to human beings (group 1) [38]. Biomass burning seems to contribute with the most part of the benzo[a]pyrene present in the environment [39]. The increase of this specific PAH in the pos-opening channel period may be associated with the higher input of terrestrial OM including some from biomass burning from

(a)

(b)

(c)

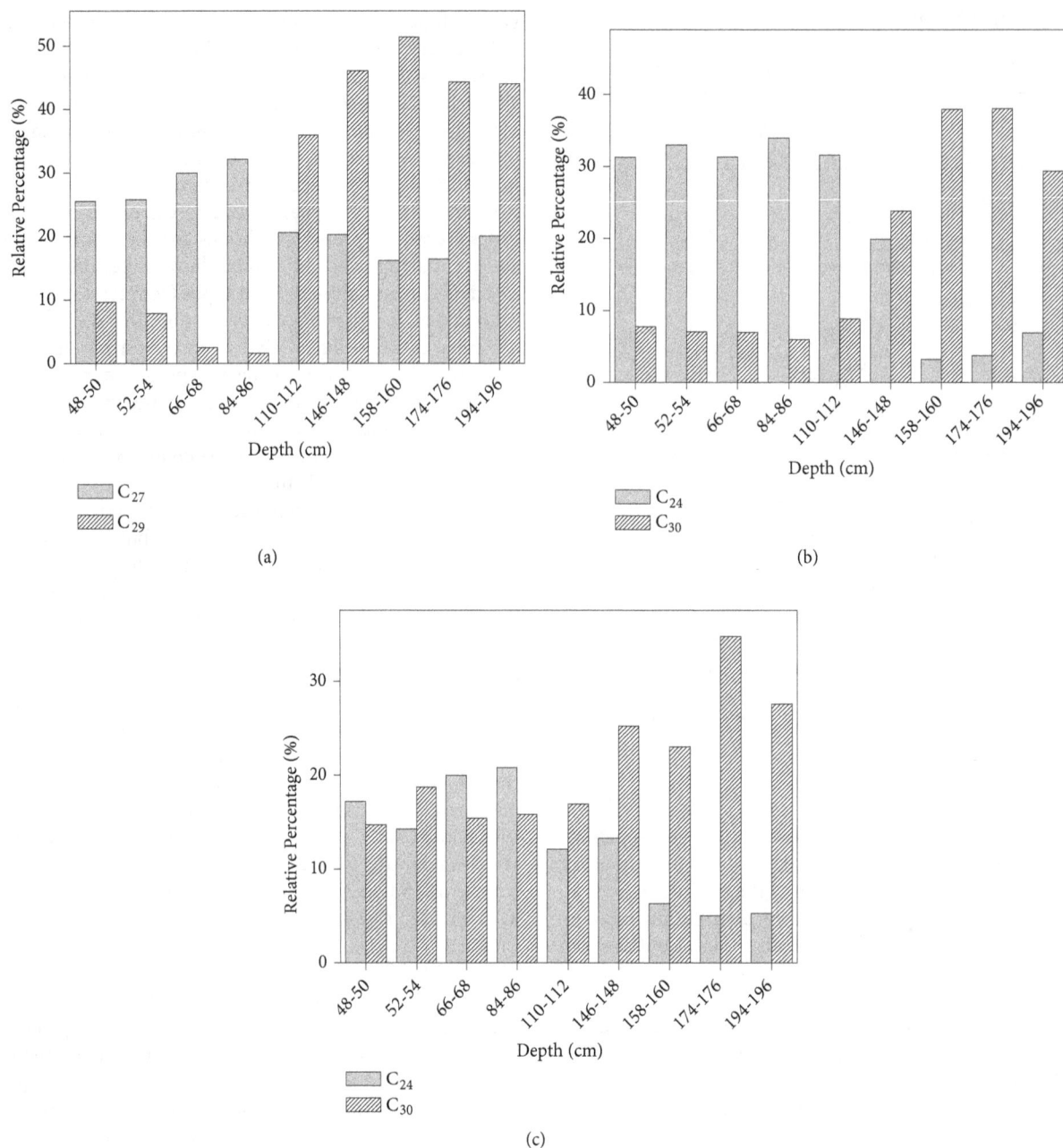

FIGURE 7: Variations of the relative abundances of n-alkanes C_{27} and C_{29} (a), n-alkanols C_{24} and C_{30} (b), and n-fatty acids C_{24} and C_{30} (c), along the sedimentary core collected at Valo Grande Channel in Cananéia-Iguape Estuarine-Lagunar System. Correspondence of sediment depth and year: 48-50 cm (1976), 52-54 cm (1972), 66-68 cm (1956), 84-86 cm (1925), 110-112 cm (1887), 146-148 cm (1816), 158-160 cm (1796), 174-176 cm (1776), 194-196 cm (1725).

agricultural areas of the Vale do Ribeira region. However, quantitative analysis of PAH is essential for evaluating the presence of this class in the environment studied.

4. Conclusions

In comparison with the traditional GC-MS technique, the alternative GC-HRTofMS one allows characterizing simultaneously a variety of geochemical biomarkers with reduction in both sample preparation time and number of chromatographic runs. The approach based on extraction with organic solvent and subsequent derivatization as well as in-situ derivatization thermal desorption can be used for such finality and deliver equivalent results. In particular, thermal desorption method requires few milligrams of sediment and no organic solvents. The geochemical biomarkers distributions along the sedimentary core collected at Valo Grande Channel revealed environmental changes occurred

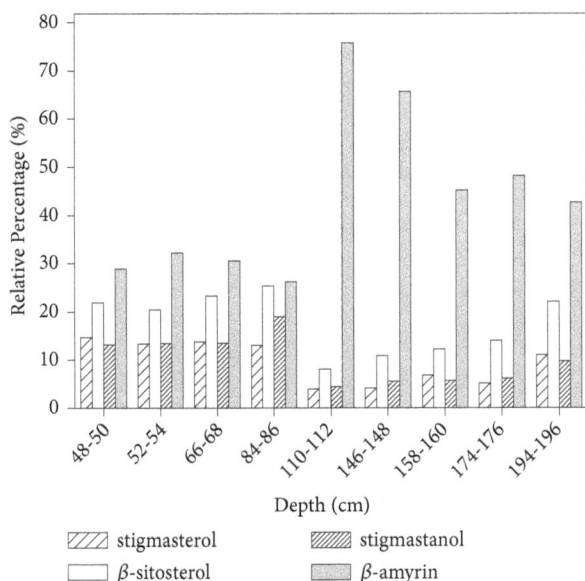

FIGURE 8: Variations in the relative abundances of some sterols and β-amyrin along the sedimentary core collected at Valo Grande Channel in Cananéia-Iguape Estuarine-Lagunar System. Correspondence of sediment depth and year: 48-50 cm (1976), 52-54 cm (1972), 66-68 cm (1956), 84-86 cm (1925), 110-112 cm (1887), 146-148 cm (1816), 158-160 cm (1796), 174-176 cm (1776), 194-196 cm (1725).

in the Cananéia-Iguape Estuarine-Lagunar System history.

Acknowledgments

The authors would like to thank FAPESP (scholarship 2012/21395-0) for sponsoring this study.

References

[1] P. Regnier, P. Friedlingstein, P. Ciais et al., "Anthropogenic perturbation of the carbon fluxes from land to ocean," *Nature Geoscience*, vol. 6, no. 8, pp. 597–607, 2013.

[2] M. A. de Abreu-Mota, C. A. de Moura Barboza, M. C. Bícego, and C. C. Martins, "Sedimentary biomarkers along a contamination gradient in a human-impacted sub-estuary in Southern Brazil: a multi-parameter approach based on spatial and seasonal variability," *Chemosphere*, vol. 103, pp. 156–163, 2014.

[3] L. Huang, S. M. Chernyak, and S. A. Batterman, "PAHs (polycyclic aromatic hydrocarbons), nitro-PAHs, and hopane and sterane biomarkers in sediments of southern Lake Michigan, USA," *Science of the Total Environment*, vol. 487, no. 1, pp. 173–186, 2014.

[4] S. S. O'Reilly, M. T. Szpak, P. V. Flanagan et al., "Biomarkers reveal the effects of hydrography on the sources and fate of marine and terrestrial organic matter in the western Irish Sea," *Estuarine, Coastal and Shelf Science*, vol. 136, pp. 157–171, 2014.

[5] D. J. Strong, R. Flecker, P. J. Valdes et al., "Organic matter dis-

[6] Y. Lu and P. A. Meyers, "Sediment lipid biomarkers as recorders of the contamination and cultural eutrophication of Lake Erie, 1909-2003," *Organic Geochemistry*, vol. 40, no. 8, pp. 912–921, 2009.

[7] B. G. Rodrigues Alves, A. Ziggiatti Güth, M. Caruso Bícego, S. Airton Gaeta, and P. Y. Gomes Sumida, "Benthic community structure and organic matter variation in response to oceanographic events on the Brazilian SE inner shelf," *Continental Shelf Research*, vol. 85, pp. 106–116, 2014.

[8] S. Christodoulou, J.-C. Marty, J.-C. Miquel, J. K. Volkman, and J.-F. Rontani, "Use of lipids and their degradation products as biomarkers for carbon cycling in the northwestern Mediterranean Sea," *Marine Chemistry*, vol. 113, no. 1-2, pp. 25–40, 2009.

[9] R. A. Andersson and P. A. Meyers, "Effect of climate change on delivery and degradation of lipid biomarkers in a Holocene peat sequence in the Eastern European Russian Arctic," *Organic Geochemistry*, vol. 53, no. 1, pp. 63–72, 2012.

[10] T. R. Silva, S. R. P. Lopes, G. Spörl, B. A. Knoppers, and D. A. Azevedo, "Evaluation of anthropogenic inputs of hydrocarbons in sediment cores from a tropical Brazilian estuarine system," *Microchemical Journal*, vol. 109, pp. 178–188, 2013.

[11] W. Prus, M. J. Fabiańska, and R. Łabno, "Geochemical markers of soil anthropogenic contaminants in polar scientific stations nearby (Antarctica, King George Island)," *Science of the Total Environment*, vol. 518-519, pp. 266–279, 2015.

[12] G. A. Bataglion, E. Meurer, A. C. R. de Albergaria-Barbosa, M. C. Bícego, R. R. Weber, and M. N. Eberlin, "Determination of geochemically important sterols and triterpenols in sediments using ultrahigh-performance liquid chromatography tandem mass spectrometry (UHPLC–MS/MS)," *Analytical Chemistry*, vol. 87, no. 15, pp. 7771–7778, 2015.

[13] G. A. Bataglion, H. H. F. Koolen, R. R. Weber, and M. N. Eberlin, "Quantification of sterol and triterpenol biomarkers in sediments of the Cananéia-Iguape estuarine-lagoonal system (Brazil) by UHPLC-MS/MS," *International Journal of Analytical Chemistry*, vol. 2016, Article ID 8361375, 8 pages, 2016.

[14] P. M. Medeiros and B. R. T. Simoneit, "Multi-biomarker characterization of sedimentary organic carbon in small rivers draining the Northwestern United States," *Organic Geochemistry*, vol. 39, no. 1, pp. 52–74, 2008.

[15] R. S. Carreira, M. P. Araújo, T. L. F. Costa, G. Spörl, and B. A. Knoppers, "Lipids in the sedimentary record as markers of the sources and deposition of organic matter in a tropical Brazilian estuarine-lagoon system," *Marine Chemistry*, vol. 127, no. 1–4, pp. 1–11, 2011.

[16] H. H. F. Koolen, R. F. Swarthout, R. K. Nelson et al., "Unprecedented insights into the chemical complexity of coal tar from comprehensive two-dimensional gas chromatography mass spectrometry and direct infusion fourier transform ion cyclotron resonance mass spectrometry," *Energy & Fuels*, vol. 29, no. 2, pp. 641–648, 2015.

[17] H. K. White, L. Xu, P. Hartmann, J. G. Quinn, and C. M. Reddy, "Unresolved complex mixture (UCM) in coastal environments is derived from fossil sources," *Environmental Science & Technology*, vol. 47, no. 2, pp. 726–731, 2013.

[18] A. T. Lebedev, O. V. Polyakova, D. M. Mazur, and V. B. Artaev, "The benefits of high resolution mass spectrometry in environmental analysis," *Analyst*, vol. 138, no. 22, pp. 6946–6953, 2013.

tribution in the modern sediments of the Pearl River Estuary," *Organic Geochemistry*, vol. 49, pp. 68–82, 2012.

[19] H. H. F. Koolen, A. F. Gomes, L. G. M. de Moura et al., "Integrative mass spectrometry strategy for fingerprinting and tentative structural characterization of asphaltenes," *Fuel*, vol. 220, pp. 717–724, 2018.

[20] J. Castro-Jiménez, G. Mariani, I. Vives et al., "Atmospheric concentrations, occurrence and deposition of persistent organic pollutants (POPs) in a Mediterranean coastal site (Etang de Thau, France)," *Environmental Pollution*, vol. 159, no. 7, pp. 1948–1956, 2011.

[21] F. Hernández, T. Portolés, M. Ibáñez et al., "Use of time-of-flight mass spectrometry for large screening of organic pollutants in surface waters and soils from a rice production area in Colombia," *Science of the Total Environment*, vol. 439, pp. 249–259, 2012.

[22] S. Hashimoto, Y. Zushi, A. Fushimi, Y. Takazawa, K. Tanabe, and Y. Shibata, "Selective extraction of halogenated compounds from data measured by comprehensive multidimensional gas chromatography/high resolution time-of-flight mass spectrometry for non-target analysis of environmental and biological samples," *Journal of Chromatography A*, vol. 1282, pp. 183–189, 2013.

[23] G. Mascolo, G. Bagnuolo, B. De Tommaso, and V. Uricchio, "Direct analysis of polychlorinated biphenyls in heavily contaminated soils by thermal desorption/gas chromatography/mass spectrometry," *International Journal of Environmental Analytical Chemistry*, vol. 93, no. 9, pp. 1030–1042, 2013.

[24] J. Tolu, L. Gerber, J.-F. Boily, and R. Bindler, "High-throughput characterization of sediment organic matter by pyrolysis-gas chromatography/mass spectrometry and multivariate curve resolution: A promising analytical tool in (paleo)limnology," *Analytica Chimica Acta*, vol. 880, pp. 93–102, 2015.

[25] B. R. T. Simoneit, "A review of current applications of mass spectrometry for biomarker/molecular tracer elucidations," *Mass Spectrometry Reviews*, vol. 24, no. 5, pp. 719–765, 2005.

[26] M. M. de Mahiques, R. C. L. Figueira, A. B. Salaroli, D. P. V. Alves, and C. Gonçalves, "150 years of anthropogenic metal input in a Biosphere Reserve: the case study of the Cananéia-Iguape coastal system, Southeastern Brazil," *Environmental Earth Sciences*, vol. 68, no. 4, pp. 1073–1087, 2013.

[27] A. De Junet, I. Basile-Doelsch, D. Borschneck et al., "Characterisation of organic matter from organo-mineral complexes in an Andosol from Reunion Island," *Journal of Analytical and Applied Pyrolysis*, vol. 99, pp. 92–100, 2013.

[28] J. A. González-Pérez, G. Almendros, J. M. De La Rosa, and F. J. González-Vila, "Appraisal of polycyclic aromatic hydrocarbons (PAHs) in environmental matrices by analytical pyrolysis (Py-GC/MS)," *Journal of Analytical and Applied Pyrolysis*, vol. 109, pp. 1–8, 2014.

[29] S. G. Tuncel and T. Topal, "Polycyclic aromatic hydrocarbons (PAHs) in sea sediments of the Turkish Mediterranean coast, composition and sources," *Environmental Science and Pollution Research*, vol. 22, no. 6, pp. 4213–4221, 2015.

[30] M. Frena, G. A. Bataglion, S. S. Sandini, K. N. Kuroshima, M. N. Eberlin, and L. A. S. Madureira, "Distribution and sources of aliphatic and polycyclic aromatic hydrocarbons in surface sediments of itajaí-Açu estuarine system in Brazil," *Journal of the Brazilian Chemical Society*, vol. 28, no. 4, pp. 603–614, 2017.

[31] L. Gao, J. Hou, J. Toney, D. MacDonald, and Y. Huang, "Mathematical modeling of the aquatic macrophyte inputs of mid-chain n-alkyl lipids to lake sediments: Implications for interpreting compound specific hydrogen isotopic records," *Geochimica et Cosmochimica Acta*, vol. 75, pp. 3781–3791, 2011.

[32] C. A. Silva and L. A. S. Madureira, "Source correlation of biomarkers in a mangrove ecosystem on Santa Catarina Island in southern brazil," *Anais da Academia Brasileira de Ciências*, vol. 84, no. 3, pp. 589–604, 2012.

[33] K. J. Ficken, B. Li, D. L. Swain, and G. Eglinton, "An n-alkane proxy for the sedimentary input of submerged/floating freshwater aquatic macrophytes," *Organic Geochemistry*, vol. 31, no. 7-8, pp. 745–749, 2000.

[34] B. Aichner, U. Herzschuh, and H. Wilkes, "Influence of aquatic macrophytes on the stable carbon isotopic signatures of sedimentary organic matter in lakes on the Tibetan Plateau," *Organic Geochemistry*, vol. 41, no. 7, pp. 706–718, 2010.

[35] R. K. Ranjan, J. Routh, J. Val Klump, and A. L. Ramanathan, "Sediment biomarker profiles trace organic matter input in the Pichavaram mangrove complex, southeastern India," *Marine Chemistry*, vol. 171, pp. 44–57, 2015.

[36] R. Jaffé, R. Mead, M. E. Hernandez, M. C. Peralba, and O. A. DiGuida, "Origin and transport of sedimentary organic matter in two subtropical estuaries: a comparative, biomarker-based study," *Organic Geochemistry*, vol. 32, no. 4, pp. 507–526, 2001.

[37] S. Boitsov, V. Petrova, H. K. B. Jensen, A. Kursheva, I. Litvinenko, and J. Klungsøyr, "Sources of polycyclic aromatic hydrocarbons in marine sediments from southern and northern areas of the Norwegian continental shelf," *Marine Environmental Research*, vol. 87-88, pp. 73–84, 2013.

[38] X.-T. Wang, Y. Miao, Y. Zhang, Y.-C. Li, M.-H. Wu, and G. Yu, "Polycyclic aromatic hydrocarbons (PAHs) in urban soils of the megacity Shanghai: Occurrence, source apportionment and potential human health risk," *Science of the Total Environment*, vol. 447, pp. 80–89, 2013.

[39] C. A. Belis, J. Cancelinha, M. Duane et al., "Sources for PM air pollution in the Po Plain, Italy: I. Critical comparison of methods for estimating biomass burning contributions to benzo(a)pyrene," *Atmospheric Environment*, vol. 45, no. 39, pp. 7266–7275, 2011.

12

Authenticity Detection of Black Rice by Near-Infrared Spectroscopy and Support Vector Data Description

Hui Chen,[1,2] **Chao Tan** ⓘ**,**[1] **and Zan Lin**[1,3]

[1]*Key Lab of Process Analysis and Control of Sichuan Universities, Yibin University, Yibin, Sichuan 644000, China*
[2]*Hospital, Yibin University, Yibin, Sichuan 644000, China*
[3]*The First Affiliated Hospital, Chongqing Medical University, Chongqing 400016, China*

Correspondence should be addressed to Chao Tan; chaotan1112@163.com

Academic Editor: Richard G. Brereton

Black rice is an important rice species in Southeast Asia. It is a common phenomenon to pass low-priced black rice off as high-priced ones for economic benefit, especially in some remote towns. There is increasing need for the development of fast, easy-to-use, and low-cost analytical methods for authenticity detection. The feasibility to utilize near-infrared (NIR) spectroscopy and support vector data description (SVDD) for such a goal is explored. Principal component analysis (PCA) is used for exploratory analysis and feature extraction. Another two data description methods, i.e., k-nearest neighbor data description (KNNDD) and GAUSS method, are used as the reference. A total of 142 samples from three brands were collected for spectral analysis. Each time, the samples of a brand serve as the target class whereas other samples serve as the outlier class. Based on both the first two principal components (PCs) and original variables, three types of data descriptions were constructed. On average, the optimized SVDD model achieves acceptable performance, i.e., a specificity of 100% and a sensitivity of 94.2% on the independent test set with tight boundary. It indicates that SVDD combined with NIR is feasible and effective for authenticity detection of black rice.

1. Introduction

Black rice is an economically important special rice species and has been consumed for a long time in Southeast Asia including China [1–3]. Many researches have showed that black rice has considerably strong free-radical scavenging and antioxidation effects, as well as other biological effects of its extracts such as antimutagenic and anticarcinogenic [4, 5]. Black rice quality in terms of nutrition is also valuable for its protein content and the balance of essential amino acids. In fact, black rice is also a mixture of various carbohydrates. There exist varying amounts of nutrient in different kinds of black rice because of genetic and environmental factors. In market, there exist many brands of black rice. The quality and price of them vary greatly and renowned brands have higher price. However, illegal tradesman often passes low-priced black rice off as high-priced ones for economic benefit, especially in some remote towns.

How to discriminate different types of black rice is interesting. Up to now, it is mainly dependent on human senses. More objective and novel methods are maybe based on complex instruments such as high performance liquid chromatography or mass spectroscopy (MS) [6]. In recent years, molecular spectroscopy has drawn more attention and proved to be a powerful tool for authenticity detection [7–9]. In particular, near-infrared (NIR) spectroscopy becomes the most widely used technique in various fields including cigarettes [10], food [11], textile [12], medicine [13], and drug [14]. It is capable of rapidly obtaining a vector/matrix signal of a complex sample and therefore provides the chance of executing a in-depth qualitative or quantitative analysis. Detection of food authenticity is a important task in food analysis and aims to answer the question on which class a particular sample belongs to by its spectral signal. Often, it can be realized by comparing spectra of a specimen to be identified with spectra of "known" or "standard." As for NIR spectroscopy, however, spectral signals for complex food systems are characterized by peak overlapping and poor resolution. So, an appropriate chemometric model is indispensable for a NIR-based application.

For the perspective of modeling, chemometrics involving qualitative tasks can be divided into two categories: classification and one-class classification, i.e., data description [15]. Classification is vey often considered as a synonym of discriminant analysis methods since they assign a new sample to one of a set of predefined classes. The corresponding classifier is trained on a training set. Data description differs in one essential aspect from the conventional classification since it is assumed that only information on a single class is available. Data description problems are common in the real world where positive objects are widely available but negative ones are maybe hard, expensive, or even impossible to gather [16]. In the literature, three main approaches can be distinguished: the density estimation, the boundary methods, and the reconstruction methods [17]. General demand of any authentication problem is that a genuine class, i.e., a target class, must be known [18, 19]. The target class is always unique for a specific authentication problem. Any other objects, or classes of objects, that are not members of the target class are considered as outliers. This also means that just samples of the target class can be utilized and that no information on the other classes is present. For data description, the boundary surrounding the target class has to be estimated from available data, such that it accepts as much of the target samples as possible and minimizes the error of accepting outlier. Up to now, much effort has been expended to develop classification algorithms, and the concept of data description is also of interest and noticeable [20–23], especially in the cases where it is impossible to meaningfully define all of the classes and obtain fully representative samples. In food authenticity, the interest is focused on a single target class so as to verify compliance of samples with the features of that class, and a data description approach should be adopted to build an enclosed boundary around the target class.

The present work focuses on exploring the feasibility to utilize near-infrared (NIR) spectroscopy and support vector data description (SVDD) for authenticity diction of black rice. Principal component analysis (PCA) is used for exploratory analysis and feature extraction. Another two data description methods, i.e., k-nearest neighbor data description (KNNDD) and GAUSS method, are used as the reference. A total of 142 samples from three brands were collected for spectral analysis. All spectra were preprocessed beforehand by standard normal transformation (SNV). Each time, the samples of a brand serve as the target class whereas other samples serve as the outlier class. Based on both the first two principal components (PCs) and original variables, three types of data descriptions were constructed. On average, the optimized SVDD model achieves acceptable performance, i.e., a specificity of 100% and a sensitivity of 94.2% on the independent test set with tight boundary. The effect of training set size and the parameter of kernel width have also been discussed. It indicates that SVDD combined with NIR is feasible and effective for authenticity detection of black rice.

2. Theory and Methods

Many methods have been developed to solve the one-class or data description problem and they can be divided into three main categories: density, boundary, and reconstruction methods. Here, three algorithms, i.e., support vector data description, Nearest Neighbor Method, and Gaussian Method, are introduced and used for experiments, among which the first two are boundary methods and the last one belongs to density method.

2.1. Support Vector Data Description (SVDD). SVDD is a novel algorithm for one-class classification problems, which has been proposed by Tax [15], inspired by the idea of the support vector machines. It focuses on finding a minimum hypersphere around the target class. The hypersphere can be used to decide whether new objects are targets or outliers. Such a sphere is characterized only by center $\mathbf{a}$ and radius R. When seeking sphere, it needs to minimize the volume of the sphere by minimizing R^2 and demand that the sphere covers as many training samples as possible. Given the training set $\{\mathbf{x}_i,\ i = 1, 2, \ldots, N\}$, the task in SVDD is to minimize error function:

$$\min \quad L(R, \mathbf{a}, \zeta_i) = R^2 + C \sum_i \zeta_i \qquad (1)$$

$$\text{s.t.} \quad \|\mathbf{x}_i - \mathbf{a}\| \le R^2 + \zeta_i, \quad \zeta_i \ge 0 \ \forall i \qquad (2)$$

where $\mathbf{a}$ and R are the center and the radius of the hypersphere, respectively; C is the penalty factor which regulates the hyperspherical volume and error, i.e., the number of target objects rejected; ζ_i is a slack variable for allowable error limitation. Almost all objects are within the sphere. This optimization problem can be solved by Lagrange multiplier method [24].

Because the target class is not spherically distributed in most cases, some traditional decision rules may not work well. To make a more effective and flexible decision, the original data can be implicitly transformed to a higher dimension by the so-called kernel function $K(\mathbf{x}_i, \mathbf{x}_j)$. Several kernel functions including linear, polynomial, Gaussian, radial basis function (RBF) are available [25, 26]. In this work, the RBF kernel, the most commonly used kernel in machine learning, was used. The form of RBF kernel is

$$K(\mathbf{x}_i, \mathbf{x}_j) = \exp\left(-\frac{\|\mathbf{x}_i - \mathbf{x}_j\|^2}{\sigma^2}\right) \qquad (3)$$

where σ is a key parameter for controlling the boundary tightness.

2.2. Nearest Neighbor Method. The most straightforward and simplest method to obtain a one-class model is to estimate the density of the training set. Unfortunately, it often requires a large number of samples to avoid the curse of dimensionality. Instead of estimating whole probability densities, an indication of the resemblance can also be acquired by comparing distances. Nearest neighbor method can be derived from a local density estimation [27]. It avoids the explicit density estimation by only using distances to the first nearest neighbor. In the process of density estimation, a cell, often an hypersphere in d-dimension space, is centered around the test object $\mathbf{z}$. The cell volume is grown until it contains k objects from the training set. The local density can be estimated by

$$p_{NN}(\mathbf{z}) = \frac{k/N}{V_k \|\mathbf{z} - NN_k^{tr}(\mathbf{z})\|} \tag{4}$$

where $NN_k^{tr}(\mathbf{z})$ and V_k are the k nearest neighbors of $\mathbf{z}$ in the training set and the volume of the cell containing this object. Later, we will use KNNDD to denote this method.

For an unknown test object $\mathbf{z}$, the distance from it to its nearest neighbor in the training set $NN^{tr}(\mathbf{z})$ is compared with the distance from $NN^{tr}(\mathbf{z})$ to its nearest neighbor. The test object $\mathbf{z}$ can be accepted when its local density is larger or equal to the density of the nearest neighbor. It seems to be very useful for distributions characterized by fast decaying probabilities. Obviously, the method can easily be generalized to a larger number of neighbors k. That is, instead of taking the first nearest neighbor into account, the kth neighbor should be considered.

2.3. Gaussian Method. When a proper probability model is assumed and the sample size is sufficient, density method is advantageous for one-class problem. With the optimization of the threshold, a minimum volume can be automatically found for the given probability density. When only a little amount of samples is available, the simplest model is the unimodal Gaussian/Normal distribution. It fits a probability density model as follows:

$$p_N(\mathbf{x}) \frac{1}{(2\pi)^{d/2} |\Sigma|^{0.5}} \exp\left\{-\frac{1}{2}(\mathbf{x} - \boldsymbol{\mu})^T \Sigma^{-1}(\mathbf{x} - \boldsymbol{\mu})\right\} \tag{5}$$

where $\boldsymbol{\mu}$ is the mean and Σ is the covariance matrix. Both should be estimated from the training set. For d dimensional data, the number of the parameters is

$$d + \frac{1}{2}d(d-1) \tag{6}$$

The method imposes a strict unimodal and convex density model on the data. The main computational effort is maybe the inversion of the covariance matrix. In case of badly scaled data or data with singular directions, it is difficult to calculate the inverse of Σ and it can be approximated by the pseudoinverse $\Sigma^+ = \Sigma^T(\Sigma\Sigma^T)^{-1}$ or by introducing regularization (adding a small constant λ to the diagonal, i.e., $\Sigma' = \Sigma + \lambda\mathbf{I}$). In the last case, the user needs to supply a parameter λ. This is also the only magic parameter that requires a user to provide.

Finally, a threshold on the probability density needs to be set for distinguishing between target and outlier data. Accepting 95% of the objects requires a threshold on the Mahanalobis distance

$$(\mathbf{x} - \boldsymbol{\mu})^T \Sigma^{-1}(\mathbf{x} - \boldsymbol{\mu}) \tag{7}$$

of

$$\theta_N = \left(\chi_d^2\right)^{-1}(0.95) \tag{8}$$

where $\left(\chi_d^2\right)^{-1}$ is the inverse χ_d^2 with d degrees of freedom. This method is expected to work effectively only if the data is unimodal and convex. To obtain a more flexible density method, it can be extended to a mixture of Gaussians. Later, we will use GAUSS to denote this method.

3. Experimental

3.1. Sample Preparation. A total of 142 samples/bag of black rice of three brands were purchased from local supermarkets in China. They were from different supplier and let us mark them as A, B, and C brands. These samples were collected from three batches of A, two batches of B, and three batches of C but different packages. For A or C, forty-eight bags of rice were sampled, sixteen bags for each batch; For B, forty-six bags of rice were sampled, twenty-three bags for each batch. In total, the number of samples belonging to A, B, and C are 48, 46, and 48, respectively. The time it takes to collect the sample is about six months. The samples of each brand could serve as the target class whereas other samples acted as the outlier class. All samples were stored in the laboratory kept at 25°C for more than 7 days in order to achieve a temperature balance. To reduce the effect of environment, the NIR spectra of all samples were recorded on the same day.

3.2. Spectral Measurement and Preprocessing. Spectra of different samples collected on an Antaris II FT-NIR spectrometer (Thermo Scientific Co./USA) were equipped with an integrating sphere module, a rotating sample cup, and a InGaAs detector, as well as a tungsten lamp as the light source. The sample was poured into a standard sample cup with a 50 mm diameter and the height was controlled on about **30 mm** for preventing light leak. An internal gold reference was used for automatic background collection. A specific sample cup spinner accessory for the integrating sphere sampling module that allows multipoint reflection measurements of heterogeneous solids such as powders, granules, and pellets, was used for obtaining NIR spectra of high quality. In this way, the final spectrum is the average of the spectra collected at different locations, which can reduce the effect of heterogeneity of solids to some extent.

The NIR spectrum was measured in the region of $10,000–4000\ cm^{-1}$ with 32 scans at a resolution of $3.856\ cm^{-1}$. Each spectrum contains 1557 data points. The experimental temperature and the related humidity were controlled around 25°C and 60%, respectively. Preprocessing of spectra is often of great importance if reasonable results need to be obtained whether it is concerned with qualitative or quantitative tasks. Several methods of preprocessing were attempted. In comparison with other preprocessing methods, standard normal transformation (SNV) achieved a satisfactory performance without the need of a reference spectrum and user decision for the computation. So, all spectra were preprocessed by SNV. The spectral measurement was controlled by the Result software [28]. DD toolbox was used for one-class classifier

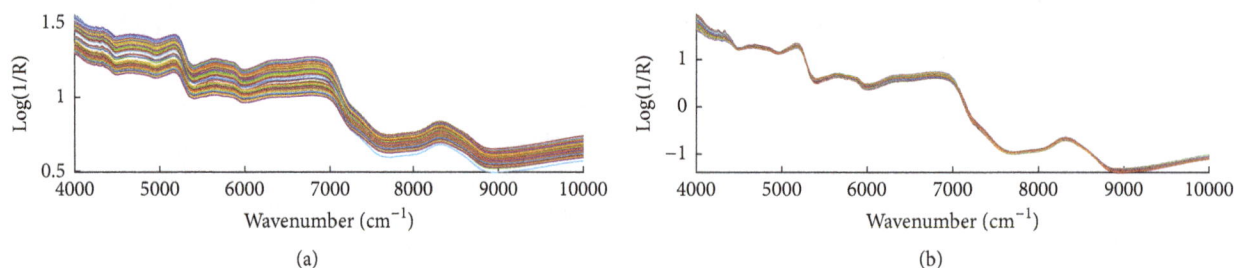

FIGURE 1: Original near-infrared (NIR) spectra (a) and all the preprocessed spectra (b) by standard normal transformation (SNV).

modeling [15]. All calculation was made on MATLAB 2015b for Windows.

4. Results and Discussions

4.1. NIR Spectral Analysis. Figure 1 shows the NIR spectra and all the preprocessed spectra of black rice samples by SNV. Seen from Figure 1, the spectra of three types of black rice share very similar absorbance patterns in the range of 4000-11000 cm^{-1}. They can hardly be distinguished just by naked eyes. General features of a NIR spectrum of solid samples include a multiplicative response to changes in particle size. SNV treatment autoscales each spectrum based on calculating the mean and standard deviation between the densities. It is also clear in Figure 1 that, by preprocessing, some additive and multiplicative effects have been removed.

It is well known that major components of black rice are complex molecules from the polymerization of monomers such as amino acids or carbohydrates. Each monomer exhibits specific chemical groups such as carboxylic and amine functions in amino acids. As each chemical group may absorb the infrared region light, it appears useful to clearly identify the characteristic NIR bands of these groups. Because NIR spectrum corresponds to molecular responses of the overtone and combination bands, for each fundamental absorption band, there exists several overtones with decreasing intensity corresponding to the increasing multiple or transition number. All the bands can form a myriad of combination bands with intensities increasing as frequency decreases. NIR band intensities are much weaker than their corresponding mid-infrared fundamentals by a factor of 10-100. In Figure 1, two strong bands at 5175 cm^{-1} and 6930 cm^{-1} result from the absorbance of water, among which the peaks around 5175 cm^{-1} are the combination of asymmetric stretching and bending vibration of H$_2$O. The band of 8200-8600 cm^{-1} can be attributed to the second overtones of C-H stretching in various groups. The wider bands in 6100–7000 cm^{-1} are mainly caused by the overlapping of the first overtones of O-H and N-H stretching. The two peaks at 4266 cm^{-1} and 4335 cm^{-1}, which can be attributed to C-H stretching and C-H deformation, are very stable and carry much useful information. However, accurate assignments of each peak were maybe difficult due to low resolution and baseline shift; therefore, it is necessary to resort to chemometric methods to extract the useful information from spectra for identification purposes.

Furthermore, one of the most interesting applications of NIR technique in the food analysis is total quality evaluation as it can provide fingerprint information of a sample. Different brands of black rice mean different balances/ratios of diverse chemical constituents and physicochemical properties, rather than simple amount of each constituent. NIR spectra contain rich information on chemical constituents and physicochemical properties. Although the quality of black rice is generally assessed by sensory evaluation, its taste is actually a function of chemical constituents such as protein, moisture, amylose, fatty acid, and minerals. Therefore, an overall evaluation is preferred based on NIR spectroscopy.

Principal component analysis (PCA), the most widespread multivariate tool, was used for an exploratory analysis and dimensional reduction. Unlike other applications, the main goal of the present work using PCA was to map the original data into its principal component score space (i.e., the first two), based on which the subsequent modeling was carried out. So, all samples were considered as a whole for PCA and mean-centering pretreatment. By computation, the first two PCs explain 79.4% and 18.4% of the total variances, respectively, and they may contain most of the useful information in the original spectra. Because of this, we decided to use the first two components as the input of subsequent data description methods.

4.2. Authenticity Detection by Data Description. Given a dataset, in general, the selection of a representative training set upon which training the prediction model is performed is very important. For this purpose, in our work, the Kennard and Stone (KS) algorithm [29] was first used to rank all samples of each class in the dataset under consideration, thereby producing three sequences (A, B, and C). The KS algorithm consists of two main steps: taking the pair of samples between which the Euclidean distance of x-vectors (predictor) is largest, and then sequentially selecting a sample to maximize the Euclidean distances between x-vectors of already selected samples and the remaining samples. This process is repeated until all samples are picked out. The former samples are more representative than the latter one. When A class served as the target class, only the first thirty samples in A sequence were used as the training set for constructing data description. The remaining samples in A sequence and all samples in B and C sequences were used as the test set (the same partition of the sample set for the cases using B or C as the target class). Based on the first

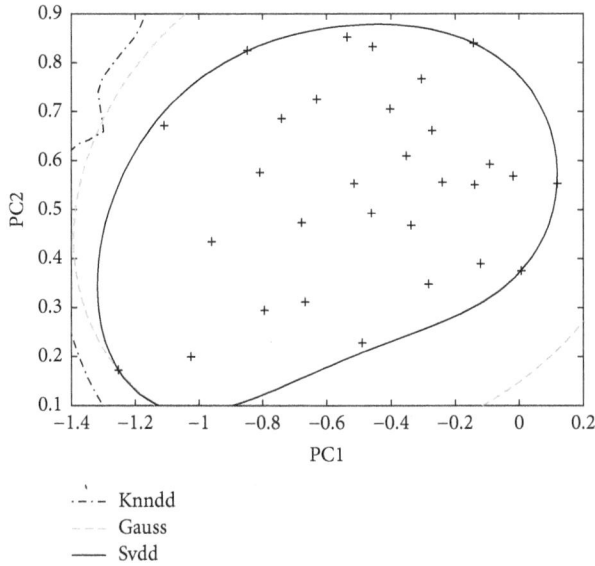

··-·- Knndd
······· Gauss
——— Svdd

FIGURE 2: Data description boundary of class A on the first two-principal-component space based on the training set.

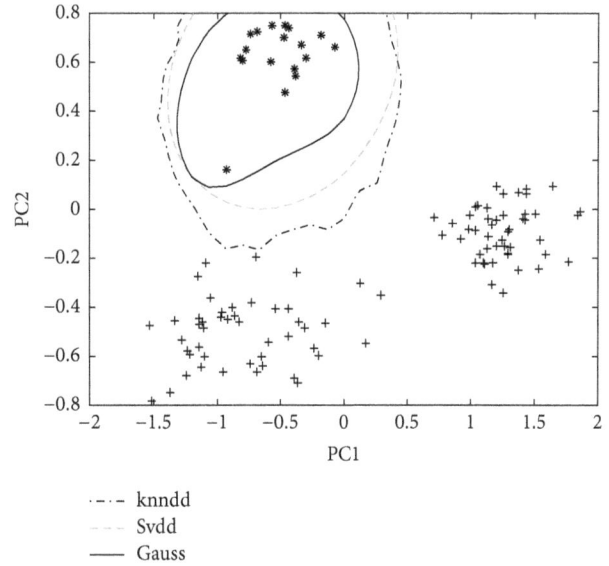

·-·-· knndd
······· Svdd
——— Gauss

FIGURE 4: Application of the data description models of class B on the test set.

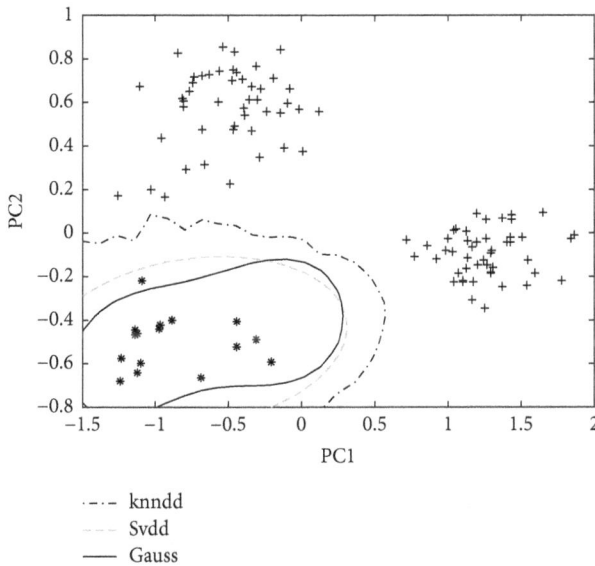

·-·-· knndd
······· Svdd
——— Gauss

FIGURE 3: Application of the data description models of class A on the test set.

two PCs of the training set, three types of data descriptions mentioned above, i.e., SVDD, KNNDD, and GAUSS, were constructed. SVDD used the Gaussian kernel. Figure 2 gives the optimized data description boundary of class A based on the training set. It seems that the boundary of SVDD is tightest. All the descriptions differ from conventional classification because they always obtain a closed boundary around one of the target classes. Unlike density methods such as GAUSS, SVDD does not require a strict representative sampling of the target class; a sampling containing extreme objects is also acceptable. This can be found explicitly in the error definition of SVDD, which minimizes the volume of the description plus the sum of slack variables for objects outside

the description. A conventional classifier, on the contrary, distinguishes between two/multiple classes without focusing on any of the classes and aims to minimize the probability of overall error. It is expected to perform very poorly when just the target class is available or the dataset is relatively small. Food validation or authenticity detection is often the case.

Figure 3 shows the application of the data description models of class A on the test set. Only one target sample was identified as outlier by SVDD. Even if the KNNDD and GAUSS correctly identified all samples, the false positive would increase when more test samples were used in the future. Similarly, Classes B and C were considered as the target class, and three corresponding data descriptions were constructed. Figure 4 shows the application of the data description models of class B on the test sets. Now, all the models correctly identified the target samples and the corresponding outliers but the SVDD use the tightest boundary, maybe implying better generalization ability. Both KNNDD and GAUSS produce looser borders. It should be noted that each time the so-called "fake" black rice is actually simulated by the samples from nontarget class.

The character of the SVDD heavily depends on the width parameter of the Gaussian kernel, which is very crucial as it can provide different prediction performance and leads to overfitting problem. Several previous studies have reported how to optimize SVDD [30]. The penalty term is sample rejection rate, i.e., the approximate proportion of samples misclassified in a training set. The other tunable parameter is kernel width. A large width can lead to a less complicated boundary and a relatively large width (compared to the maximum distance between samples in training set) could lead to a rigid hypersphere. In this work, based on the average nearest neighbor distance in the dataset, one can distinguish three types of cases: very small, very large, and intermediate values. By changing the value, the description ranges from

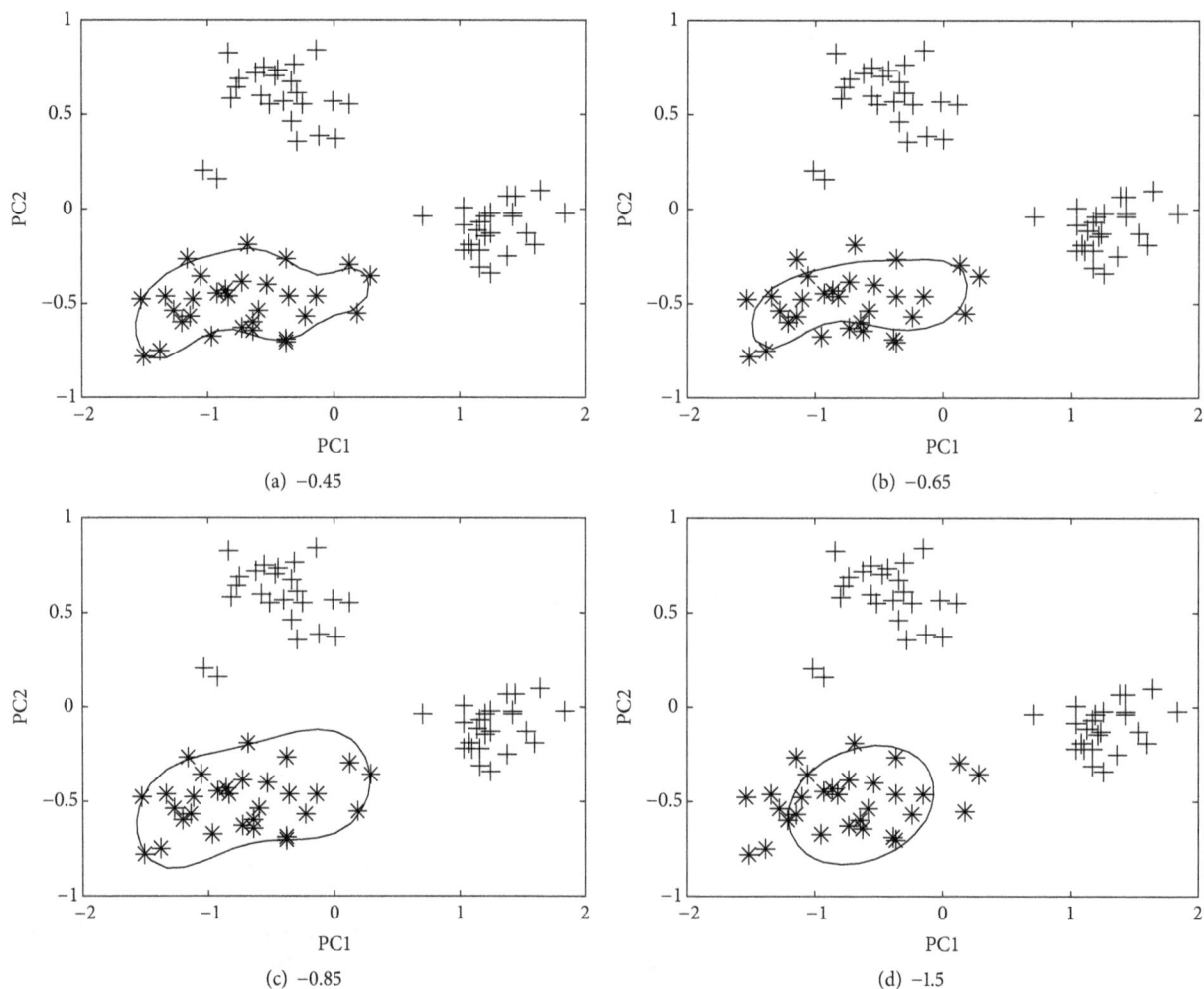

FIGURE 5: The influence of the kernel parameter on the boundary of support vector data description (SVDD) using class A as the target class (classification error on the target class is set as 0.1).

Parzen density estimation, via a mixture of Gaussian to the rigid hypersphere, can be observed in Figure 5, which shows the influence of the kernel parameter on the boundary of SVDD when using class A as the target class. The boundary of SVDD seems to be sensitive to the kernel parameter. With the increase of the width, the boundary undergoes a complex change; it gradually achieves the optimum and then gets worse. For different cases, the number of support vectors is also different. In order to facilitate the comparison, Figure 6 gives a similar ensemble plot of the influence of the kernel parameter on the boundary. Based on the shape and the compactness of the edge of the description, the optimal width parameter is 0.85 for this case A. Such a boundary contains all the target samples, among which six samples are just on the edge, and the shape is also simple. Also, one important advantage of SVDD over some traditional methods is that the classifier does not require that the data follow a normal distribution. However, there exist some alternative procedures for optimizing kernel width such as cross-validation, bootstrapping, and the consistency evaluation of the classifier using only the error of the nontarget class [15, 30, 31].

Taking the first case as an example (A as the target class), instead of the PCs, the original spectral variables were used as the independent variables for constructing data descriptions. On the independent test set, all these models including SVDD, KNNDD, and GAUSS achieved a specificity of 100% (the ration of outliers that were rejected), while the corresponding sensitivity, i.e., the ratio of the target class that was accepted, is 100% for GUASS and 94.4% for both SVDD and KNNDD, despite whether PCs or original variables are used. It indicates that using PCs or original variables does not make substantial difference. However, using all features is likely to result in overfitting, while using PCs will likely reduce overfitting. Also, using PCs makes the computation to be faster and to be more convenient for visual purposes. When B or C is the target class, the corresponding specificity and sensitivity have also been summarized in Table 1. On average, the SVDD achieves best prediction, with the specificity of 100% and the sensitivity of 94.2%.

On the whole, the data description, especially SVDD, achieved an acceptable sensitivity and specificity for the so-called small-sample problem. Such a procedure is maybe

TABLE 1: Summary of the performance of different models.

Target class	GAUSS		KNNDD		SVDD	
	SPE	SEN	SPE	SEN	SPE	SEN
A	100%	100%	100%	94.4%	100%	94.4%
B	96.8%	87.5%	96.8%%	93.8%	100%	93.8%
C	97.8%	88.9%	98.9%	88.9%	100%	94.4%
Average	98.2%	92.1%	98.5%	92.3%	100%	94.2%

Note. SPE and SEN denote the specificity and sensitivity, respectively.

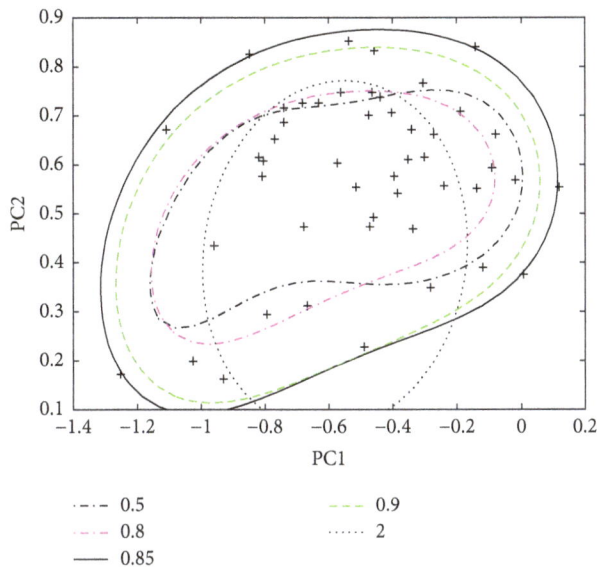

FIGURE 6: Ensemble of the influence of the kernel parameter on the boundary of support vector data description (SVDD) on the same plot.

potential tool for authenticity detection of various foods including black rice.

5. Conclusions

The work reveals that NIR spectroscopy combined with support vector data description is feasible and advantageous to implement authenticity detection of black rice. It can serve as an alternative to laborious, time-consuming, wet chemical methods and sensory analysis of human. However for obtaining more reliable results, more samples need to be collected, which remains our next work.

Acknowledgments

This work was supported by the National Natural Science Foundation of China (21375118, J1310041), Scientific Research Foundation of Sichuan Provincial Education Department of China (17TD0048), Scientific Research Foundation of Yibin University (2017ZD05), the Applied Basic Research Programs of Science and Technology Department of Sichuan Province of China (2018JY0504), and Opening Fund of Key Lab of Process Analysis and Control of Sichuan Universities of China (2015006, 2016002).

References

[1] K. Tananuwong and W. Tewaruth, "Extraction and application of antioxidants from black glutinous rice," *LWT- Food Science and Technology*, vol. 43, no. 3, pp. 476–481, 2010.

[2] C. Hu, J. Zawistowski, W. Ling, and D. D. Kitts, "Black rice (Oryza sativa L. indica) pigmented fraction suppresses both reactive oxygen species and nitric oxide in chemical and biological model systems," *Journal of Agricultural and Food Chemistry*, vol. 51, no. 18, pp. 5271–5277, 2003.

[3] C. J. Liu, H. M. Wang, C. L. Liu, L. Wang, and X. M. Meng, "uncertainty evaluation Of toatal phenol determination on black rice by spectrophotometry," *China Food Additives*, vol. 2, pp. 172–175, 2017.

[4] J. W. Hyun and H. S. Chung, "Cyanidin and malvidin from *Oryza sativa* cv. heugjinjubyeo mediate cytotoxicity against human monocytic leukemia cells by arrest of G(2)/M phase and induction of apoptosis," *Journal of Agricultural and Food Chemistry*, vol. 52, no. 8, pp. 2213–2217, 2004.

[5] Z. F. Liang, W. J. Wang, X. Y. Wang, and Y. Fang, "Determination of trace elements content of the same ogill of black rice and ordinary rice," *Chemical Engineering*, vol. 11, pp. 27–31, 2015.

[6] B. Zhang, Z. Q. Rong, Y. Shi, J. G. Wu, and C. H. Shi, "Prediction of the amino acid composition in brown rice using different sample status by near-infrared reflectance spectroscopy," *Food Chemistry*, vol. 127, no. 1, pp. 275–281, 2011.

[7] D. Cozzolino, "Near Infrared Spectroscopy and Food Authenticity," *Advances in Food Traceability Techniques and Technologies: Improving Quality Throughout the Food Chain*, pp. 119–136, 2016.

[8] E. Domingo, A. A. Tirelli, C. A. Nunes, M. C. Guerreiro, and S. M. Pinto, "Melamine detection in milk using vibrational spectroscopy and chemometrics analysis: a review," *Food Research International*, vol. 60, pp. 131–139, 2014.

[9] H. Chen, C. Tan, Z. Lin, and T. Wu, "Detection of melamine adulteration in milk by near-infrared spectroscopy and one-class partial least squares," *Spectrochimica Acta Part A: Molecular and Biomolecular Spectroscopy*, vol. 173, pp. 832–836, 2017.

[10] C. Tan, M. Li, and X. Qin, "Study of the feasibility of distinguishing cigarettes of different brands using an Adaboost algorithm and near-infrared spectroscopy," *Analytical and Bioanalytical Chemistry*, vol. 389, no. 2, pp. 667–674, 2007.

[11] J. Zhao, H. Lin, Q. Chen, X. Huang, Z. Sun, and F. Zhou, "Identification of egg's freshness using NIR and support vector data description," *Journal of Food Engineering*, vol. 98, no. 4, pp. 408–414, 2010.

[12] C. Ruckebusch, F. Orhan, A. Durand, T. Boubellouta, and J. P. Huvenne, "Quantitative analysis of cotton-polyester textile blends from near-infrared spectra," *Applied Spectroscopy*, vol. 60, no. 5, pp. 539–544, 2006.

[13] H. Chen, Z. Lin, H. Wu, L. Wang, T. Wu, and C. Tan, "Diagnosis of colorectal cancer by near-infrared optical fiber spectroscopy and random forest," *Spectrochimica Acta Part A: Molecular and Biomolecular Spectroscopy*, vol. 135, pp. 185–191, 2015.

[14] K. Dégardin, A. Guillemain, N. V. Guerreiro, and Y. Roggo, "Near infrared spectroscopy for counterfeit detection using a large database of pharmaceutical tablets," *Journal of Pharmaceutical and Biomedical Analysis*, vol. 128, pp. 89–97, 2016.

[15] D. M. Tax, *One-class classification*, Delft University of Technology, Delft, The Netherlands, 2001.

[16] B. Krawczyk and M. Woźniak, "Diversity measures for one-class classifier ensembles," *Neurocomputing*, vol. 126, pp. 36–44, 2014.

[17] O. Mazhelis, "One-Class Classifiers: A Review and Analysis of Suitability in the Context of Mobile-Masquerader Detection," *Advances in end-user data-mining techniques*, vol. 30, pp. 39–47, 2006.

[18] P. Oliveri, "Class-modelling in food analytical chemistry: Development, sampling, optimisation and validation issues - A tutorial," *Analytica Chimica Acta*, vol. 982, pp. 9–19, 2017.

[19] P. Oliveri and G. Downey, "Multivariate class modeling for the verification of food-authenticity claims," *TrAC - Trends in Analytical Chemistry*, vol. 35, pp. 74–86, 2012.

[20] M. Forina, C. Armanino, R. Leardi, and G. Drava, "A class-modelling technique based on potential functions," *Journal of Chemometrics*, vol. 5, no. 5, pp. 435–453, 1991.

[21] O. Y. Rodionova, P. Oliveri, and A. L. Pomerantsev, "Rigorous and compliant approaches to one-class classification," *Chemometrics and Intelligent Laboratory Systems*, vol. 159, pp. 89–96, 2016.

[22] L. Xu, S. Yan, C. Cai, and X. Yu, "One-class partial least squares (OCPLS) classifier," *Chemometrics and Intelligent Laboratory Systems*, vol. 126, pp. 1–5, 2013.

[23] R. G. Brereton, "One-class classifiers," *Journal of Chemometrics*, vol. 25, no. 5, pp. 225–246, 2011.

[24] Z. S. Pan, B. Chen, Z. M. Miao, and G. Q. Ni, "Overview of study on one-class classifier," *Acta Electronica Sinica*, vol. 37, pp. 2496–2503, 2009.

[25] F. S. Uslu, H. Binol, M. Ilarslan, and A. Bal, "Improving SVDD classification performance on hyperspectral images via correlation based ensemble technique," *Optics and Lasers in Engineering*, vol. 89, pp. 169–177, 2016.

[26] L. Duan, M. Xie, T. Bai, and J. Wang, "A new support vector data description method for machinery fault diagnosis with unbalanced datasets," *Expert Systems with Applications*, vol. 64, pp. 239–246, 2016.

[27] D. Tax and R. Duin, "Data description in subspaces," in *Proceedings of the 15th International Conference on Pattern Recognition*, pp. 672–675, Barcelona, Spain, 2002.

[28] Thermo Scientific, Result Integration Software user Guide.

[29] R. W. Kennard and L. A. Stone, "Computer aided design of experiments," *Technometrics*, vol. 11, no. 1, pp. 137–148, 1969.

[30] S. Kittiwachana, D. L. S. Ferreira, G. R. Lloyd et al., "One class classifiers for process monitoring illustrated by the application to online HPLC of a continuous process," *Journal of Chemometrics*, vol. 24, no. 3-4, pp. 96–110, 2010.

[31] L. S. Zhong and C. R. Hou, "Fault monitoring of industrial process based on independent component and support vector description (IC-SVDD," *Computers and Applied Chemistry*, vol. 34, pp. 285–290, 2017.

13

Method Development for Assessing Carbamazepine, Caffeine, and Atrazine in Water Sources from the Brazilian Federal District using UPLC-QTOF/MS

Fernando F. Sodré [iD][1] **and Cínthia M. P. Cavalcanti**[2]

[1]*Institute of Chemistry, University of Brasília, Brasília 70910-000, Brazil*
[2]*Environmental Sanitation Company of the Federal District (CAESB), Águas Claras 71928-720, Brazil*

Correspondence should be addressed to Fernando F. Sodré; ffsodre@unb.br

Academic Editor: Eladia M. Pena-Mendez

About 3.0 million people living under a typical tropical savannah climate in the Brazilian Federal District (FD) have faced an unprecedented water crisis. Considering the need for indirect reuse of wastewater for public supply, this work aimed to investigate FD water sources regarding the presence and risks of three contaminants of emerging concern: caffeine, carbamazepine, and atrazine. Samples from two current water sources (Descoberto and Santa-Maria Lakes) and two future water sources of the FD (Paranoá and Corumbá Lakes) were analyzed by solid-phase extraction followed by liquid chromatography coupled to hybrid quadrupole-time-of-flight mass spectrometry (UPLC-QTOF/MS). Method precision and accuracy were satisfactory and limits of quantification ranged from 0.37 to 0.54 ng/L. Higher concentrations were observed for caffeine in the future water sources (39 to 180 ng/L) followed by carbamazepine (5.4 to 25 ng/L) and atrazine (3.9 to 15 ng/L). The less-impacted water sources, in current use in the FD, present caffeine concentrations ranging from 4.8 to 32 ng/L and atrazine levels varying between 2.4 and 5.5 ng/L. Carbamazepine was not detected in these reservoirs. Environmental risk assessment indicates a possible risk for carbamazepine and atrazine, evidencing the need for further studies. No human health risk was depicted within the results.

1. Introduction

The capital of Brazil, Brasília, is located in the Brazilian Federal District (FD) under a typical tropical savannah climate with distinct periods of precipitation and humidity. The winter is dry with approximately 120 days without rainfall, resulting in severe problems related to water scarcity and rationing. The most important drinking water systems of the FD (Descoberto Lake and Santa-Maria Lake production systems) have become insufficient to supply about 3.0 million people living in the region. Thus, several actions have already been taken by the Environmental Sanitation Company to improve water availability, such as the use of alternative low-flow water intakes, the constant policing of the water sources, and the minimization of water losses during production processes.

As a result of low rainfall rates for three consecutive years, combined with a lower water recharge and an intense water use, the region is experiencing the largest water crisis in its history. To alleviate this problem, several long-term alternatives were evaluated, two of which were selected for the expansion of the water supply system: the use of Corumbá and Paranoá Lakes as water sources. The former is located beyond the borders of the FD and receives effluents from wastewater treatment plants (WWTPs), either directly or through its tributaries, while Paranoá Lake is an urban water system that receives effluents from two important WWTPs of the FD, as well as urban drainage waters and contaminated waters from tributaries, some of them running through densely populated areas.

Under this new reality, the indirect reuse of water is significant [1] and WWTPs become the most promising sources of water to be recycled [2]. Therefore, monitoring of water quality should incorporate aquatic parameters of emerging concern related to the lifestyle of the modern human societies [3, 4]. In this context, emerging contaminants are of particular interest, as they arise in the environment through a variety of routes [5], may be refractory to conventional treatment methods [6–8], may promote adverse reproductive and developmental effects [9–11], and have been widely used as environmental tracers of several human-related substances [12, 13].

As pointed out by Snyder and Bennoti [14], the implementation of water reuse operations may experience resistance from public opinion mainly due to the presence of emerging contaminants such as pharmaceuticals, personal care products, endocrine disruptor chemicals, and pesticides. Thus, it is necessary to generate reliable information regarding different aspects of this issue, such as the removal efficiency by conventional and advanced drinking and wastewater treatment processes, the impact of wastewater discharge on the presence of such contaminants in drinking water supplies, and the natural attenuation of such contaminants in the environment [14].

Although Descoberto and Santa-Maria Lakes are the main water sources in the FD, most of the work involving the presence of emerging contaminants has been carried out in the waters of Paranoá Lake considering its historical importance and future multiple-use possibilities. Abbt-Braun et al. [15] investigated the presence of pesticides, sweeteners, and perfluorinated substances in Paranoá Lake and observed that high concentrations (nearby WWTPs) decreased through natural processes such as dilution, photolysis, and degradation until reaching the regions near the lake dam where water will be used for public supply purposes. The authors also show that only caffeine and iopromide present concentrations higher than 100 ng/L in the site of the future raw water withdrawal. Da Costa et al [16] investigated the occurrence of emergent contaminants at five sampling points along Paranoá Lake, including the point at the lake dam, where they observed lower concentrations of caffeine, bezafibrate, bisphenol A, diethyl phthalate, and nonylphenol compared to the other points located in the four branches of the lake.

Descoberto Lake was previously investigated for the presence of caffeine, atrazine, atenolol, and DEET, an active ingredient in insect repellents [17]. The concentration of these contaminants varied between 2.6 (DEET) and 10 ng/L (caffeine) being considerably lower than those found in Paranoá Lake.

In Brazil, the monitoring of contaminants in drinking water became mandatory only in 1977, with the publication of BSB Ordinance No. 56/1977, which recommends periodic determinations of 10 inorganic contaminants, 12 pesticides, and 14 organoleptic parameters. After successive revisions over time, water quality standards were gradually increased. Nowadays, determinations of 15 inorganic contaminants, 15 organic substances, 27 pesticides, 7 disinfectants and their by-products, and more than 21 organoleptic parameters are required, through Annex XX of Consolidation Ordinance No. 05/2017, published by the Brazilian Ministry of Health. Within the contaminants investigated in the present work,

only atrazine is considered in the Brazilian legislation for raw and treated waters. For the other substances, i.e., caffeine and carbamazepine, there are still no initiatives in Brazil for their inclusion in a monitoring program of national proportions.

This work was motivated by the importance of emerging contaminants in situations of indirect water reuse and by the limited amount of information regarding this class of contaminants in Brazil, especially in the FD. Thus, the present work aimed to develop and apply a method based on the solid-phase extraction of caffeine, atrazine, and carbamazepine followed by the quantification by ultra-efficiency liquid chromatography coupled to a high-resolution hybrid mass spectrometer (quadrupole-time-of-flight). These chemicals were selected due to their use as tracers of anthropogenic discharges in surface waters [18–20].

2. Materials and Methods

2.1. Chemicals. Methanol and acetone (HPLC grade) were obtained from Scharlau Chemie SA (Spain). Formic acid, acquired from Sigma-Aldrich (USA), was used as mobile phase additive. Ultrapure water was produced in a Milli-Q Academic system (Millipore, USA). Caffeine (98%, CAS 58-08-2) and atrazine (99%, CAS 1912-24-9) were purchased from Fluka Analytical (USA), whereas carbamazepine (99%, CAS 298-46-4) was obtained from Sigma-Aldrich (USA). Stock solutions (200 mg L^{-1}), prepared by the solubilization of appropriate amounts of each solid standard in methanol, were kept in amber glass bottles at -10°C. A mixed stock solution containing 400 μg L^{-1} of each standard was used during the optimization of instrumental parameters. For quantification, mixed stock and working solutions were prepared weekly in methanol and kept under refrigerated conditions when not in use. A standard solution (Sciex, Canada) containing a mixture of cesium iodide (CsI, m/z 132.9054) and the synthetic peptide ALILTLVS (m/z 829.5398) was used for mass calibration and tuning.

2.2. Study Area and Sampling. Figure 1 shows the location of the sampling points selected in the present work. The sampling points DL and SL were established at the water intakes of Descoberto and Santa-Maria Lakes, respectively. The water supply systems of both water sources account for approximately 86% of the drinking water production for the population of the FD.

Although located in an environmental protection area, the Descoberto Lake basin suffers from several problems such as invasions of protected areas, high population densities, agricultural activities, and siltation. Santa-Maria Lake is considered the most protected water source of the FD due to its restricted access through the National Park of Brasília. Sampling points were also established in the tributaries of Santa-Maria Lake: Santa-Maria River (SR), Milho-Cozido Stream (MC), and Vargem-Grande Stream (VG).

Two sampling points were established in Paranoá Lake. This urban artificial lake was built in 1959 to generate electric power and to improve the microclimate of the future capital of Brazil, Brasília, but is currently used for recreation, sports, tourism, and fishing. Paranoá Lake also receives urban

Method Development for Assessing Carbamazepine, Caffeine, and Atrazine in Water Sources...

123

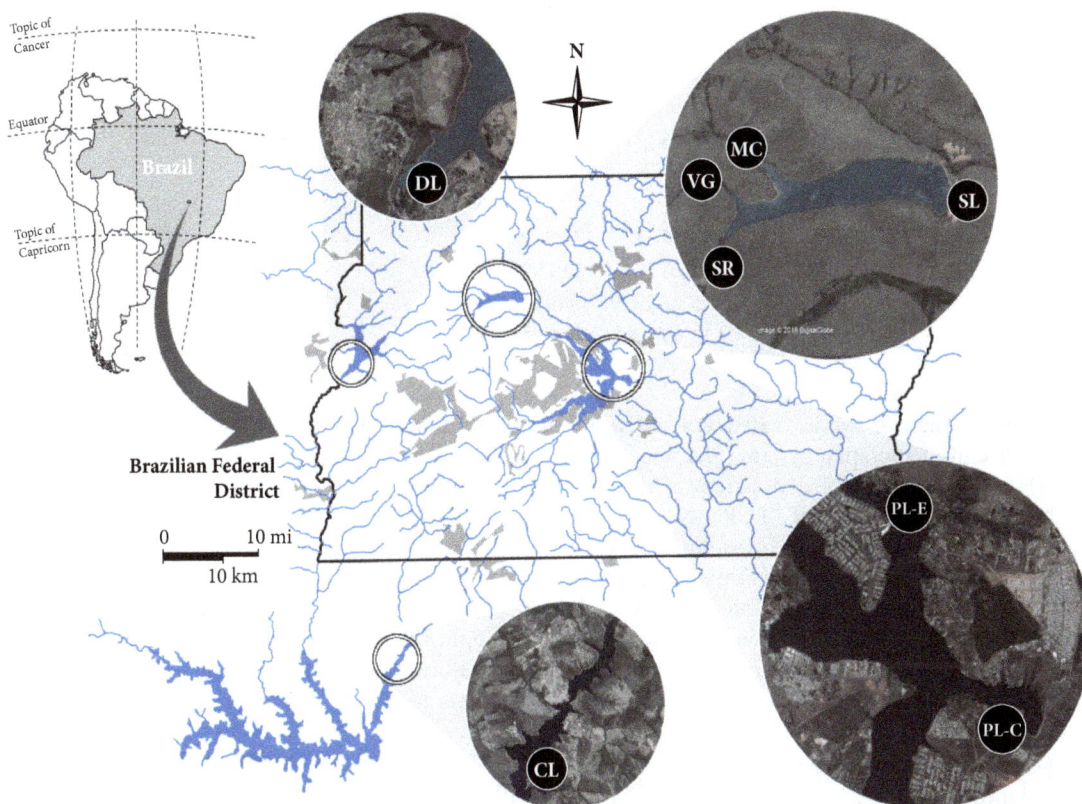

FIGURE 1: Map showing the Federal District in Brazil and the sampling points selected. DL: Descoberto Lake, SL: Santa-Maria Lake, SR: Santa-Maria River, VG: Vargem-Grande Stream, MC: Milho-Cozido Stream, CL: Corumbá Lake, PL-C: Paranoá Lake (conventional uptake), and PL-E: Paranoá Lake (emergency uptake).

drainage waters, effluents from two wastewater treatment plants, and other diffuse contributions [15, 21]. The sampling point PL-E is located in the water intake of the emergency water treatment plant (WTP), in operation since October 2017, while the sampling point PL-C is located at the water intake of the conventional WTP that will be in permanent operation in the near future.

Finally, the sampling point CL is located in one of the branches of Corumbá Lake, an artificial lake formed for electric power generation. This lake faces similar problems to those suffered by Descoberto Lake.

A total of 25 samples were collected in different sampling campaigns. Most of them (60%) were from Paranoá Lake. In this case, nine samples were obtained from the conventional water intake point whereas six were obtained at the emergency intake point. Three samples were from Descoberto Lake, three were from Santa-Maria Lake, and one sample was from Corumbá Lake. The remaining three samples were collected in three tributaries of the Santa-Maria Lake. Table 1 shows details on the samples and sampling periods.

Samples from different depths were collected using a Van Dorn water sampler and transferred to amber glass bottles (1 L). Surface samples were obtained directly into glass bottles. All bottles were previously cleaned in the laboratory and rinsed with the sampled water on site. Samples were transported to the laboratory on ice and preserved at 4°C until further preparation steps.

2.3. Sample Preparation. In the laboratory, samples were first passed through one or more glass fiber filters (GF-3, 0.7 μm, 47 mm diameter, Macherey-Nagel, Germany) and then through 0.45 μm pore-sized nitrocellulose membranes (47 mm diameter, Scharlau, Spain). Solid-phase extraction (SPE) of the analytes was carried out using a procedure described elsewhere [22]. Briefly, filtered samples (1 L) were passed through 200 mg SPE cartridges (HLB Oasis, Waters, USA) fitted to a lab-made extraction system [23] and to a peristaltic pump (Minipuls Evolution, Gilson, USA). Cartridges were previously conditioned using two aliquots (5 mL) of methanol followed by two aliquots (5 mL) of ultrapure water. Samples were passed through the solid phase at a flow rate of 5 mL min^{-1}. After extraction, the cartridges were centrifuged (1 min, 4000 rpm) to remove the excess of water and dried with a gentle flow of N_2. Analytes were eluted into precleaned glass tubes in a 12-port Prep Sep vacuum manifold (Visiprep DL, Supelco, USA) using two aliquots (3 mL) of methanol followed by another aliquot of a mixture of methanol and acetone (1:1, v/v). The eluates were transferred to evaporation flasks for concentration in a 12-vessel parallel evaporator (Syncore Analyst, Büchi, France) to a final volume of 1.0 mL.

TABLE 1: Characteristics of the investigated samples and acronyms used in the present work.

Aquatic system	Acronym	Coordinates	Sampling depth (m)	Season
Paranoá Lake (Conventional)	PL-C	15°47'36.9"S 47°47'22.9"W	1, 5 and 10	Rainy and dry
Paranoá Lake (Emergency)	PL-E	15°44'37.2"S 47°49'51.9"W	1, 5 and 10	Rainy
Corumbá Lake	CL	16°12'26.7"S 48°09'55.2"W	0	Dry
Descoberto Lake	DL	15°46'41.5"S 48°13'52.9"W	9 and 16	Rainy (only 16 m) and dry
Santa-Maria Lake	SL	15°40'33.2"S 47°57'19.6"W	9 and 16	Rainy (only 16 m) and dry
Santa-Maria River	SR	15°40'58.1"S 48°01'09.8"W	0	Dry
Vargem-Grande Stream	VG	15°40'23.4"S 48°01'11.3"W	0	Dry
Milho-Cozido Stream	MC	15°40'11.8"S 48°00'24.0"W	0	Dry

2.4. Instrumentation.

Analyses were carried out using an Expert Ultra LC100 chromatographic system (Eksigent Technologies, USA) consisting of a binary pump, a vacuum degasser, a thermostated autosampler (LC100-XL), and a thermostated column oven, coupled to a hybrid quadrupole-time-of-flight tandem mass spectrometer (TripleTOF 5600+, Sciex, Canada) with a DuoSpray ion source interface operated in the electrospray ionization (ESI) mode. Nitrogen was produced by a high-purity generator (Genius 3010, Peak Scientific, USA) and used as source gas.

2.5. Chromatographic Separation.

Separation was performed using a C18 column (Kinetex 2,6 μm EVO, 100 Å, 50 × 4.6 mm) obtained from Phenomenex (USA) at 40°C with gradient elution under a flow rate of 0.80 mL/min. Formic acid solutions (0.1% v/v) prepared in ultrapure water and in methanol were used as mobile phase solvents. The gradient was achieved after 3.5 min by keeping the relative methanol concentration at 15% (initial condition) for 0.5 min, followed by an increase to 50% in 0.5 min, and held constant for 2.5 min. After the elution of analytes, the column was washed with 95% methanol for 1.5 min, readjusted to the initial conditions, and reequilibrated for 2 min. During analyses, the autosampler was kept under 8°C and the injection volume was 2.0 μL.

2.6. Mass Spectrometry Conditions.

Successive injections of a 100 μg/L caffeine solution were performed to optimize ESI gas parameters in a 2^3 factorial design. Auxiliary ion source gas 1 (GS1) and ion source gas 2 (GS2) were used as nebulizer and drying gases, respectively, at a back pressure of 60 psi for both parameters. Curtain gas (CUR) was 40 psi. The ESI interface was operated in the positive mode at 650°C with a capillary voltage of 5500 V. Under these conditions the highest sensitivity without in-source fragmentation was observed.

Data acquisition was performed using the high-resolution multiple reaction monitoring (HR-MRM) mode. Firstly, precursor ions were selected by direct infusions of a 0.1% formic acid solution prepared in methanol containing 400 μg/L of each analyte using a declustering potential (DP) of 100 V and a collision energy (CE) of 10 eV. Then, after preliminary direct infusions, product-ions and definitive DP and CE were obtained for each analyte by chromatographic injections of a 100 μg/L mix solution using a 2^2 factorial design. Table 2 shows the HR-MRM parameters obtained for each analyte and their chromatographic retention time.

The mass spectrometry system was firstly calibrated using the CsI/ALILTLVS solution under direct infusion. Then, a CMZ solution was used for routine mass tuning on a daily basis. During the analyses, a mix solution was injected every five chromatographic runs for mass calibration. An error up to 2 ppm was considered acceptable. Formic acid was added to all working solutions to favor positive ionization of the target compounds.

2.7. Validation.

Analytical curves were tested for the homogeneity of variances by the Cochran test. Outliers were verified by the Grubbs test. For the heteroscedastic data, a weighted least squares regression method was performed using weighting factors that produced the lowest sum of the relative errors [24]. Repeatability was based on the analysis of five-point analytical curves performed by the same analyst on the same day. The matrix effect was evaluated for each compound at three concentration levels (5, 95, and 190 μg/L, on-column) using two replicates for each level. It was estimated by comparing the slopes of response curves made in solvent (methanol) and in extracts of a Paranoá Lake sample collected at 1 m depth. Extraction efficiency (in %) was also achieved for another Paranoá Lake sample.

2.8. Risk Assessment.

The environmental risk was assessed by calculating the risk quotient (RQ_{env}) based on the MEC/PNEC ratio, where MEC stands for the measured environmental concentrations obtained in the surface water samples. RQ_{env} was calculated considering the most restrictive Predicted No-Effect Concentration (PNEC) found in the literature for each investigated contaminant. Human health risks were also evaluated by risk quotients (RQ_{hum}) comparing the target chemicals concentrations with water quality criteria (WQC) calculated using

$$WQC = \frac{ADI \times P \times BW}{C}, \quad (1)$$

where ADI is the acceptable daily intake, in mg/kg, P is the allocation factor considering the percentage of the contaminant ingested via water consumption, BW is the body weight, and C is the daily water consumption. Default values for P (20%), BW (60 kg), and C (2 L) were used considering water consumption for an adult according to the Guidelines for Drinking-Water Quality [25].

TABLE 2: Acquisition parameters used in the UPLC-QTO/MS system.

Analyte	Formula	Exact mass (Da)	Precursor ion (Da)	Product-ion (Da)	DP (V)	CE (eV)	RT (min)
CAF	$C_8H_{10}N_4O_2$	194.08037	195.0877	138.0662[a] 195.0877 110.0349	100	25	1.74
CMZ	$C_{15}H_{12}N_2O$	236.094963	237.1022	194.0964[a] 237.1022 192.0808	70	30	2.27
ATZ	$C_8H_{14}ClN_5$	215.093773	216.1010	174.0541[a] 216.1010 104.0010	100	30	2.55

[a]Transitions used for quantification. DP: declustering potential, CE: collision energy, EP: entrance potential, CEP: collision cell entrance potential, CXP: collision cell exit potential, RT: retention time.

3. Results and Discussion

3.1. Linearity and Limit of Quantification. For all analytes, the five-point analytical curves were heteroscedastic according to the Cochran test. Also, no outliers were depicted using the Grubbs test. The best weighting factor for caffeine and atrazine was $1/\sigma$, while for carbamazepine a weighting factor of $1/\sqrt{y}$ provided the lowest sum of the relative errors. Under these conditions, correlation coefficients were significant and higher than 0.99, as can be seen in Table 3.

The limit of quantification (on-column) was admitted as the lowest concentration of the analytical curves for all analytes, i.e., 0.48 μg/L. Under these conditions, signal-to-noise ratios (S/N) were higher than 10.

3.2. Precision. Table 4 shows the precision obtained during the construction of weighted analytical curves by external calibration.

Precision was considered satisfactory for all analytes since coefficients of variance below 5% were observed, with the exception of caffeine, where values of 6% were obtained for the highest concentration levels. During the experiments, it was noticed that the standard deviation of the analytical curves was influenced by the ambient temperature, the cleaning and maintenance of the mass spectrometer orifice plate, the stabilization of the analytical signals, and mainly the constant calibration of the exact mass, which must be done frequently throughout the analysis. Thus, results shown in Table 2 were obtained under constant room temperature (20°C), after adequate cleaning of the source and other components of the mass spectrometry system and by periodic injections of the mass calibration solution. The automatic integration of the peaks also influenced the precision, being necessary to check and adjust the baseline manually, mainly for low concentrations. In this sense, higher precision was obtained by parameterizing the noise reduction by 100%.

3.3. Accuracy. The matrix effect (ME) was evaluated by plotting two curves: the first one obtained by the analysis of three solutions containing increasing concentrations of the analytes in methanol and the second one with the same concentrations of the analytes in a sample matrix, i.e., an extract of a Paranoá Lake sample collected at 1 m depth. Both curves were plotted with three concentration levels due to the small amount of sample extract available. Figure 2 portrays linear correlations obtained during ME experiments.

For all analytes, the matrix effect was manifested in order to attenuate the analyte response with a tendency to under-estimate higher concentrations. Caffeine suffered less influence of the matrix (14% attenuation), followed by atrazine (18%) and carbamazepine (19%). The slopes were statically compared using Student's t-test based on the standard errors of the regressions [26]. Experimental t-values for caffeine, carbamazepine, and atrazine were 3.713, 1.113, and 2.048, respectively, being below the critical value of 4.303 (95% confidence interval) and indicating that the matrix effect was not significative. For caffeine, more intense responses were observed for the in-extract curve in comparison to the in-solvent one. Table 5 shows that there was satisfactory recovery for caffeine at the three concentrations investigated. Satisfactory recovery rates at high concentrations (190 ng/L) are important for caffeine, considering that higher levels of this substance are expected in Brazilian natural waters compared to the other tested analytes [17, 27, 28].

It is observed in Figure 2 that the central points of the curves in solvent and in extract for carbamazepine and atrazine were similar, whereas maximum and minimum concentrations differ leading to the attenuation of the sensitivity in the curves plotted in extract. Considering only the lowest concentrations, i.e., compatible with the environmental levels commonly found in the region [15], a 13% attenuation was estimated for carbamazepine. For atrazine, there was no change in attenuation at the lower concentration levels. However, for both analytes Table 5 shows more satisfactory recoveries for the lower concentration levels, making the results achieved also satisfactory.

Accuracy was also assessed by a recovery test for extraction efficiency. In this case, a sample of Paranoá Lake was enriched with 55 ng/L of each analyte and submitted to extraction and analysis using the weighted analytical curves described in Table 4. Table 5 shows satisfactory recoveries ranging from 78±5 to 112±7%.

TABLE 3: Work range and linearity of the external calibration curves.

Analyte	Work range (μg/L)	Weight	R^2	LOQ (μg/L)	S/N at LOQ
CAF	0.48 to 300	$1/\sigma$	0,99	0.48	13.7
CMZ	0.48 to 300	$1/\sqrt{y}$	0,99	0.48	20.6
ATZ	0.48 to 300	$1/\sigma$	0,99	0.48	16.6

TABLE 4: Coefficients of variance for intraday analysis of mixed solutions of the analytes.

Concentration[a]	Precision (%)[b]		
	Atrazine	Caffeine	Carbamazepine
0.48	1	5	3
2.4	4	2	4
12	3	5	2
60	2	6	5
300	1	6	5

[a]Concentrations in column (μg/L) used the for the construction of analytical curves. [b]Peak areas for six replicates

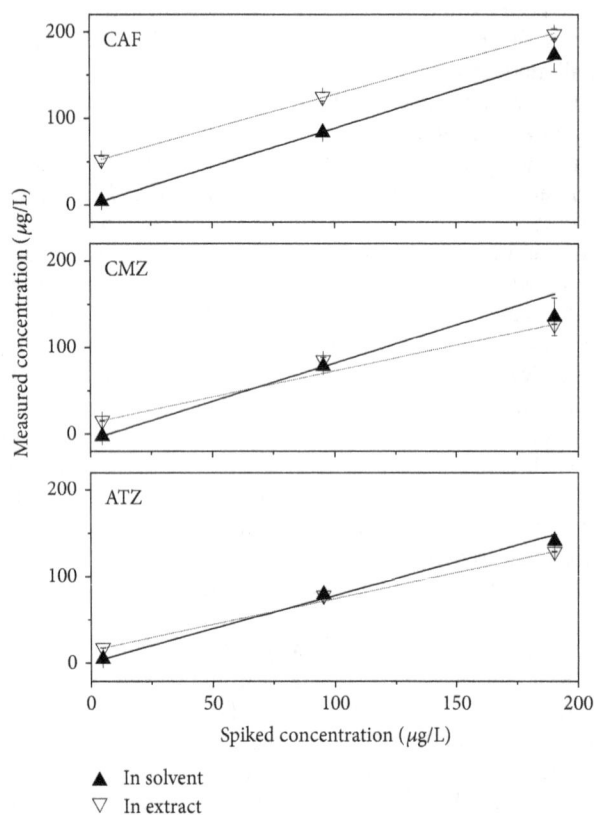

FIGURE 2: Matrix effects on the analytical response of the investigated contaminants.

3.4. *Emerging Contaminants in Water Sources.* Method limits of quantification (LOQm) were expressed by the instrument LOQ (Table 3) multiplied by the extraction recovery (Table 5) and divided by the preconcentration factor of 1000 times. Values for caffeine, carbamazepine, and atrazine were 0.49, 0.37, and 0.54 ng/L, respectively. Prior to analysis, extraction blank controls, obtained by the analysis of 1 L of ultrapure water, in triplicate, revealed the presence of 4.4 ng/L of caffeine. As solvent blank controls did not reveal the presence of any analyte, this interfering concentration was considered in the calculation of the final concentrations of caffeine in the samples.

Table 6 shows the concentrations of caffeine, carbamazepine, and atrazine in the water sources in current use in the FD.

Only caffeine and atrazine were detected in Descoberto and Santa-Maria Lakes in all samples investigated. The concentration of the former was higher varying between 10 and 32 ng/L in Descoberto Lake and from 4.8 to 10 ng/L in Santa-Maria Lake.

Higher levels of caffeine in the Descoberto Lake are expected due to the occupation of adjacent areas by condominiums and by the increasing population density observed in the region in the last years. These factors may contribute to the presence of caffeine, a known tracer of human activities [20, 29, 30]. However, the presence of caffeine in Santa-Maria Lake was not expected, even under lower concentrations, since this compartment is under restricted access within the borders of the National Park of Brasília. In view of these results, we sought to investigate samples from three tributaries from this reservoir in an attempt to trace possible sources of contamination. Results revealed significantly higher concentrations (75 to 123 ng/L) of caffeine in these streams (Table 6) compared to those determined in Santa-Maria Lake. Therefore, there is evidence of potentially contaminating human activities in the drainage areas of the tributaries. The possible causes of the presence of caffeine in these streams have not been identified yet, but high levels of coliforms in these tributaries (data not shown) suggest a common source for chemical and biological contamination.

Atrazine levels varied between 2.4 and 5.5 ng/L in both compartments and probably arise due to minor diffuse sources related to agricultural activities in the surrounding areas. No significant differences were observed when results from different seasons were compared indicating that pollution processes may be stable over the year.

Table 7 shows the levels of the emerging contaminants investigated in the future sources of water in the FD.

TABLE 5: Percentage of recovery obtained for spiked extracts and for a Paranoá Lake sample spiked with all analytes.

Samples	Recovery (%)		
	Caffeine	Carbamazepine	Atrazine
Extract[a] spiked with 5.0 μg/L	83±11	79±9	107±6
Extract[a] spiked with 95 μg/L	81±8	77±9	74±7
Extract[a] spiked with 190 μg/L	80±8	62±8	64±6
Natural water[b] spiked with 55 ng/L	102±6	78±5	112±7

[b]Extracts of a Paranoá Lake sample (1 m) obtained after solid phase extraction. [a]Filtered (0.45 um) Paranoá Lake sample.

TABLE 6: Concentrations of caffeine, carbamazepine, and atrazine in the current water sources of the Brazilian Federal District and in selected tributaries.

Analytes	Concentration (ng/L)						
	DL		SL		SR	VG	MC
	9 m	16 m	9 m	16 m	0 m	0 m	0 m
CAF	13 (D)	32 (R) 10 (D)	4.8 (D)	10 (R) 7.0 (D)	83 (D)	75 (D)	123 (D)
CMZ	ND (D)	ND (R) ND (D)	ND (D)	ND (R) ND (D)	ND (D)	ND (D)	ND (D)
ATZ	5.5 (D)	2.8 (R) 4.8 (D)	3.4 (D)	2.4 (R) 2.9 (D)	ND (D)	ND (D)	ND (D)

CAF: Caffeine, CMZ: Carbamazepine, ATZ: Atrazine, DL: Descoberto Lake, SL: Santa-Maria Lake, SR: Santa-Maria River, VG: Vargem-Grande Stream, MC: Milho-Cozido Stream, ND: Not detected, R: Rainy season, D: Dry season

TABLE 7: Concentrations of caffeine, carbamazepine, and atrazine in the future water sources of the Brazilian Federal District.

Analyte	Concentration (ng/L)						
	CL	PL-C			PL-E		
	0 m	1 m	5 m	10 m	1 m	5 m	10 m
CAF	149 (D)	77 (R) 81 (R) 58 (D)	77 (R) 61 (R) 43 (D)	50 (R) 39 (R) 56 (D)	49 (R) 103 (R)	121 (R) 80 (R)	180 (R) 59 (R)
CMZ	8.5 (D)	17 (R) 15 (R) 15 (D)	25 (R) 15 (R) 18 (D)	15 (R) 13 (R) 12 (D)	5.4 (R) 8.9 (R)	21 (R) 11 (R)	9.0 (R) 10 (R)
ATZ	9.3 (D)	9.4 (R) 13 (R) 10 (D)	13 (R) 11 (R) 9.0 (D)	5.8 (R) 7.7 (R) 8.0 (D)	3.9 (R) 6.9 (R)	15 (R) 7.0 (R)	7.6 (R) 6.8 (R)

CAF: Caffeine, CMZ: Carbamazepine, ATZ: Atrazine, CL: Corumbá Lake, PL-C: Paranoá Lake (Conventional uptake), PL-E: Paranoá Lake (Emergency uptake), R: Rainy season, D: Dry season

The three analytes investigated were found in all samples of the Corumbá and Paranoá Lakes. Concentrations of caffeine were higher in both lakes, followed by carbamazepine and atrazine. As expected, concentrations were also consistently higher in these reservoirs compared to the current water sources of the FD. The presence of contaminants in samples from different depths indicates the vertical mixture of the waters of the Paranoá Lake. However, no further tendency was depicted within the samples investigated.

No significant differences were observed in the results considering both sampling points of Paranoá Lake. Caffeine Levels varied between 39 and 180 ng/L in Paranoá Lake considering all investigated samples, corroborating previous reports regarding such contaminant in the lake. In the sampling point PL-C, Abbt-Braun et al. [31] report caffeine concentrations varying from 28 to 193 ng/L during sampling campaigns carried out in 2010. Our results, from samples collected in 2017, show that the sources of contamination remained stable over the last few years in the surroundings of the lake. In the emergency point (PL-E), previous studies revealed caffeine levels varying between 29 and 138 ng/L [16, 31]. In Corumbá Lake, a similar concentration was depicted in comparison with the results portrayed for Paranoá Lake, suggesting similar degrees of anthropic influence in both reservoirs.

Carbamazepine concentrations ranged from 5.4 to 25 ng/L in the samples from Paranoá Lake. For Corumbá Lake, a concentration of 8.5 ng/L was obtained. Again, no significative differences were observed between sampling points, seasons, and depths investigated. A previous report also found carbamazepine in both sampling points of Paranoá Lake in concentrations varying from <5 to 16 ng/L. Atrazine

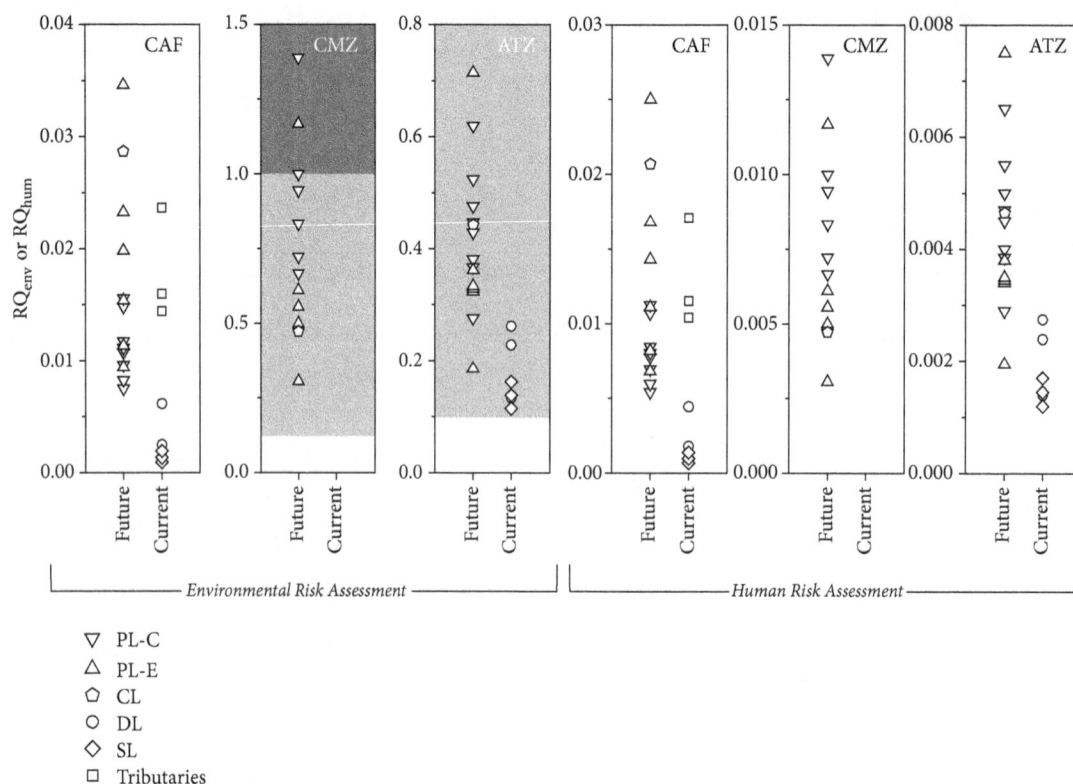

FIGURE 3: Environmental and human risk assessments of selected emerging contaminants in future and current water sources of the Brazilian Federal District. Dark and light grey regions indicate risk and possible risk, respectively. Blank regions indicate no risk.

levels, varying from 3.9 to 15 ng/L, were also similar to previous results corroborating with the scenario of contamination that has remained stable since 2010 in the FD.

3.5. Risk Assessment.
Figure 3 portrays risk quotients for environmental and human risk assessment considering the presence of caffeine, carbamazepine, and atrazine in the water sources of the FD.

For the environmental risk assessment, risk quotients (RQ_{env}) were calculated as suggested by the Technical Guidance Document on Risk Assessment of the European Commission [32], where the measured environmental concentrations (MEC), portrayed in Tables 6 and 7, are evaluated against previously reported PNEC values. The PNEC can be described as the concentration limit at which harmful effects on organisms will most likely not occur. For aquatic systems, a PNEC should be derived that, if not exceeded, ensures an overall protection of the environment. In the present work, the most restrictive PNECs, representing the worst-case scenario, were selected in literature for carbamazepine (18 ng/L) [33], atrazine (21 ng/L) [34], and caffeine (5200 ng/L) [35].

As PNEC is an estimate, a restrictive demarcation of what is "acceptable" or "not acceptable" is not possible for MEC values below or above this parameter, respectively. Therefore, for a more realistic risk assessment, it is considered that RQ greater than 1 may imply risk while values lower than 0.1 indicate no risk. Intermediate values indicate possible risk as well as the need for further studies [28, 35]. Figure 3

shows no ecological risks for caffeine in both current and future water sources of the FD. Higher RQ_{env} is noticed for carbamazepine, with three samples from Paranoá Lake presenting environmental risk. The remaining samples are classified in a situation of possible risks. Considering that carbamazepine was not detected in the current water sources of the FD, no risk was depicted. For atrazine, all samples investigated in the present work were in the range of possible risks.

For human risk assessment, it was also considered a worst-case scenario where removal efficiency during drinking water treatment processes was null. Water quality criteria were derived considering ADI data available in the literature for carbamazepine (300 ng/kg) [6, 36] and caffeine (1200 ng/kg) [37]. Using (1), these values provide WQCs of 1800 and 7200 ng/L for carbamazepine and caffeine, respectively. For atrazine, a WQC of 2000 ng/L, corresponding to the drinking water standard proposed by the Brazilian Ministry of Health [38], was considered. Human risk quotients (RQ_{hum}) portrayed in Figure 3 were based on MEC/WQC ratios and indicate no risk for all analytes investigated in this work, to the best of our knowledge.

4. Conclusions

A method based on the solid-phase extraction followed by quantification using liquid chromatography coupled to high-resolution hybrid mass spectrometry (UPLC-QTOF/MS) was developed and applied for the quantification of caffeine, carbamazepine, and atrazine in water sources of the Brazilian Federal District.

Accuracy was considered satisfactory considering matrix effects, as well as recoveries experiments, carried out with samples collected in Paranoá Lake. Precision was also satisfactory under weighted least squares regressions using the most appropriated weighting factors.

Concentrations of the investigated analytes were consistently higher in the future water sources as they receive urban drainage waters, effluents from wastewater treatment plants, and other diffuse contributions. As a result, possible environmental risks were depicted for carbamazepine in the future water sources. Atrazine levels in all water sources were also in a range of possible environmental risks. No risk for human health was estimated based on the worst-case scenario where removal in water treatment plants is not achieved.

Our results point towards a crucial role of indirect water reuse in situations of water scarcity and rationing. Receiving waters may contain several contaminants of recent concern that should be investigated to ensure the safe use of water for different purposes. Although risks to human health have not been evidenced in this work, our results may be useful for constructing a reliable contamination scenario to other alternative and more complete risk assessment models.

Acknowledgments

This research was supported by the Brazilian Funding for Innovation and Research (FINEP 01.13.0470.00) and by the Federal District Research Foundation (FAPDF 193.000.714/2016). The authors thank the Environmental Sanitation Company of the Federal District (CAESB) for the operational support during the sampling and the Analytical Center of the Institute of Chemistry at the University of Brasília for providing access to the UPLC-QTOF/MS used in this work.

References

[1] J. A. Soller, M. H. Nellor, C. J. Cruz, and E. McDonald, "Human health risk associated with direct potable reuse-evaluation through quantitative relative risk assessment case studies," *Environmental Science: Water Research and Technology*, vol. 1, no. 5, pp. 679–688, 2015.

[2] M. Brienza, M. Mahdi Ahmed, A. Escande et al., "Use of solar advanced oxidation processes for wastewater treatment: Follow-up on degradation products, acute toxicity, genotoxicity and estrogenicity," *Chemosphere*, vol. 148, pp. 473–480, 2016.

[3] S. D. Richardson, "Water analysis: emerging contaminants and current issues," *Analytical Chemistry*, vol. 75, no. 12, pp. 2831–2857, 2003.

[4] E. T. Furlong, A. L. Batt, S. T. Glassmeyer et al., "Nationwide reconnaissance of contaminants of emerging concern in source and treated drinking waters of the United States: Pharmaceuticals," *Science of the Total Environment*, vol. 579, pp. 1629–1642, 2017.

[5] M. Raghav, S. Eden, K. Mitchell, and B. Witte, "Contaminants of emerging concern in water," in *Water Resources Research Center College of Agriculture and Life Sciences*, Arroyo, 2013.

[6] V. de Jesus Gaffney, C. M. M. Almeida, A. Rodrigues, E. Ferreira, M. J. Benoliel, and V. V. Cardoso, "Occurrence of pharmaceuticals in a water supply system and related human health risk assessment," *Water Research*, vol. 72, pp. 199–208, 2015.

[7] S. Tewari, R. Jindal, Y. L. Kho, S. Eo, and K. Choi, "Major pharmaceutical residues in wastewater treatment plants and receiving waters in Bangkok, Thailand, and associated ecological risks," *Chemosphere*, vol. 91, no. 5, pp. 697–704, 2013.

[8] A. Joss, S. Zabczynski, A. Göbel et al., "Biological degradation of pharmaceuticals in municipal wastewater treatment: Proposing a classification scheme," *Water Research*, vol. 40, no. 8, pp. 1686–1696, 2006.

[9] V. L. Flöter, G. Galateanu, R. W. Fürst et al., "Sex-specific effects of low-dose gestational estradiol-17β exposure on bone development in porcine offspring," *Toxicology*, vol. 366-367, pp. 60–67, 2016.

[10] M. Parolini, S. Magni, S. Castiglioni, and A. Binelli, "Genotoxic effects induced by the exposure to an environmental mixture of illicit drugs to the zebra mussel," *Ecotoxicology and Environmental Safety*, vol. 132, pp. 26–30, 2016.

[11] K. A. Kidd, P. J. Blanchfield, K. H. Mills et al., "Collapse of a fish population after exposure to a synthetic estrogen," *Proceedings of the National Acadamy of Sciences of the United States of America*, vol. 104, no. 21, pp. 8897–8901, 2007.

[12] C. C. Montagner, G. A. Umbuzeiro, C. Pasquini, and W. F. Jardim, "Caffeine as an indicator of estrogenic activity in source water," *Environmental Science: Processes & Impacts*, vol. 16, no. 8, pp. 1866–1869, 2014.

[13] Z.-F. Chen, G.-G. Ying, Y.-S. Liu et al., "Triclosan as a surrogate for household biocides: An investigation into biocides in aquatic environments of a highly urbanized region," *Water Research*, vol. 58, pp. 269–279, 2014.

[14] S. A. Snyder and M. J. Benotti, "Endocrine disruptors and pharmaceuticals: Implications for water sustainability," *Water Science and Technology*, vol. 61, no. 1, pp. 145–154, 2010.

[15] G. Abbt-Braun, H. Börnick, C. C. S. Brandão et al., "Warer quality of tropical reservoirs in a changing world - the case of Lake Paranoá, Brasília, Brazil," in *Integrated Water Resource Management in Brazil Chapter*, C. Lorz, F. Makeschin, and H. Weiss, Eds., p. 152, IWA Publishing, London, 2014.

[16] N. Y. Mar da Costa, G. R. Boaventura, D. S. Mulholland et al., "Biogeochemical mechanisms controlling trophic state and micropollutant concentrations in a tropical artificial lake," *Environmental Earth Sciences*, vol. 75, no. 10, 2016.

[17] F. F. Sodré, J. S. Santana, T. R. Sampaio, and C. C. S. Brandão, "Seasonal and spatial distribution of caffeine, atrazine, atenolol and DEET in surface and drinking waters from the Brazilian Federal District," *Journal of the Brazilian Chemical Society*, vol. 29, no. 9, pp. 1854–1865, 2018.

[18] Y. Vystavna, P. Le Coustumer, and F. Huneau, "Monitoring of trace metals and pharmaceuticals as anthropogenic and socioeconomic indicators of urban and industrial impact on surface waters," *Environmental Modeling & Assessment*, vol. 185, no. 4, pp. 3581–3601, 2013.

[19] D. D. Bussan, C. A. Ochs, C. R. Jackson, T. Anumol, S. A. Snyder, and J. V. Cizdziel, "Concentrations of select dissolved

trace elements and anthropogenic organic compounds in the Mississippi River and major tributaries during the summer of 2012 and 2013," *Environmental Modeling & Assessment*, vol. 189, no. 2, 2017.

[20] S. Froehner, D. B. Souza, K. S. Machado, and E. C. Da Rosa, "Tracking anthropogenic inputs in Barigui River, Brazil using biomarkers," *Water, Air, & Soil Pollution*, vol. 210, no. 1-4, pp. 33–41, 2010.

[21] C. Lorz, G. Abbt-Braun, F. Bakker et al., "Challenges of an integrated water resource management for the Distrito Federal, Western Central Brazil: Climate, land-use and water resources," *Environmental Earth Sciences*, vol. 65, no. 5, pp. 1575–1586, 2012.

[22] M. B. Campanha, A. T. Awan, D. N. R. de Sousa, G. M. Grosseli, A. A. Mozeto, and P. S. Fadini, "A 3-year study on occurrence of emerging contaminants in an urban stream of São Paulo State of Southeast Brazil," *Environmental Science and Pollution Research*, vol. 22, no. 10, pp. 7936–7947, 2015.

[23] F. F. Sodré, M. A. F. Locatelli, and W. F. Jardim, "An in-line clean system for the solid-phase extraction of emerging contaminants in natural waters," *Química Nova*, vol. 33, pp. 216–219, 2010 (Portuguese).

[24] A. M. Almeida, M. M. Castel-Branco, and A. C. Falcão, "Linear regression for calibration lines revisited: weighting schemes for bioanalytical methods," *Journal of Chromatography B*, vol. 774, no. 2, pp. 215–222, 2002.

[25] WHO, *Guidelines for Drinking-Water Quality*, World Health Organization Press, Geneva, 2011.

[26] J. M. Andrade and M. G. Estévez-Pérez, "Statistical comparison of the slopes of two regression lines: A tutorial," *Analytica Chimica Acta*, vol. 838, pp. 1–12, 2014.

[27] K. C. Machado, M. T. Grassi, C. Vidal et al., "A preliminary nationwide survey of the presence of emerging contaminants in drinking and source waters in Brazil," *Science of the Total Environment*, vol. 572, pp. 138–146, 2016.

[28] F. F. Sodré, P. M. Dutra, and V. P. Dos Santos, "Pharmaceuticals and personal care products as emerging micropollutants in Brazilian surface waters: a preliminary snapshot on environmental contamination and risks," *Eclética Química Journal*, vol. 43, no. 1SI, pp. 22–34, 2018.

[29] I. J. Buerge, T. Poiger, M. D. Müller, and H. Buser, "Caffeine, an anthropogenic marker for wastewater contamination of surface waters," *Environmental Science & Technology*, vol. 37, no. 4, pp. 691–700, 2003.

[30] F. F. Sodré, M. A. F. Locatelli, and W. F. Jardim, "Occurrence of emerging contaminants in Brazilian drinking waters: A sewage-to-tap issue," *Water, Air, & Soil Pollution*, vol. 206, no. 1-4, pp. 57–67, 2010.

[31] G. Abbt-Braun, E. Worch, H. B et al., *Progress Report of the Brazilian-German IWRM Project IWAS-Água DF*, Dresden University of Technology, Brasília, 2012.

[32] J. De Bruijn, B. G. Hansen, S. Johansson et al., *Technical Guidance Document on Risk Assessment. Part II: Environmental Risk Assessment*, Luxembourg, 2003.

[33] B. I. Escher, R. Baumgartner, M. Koller, K. Treyer, J. Lienert, and C. S. McArdell, "Environmental toxicology and risk assessment of pharmaceuticals from hospital wastewater," *Water Research*, vol. 45, no. 1, pp. 75–92, 2011.

[34] A. E. Girling, L. Tattersfield, G. C. Mitchell et al., "Derivation of predicted no-effect concentrations for lindane, 3,4-dichloroaniline, atrazine, and copper," *Ecotoxicology and Environmental Safety*, vol. 46, no. 2, pp. 148–162, 2000.

[35] K. Komori, Y. Suzuki, M. Minamiyama, and A. Harada, "Occurrence of selected pharmaceuticals in river water in Japan and assessment of their environmental risk," *Environmental Modeling & Assessment*, vol. 185, no. 6, pp. 4529–4536, 2013.

[36] H. W. Leung, L. Jin, S. Wei et al., "Pharmaceuticals in tap water: Human health risk assessment and proposed monitoring framework in China," *Environmental Health Perspectives*, vol. 121, no. 7, pp. 839–846, 2013.

[37] R. S. Prosser and P. K. Sibley, "Human health risk assessment of pharmaceuticals and personal care products in plant tissue due to biosolids and manure amendments, and wastewater irrigation," *Environment International*, vol. 75, pp. 223–233, 2015.

[38] Brazil, *Brazilian Drinking Water Standards (Ordinance 2914/2011)*, Ministry of Health, Brasília, 2011.

14

Development of a Solid-Phase Extraction (SPE) Cartridge based on Chitosan-Metal Oxide Nanoparticles (Ch-MO NPs) for Extraction of Pesticides from Water and Determination by HPLC

Mohamed E. I. Badawy⬡, Mahmoud A. M. El-Nouby, and Abd El-Salam M. Marei

Department of Pesticide Chemistry and Technology, Faculty of Agriculture, Alexandria University, El-Shatby, Alexandria 21545, Egypt

Correspondence should be addressed to Mohamed E. I. Badawy; m_eltaher@yahoo.com

Academic Editor: Valentina Venuti

The present study aims to prepare two new types of chitosan-metal oxide nanoparticles (Ch-MO NPs), namely, chitosan-copper oxide nanoparticles (Ch-CuO NPs) and chitosan-zinc oxide nanoparticles (Ch-ZnO NPs), using sol-gel precipitation mechanism, and test them new as adsorbent materials for extraction and clean-up of different pesticides from water. The design of core-shell was implemented by metal oxide core with chitosan as a hard shell after crosslinking mechanism by glutaraldehyde and then epichlorohydrin. The characterizations of the prepared nanoparticles were investigated using Fourier transform infrared spectrometry (FT-IR), zeta potential, scanning electron microscopy (SEM), transmission electron microscope (TEM), and X-ray diffraction (XRD). FT-IR confirmed the interaction between chitosan, metal oxide, and crosslinking mechanism. SEM and TEM explained that the nanoparticles have a spherical morphology and nanosize of 93.74 and 97.95 nm for Ch-CuO NPs and Ch-ZnO NPs, respectively. Factorial experimental design was applied to study the effect of pH, concentration of pesticide, agitation time, and temperature on the efficiency of adsorption of pesticides from water samples. The results indicated that optimum conditions were pH of 7, temperature of 25°C, and agitation time of 25 min. The SPE cartridges were then packed with Ch-MO NPs, and seven pesticides of abamectin, diazinon, fenamiphos, imidacloprid, lambda-cyhalothrin, methomyl, and thiophanate-methyl were extracted from water samples and determined by HPLC. The extraction efficiency of Ch-ZnO NPs was higher than Ch-CuO NPs, but both removed a larger amount of most of tested pesticides than the standard ODS cartridge (C18). The results showed that this method achieves rapid and simple extraction in small quantities of adsorbents (Ch-MO NPs) and solvents. In addition, the method is highly sensitive to pesticides and has a high recovery rate.

1. Introduction

Pesticides are widely used in agricultural production to prevent or control pests, diseases, weeds, and other plant pathogens in an effort to minimize or eliminate yield losses and maintain high quality of products [1, 2]. Widespread uses of pesticides with all groups such as organochlorines, organophosphorus, carbamates, pyrethroids, and neonicotinoids have resulted in extensive contamination of water, atmosphere, and soil as well as agricultural products which eventually lead to food safety issues [3]. Water contamination with pesticides is considered a serious problem and may

pose a risk to human health such as acute neurological toxicity, neurodevelopmental impairment, cancer, allergies, neurological disorders, and reproductive disorders [2, 4–6].

Different analytical techniques have been used for sample preparation and clean-up with differentiation of sensitivity and selectivity [7], which include liquid-liquid extraction (LLE) [8], solid-phase extraction (SPE) [9], solid-phase microextraction (SPME) [10], dispersible solid-phase extraction (d-SPE), headspace solid-phase extraction [11], and stir bar sorptive extraction (SBSE) [12]. SPE was introduced in the early 1970s to avoid and minimize the disadvantage of LLE technique. The SPE is a superior extraction and clean-up method that uses a solid phase and a liquid phase to separate the analyte from the sample without impurities before analysis by dint of speed, less usage of organic solvent, low cost, and ability to obtain a higher preconcentration factor [13, 14]. Recently, advanced materials for SPE extraction have been investigated with separation by liquid chromatography and ultraviolet absorption detection (HPLC/UV) [15–17].

Some of the most common sorbents in SPE are generally similar to those in column liquid chromatography such as the primary secondary amine (PSA), octadecyl-siloxane (C_{18}), graphitized carbon black (GCB), alumina, and florisil. PSA is normally used in the d-SPE to remove interferences, such as free fatty acids, sugars, and other nonpolar compounds from the sample. However, the most commercial stationary phase used in SPE is octadecyl-siloxane (C_{18}) used in the reversed phase to extract the nonpolar compounds like pesticides [18, 19].

Recently, the biopolymer materials have been shown to be of low cost and good efficiency in removal of various contaminants from aqueous media. Among these biopolymers, chitosan (poly-β-(1⟶4)-2-amino-2-deoxy-D-glucose) [20, 21] has been considered to be one of the most promising and applicable materials in adsorption applications [22]. The existence of the two functional groups of hydroxyl (-OH) and amino (-NH_2) in its molecular structure contributes to many possible adsorptions and gives highly powerful removal capacity of dyes, metal ions, phenols, pharmaceuticals drugs, and other pollutants including the pesticides from environment and wastewater [23].

Metal oxide particles have been used in many functions [24] in various polymeric materials to improve the permanence of the polymeric products [25]. In addition, the nanoparticles of these products could increase the stiffness, toughness, and service life of polymeric materials [26]. The chitosan-metal oxide complexes in nanostructure form have been extensively modified to improve the adsorption capacity of chitosan molecule because of their limited size and a high density in their corner or edge surface sites [6]. Dehaghi and coauthors synthesized the chitosan-ZnO nanoparticles (Ch-ZnO NPs) for adsorption applications in the removal of pesticide pollutants [25]. They found that the 0.5 g of the Ch-ZnO NPs, in room temperature and pH 7, removed 99% of permethrin insecticide solution (0.1 mg/L). Copper-coated chitosan nanocomposite (Ch-Cu) was prepared and used for adsorption of parathion and methyl parathion insecticide in the batch mode. The maximum adsorption capacity of malathion was found to be 322.6 mg/g at an optimum pH of 2.0. The adsorbent was found to remove malathion completely from the spiked concentration of 2 mg/L in one min in the agricultural run-off samples [24].

Chemical modification promotes crosslinking of the polymer chains. This process consists of joining polymer chains with the help of high reactivity chemicals called crosslinking agents, generating polymer networks. This modification type is only possible by the presence of functional groups of high reactivity in the structure of these polymers. Most notably, glutaraldehyde and epichlorohydrin as crosslinking agents considerably improve the mechanical strength, the hardness of the chitosan particles, and the

chemical stability in acidic media [27, 28]. Epichlorohydrin was selected as a convenient base catalyzed crosslinking agent. An advantage of epichlorohydrin is that it does not eliminate the cationic amine function of chitosan, but it reacts with hydroxyl groups in chitosan. Glutaraldehyde has been used more frequently since it is less expensive, nontoxic, and highly soluble in aqueous solution. It is a dialdehyde whose aldehydic groups are highly reactive and can form covalent bonds with functional groups such as primary amine by Schiff base suggesting that the conjugated aldehyde moieties in the polymers yield more stable reaction products [29–31].

In the current study, new chitosan-metal oxide nanoparticles (Ch-MO NPs) including chitosan-CuO nanoparticles (Ch-CuO NPs) and chitosan-ZnO nanoparticles (Ch-ZnO NPs) were prepared through the crosslinking mechanism by glutaraldehyde and then epichlorohydrin. The nanoparticles were used as a stationary phase in the preparation of SPE cartridge. The SPE cartridges were used in extraction and cleanup of pesticides from water samples. The efficiency of the prepared cartridge of adsorption or retention of the different pesticides including abamectin, diazinon, fenamiphos, imidacloprid, lambda-cyhalothrin, methomyl, and thiophanate methyl was tested at three concentrations of each pesticide. The targeted pesticides are known to have been extensively used in agriculture in Egypt. The pesticide residues were determined by HPLC system. This protocol addresses the detection of trace amounts of these pesticides in water and optimizes the conditions for SPE technique compared with the commercial SPE of Supelco Sigma product.

2. Materials and Methods

2.1. Chemicals. Low molecular weight of acid-soluble chitosan (3.60×10^5 Da and 88% degree of deacetylation), glutaraldehyde (50%), epichlorohydrin (99%), toluene, dimethylformamide, and ethyl acetate were purchased from Sigma-Aldrich Co (St. Louis, Missouri, USA). HPLC-grade of acetonitrile, methanol, and water were purchased from Carlo-Erba Reagents SAS, Co (Chaussee du Vexin, 27100 Val-de-Reuil, France). Zinc oxide (ZnO), red copper (I) oxide (Cu_2O), acetic acid, nitric acid, and sodium hydroxide were purchased from El-Gomhoria for pharmaceutical and chemicals Co (Adeb Ishak St, Manshia, Alexandria, Egypt) and used without further purification.

2.2. Technical Pesticides. Technical grade of abamectin (96% purity) was purchased from Merck and Co., Inc., (Kenilworth, New Jersey, USA). Chlorpyrifos methyl (97%) was purchased from Dow Chemical Co., (Midland, Michigan, USA). Diazinon (90%) was purchased from Syngenta International AG Co, (Schwarzwaldallee 215, 4002 Basel, Switzerland). Fenamiphos (90%) was purchased from Miles Inc, Co, (8400 Hawthorn Road, Stilwell, Kansas City, USA). Imidacloprid (96%) was purchased from Bayer AG Co (51368 Leverkusen, Germany). Lambda-cyhalothrin (97%) was purchased from Syngenta International AG Co (Schwarzwaldallee 215, 4002 Basel, Switzerland). Methomyl (98%) was purchased from E.I. du Pont de Nemours and Co (Wilmington, Delaware 19805, USA) and thiophanate-methyl (94%)

was purchased from Pennwalt Ltd, Co, (D-221, M.I.D.C, T.T.C. Industrial Area, Thane Belapur Road, Nerul, Navi Mumbai, Maharashtra, India). The chemical structures of these pesticides are shown in Figure S1.

2.3. Instruments and Equipment.

High-Performance Liquid Chromatography (HPLC) Agilent technology infinity 1260 (Germany) equipped with an Agilent variable wavelength ultraviolet detector (VWD) was used. The system consists of a quaternary gradient solvent pump to control the flow rate of the mobile phase and an autosampler for automatic injection with a $100\,\mu L$ sample loop, a vacuum degasser, and a column oven (5-80°C). Separation was performed on ZORBAX Eclipse Plus C18 analytical column (250 × 4.6 mm id, $5\,\mu m$ particle size). Data were managed using HP Chemstation software. Perkin Elmer FT-IR Spectrophotometer L160000A with detector $LiTaO_3$, PerkinElmer, Inc, (Waltham, Massachusetts, USA); Malvern Zeta-Nano-sizer, using Laser Doppler Micro-Electrophoresis Malvern instrument Ltd Co (Enigma Business Park, Grove wood Road, Malvern WR14 1XZ, UK); UV-visible Spectrophotometer Alpha 1502 (Laxco, Inc., Bothell, WA 98021, USA); scanning electron microscope (SEM) JSM5300, JEOL Ltd, (Akishima, Tokyo, Japan); transmission electron microscope (TEM) JEOL JEM-1400 (USA); Bruker's X-ray diffraction (XRD, USA); ultrasonic homogenizer HD 2070 with HF generator (GM 2070), ultrasonic converter UW 2070, booster horn (SH 213 G), and probe microtip MS 73, Ø 3 mm, BANDELIN electronic GmbH & Co. (KG. Heinrichstraße, Berlin, Germany); hotplate with magnetic stirrer, IKA-Werke GmbH & Co (Breisgau-Hochschwarzwald, Germany); oven, Heraeus Co (KG-Hanau, Germany); and electric balances three and four digits, BL-410SLCD, Setra systems Inc, (59 Swanson Rd, Boxborough, MA 01719, USA) were used.

2.4. Preparation of Chitosan-Metal Oxide Nanoparticles (Ch-MO NPs).

Ch-MO NPs including chitosan-copper oxide (Ch-CuO) and chitosan-zinc oxide (Ch-ZnO) nanoparticles were prepared according to the method of Dehaghi and others with minor modifications [25]. A weight (4 g) of chitosan was dissolved in 100 mL aqueous acetic acid solution (1%, v/v) and stirred for 2 h using magnetic stirrer (solution A). The desired amount of metal oxide (1 mol metal ions per 1 mol amino group of chitosan) was added to the solution. In the case of Ch-Cu complex, Cu_2O (7.09 g) was dissolved in 20 mL diluted nitric acid (2%, v/v) (solution B); however, in the case of Ch-Zn complex, ZnO (8 g) was dissolved in 10 mL concentrated nitric acid (solution C). Solution B or C was added dropwise to the solution A using a syringe under continuous stirring for 2 h until the metal ions conjugated with a chitosan polymer. After that, 12 mL of glutaraldehyde (50%, v/v), as a first crosslinking agent, was added dropwise to the mixture under stirring, followed by addition of 8 mL epichlorohydrin (99%) as a second crosslinking agent, under continuous stirring. The pH was adjusted to 10 by NaOH (1N) dropwise by syringe under stirring. The reaction mixture was then sonicated for 15 min at a sonication power of 10 kHz and pulses or cycles (9 cycle /sec). Finally, the solution was stored in a water bath at 60°C for 3 h until precipitation. The precipitate was filtered, washed with distilled water, and dried at 70°C for 3 h.

2.5. Characterizations of Ch-MO NPs

2.5.1. Scanning Electron Microscope (SEM).
The samples of Ch-MO NPs were investigated using a JEOL SEM with a magnification of 20000x and acceleration voltage of 19 kV. The dry particles were suspended in ethyl alcohol by sonication in dismantling the assembled particles. After that, the particles were mounted on metal stubs with double-sided tape, sputtered with gold, and viewed in an SEM.

2.5.2. Transmission Electron Microscope (TEM).
TEM observation was performed on a JEOL JEM-1400 electron microscope (USA) at accelerating voltage of 120 kV. Specimens for TEM measurements were prepared by depositing a drop of colloid solution on a 400 mesh copper grid coated by an amorphous carbon film and evaporating the solvent in air at room temperature.

2.5.3. X-Ray Diffraction (XRD).
X-ray diffractograms on powder samples were obtained using a Bruker's X-ray diffraction (USA) with Cu tube radiation ($k = 1.54184$ Å), a graphite monochromator and Lynxeye detector at 30 kV, and a current of 10 mA. The diffractometer was controlled and operated by a PC computer with the DIFFRAC.SUITE™ software package. Measurements were taken over an angular range of $0.99° \leq 2\theta \leq 89.99°$ with a scanning step of 0.05 and a fixed counting time of 10 s. Divergence, scattered, and receiving radiation slits were 1°, 1°, and 0.2 mm, respectively.

2.5.4. Zeta Potential.
The surface charge of Ch-MO NPs was investigated by a Malvern Zeta-Nano-sizer instrument. The fixed weight (0.1gm) of the prepared particles was suspended in glycerol (50%) in isopropanol (v/v) and then they were sonicated for 30 min. The suspension was transferred to zeta potential cell [32].

2.5.5. FT-IR Spectroscopy.
The functional groups of Ch-MO NPs was analyzed by FT-IR spectroscopy with KBr discs (5 mg of Ch-MO NPs and 100 mg KBr pellets), in the range from 4000 to $400\,cm^{-1}$, with a resolution of $4.0\,cm^{-1}$ on a Perkin Elmer 1600 FT-IR Spectrophotometer (USA) [20].

2.6. Kinetic Study.
The preliminary study was conducted to investigate the influence of some factors (pH of the solution, temperature, and agitation time) on the adsorption efficiency of imidacloprid (as a pesticide example) on Ch-CuO NPs using full factorial design in MINITAB® software v17.1.0, 2002 (Minitab Inc, Co, Pine Hall Rd, State College, PA 16801-3008, USA). The three factors were tested at three levels including low level, high level, and medium level, coded as -1, +1, and 0, respectively. The minimum number of experimental runs that have to be carried out for two levels with three factors design is $2^3 = 8$ runs plus 1 run at a center point. The experiments were carried out using 100 mg of each type of nanoparticles suspended in 25 mL of imidacloprid solution (25 mg/L) at 10, 25, and 40°C, pH 5, 7, and 9, and different agitation times

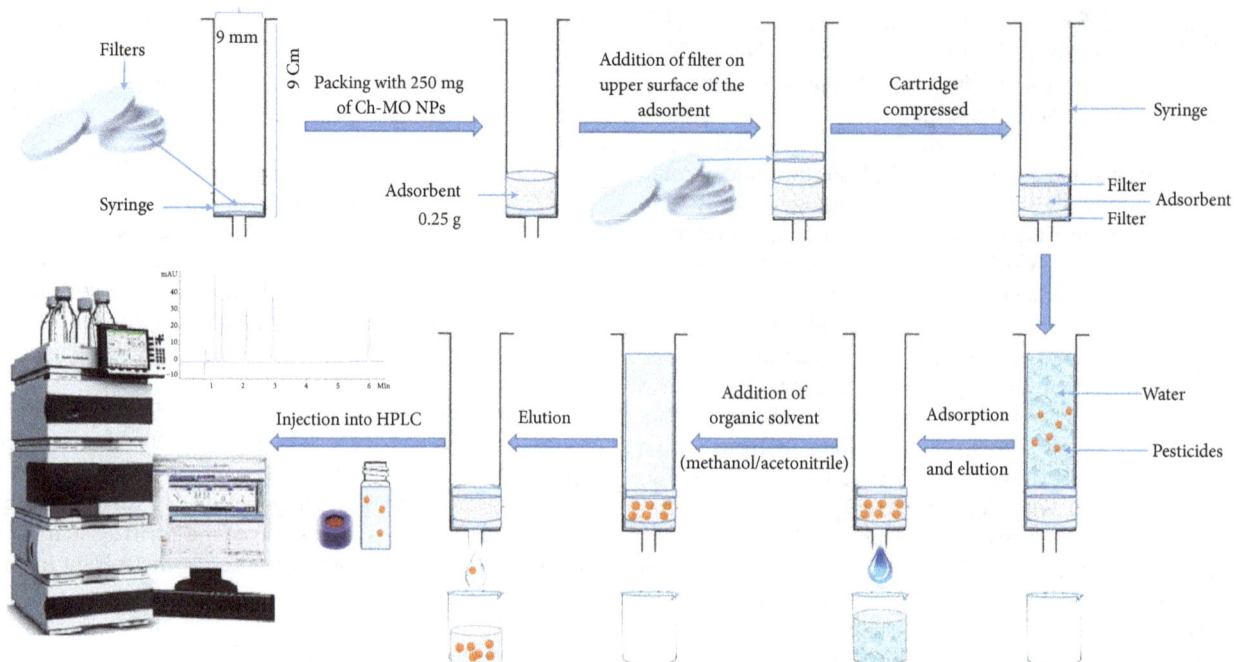

FIGURE 1: A schematic diagram shows extraction and clean-up of pesticides using SPE cartridge packed with Ch-MO NPs (Ch-CuO NPs and Ch-ZnO NPs). This figure is reproduced from Badawy et al. (2018).

(10, 25 and 40 min) with shaking at 150 rpm. The blank samples were added and placed in the same shaker to avoid loss of evaporation of pesticide or solvent. After each time with different experiments, the eluent was determined by HPLC [2, 25, 33].

2.7. Solid-Phase Extraction (SPE) of Different Pesticides by Ch-MO NPs. The prepared nanoparticles were studied as solid matrix materials in SPE cartridge. The SPE cartridge was performed using a plastic syringe column of 0.9 cm diameter and 9 cm in length (Figure 1). The column was filled up without gaps by compressing a frit on the bottom and then adding 0.25 g of each Ch-MO NPs and stopcock frit on the upper [34]. We compared these cartridges with the ODS (C18, Supelco) cartridge as it is the most common material used in extraction and clean-up of pesticide residues. Three different concentrations (10, 50, and 100 mg/L) of each pesticide (abamectin, diazinon, fenamiphos, imidacloprid, lambda-cyhalothrin, methomyl, and thiophanate-methyl) were prepared by dissolving the tested pesticide in a minimum volume of methanol and then completed to the final volume of 20 mL with water. The prepared solutions were allowed to pass through the SPE cartridge. After that, the adsorbed amount of each pesticide was eluted by 5 mL of acetonitrile/methanol (1:1, v/v).

2.8. HPLC Analysis. The water phase (effluent) and organic phase (eluent) were collected from SPE cartridge and injected into HPLC. The summary of the optimum conditions for chromatographic analysis of each pesticides is presented in Table S1. For analysis calibration, five standard solutions of each pesticide were prepared by dissolving weighed amount

in the mobile phase used for each pesticide, and different quantities (0.0125-0.15 μg/mL) were injected into HPLC. Calibration curves were constructed by plotting the peak areas of compound against the amount injected in μg. Regression analysis of the data (n = 5) for each calibration curve gave the values of slope, along with the intercept and correlation coefficient. Calibration curves were used for the quantification of the pesticides in water samples. The limit of detection (LOD) and limit of quantification (LOQ) for each pesticide were calculated. The LOD is the lowest concentration of the analyte in a sample that can still be detected by the analytical method but should not be quantified as an appropriate value. However, the LOQ is the lowest concentration of the sample that can still be quantitatively detected with acceptable precision and accuracy [35]. LOD was defined as $3\sigma/S$ and LOQ was defined as $10\sigma/S$, where σ is the standard deviation and S is the slope of the calibration curve [36].

2.9. Statistical Analysis. The statistical analysis was performed using the SPSS 25.0 software (Statistical Package for Social Sciences, USA). Analysis of variance (ANOVA) of the data was conducted, and means property values were separated by Student-Newman-Keuls (SNK) test. Differences were considered significant at $p \leq 0.05$. The statistical analysis of adsorption kinetics was investigated by full factorial design using a MINITAB® software v17.1.0, 2002 (Minitab Inc, Co., Pine Hall Rd, State College, PA 16801-3008, USA).

3. Results and Discussion

3.1. Preparation of Ch-MO NPs. The Ch-MO NPs were synthesized through combining the sol-gel precipitation and

TABLE 1: Reaction conditions and characterizations of chitosan-metal oxide nanoparticles (Ch-MO NPs).

Product code	Reaction components	Mole ratio	Product color	Yield (%)	Particles diameter (nm) ± SE	Zeta-potential (mV)
Ch-CuO NPs	Chitosan: Cu$_2$O: Glutaraldehyde: Epichlorohydrin	1:2:2:3	Yellowish-dark	85.29	93.74±5.70	+0.516
Ch-ZnO NPs	Chitosan: ZnO: Glutaraldehyde: Epichlorohydrin	1:4:2:3	Yellowish	91.67	97.95±9.46	+0.086

crosslinking mechanism [27] as illustrated in Figure S2. Monodispersed metal oxide particles were coated by chitosan as the uniform of core or shell layer. They were then sequentially crosslinked with glutaraldehyde and epichlorohydrin. Firstly, glutaraldehyde forms the hard-spherical shape of particles through reaction with the amino groups of chitosan. In the second stage, the epichlorohydrin reacted with the hydroxyl groups to give more hardness for particles and reduce the hydrophilicity of chitosan. The final product was precipitated by aqueous solution of NaOH (1N). The yields were 85.29% and 91.67% for Ch-CuO NPs and Ch-ZnO NPs, respectively, with a yellowish and dark yellowish color, respectively (Table 1).

Many research articles prepared and characterized polymer-supported metals and metal oxide nanoparticles including chitosan-ZnO and chitosan-CuO, and some of which suggested the previous mechanism of the particle formation [26, 37]. For example, Shrifian-Esfahni et al. prepared and characterized Fe$_3$O$_4$/chitosan core-shell and the mechanism investigated hydrogen-bonding formation. In addition, the authors indicated the unbonded hydroxyl groups with partial positive charges surrounding nanoparticle [37]. Therefore, we completed this reaction in our study by crosslinking agent to cover the reactive functional groups (amino and hydroxyl). Recently, we prepared chitosan-siloxane magnetic nanoparticles from Fe$_3$O$_4$ functionalized by siloxane derivatives followed by coating with chitosan through a crosslinking mechanism using glutaraldehyde and epichlorohydrin [34].

3.2. Characterizations of Ch-MO NPs

3.2.1. Scanning Electron Microscope (SEM). The SEM was used to investigate the surface morphology and particle size of Ch-CuO NPs and Ch-ZnO NPs as shown in Figures 2(a) and 2(b), respectively. The particles in nanocomposites were found with almost spherical morphology with aggregations of the nanoparticles. Nanoparticles were measured with an average size of 93.74 and 97.95 nm for Ch-CuO NPs and Ch-ZnO NPs, respectively (Table 1). Dehaghi and coauthors prepared Ch-ZnO NPs without crosslinking reaction and they found that the particles size was in a arrange of 58 nm [25]. However, Manikanndan and others prepared the Ch-Cu complex without crosslinking reactions with an average size ranging from 20 to 30 nm [38]. Gouda and Hebeish loaded CuO NPs into chitosan by using drops of H$_2$O$_2$ (30%) and then stirring with a high-speed homogenizer at 10000 rpm for 30 min. The corresponding CuO/chitosan nanocomposite

formed was characterized by using transmission electron microscope (TEM) images and they presented a very homogeneous morphology with a quite uniform particle size distribution and a rather spherical shape [39]. The particle size diameters obtained were 10 nm for chitosan nanoparticle and 20 nm for CuO/chitosan nanocomposite.

3.2.2. Transmission Electron Microscope (TEM). TEM photographs of Ch-CuO NPs and Ch-ZnO NPs are presented in Figures 2(c) and 2(d), respectively. It is evident that the particles are formed with average sizes ranging from 75 to 100 nm. In addition, the nanoparticles of both products showed high agglomeration of smaller size nanoparticles and their surface was rough and porous because metal oxide particles were wrapped by chitosan matrix.

3.2.3. X-Ray Powder Diffraction (XRD). The X-ray diffraction patterns of Ch-MO NPs are shown in Figure 3. Figure 3(a) shows the characteristic peaks at $2\theta \sim 10°$ and $2\theta \sim 20°$, due to inter- and intramolecular hydrogen bonds in chitosan molecule [40, 41]. However, these two peaks are very weak in the spectra of Ch-CuO NPs and Ch-ZnO NPs (Figures 3(b) and 3(c), respectively), which suggest a low crystallinity and an amorphous nature of the products. The weak peaks reflect great disarray in chain alignment of chitosan with the production of new peaks identifying zinc oxide and copper oxide. The X-ray diffraction patterns of Ch-CuO NPs (Figure 3(b)) demonstrated diffraction angles of 23.58°, 26.08°, 29.98°, 33.67°, 39.87°, 53.35°, and 77.80°, which correspond to the characteristic face centered CuO core with counts index (260), (415), (240), (458), (255), (149), and (110), respectively [42, 43]. The diffraction angles observed at 10.86° and 20.34° corresponding to count indexes (134) and (250), respectively, refer to the chitosan shell. The main peaks of Ch-ZnO NPs (Figure 3(c)) were at $2\theta = 30.91°$, 33.55°, 35.42°, 46.71°, 55.80°, 62.08°, 67.22°, and 68.28°, which correspond to the (1159), (1023), (1563), (391), (566), (449), (411), and (258) crystal planes, respectively. These peaks are consistent with the database in Joint Committee on Powder Diffraction Standards for ZnO (JCPDS file, PDF No. 36-1451) [44]. In addition, two smaller peaks at $2\theta = 76.31°$ and 88.84° corresponding to the count (157) and (170), respectively, were also observed. The diffraction angles observed at 10.98° and 20.76° corresponding to count indexes (211) and (289), respectively, refer to the chitosan shell.

3.2.4. Zeta Potential. Zeta potential is the surface charge value and it is a key indicator of the stability of colloidal

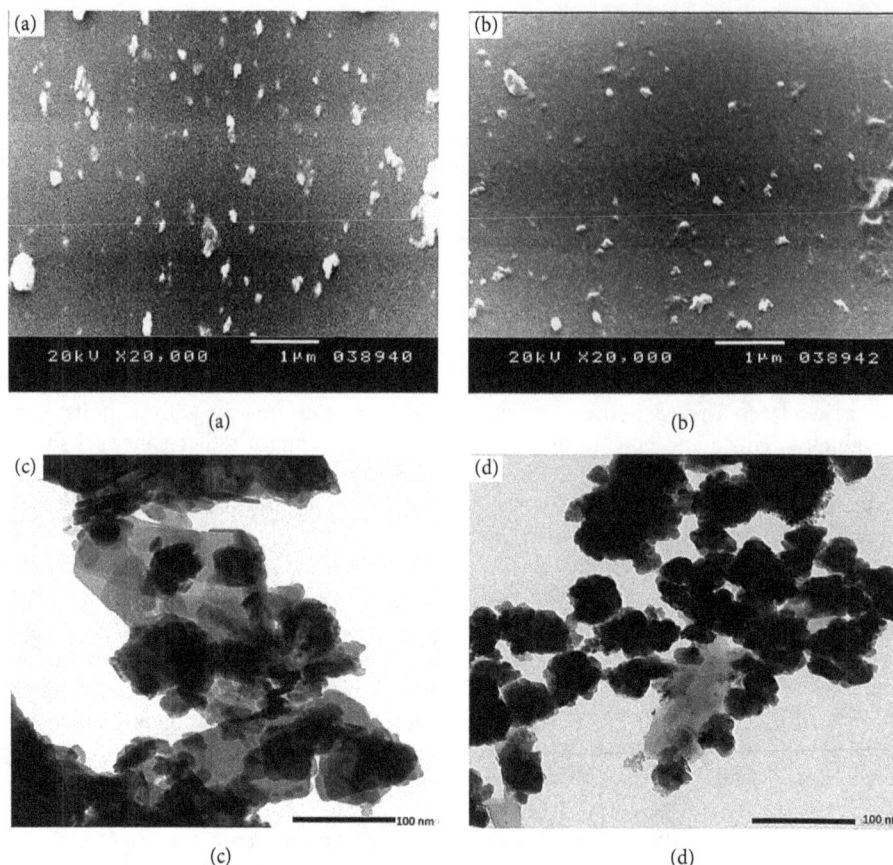

FIGURE 2: Electron microscopy images of Ch-MO NPs. **(a)**, **(b)** The SEM of Ch-CuO NPs and Ch-ZnO NPs, respectively. **(c)**, **(d)** The TEM of Ch-CuO NPs and Ch-ZnO NPs, respectively. Scale bar for SEM measurements was 1 μm and magnification x20000 at 20 Kv. Scale bar for TEM measurements was 100 nm and magnification x40000 at 20 Kv.

dispersions. The magnitude of the zeta potential indicates the degree of electrostatic repulsion between charged particles in a dispersion. For molecules and particles that are small enough, a high zeta potential will confer stability; i.e., the solution or dispersion will resist aggregation [32, 45]. In the present study, the values were +0.516 mV for Ch-CuO NPs and +0.086 mV for Ch-ZnO NPs (Table 1 and Figure S3), indicating a rapid coagulation or flocculation of particles in suspension at pH 7 and 25°C. It can be noted that the nanoparticles of Ch-CuO NPs have a higher charge ($\approx$ 5-fold) than Ch-ZnO NPs. The positive charge of zeta potential values obtained refers to the surface charge of the particles. The previous study reported that the Ch-Cu complex has a negative charge (-29 mv) [38]. However, the Ch-Zn complex had a positive charge (+26.6) [46]. The low surface charge of the prepared nanoparticles (Ch-CuO and Ch-ZnO) may be due to the crosslinking reaction that blocked the hydroxyl and amino functional groups. The glutaraldehyde blocks the amino groups of chitosan while the hydroxyl groups were blocked by epichlorohydrin [29, 47, 48].

3.2.5. FT-IR. The FT-IR spectra of chitosan and Ch-MO NPs are shown in Figure 4. The spectrum of pure chitosan exhibits bands at 3436 cm^{-1} due to the stretching vibration mode of –OH and -NH$_2$ groups. The peak at 2924 cm^{-1} is a type of C-H stretching vibration, while the band at 1655 cm^{-1} is due to the amide I group (C-O stretching along with N-H deformation mode). A band at 1590 cm^{-1} is attributed to the NH$_2$ group due to N-H deformation, while a band at 1419 cm^{-1} is due to C-N axial deformation (amine group band). In addition, the peak at 1380 cm^{-1} peak is due to the COO$^-$ group in carboxylic acid salt, and the band at 1160 cm^{-1} is assigned to the special broad peak of β (1–4) glucosidic bond in polysaccharide unit. However, the peak at 1080 cm^{-1} is attributed to the stretching vibration mode of the hydroxyl group, 989-1060 cm^{-1} stretching vibrations of C-O-C in glucose units [20].

The FT-IR spectrum of Ch-ZnO NPs exhibits band at 3401 cm^{-1} due to the combination between -OH and -NH$_2$ groups. The peak at 2932 cm^{-1} is a typical of C-H stretch vibration. The band at 1657 cm^{-1} is due to the rest of amide I group while a band at 1553 cm^{-1} is attributed to the NH$_2$ group due to N-H deformation. The peak at 1407 cm^{-1} is due to C-N axial deformation (amine group band). In addition, the band at 1067 cm^{-1} is attributed to the stretching vibration mode of the hydroxyl group and the band at 682 cm^{-1} ascribed to the vibration of O-Zn-O core groups.

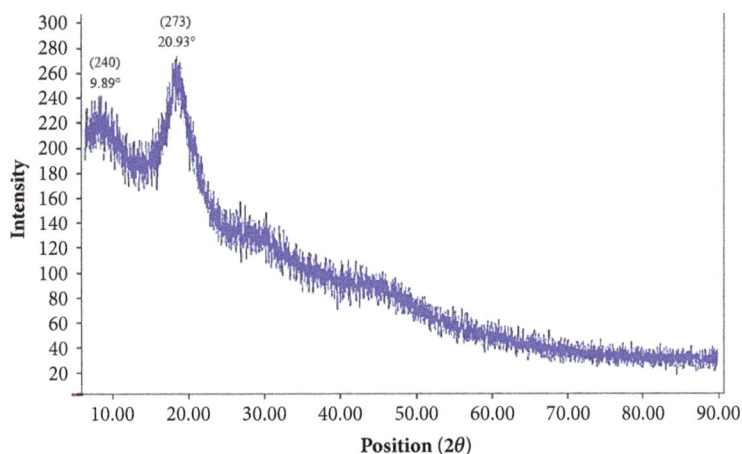

(a)

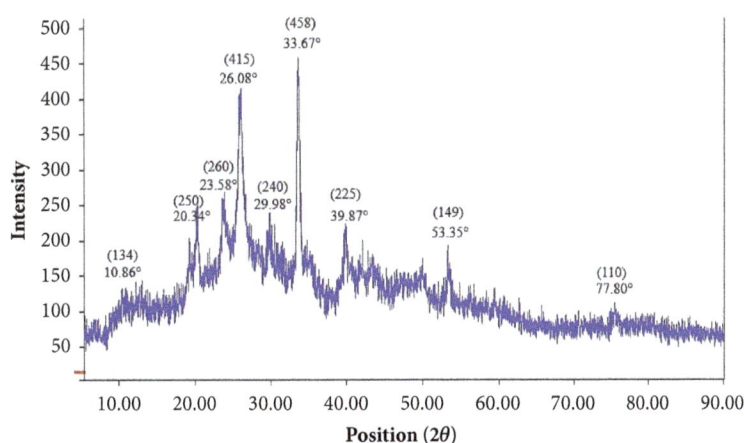

(b)

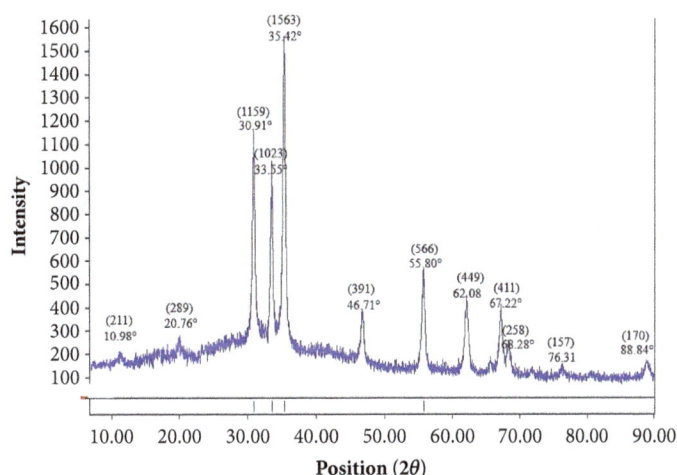

(c)

Figure 3: X-ray diffraction (XRD) patterns of chitosan (a), Ch-CuO NPs (b), and Ch-ZnO NPs (c).

The spectrum of Ch-CuO NPs exhibits bands at $3390 \, cm^{-1}$ due to the combination between -OH and -NH$_2$ groups. The peak at $2924 \, cm^{-1}$ indicates a C-H stretching vibration. A band at $1583 \, cm^{-1}$ is attributed to the NH$_2$ group due to N-H deformation, and $1410 \, cm^{-1}$ peak is due to C-N axial deformation (amine group band). A band at $1380 \, cm^{-1}$ is due to the COO- group in carboxylic acid salt, while the peak at $1070 \, cm^{-1}$ is attributed to the stretching vibration mode of the hydroxyl group. The band at $682 \, cm^{-1}$ is attributed to the vibration of O-Cu-O core groups. However, the peak at 493 is ascribed to Cu-O bond vibration.

TABLE 2: Experimental design using Minitab software and standardized effects of temperature, pH, and time on the adsorption of imidacloprid insecticide at 25 mg/L on Ch-CuO NPs.

Run order	Temperature (°C)	pH	Time (min)	Adsorption (%) ± SE
1	10	5	10	12.18±0.58
2	40	5	10	31.86±1.16
3	10	9	10	62.21±0.62
4	40	9	10	84.24±0.78
5	10	5	40	19.23±1.77
6	40	5	40	27.93±2.01
7	10	9	40	92.91±1.72
8	40	9	40	100.00±0.00
9	25	7	25	87.43±0.98

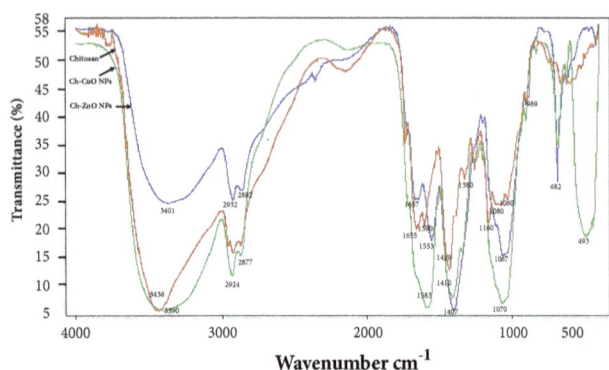

FIGURE 4: FT-IR spectra of chitosan (A), chitosan-copper oxide nanoparticles (Ch-CuO NPs), and chitosan-zinc oxide nanoparticles (Ch-ZnO NPs).

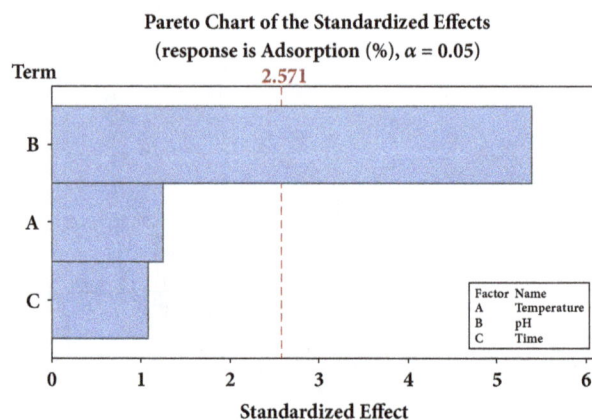

FIGURE 5: Pareto Chart of the standardized effects of pH, temperature, and time of adsorption (response is adsorption (%), $\alpha = 0.05$).

In comparison with chitosan, the broader and stronger peak shifted considerably to lower wave number at $3390 \, cm^{-1}$ in Ch-CuO NPs and $3401 \, cm^{-1}$ in Ch-ZnO NPs, which indicates strong attachment of metal oxide to the amide groups of chitosan molecules (Figure 4). The absorption peaks at 2877-2924 in Ch-MO NPs are due to asymmetric stretching of CH_2 and CH_3 of chitosan polymer and the overlapping with -NH. The absorption peaks at 1583 and $1070 \, cm^{-1}$ in the spectrum of Ch-CuO NPs are attributed to bending vibration of the -NH group and the C-O stretching group but these peaks were observed at 1553 and $1067 \, cm^{-1}$ in spectrum of Ch-ZnO NPs. New broad absorption bands at 682 and $400 \, cm^{-1}$ were found in the FT-IR spectra of Ch-MO NPs which were ascribed to the vibration of O-Cu-O and O-Zn-O groups [49, 50].

3.3. Kinetic Studies of Adsorption Efficiency of Pesticides by Ch-MO NPs. Three factors (pH, temperature, and agitation time) were studied on the efficiency of Ch-CuO NPs in the adsorption of imidacloprid insecticide at 25 mg/L. The full factorial design was used in terms of the experimental runs, and the experimental data are shown in Table 2. The results indicate that the pH values of 7 and 9 showed the most significant effect on the adsorption efficiency of imidacloprid with 62.21, 84.24, 92.91, 100, and 87.43 for run 3, 4, 7, 8, and

9, respectively. To investigate the main effect of all factors, the adsorption efficiency was studied using the Pareto chart and the result is shown in Figure 5. The most affecting factor is the pH followed by temperature and then agitation time. The Pareto chart provides a clear visualization of the factor effects and indicates that the pH has the most significant effect on the adsorption at $\alpha = 0.05$; however, the temperature and agitation time did not show values lower than the reference line (2.571 at $\alpha = 0.05$) [2, 25]. From this analysis, the adsorption (%) can be calculated or predicted according to the following model (1).

$$\text{Adsorption (\%)} = -73.3 + 0.479 \, \text{Temperature}$$

$$+ 15.51 \, \text{pH} + 0.413 \, \text{Time}$$

$$\text{S} = 16.28 \text{ and}$$

$$\text{R}^2 = 86.40\%$$

(1)

It can be noted that the three factors have a positive sign that means that the adsorption will be increased with an increase in each factor. The factor has a greater correlation factor denoting the great effects. Therefore, the pH has a great effect (coefficient = 15.51) on the adsorption followed in the descending order by temperature (coefficient = 0.479)

TABLE 3: Statistical data from regression analysis of different pesticides obtained from the study with analytical HPLC methods.

Pesticide	Rt (min) ± SD	A_s ± SD	Regression equation	R^2	LOD (μg/mL)	LOQ (μg/mL)
Abamectin	7.999 ± 0.01	0.871 ± 0.00	y = 4523.45190x-2.70225	0.9998	0.023	0.077
Diazinon	7.975 ± 0.00	0.870 ± 0.01	y = 1177.60010x+0.42100	0.9999	0.046	0.154
Fenamiphos	3.374 ± 0.01	0.885 ± 0.01	y = 3214.11453x+0.89949	0.9997	0.002	0.006
Imidacloprid	3.647 ± 0.00	0.853 ± 0.04	y = 4728.25710x+0.794634	0.9998	0.020	0.066
Lambda-cyhalothrin	10.761 ± 0.05	0.923 ± 0.05	y = 2874.16095x+0.431849	0.9999	0.012	0.040
Methomyl	2.795 ± 0.03	0.953 ± 0.00	y = 4972.13330x+3.61685	0.9997	0.018	0.059
Thiophanate-methyl	4.566 ± 0.01	1.070 ± 0.00	y = 3412.34475x+11.24269	0.9997	0.024	0.081

Rt: retention time. A_s: peak asymmetry factor. R^2: linear correlation coefficient. LOD: limit of detection. LOQ: limit of quantification.

and then the agitation time (coefficient = 0.413). In addition, three-dimensional response surface plots are presented in Figure S4. These plots provide useful information about the behavior of the system within the experimental design, which was used to understand the main and interactive effects of the factors. The effect of pH, temperature, and agitation time on pesticides adsorption percentage was shown at initial concentration in Figure S4 right. The results indicated that the adsorption or retention percentage increased with increasing of the pH and temperature, but the optimum adsorption percentage was observed at pH 7 and temperature of 25°C. These results are consistent with the previous study, which reported that the removal rate of pyrethrin increased by an increase of pH to 8 [25]. The adsorption ratio increased at pH increase and induction time from 10 to 40 min, but the optimal adsorption was performed at pH 7 and after 25 minutes. However, the effect of time and temperature has proved the previous theory that confirmed that optimal temperature and induction time are from 25°C to 40°C and 25 to 40 minutes, respectively at the top of the surface plot curve. The contour plots shown in Figure S4 indicate the interaction between the pH and temperature and confirmed that the optimum adsorption was found at pH ranging from 6.5 to 9 with the optimal temperature from 25 to 40°C.

3.4. SPE of Pesticides Using Ch-MO NPs and HPLC Analysis. HPLC analytical methods for the tested pesticides were validated by calculating regression equation, correlation coefficient (R^2), peak asymmetry factor (A_s), LOD, and LOQ for each pesticide and the data are presented in Table 3. The values of R^2 obtained for the regression lines demonstrate the excellent relationship between peak area and the injected amount of all pesticides ($R^2 \geq 0.999$). The LOD of the pesticides determined by HPLC ranged from 0.002 to 0.046 μg/mL and the LOQ was in the range of 0.006 to 0.154 μg/mL. The asymmetry factor (A_s) is an indication for the peak tailing [51, 52] being in the range of 0.870 to 1.070.

The efficacy data of Ch-MO NPs (250 mg) in extraction and removal of pesticides from water samples at three levels (10, 50, and 100 mg/L) is presented in Tables 4 and 5 for Ch-CuO NPs and Ch-ZnO NPs, respectively, and compared to the standard ODS cartridge (Supelco) (Table 6). The data are presented as a percentage of that extracted by methanol: acetonitrile (50:50) and that found in water phase. It can be noted that the removal percentages were

decreased with the increase of the concentration. Table 4 shows the results of cartridge loaded with Ch-CuO NPs. All pesticides were adsorbed into the Ch-CuO NPs with high percentages compared to the amount remaining in the water phase. Lambda-cyhalothrin was the highest in removal (98.93, 95.19, and 92.66% at 10, 50, and 100 mg/L, respectively) followed in the descending order by abamectin (98.02, 94.34, and 92.31% at 10, 50, and 100 mg/L, respectively). However, there is no significant difference between both insecticides. Fenamiphos showed 95.33, 93.28, and 90.44% and then imidacloprid with 93.78, 90.39, and 72.91% at 10, 50, and 100 mg/L, respectively. However, methomyl and thiophanate-methyl showed moderate values (63.85-84.75%). Diazinon was the lowest pesticide among all the tested pesticides in removal percentages (70.15, 34.21, and 21.44% at 10, 50, and 100 mg/L, respectively). Ch-CuO NPs demonstrated that no amount of lambda-cyhalothrin was found in water at any of the tested concentrations. This finding may be due to the fact that the lambda-cyhalothrin has a very low solubility in water and a highest octanol-water partition coefficient value compared to the other tested pesticides [53], followed in the descending order by imidacloprid, thiophanate-methyl, fenamiphos, and abamectin. However, methomyl indicated high percentages in water (20.55, 25.00, and 33.37% at 10, 50, and 100 mg/L, respectively). This is may be due to the high solubility of this compound in the water [54].

All pesticides were also adsorbed into the Ch-ZnO NPs with high percentage compared to that found in the water phase and lambda-cyhalothrin was the highest in removal with 99.09, 98.00, 94.47% at 10, 50, and 100 mg/L, respectively (Table 5), followed in the descending order by abamectin, fenamiphos, and imidacloprid. However, diazinon and thiophanate-methyl showed moderate values (60.10-94.28%). Methomyl was the lowest pesticide among all tested pesticides (41.40, 38.51, and 36.62% at 10, 50, and 100 mg/L, respectively). These particles proved that the insecticide lambda-cyhalothrin was not detected in water at any of the tested concentrations. However, methomyl showed high percentages in water (18.09, 57.82, and 62.59% at 10, 50, and 100 mg/L, respectively).

Table 6 shows the recovery of pesticides at 10, 50, and 100 mg/L from water using the standard SPE cartridge (C_{18}) obtained from Supelco. Diazinon, fenamiphos, and thiophanate-methyl were the most pesticides extracted from this type of cartridge in all tested concentrations. However,

TABLE 4: Efficiency of Ch-CuO NPs in adsorption of different pesticides using solid phase extraction cartridge technique.

Pesticides	Removal (%) ± SE at three levels of concentration (mg/L) (Extracted in methanol / acetonitrile)			Active ingredient (%) found in water ± SE (Remaining in water sample)			Total found (%) * ± SE		
	10	50	100	10	50	100	10	50	100
Abamectin	$98.02^a \pm 3.41$	$94.74^a \pm 1.02$	$92.31^a \pm 0.23$	$0.00^f \pm 0.00$	$2.53^e \pm 0.47$	$3.94^e \pm 0.09$	$98.02^a \pm 4.55$	$97.27^a \pm 1.35$	$96.25^a \pm 0.13$
Diazinon	$70.15^b \pm 1.46$	$34.21^e \pm 1.10$	$21.44^f \pm 0.25$	$18.89^b \pm 1.06$	$19.05^b \pm 1.03$	$23.81^b \pm 0.60$	$89.04^a \pm 2.01$	$53.27^c \pm 1.98$	$45.25^c \pm 0.72$
Fenamiphos	$95.33^a \pm 1.69$	$93.28^{ab} \pm 0.99$	$90.44^a \pm 1.04$	$4.03^d \pm 0.18$	$4.67^d \pm 0.15$	$7.31^d \pm 0.09$	$99.36^a \pm 1.59$	$97.94^a \pm 0.63$	$97.76^a \pm 0.32$
Imidacloprid	$93.78^a \pm 0.45$	$90.39^b \pm 0.61$	$72.91^d \pm 0.30$	$5.80^c \pm 0.28$	$8.16^c \pm 0.09$	$25.96^b \pm 0.96$	$99.58^a \pm 0.22$	$99.75^a \pm 0.49$	$98.87^a \pm 0.58$
Lambda-cyhalothrin	$98.44^a \pm 1.01$	$95.14^a \pm 0.41$	$92.66^a \pm 0.07$	$0.00^f \pm 0.00$	$0.00^f \pm 0.00$	$0.00^f \pm 0.00$	$98.44^a \pm 0.58$	$95.14^a \pm 0.14$	$92.66^b \pm 0.02$
Methomyl	$77.15^b \pm 0.28$	$70.17^d \pm 0.00$	$63.85^e \pm 0.39$	$20.55^a \pm 0.27$	$25.00^a \pm 0.65$	$33.37^a \pm 2.34$	$97.70^a \pm 0.48$	$95.16^a \pm 0.57$	$97.22^a \pm 0.79$
Thiophanate-methyl	$84.75^b \pm 1.82$	$78.91^c \pm 0.89$	$74.62^c \pm 0.22$	$5.14^e \pm 0.11$	$8.19^c \pm 0.04$	$22.47^c \pm 0.10$	$89.89^a \pm 0.92$	$87.10^b \pm 0.86$	$97.09^a \pm 0.16$

* The values lower than 100% mean the nonextracted amount of pesticide from Ch-CuO NPs. Values are mean of three replicates and are given as mean ± standard error. Different letters in the same column indicate significant differences according to Student-Newman-Keuls (SNK) test ($P \leq 0.05$).

TABLE 5: Efficiency of Ch-ZnO NPs in adsorption of different pesticides using solid phase extraction cartridge technique.

Pesticides	Removal (%) ± SE at three levels of concentration (mg/L) (Extracted in methanol / acetonitrile)			Active ingredient (%) found in water ± SE (Remaining in water sample)			Total found (%)* ± SE		
	10	50	100	10	50	100	10	50	100
Abamectin	$98.72^a \pm 5.31$	$93.15^a \pm 0.65$	$92.63^a \pm 0.66$	$0.00^e \pm 0.00$	$1.84^e \pm 0.18$	$2.51^e \pm 0.06$	$98.72^a \pm 4.05$	$94.99^b \pm 0.26$	$95.15^b \pm 0.57$
Diazinon	$94.28^b \pm 1.53$	$76.12^b \pm 1.14$	$72.55^c \pm 1.22$	$5.25^b \pm 0.47$	$18.08^b \pm 0.25$	$23.01^b \pm 0.36$	$99.54^a \pm 1.72$	$94.15^b \pm 0.42$	$95.56^b \pm 0.89$
Fenamiphos	$95.21^b \pm 3.53$	$93.33^a \pm 0.92$	$87.20^b \pm 0.44$	$4.34^c \pm 0.36$	$4.42^d \pm 0.13$	$7.52^{de} \pm 0.04$	$99.55^a \pm 2.79$	$97.75^{ab} \pm 0.68$	$94.72^b \pm 0.40$
Imidacloprid	$96.90^{ab} \pm 0.35$	$97.76^a \pm 0.68$	$88.47^b \pm 0.61$	$1.69^d \pm 0.16$	$2.21^e \pm 0.16$	$10.42^{cd} \pm 0.07$	$99.58^a \pm 0.22$	$99.97^a \pm 0.76$	$98.88^a \pm 0.27$
Lambda-cyhalothrin	$99.09^a \pm 0.78$	$98.00^a \pm 1.61$	$94.47^c \pm 0.41$	$0.00^e \pm 0.00$	$0.00^f \pm 0.00$	$0.00^f \pm 0.00$	$99.09^a \pm 0.55$	$98.00^{ab} \pm 1.00$	$94.47^b \pm 0.20$
Methomyl	$41.47^d \pm 1.08$	$38.51^c \pm 0.31$	$36.62^f \pm 0.56$	$18.09^a \pm 0.27$	$57.82^a \pm 0.25$	$62.59^a \pm 0.33$	$59.56^b \pm 1.18$	$96.33^{ab} \pm 0.09$	$99.21^a \pm 0.78$
Thiophanate-methyl	$90.62^c \pm 0.86$	$60.60^b \pm 0.52$	$60.10^e \pm 0.22$	$3.49^c \pm 0.04$	$6.66^c \pm 0.14$	$12.34^c \pm 0.07$	$94.11^a \pm 0.88$	$67.26c \pm 0.38$	$72.44^c \pm 0.18$

* The values lower than 100% mean the nonextracted amount of pesticide from Ch-ZnO NPs. Values are mean of three replicates and are given as mean ± standard error. Different letters in the same column indicate significant differences according to Student-Newman-Keuls (SNK) test ($P \leq 0.05$).

TABLE 6: Efficiency of standard ODS cartridge (Supelco) in adsorption of different pesticides using SPE technique.

Pesticides	Removal efficiency (%) ± SE at three levels of concentration (mg/L) (Extracted in methanol / acetonitrile)			Active ingredient found in water (%) ± SE (Remaining in water sample)			Total found (%)* ± SE		
	10	50	100	10	50	100	10	50	100
Abamectin	$97.59^a \pm 2.51$	$95.4^a4 \pm 0.48$	$48.11^c \pm 0.17$	$00.00^e \pm .00$	$4.28^e \pm 0.43$	$11.86^d \pm 0.65$	$97.59^a \pm 2.51$	$99.72^a \pm 0.45$	$59.97^c \pm 0.34$
Diazinon	$99.36^a \pm 2.05$	$96.28^a \pm 0.43$	$87.65^a \pm 0.28$	$00.00^e \pm 0.00$	$2.00^f \pm 0.04$	$7.45^e \pm 0.67$	$99.36^a \pm 2.05$	$98.32^a \pm 0.42$	$95.10^a \pm 0.47$
Fenamiphos	$84.20^b \pm 3.04$	$78.28^b \pm 0.46$	$78.60^b \pm 0.41$	$14.45^a \pm 0.65$	$16.54^b \pm 0.29$	$16.96^c \pm 0.25$	$98.65^a \pm 1.84$	$94.82^a \pm 0.56$	$95.56^a \pm 0.45$
Imidacloprid	$80.16^b \pm 1.03$	$51.26^c \pm 0.45$	$31.20^d \pm 1.19$	$8.11^d \pm 0.11$	$13.90^c \pm 0.14$	$36.84^a \pm 0.23$	$88.27^a \pm 0.98$	$65.16^c \pm 0.34$	$68.04^b \pm 0.71$
Lambda-cyhalothrin	$93.88^a \pm 1.21$	$72.05^b \pm 2.46$	$51.70^{9c} \pm 0.55$	$00.00^e \pm .00$	$7.42^d \pm 0.34$	$10.64^d \pm 0.65$	$93.88^a \pm 1.21$	$79.47^b \pm 1.49$	$62.43^b \pm 0.60$
Methomyl	$40.37^d \pm 0.63$	$28.20^d \pm 0.46$	$23.35^d \pm 1.08$	$11.87^c \pm 0.87$	$13.99^c \pm 0.87$	$22.98^b \pm 0.98$	$52.24^b \pm 0.76$	$42.19^d \pm 0.63$	$46.33^d \pm 0.96$
Thiophanate-methyl	$78.98^c \pm 4.26$	$75.30^b \pm 0.40$	$74.28^b \pm 0.22$	$13.07^b \pm 0.00$	$19.65^a \pm 0.65$	$24.67^b \pm 0.83$	$92.05^a \pm 4.26$	$94.95^a \pm 0.53$	$98.95^a \pm 0.52$

* The values lower than 100% mean the nonextracted amount of pesticide from standard solid phase extraction cartridge. Values are mean of three replicates and are given as mean ± standard error. Different letters in the same column indicate significant differences according to Student-Newman-Keuls (SNK) test ($P \leq 0.05$).

TABLE 7: Enrichment factor (EF) of Ch-Si MNPs for adsorption of different pesticides from water sample.

Pesticides	EF ± SE of Ch-MO NPs at three levels of pesticide concentrations (μg/mL)											
	10	50	100	Mean ± SE	10	50	100	Mean ± SE	10	50	100	Mean ± SE
	Ch-CuO NPs				Ch-ZnO NPs				ODS (Supelco)			
Abamectin	8.22	9.47	8.51	$8.73^a \pm 0.31$	8.28	9.31	8.54	$8.71^a \pm 0.26$	8.19	9.58	4.43	$7.40^a \pm 1.26$
Diazinon	5.24	2.74	1.75	$3.24^b \pm 0.85$	7.04	6.10	5.93	$6.36^{ab} \pm 0.28$	7.42	7.71	7.17	$7.43^a \pm 0.13$
Fenamiphos	7.56	7.35	7.24	$7.38^a \pm 0.08$	7.55	7.35	6.98	$7.29^{ab} \pm 0.14$	6.68	6.16	6.28	$6.37^b \pm 0.13$
Imidacloprid	7.39	7.60	5.12	$6.70^{ab} \pm 0.65$	7.64	8.22	6.22	$7.36^{ab} \pm 0.49$	6.32	4.59	2.19	$4.37^c \pm 0.98$
Lambda-cyhalothrin	7.87	10.80	7.31	$8.66^a \pm 0.89$	7.93	11.13	7.45	$8.83^a \pm 0.95$	7.37	8.18	4.08	$6.54^b \pm 1.03$
Methomyl	9.34	5.64	4.31	$6.43^{ab} \pm 1.24$	5.02	3.10	2.47	$3.53^c \pm 0.63$	4.89	2.27	1.58	$2.91^d \pm 0.83$
Thiophanate-methyl	6.76	6.32	5.97	$6.35^{ab} 0.19$	7.23	4.85	4.81	$5.63^{bc} \pm 0.66$	6.30	6.03	5.94	$6.09^b \pm 0.09$

Values are mean of three replicates and are given as mean ± standard error. Different letters in the same column indicate significant differences according to Student-Newman-Keuls (SNK) test ($P \leq 0.05$).

methomyl is still less compared to others. It can be observed that the standard SPE cartridge (C_{18}) showed a disparity in extraction efficiency and was the least cartridge compared with Ch-CuO NPs and Ch-ZnO NPs in the recovery of most tested pesticides including abamectin (recovery of 48.11-97.59%), fenamiphos (recovery of 78.60-84.20%), imidacloprid (recovery of 31.20-80.16%), lambda-cyhalothrin (recovery of 51.70-93.88%), and methomyl (recovery of 23.35-40.37%). Unfortunately, the SPE has certain limitations, primarily related to low recovery, i.e., slightly lower sensitivity, in cases where the SPE column is blocked (blocking the absorption centers by the sample's solid and organic components) [55].

The enrichment factor (EF) of the prepared and standard cartridges is shown in Table 7. EF can be defined as the concentration of the analyte in organic phase to the original concentration in the aqueous phase. The results showed that the EF of Ch-CuO NPs ranged from 3.24 for diazinon to 8.73 for abamectin. However, there is no significant difference among the other pesticides. The EF of Ch-ZnO NPs ranged from 3.53 for methomyl to 8.83 for lambda-cyhalothrin. It can be noted that the EF values of the prepared cartridges were higher than the standard ODS (C_{18}), which had a range of 2.91-7.43.

SPE became one of the most widely used treatment methods for various samples [56, 57]. This technology has many advantages, including high enrichment factor, easy operation, high recovery, rapid phase separation, low cost, low consumption of organic solvents, and effective matrix interference [58]. In the SPE process, the synthesis of adsorbents is the fundamental issue since the type and amount of absorbance largely determine selectivity, sensitivity, and full recovery. In general, properties with large surface areas, active surface locations, and a short propagation path can provide a significant number of improvements in extraction kinetics [59]. Compared with conventional adsorbents, nanoscale metal oxides have attracted more interest from researchers in recent years, given their high surface area and rapid absorption kinetics. Several results confirmed that the Ch-MO NPs were high adsorbent materials and used in SPE technique for extraction and removal of different pollutants [24, 25]. Ch-Zn was prepared and applied for removal

of permethrin at optimum conditions, including adsorbent dose, agitating time, the initial concentration of pesticide, and pH on the adsorption [25]. The results indicated that the weight of 0.5 g of the bionanocomposite, at room temperature and pH 7, removed 99% of permethrin solution (25 mL, 0.1 mg L) using UV spectrophotometer at 272 nm. Copper-coated chitosan nanocomposite (Ch-Cu) was found to have high adsorption efficiency for parathion and methyl parathion, and maximum adsorption capacity of parathion was found to be 322.60 mg/g at an optimum pH of 2.0 [24]. This could be attributed to the inherent alkalinity of the adsorbent. In addition, high adsorption value of malathion could be explained by acidic hydrolysis of malathion to dithiophosphate followed by complexation of copper to form Cu (II) dithiophosphate. Ch-AgO NPs composite beads were also optimized to remove maximum permethrin as the model pesticide, with the amount of sorbent, agitating time, initial concentration of pesticide, and pH parameters [2]. In optimum conditions, room temperature and pH 7, the Ch-AgO NPs beads recovered 99% of permethrin solution (0.10 mg/L) using UV spectrophotometer compared to 50% with the pure chitosan.

3.5. Adsorption Isotherm Study. Adsorption isotherm models are important to determine the efficiency of the adsorption process. Adsorption isotherms illustrate the connection between the amount of adsorbed component per adsorbent weight and the concentration of the contaminated components in the solution. Determination of the adsorption parameters provides useful information, which can improve the adsorption efficiency of the systems. In the present study, the adsorption percentages were applied in Freundlich (1) and Langmuir (3) isotherm models as follows to predict which model is fit.

$$q = K_f C^{1/n} \tag{2}$$

$$q = \frac{q_{max} K_l C}{1 + K_l C} \tag{3}$$

where q is adsorption capacity (μg/g); K_f is Freundlich isotherm constant (μg/g); C is concentration of the analyte (adsorbate) in the solution at equilibrium (μg/mL), n is adsorption intensity; q_{max} is maximum adsorption monolayer capacity (μg/g); and K_l is Langmuir isotherm constant (mL/μg).

By analyzing the linear correlation coefficient (R^2) obtained, it is possible to identify the isotherm model that best represents the experimental data of this study [60]. From the values of R^2 obtained (Table S2) for the Ch-MO NPs, it is possible to conclude that both of Langmuir and Freundlich isotherms are fit to this study with $R^2 > 0.92$. When the experimental data follows the Langmuir model, this assumes that a monomolecular layer is formed when adsorption takes place without any interaction between the adsorbed molecules. However, the data follows the Freundlich isotherm, which means that the adsorption process takes place on heterogeneous surfaces and adsorption capacity is related to the concentration of the analyte at equilibrium [61]. The maximum adsorption capacity (q_{max}) of Ch-MO NPs was observed for all the tested pesticides. The Ch-CuO NPs and Ch-ZnO NPs showed the highest adsorption capacities (2.50×10^4 and 1.00×10^5 μg/g, respectively) for thiophanate-methyl compared to 1.00×10^4 μg/g by using ODS (C_{18}). However, the insecticide methomyl showed a low q_{max} on Ch-CuO NPs and Ch-ZnO NPs (2.00×10^3, 1.00×10^3 μg/g, respectively) compared to 2.86×10^2 by using ODS (C_{18}).

4. Conclusion

Novel Ch-MO NPs, stationary phases for SPE technique, were prepared and characterized by FT-IR, SEM, TEM, XRD, and Zeta-Nano-sizer. The chromatographic retention behaviors of seven pesticides on Ch-MO NPs were investigated and compared with standard ODS (C_{18} column). The factors of the pH, temperature and agitation time were studied on the efficiency of these products in adsorption or retention of imidacloprid insecticide and the results proved that the pH was the most significant factor. It was reported that the Ch-MO NPs are able to remove the selected pesticides at the optimum condition of agitation time 25 min, pH 7, and 25°C. Ch-CuO NPs and Ch-ZnO NPs exhibited high selectivity for the tested pesticides as solutes and the extracted amount by these products was more than the ODS in most cases at three levels of concentrations (10, 50 and 100 mg/L in aqueous solution). The new adsorbent nanoparticles behaved as a reversed phase retention mechanism based on hydrophobic interaction as well as inclusion interactions and weak hydrophilicity for the polar pesticides such as methomyl based on partitioning and surface adsorption process. The nanoparticles will possess great prospect in chromatographic analysis especially SPE and SPME techniques. In addition, these products are newly biocompatible, environmentally friendly, and low cost to extract and clean-up pesticides from wastewater. In future, this work will be conducted on the packing of the HPLC columns with these products as new alternatives to the current stationary phases for separation of pesticide residues.

Supplementary Materials

Figure S1 shows the chemical structures of tested pesticides (abamectin, chlorpyrifos methyl, diazinon, fenamiphos, imidacloprid, lambda-cyhalothrin, methomyl, and thiophanate-methyl). Figure S2 shows the 3D-schematic diagram for preparation mechanism of Ch-MO NPs. Figure S3 shows the zeta potential distribution graph of Ch-MO NPs. Figure S4 presents the surface plot and contour plot of the adsorption (%) of imidacloprid insecticide on Ch-CuO NPs versus temperature, pH, and agitation time. Table S1 shows a summary of the methods conditions used for determination of different pesticides by HPLC system. Table S2 indicates the parameters of the isothermal models of Ch-MO NPs for adsorption of different pesticides. *(Supplementary Materials)*

References

[1] F. Ahmadi, Y. Assadi, S. M. R. M. Hosseini, and M. Rezaee, "Determination of organophosphorus pesticides in water samples by single drop microextraction and gas chromatography-flame photometric detector," *Journal of Chromatography A*, vol. 1101, no. 1-2, pp. 307–312, 2006.

[2] B. Rahmanifar and S. Moradi Dehaghi, "Removal of organochlorine pesticides by chitosan loaded with silver oxide nanoparticles from water," *Clean Technologies and Environmental Policy*, vol. 16, no. 8, pp. 1781–1786, 2014.

[3] K. L. Howdeshell, A. K. Hotchkiss, and L. E. Gray, "Cumulative effects of antiandrogenic chemical mixtures and their relevance to human health risk assessment," *International Journal of Hygiene and Environmental Health*, vol. 220, no. 2, pp. 179–188, 2017.

[4] K.-H. Kim, E. Kabir, and S. A. Jahan, "Exposure to pesticides and the associated human health effects," *Science of the Total Environment*, vol. 575, pp. 525–535, 2017.

[5] A. M. Cimino, A. L. Boyles, K. A. Thayer, and M. J. Perry, "Effects of neonicotinoid pesticide exposure on human health: A systematic review," *Environmental Health Perspectives*, vol. 125, no. 2, pp. 155–162, 2017.

[6] K. Yoshizuka, Z. Lou, and K. Inoue, "Silver-complexed chitosan microparticles for pesticide removal," *Reactive and Functional Polymers*, vol. 44, no. 1, pp. 47–54, 2000.

[7] S. D. Zaugg, M. W. Sandstrom, S. G. Smith, and K. M. Fehlberg, "Methods of analysis by the US Geological Survey National Water Quality Laboratory; determination of pesticides in water by C-18 solid-phase extraction and capillary-column gas chromatography/mass spectrometry with selected-ion monitoring," US Geological Survey: Open-File Reports Section/ESIC, 1995.

[8] D. A. J. Murray, "Rapid micro extraction procedure for analyses of trace amounts of organic compounds in water by gas chromatography and comparisons with macro extraction methods," *Journal of Chromatography A*, vol. 177, no. 1, pp. 135–140, 1979.

[9] I. Liška, J. Krupčíik, and P. A. Leclercq, "The use of solid sorbents for direct accumulation of organic compounds from water matrices–a review of solid-phase extraction techniques," *Journal of High Resolution Chromatography*, vol. 12, no. 9, pp. 577–590, 1989.

[10] M. T. Muldoon and L. H. Stanker, "Molecularly imprinted solid phase extraction of atrazine from beef liver extracts," *Analytical Chemistry*, vol. 69, no. 5, pp. 803–808, 1997.

[11] S. M. Yousefi, F. Shemirani, and S. A. Ghorbanian, "Deep eutectic solvent magnetic bucky gels in developing dispersive solid phase extraction: Application for ultra trace analysis of organochlorine pesticides by GC-micro ECD using a large-volume injection technique," *Talanta*, vol. 168, pp. 73–81, 2017.

[12] T. A. Albanis, D. G. Hela, T. M. Sakellarides, and I. K. Konstantinou, "Monitoring of pesticide residues and their metabolites in surface and underground waters of Imathia (N. Greece) by means of solid-phase extraction disks and gas chromatography," *Journal of Chromatography A*, vol. 823, no. 1-2, pp. 59–71, 1998.

[13] T. F. Jenkins, P. H. Miyares, K. F. Myers, E. F. McCormick, and A. B. Strong, "Comparison of solid phase extraction with salting-out solvent extraction for preconcentration of nitroaromatic and nitramine explosives from water," *Analytica Chimica Acta*, vol. 289, no. 1, pp. 69–78, 1994.

[14] G.-M. Momplaisir, C. G. Rosal, E. M. Heithmar et al., "Development of a solid phase extraction method for agricultural pesticides in large-volume water samples," *Talanta*, vol. 81, no. 4-5, pp. 1380–1386, 2010.

[15] Y. S. Al-Degs, M. A. Al-Ghouti, and A. H. El-Sheikh, "Simultaneous determination of pesticides at trace levels in water using multiwalled carbon nanotubes as solid-phase extractant and multivariate calibration," *Journal of Hazardous Materials*, vol. 169, no. 1-3, pp. 128–135, 2009.

[16] L. Vidal, M.-L. Riekkola, and A. Canals, "Ionic liquid-modified materials for solid-phase extraction and separation: a review," *Analytica Chimica Acta*, vol. 715, pp. 19–41, 2012.

[17] L. Costa dos Reis, L. Vidal, and A. Canals, "Graphene oxide/Fe3O4 as sorbent for magnetic solid-phase extraction coupled with liquid chromatography to determine 2,4,6-trinitrotoluene in water samples," *Analytical and Bioanalytical Chemistry*, vol. 409, no. 10, pp. 2665–2674, 2017.

[18] A. Zwir-Ferenc and M. Biziuk, "Solid phase extraction technique - Trends, opportunities and applications," *Polish Journal of Environmental Studies*, vol. 15, no. 5, pp. 677–690, 2006.

[19] J. Pawliszyn, *Solid phase microextraction: theory and practice*, John Wiley Sons, 1997.

[20] M. E. Badawy, E. I. Rabea, N. E. Taktak, and M. A. El Nouby, "Production and Properties of Different Molecular Weights of Chitosan from Marine Shrimp Shells," *Journal of Chitin and Chitosan Science*, vol. 4, no. 1, pp. 46–54, 2016.

[21] E. I. Rabea, M. E.-T. Badawy, C. V. Stevens, G. Smagghe, and W. Steurbaut, "Chitosan as antimicrobial agent: applications and mode of action," *Biomacromolecules*, vol. 4, no. 6, pp. 1457–1465, 2003.

[22] A. Domard and M. Domard, "Chitosan: structure-properties relationship and biomedical applications," *Polymeric Biomaterials*, vol. 2, pp. 187–212, 2001.

[23] M. Masuelli and D. Renard, *Advances in Physicochemical Properties of Biopolymers (Part 2)*, BENTHAM SCIENCE PUBLISHERS, 2017.

[24] M. Jaiswal, D. Chauhan, and N. Sankararamakrishnan, "Copper chitosan nanocomposite: Synthesis, characterization, and application in removal of organophosphorous pesticide from agricultural runoff," *Environmental Science and Pollution Research*, vol. 19, no. 6, pp. 2055–2062, 2012.

[25] S. Moradi Dehaghi, B. Rahmanifar, A. M. Moradi, and P. A. Azar, "Removal of permethrin pesticide from water by chitosan-zinc oxide nanoparticles composite as an adsorbent," *Journal of Saudi Chemical Society*, vol. 18, no. 4, pp. 348–355, 2014.

[26] S. Sarkar, E. Guibal, F. Quignard, and A. K. SenGupta, "Polymer-supported metals and metal oxide nanoparticles: synthesis, characterization, and applications," *Journal of Nanoparticle Research*, vol. 14, no. 2, article 715, 2012.

[27] M. E. I. Badawy, N. E. M. Taktak, O. M. Awad, S. A. Elfiki, and N. E. A. El-Ela, "Preparation and Characterization of Biopolymers Chitosan/Alginate/Gelatin Gel Spheres Crosslinked by Glutaraldehyde," *Journal of Macromolecular Science, Part B Physics*, vol. 56, no. 6, pp. 359–372, 2017.

[28] C. Tual, E. Espuche, M. Escoubes, and A. Domard, "Transport properties of chitosan membranes: Influence of crosslinking," *Journal of Polymer Science Part B: Polymer Physics*, vol. 38, no. 11, pp. 1521–1529, 2000.

[29] W.-W. Xiong, W.-F. Wang, L. Zhao, Q. Song, and L.-M. Yuan, "Chiral separation of (R,S)-2-phenyl-1-propanol through glutaraldehyde-crosslinked chitosan membranes," *Journal of Membrane Science*, vol. 328, no. 1-2, pp. 268–272, 2009.

[30] M. Gabriel Paulraj, S. Ignacimuthu, M. R. Gandhi et al., "Comparative studies of tripolyphosphate and glutaraldehyde cross-linked chitosan-botanical pesticide nanoparticles and their agricultural applications," *International Journal of Biological Macromolecules*, vol. 104, pp. 1813–1819, 2017.

[31] W. Tong, C. Gao, and H. Möhwald, "Manipulating the properties of polyelectrolyte microcapsules by glutaraldehyde cross-linking," *Chemistry of Materials*, vol. 17, no. 18, pp. 4610–4616, 2005.

[32] S. Honary and F. Zahir, "Effect of zeta potential on the properties of nano-drug delivery systems—a review (part 1)," *Tropical Journal of Pharmaceutical Research*, vol. 12, no. 2, pp. 255–264, 2013.

[33] J. L. D. O. Arias, C. Rombaldi, S. S. Caldas, and E. G. Primel, "Alternative sorbents for the dispersive solid-phase extraction step in quick, easy, cheap, effective, rugged and safe method for extraction of pesticides from rice paddy soils with determination by liquid chromatography tandem mass spectrometry," *Journal of Chromatography A*, vol. 1360, pp. 66–75, 2014.

[34] M. E. Badawy, A. E. Marei, and M. A. El-Nouby, "Preparation and characterization of chitosan-siloxane magnetic nanoparticles for the extraction of pesticides from water and determination by HPLC," *Separation Science Plus*, vol. 1, no. 7, pp. 506–519, 2018.

[35] US Department of Health and Human Services (FDA), *Analytical Procedures And Methods Validation Chemistry, Manufacturing, And Controls Documentation*, vol. 65, 2000.

[36] A. Teasdale, D. Elder, and R. W. Nims, *ICH Quality Guidelines*, John Wiley & Sons, Inc., Hoboken, NJ, USA, 2017.

[37] A. Shrifian-Esfahni, M. T. Salehi, M. Nasr-Esfahni, and E. Ekramian, "Chitosan-modified superparamgnetic iron oxide nanoparticles: Design, fabrication, characterization and antibacterial activity," *Chemik*, vol. 69, no. 1, pp. 19–32, 2015.

[38] A. M. Muthukrishnan, "Green synthesis of copper-chitosan nanoparticles and study of its antibacterial activity," *Journal of Nanomedicine & Nanotechnology*, vol. 6, no. 1, 2015.

[39] M. Gouda and A. Hebeish, "Preparation and evaluation of CuO/chitosan nanocomposite for antibacterial finishing cotton fabric," *Journal of Industrial Textiles*, vol. 39, no. 3, pp. 203–214, 2010.

[40] K. L. Haas and K. J. Franz, "Application of metal coordination chemistry to explore and manipulate cell biology," *Chemical Reviews*, vol. 109, no. 10, pp. 4921–4960, 2009.

[41] F. S. Pereira, S. Lanfredi, E. R. P. González, D. L. da Silva Agostini, H. M. Gomes, and R. dos Santos Medeiros, "Thermal and morphological study of chitosan metal complexes," *Journal of Thermal Analysis and Calorimetry*, vol. 129, no. 1, pp. 291–301, 2017.

[42] M. S. Usman, N. A. Ibrahim, K. Shameli, N. Zainuddin, and W. M. Z. W. Yunus, "Copper nanoparticles mediated by chitosan: synthesis and characterization via chemical methods," *Molecules*, vol. 17, no. 12, pp. 14928–14936, 2012.

[43] P. Senthil Kumar, M. Selvakumar, S. Ganesh Babu, S. Induja, and S. Karuthapandian, "CuO/ZnO nanorods: An affordable efficient p-n heterojunction and morphology dependent photocatalytic activity against organic contaminants," *Journal of Alloys and Compounds*, vol. 701, pp. 562–573, 2017.

[44] L.-H. Li, J.-C. Deng, H.-R. Deng, Z.-L. Liu, and L. Xin, "Synthesis and characterization of chitosan/ZnO nanoparticle composite membranes," *Carbohydrate Research*, vol. 345, no. 8, pp. 994–998, 2010.

[45] S. Patil, A. Sandberg, E. Heckert, W. Self, and S. Seal, "Protein adsorption and cellular uptake of cerium oxide nanoparticles as a function of zeta potential," *Biomaterials*, vol. 28, no. 31, pp. 4600–4607, 2007.

[46] A. Regiel-Futyra, M. Kus-Liśkiewicz, S. Wojtyła, G. Stochel, and W. Macyk, "The quenching effect of chitosan crosslinking on ZnO nanoparticles photocatalytic activity," *RSC Advances*, vol. 5, no. 97, pp. 80089–80097, 2015.

[47] Y. Gao, K.-H. Lee, M. Oshima, and S. Motomizu, "Adsorption behavior of metal ions on cross-linked chitosan and the determination of oxoanions after pretreatment with a chitosan column," *Analytical Sciences*, vol. 16, no. 12, pp. 1303–1308, 2000.

[48] I. A. Udoetok, R. M. Dimmick, L. D. Wilson, and J. V. Headley, "Adsorption properties of cross-linked cellulose-epichlorohydrin polymers in aqueous solution," *Carbohydrate Polymers*, vol. 136, pp. 329–340, 2016.

[49] A. Bagabas, A. Alshammari, M. F. A. Aboud, and H. Kosslick, "Room-temperature synthesis of zinc oxide nanoparticles in different media and their application in cyanide photodegradation," *Nanoscale Research Letters*, vol. 8, no. 1, pp. 1–10, 2013.

[50] S. Basumallick and S. Santra, "Chitosan coated copper-oxide nano particles: A novel electro-catalyst for CO_2 reduction," *RSC Advances*, vol. 4, no. 109, pp. 63685–63690, 2014.

[51] Z. Pápai and T. L. Pap, "Determination of chromatographic peak parameters by non-linear curve fitting using statistical moments," *Analyst*, vol. 127, no. 4, pp. 494–498, 2002.

[52] G. I. K. Marei, E. I. Rabea, and M. E. Badawy, "Preparation and Characterizations of Chitosan/Citral Nanoemulsions and their Antimicrobial Activity," *Applied Food Biotechnology*, vol. 5, pp. 69–78, 2018.

[53] J. Liu, X. Lü, J. Xie, Y. Chu, C. Sun, and Q. Wang, "Adsorption of lambda-cyhalothrin and cypermethrin on two typical Chinese soils as affected by copper," *Environmental Science and Pollution Research*, vol. 16, no. 4, pp. 414–422, 2009.

[54] R. I. Krieger, P. Brutsche-Keiper, H. R. Crosby, and A. D. Krieger, "Reduction of pesticide residues of fruit using water only or plus Fit™ Fruit and Vegetable Wash," *Bulletin of Environmental Contamination and Toxicology*, vol. 70, no. 2, pp. 213–218, 2003.

[55] R. Ðurovic and T. Ðordevic, *Modern extraction techniques for pesticide residues determination in plant and soil samples Pesticides in the Modern World-Trends in Pesticides Analysis*, InTech, 2011.

[56] Ł. Rajski, A. Lozano, A. Uclés, C. Ferrer, and A. R. Fernández-Alba, "Determination of pesticide residues in high oil vegetal commodities by using various multi-residue methods and clean-ups followed by liquid chromatography tandem mass spectrometry," *Journal of Chromatography A*, vol. 1304, pp. 109–120, 2013.

[57] D. Molins-Delgado, D. García-Sillero, M. S. Díaz-Cruz, and D. Barceló, "On-line solid phase extraction-liquid chromatography-tandem mass spectrometry for insect repellent residue analysis in surface waters using atmospheric pressure photoionization," *Journal of Chromatography A*, vol. 1544, pp. 33–40, 2018.

[58] Z. Li, J. Li, Y. Wang, and Y. Wei, "Synthesis and application of surface-imprinted activated carbon sorbent for solid-phase extraction and determination of copper (II)," *Spectrochimica Acta Part A: Molecular and Biomolecular Spectroscopy*, vol. 117, pp. 422–427, 2014.

[59] R. Khorasani, K. Dindarloo Inaloo, M. Heidari, M. Behbahani, and O. Rahmanian, "Application of solvent-assisted dispersive solid phase extraction combined with flame atomic absorption spectroscopy for the determination of trace amounts of Cadmium," *Hormozgan Medical Journal*, vol. 20, no. 6, pp. 383–392, 2017.

[60] P. M. Silva, J. E. Francisco, J. C. Cajé, R. J. Cassella, and W. F. Pacheco, "A batch and fixed bed column study for fluorescein removal using chitosan modified by epichlorohydrin," *Journal of Environmental Science and Health, Part A: Toxic/Hazardous Substances and Environmental Engineering*, vol. 53, no. 1, pp. 55–64, 2017.

[61] F. Naseeruteen, N. S. A. Hamid, F. B. M. Suah, W. S. W. Ngah, and F. S. Mehamod, "Adsorption of malachite green from aqueous solution by using novel chitosan ionic liquid beads," *International Journal of Biological Macromolecules*, vol. 107, pp. 1270–1277, 2018.

15

Response in Ambient Low Temperature Plasma Ionization Compared to Electrospray and Atmospheric Pressure Chemical Ionization for Mass Spectrometry

Andreas Kiontke (ID), **Susan Billig** (ID), **and Claudia Birkemeyer** (ID)

Research Group of Mass Spectrometry at the Faculty of Chemistry and Mineralogy, University of Leipzig, Linnéstr. 3, 04103 Leipzig, Germany

Correspondence should be addressed to Claudia Birkemeyer; birkemeyer@chemie.uni-leipzig.de

Academic Editor: David M. Lubman

Modern technical evolution made mass spectrometry (MS) an absolute must for analytical chemistry in terms of application range, detection limits and speed. When it comes to mass spectrometric detection, one of the critical steps is to ionize the analyte and bring it into the gas phase. Several ionization techniques were developed for this purpose among which electrospray ionization (ESI) and atmospheric pressure chemical ionization (APCI) are two of the most frequently applied atmospheric pressure methods to ionize target compounds from liquid matrices or solutions. Moreover, recent efforts in the emerging field of "ambient" MS enable the applicability of newly developed atmospheric pressure techniques to solid matrices, greatly simplifying the analysis of samples with MS and anticipating, to ease the required or even leave out any sample preparation and enable analysis at ambient conditions, outside the instrument itself. These developments greatly extend the range of applications of modern mass spectrometry (MS). Ambient methods comprise many techniques; a particular prominent group is, however, the plasma-based methods. Although ambient MS is a rather new field of research, the interest in further developing the corresponding techniques and enhancing their performance is very strong due to their simplicity and often low cost of manufacturing. A precondition for improving the performance of such ion sources is a profound understanding how ionization works and which parameters determine signal response. Therefore, we review relevant compound characteristics for ionization with the two traditional methods ESI and APCI and compare those with one of the most frequently employed representatives of the plasma-based methods, i.e., low temperature plasma ionization. We present a detailed analysis in which compound characteristics are most beneficial for the response of aromatic nitrogen-containing compounds with these three methods and provide evidence that desorption characteristics appear to have the main common, general impact on signal response. In conclusion, our report provides a very useful resource to the optimization of instrumental conditions with respect to most important requirements of the three ionization techniques and, at the same time, for future developments in the field of ambient ionization.

1. Introduction

In the recent past, the interest in multiselective methods analyzing complex samples as quick as possible, and its components as sensitive and complete as possible, has grown tremendously. Thus, the era of the "omic" techniques evolved and a growing number of scientists are facing now the very challenging task to set up analytical methods that would be applicable to as many target compounds as possible at a time, in very different matrices. This task demands methods with a very high performance in terms of analytical resolution, selectivity and sensitivity. Therefore for this purpose, high-performance analytical detection methods such as mass spectrometry (MS) are very useful. MS is widely used for many multiselective techniques, the nowadays so-called "omic" techniques such as proteomics [1], metabolomics [2], or lipidomics [3], as to name a few. In forensics [4], drug development [5] and process monitoring [6], structural elucidation of natural substances [7], and even in the identification of counterfeits [8], MS is also the method of choice because of the rich information this technique delivers from a sample.

For MS analysis, the target molecule is converted into an ion, which subsequently needs to be transferred into the gas phase to enter the analyzer for determination of its m/z, mass-to-charge ratio. The basic principle of mass spectrometers remained almost unchanged in recent years. For example for the separation of the ions, MS analyzers such as the linear ion trap [9], the reflector TOF (time of flight) [10] and the FT-ICR (Fourier-transform ion cyclotron resonance) [11] were already established decades ago, and the introduction of the orbitrap [12] based on the work of Kingdon [13] and others, can be considered as the latest remarkable step forward. Nonetheless, the technical quality of the devices improved constantly in recent decades leading to a significantly enhanced performance of modern instruments not only in terms of sensitivity but also in that high resolution instruments became increasingly common in analytical labs.

Therefore, the main task of instrumental development nowadays seems to broaden the applicability of the method and, thus, the ionization process, in which end the target species is ionized and brought to the gas phase, became the greatest limitation of MS. However, analytical questions with their corresponding target compounds are highly diverse and the anticipated target analytes have very different prerequisites for ionization, which is why there are many different methods used. In detail, ionization requirements of small volatile molecules differ a lot from those of nonvolatile and large molecules such as proteins, and the polarity also plays a crucial role. Based on the energy that is transferred to the analyte during the ionization process, a classification into "hard" and "soft" ionization can be made [14]. Furthermore, the ionization types can be divided into vacuum methods and those under atmospheric pressure, or the order of ionization and desorption to the gas phase, i.e., if ions or neutrals are brought to the gas phase. In particular, the introduction of atmospheric pressure ionization (API) techniques can be considered a quantum leap within this context. Among API methods, electrospray ionization (ESI-MS) has become one of the most commonly employed techniques in analytical chemistry, mainly due to its broad applicability to polar and semipolar compounds and the superior selectivity which is achieved in combination with high resolution separation techniques such as liquid chromatography or capillary electrophoresis [15]. Another common API technique is the atmospheric pressure chemical ionization (APCI).

With the introduction of two ambient ionization techniques for API-MS, DESI (desorption electrospray ionization) [16] and DART (direct analysis in real time) [17], direct analysis of samples with minimal or no sample preparation became possible offering an enormous potential for saving time and resources. The introduction of ambient ionization at atmospheric pressure for mass spectrometry (AI-MS) attracted the interest of many researchers in the field and various ionization techniques have been described in recent years. Among those, plasma-based techniques including the low-temperature plasma probe (LTP) require very little resources thereby providing great potential for implementation in mobile analytical devices [18]. The ultimate objective of current research in that area is to increase the range of applicability of MS and therefore it is essential to understand

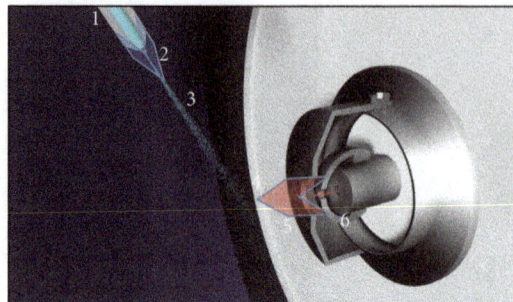

FIGURE 1: Schematic drawing of the ESI source in front of the MS inlet. 1, sprayer; 2, nebulizer gas (blue arrow); 3, spray/plume; 4, ions (red dots); 5, dry gas (red arrow); 6, MS inlet.

the different ionization mechanisms in detail. However, systematic studies on relative signal response among the different API techniques, such as the influence of the analyte and matrix characteristics on relative signal intensity, are still rare. Here, we review the available literature to compare the two most commonly employed API methods, namely ESI and APCI, with LTPI. To enable a direct comparison of these techniques, we add our review with own data revealing analyte characteristics that make their carrier particularly suitable for either of the three methods. We hope that sharing our results will help to further improve the general understanding of different API mechanisms, its common requirements but also the selectivity of the different techniques.

2. ESI and APCI – Two of the Most Common Atmospheric Pressure Ionization Techniques

2.1. ESI Mechanism. Dole et al. established the basics of ESI about 50 years ago [19, 20]. Years later, John Fenn's group, inspired by Dole's work, was able to show that using ESI, it was possible to make large, polar and even instable molecules accessible to MS [21, 22]. For ESI, the chargeability of the analyte is absolutely essential [23] which is achieved through different processes, whether by charge separation (e.g., deprotonation) or adduct formation (e.g., protonation) [24] or, less frequently, by electrolytic oxidation or reduction [25, 26]. A schematic view of an ESI-source is illustrated in Figure 1. First, the dissolved analyte is pumped through a conductive sprayer capillary. Flow rates here range from several nL/min (nanoESI) to the mL/min-range (conventional liquid chromatography, LC coupling with ESI). An electrical field ($E \approx 10^6$ V/m) is created by applying a voltage difference in the kV range (usually between 2 to 6 kV) between the sprayer capillary and the MS, which acts as a counter electrode. The respective charged species (positive or negative, depending on the applied direction of the electric field) are separated by acceleration to the MS. Depending on the vendor, either the sprayer capillary is grounded while the MS inlet is on high voltage (*needle on ground* configuration), or the sprayer capillary is on high voltage and the MS inlet on low voltage (*needle on potential*).

In the sprayer capillary, the first important process takes place, where the strong electric field leads to a charge separation by electrophoretic migration within the liquid. In the positive ion mode as an example, the cations are accelerated in direction of the MS inlet, while the anions are attracted to the inner capillary wall and can be oxidized there [27], which is reversed in negative ion mode. When the liquid sample leaves the sprayer capillary, the counteracting effect of the surface tension of the solvent on one, and the attracting force of the applied electrical field on the ions in the solution on the other side, is responsible for the formation of a cone, the so-called *Taylor-cone* named after Sir Geoffrey Taylor who was one of the first scientists describing and investigating this phenomenon [28]. The shape of the cone and thus the further formation of the so-called jet or filament, formed at the point of the highest charge density, the break-off of droplets from it and the resulting properties of these droplets all depend on the operating parameters of the mass spectrometer (e.g., needle voltage and flow rate) [29]. Basically, as soon as the surface tension of the solvent is exceeded by the electrostatic forces between the dissolved ions and the applied potential at the MS-entrance, the filament disintegrates and droplets are pinched off [30] due to instabilities and propagating waves along the filament [29]. This process is often supported by a nebulizer or sheath gas, an inert gas such as nitrogen, which encircles the ESI plume and thus diminishes the influence of surface tension. All of these droplets carry a net charge reversed to the MS electric pole. Subsequent evaporation of the solvent supported by a heated dry gas (nitrogen) causes the droplets to shrink. In succession, the charge density on the surfaces increases until the coulomb repulsion forces between the like-charged ions exceed the cohesive intermolecular forces at the so-called *Rayleigh* limit [31], where the surface tension equals the coulomb repulsion, and droplet fission occurs. The process of evaporation and splitting is repeated several times until the droplets have radii of a few nanometers.

Two theories are commonly accepted for gas-phase ion formation during the electrospray process, the *ion evaporation model* (IEM) [32] and the *charged residue model* (CRM) [19]. IEM applies mainly to low-molecular weight analytes and suggests that the resulting coulombic repulsion is strong enough to overcome surface tension and a dissolved ion is released from the droplet surface to the gas phase. It is believed that this mechanism takes place when the droplets have radii smaller than 10 nm, as the Rayleigh instability with droplet fission is preferred for larger radii [32]. The CRM on the other hand can be utilized to explain the release into the gas phase of larger analytes such as proteins [19], it is assumed that the cycles of shrinkage and fission caused by solvent evaporation ultimately end in a single ion in a solvent shell, which transfers the charge to the analyte after drying. Konermann et al. proposed an advanced model for gas phase ion formation called chain ejection model (CEM) [33]. It is assumed that this mechanism can be used for large molecules with nonpolar side chains, for example proteins that are unfolded due to an acidic solvent (e.g., a mobile phase in LC), while CRM can be used for native proteins [34]. The CEM describes an IEM-like process, in which the nonpolar

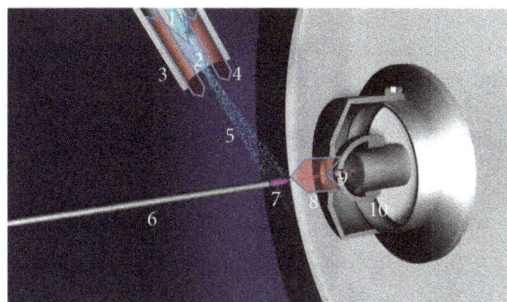

Figure 2: Schematic drawing of the APCI source in front of the MS inlet. 1: sprayer; 2: nebulizer gas (blue arrow); 3: heater; 4: auxiliary gas (red arrow); 5: vaporized spray; 6: corona needle; 7: corona discharge; 8: dry gas (red arrow); 9: ions (red dots); 10: MS inlet.

side chain migrates to the droplet surface and from there is expelled into the gas phase till the protein separates from the droplet completely [33, 35].

2.2. APCI Mechanism. Atmospheric pressure chemical ionization, APCI, is the second most important ionization method when it comes to LC-MS coupling. It is a gas phase ionization technique in which the analytes are ionized similar to chemical ionization, with the difference that the ionization takes place at atmospheric pressure and not under reduced pressure. First, ^{63}Ni was used in the ionization source [36, 37] but was soon replaced by a corona discharge since it produces comparable spectra [38] with an improved dynamic range [39], easier manufacture, use, maintenance and disposal with regard to radioactive waste. Approaches using a glow discharge were also described [40, 41].

APCI is typically used for small molecules (<1000 u) that are not polar enough for efficient electrospray ionization (ESI). Although APCI was developed earlier than ESI and according to the literature, APCI is supposed to be less vulnerable to matrix effects [42], ESI has become much more widespread. Possibly, the reason for this was the great interest in the analysis of large proteins [43], which remained inaccessible to APCI. However, APCI benefited from the rapid development and expansion of ESI and the related development of atmospheric pressure interfaces, since in the late 1980s and early 1990s all major MS device manufacturers introduced APCI sources [44]. Most instruments can accommodate an ESI *and* an APCI source, since the two sources can be easily exchanged due to their similarity. Figure 2 illustrates the general appearance of a typical APCI source.

Instead of a sprayer capillary with spray voltage, a pneumatic nebulizer with a downstream vaporizer or heater block is used here. Nitrogen is commonly used as nebulizing and auxiliary gas. Analyte and solvent are vaporized in the heater (up to 550°C) near the corona discharge needle. A high potential of 3-5 kV is applied to the needle since corona discharges generally occur at sharp-edged points or corners if the electric field is sufficiently large. Corona discharge is an uneven discharge; it acts as an electron source and the effects are produced at the electrode, i.e., a strong electric field, ionization, and the resulting glow [45]. The ionization mechanism has already been thoroughly investigated and the most important reactions of positive-mode APCI are described as follows [38, 46–49].

First, nitrogen is ionized by electrons generated by the corona discharge. The generated ion reacts with surrounding nitrogen and forms $N_4^{+\cdot}$. Although the ionization energy of nitrogen is higher than that of the analyte or solvent, ionization of nitrogen is most likely due to its high concentration (nebulizing and auxiliary gas) compared to the analyte or solvent.

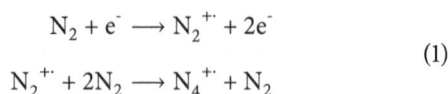

$$N_2 + e^- \longrightarrow N_2^{+\cdot} + 2e^-$$

$$N_2^{+\cdot} + 2N_2 \longrightarrow N_4^{+\cdot} + N_2 \tag{1}$$

The high-energy nitrogen ions $N_2^{+\cdot}$ and $N_4^{+\cdot}$ transfer the positive charge very quickly to the solvent or to water (as traces in surrounding gas), which is why the latter ion cannot be detected under standard conditions [38].

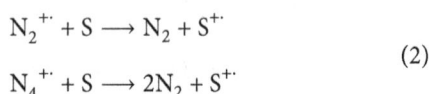

$$N_2^{+\cdot} + S \longrightarrow N_2 + S^{+\cdot}$$

$$N_4^{+\cdot} + S \longrightarrow 2N_2 + S^{+\cdot} \tag{2}$$

Considering the higher concentration of solvent molecules compared to analyte molecules, $S^{+\cdot}$ is most likely to react with other solvent molecules by hydrogen abstraction leading to the formation of protonated solvent and solvent clusters.

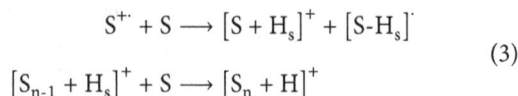

$$S^{+\cdot} + S \longrightarrow [S + H_s]^+ + [S\text{-}H_s]^{\cdot}$$

$$[S_{n-1} + H_s]^+ + S \longrightarrow [S_n + H]^+ \tag{3}$$

Finally, the analyte can be ionized by proton transfer if the gas-phase basicity of the analyte is higher than that of the solvent (or solvent cluster) [49].

$$M + [S_n + H]^+ \longrightarrow [M + H]^+ + S_n \tag{4}$$

In addition to protonation via the solvent (cluster), there are also possibilities of ionizing the analyte as a radical cation [47]. Thus, the analyte can also be ionized directly by a high-energy electron, or by charge transfer from the high-energy nitrogen species and ionized solvent, respectively.

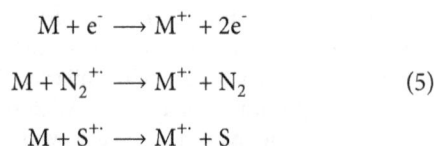

$$M + e^- \longrightarrow M^{+\cdot} + 2e^-$$

$$M + N_2^{+\cdot} \longrightarrow M^{+\cdot} + N_2 \tag{5}$$

$$M + S^{+\cdot} \longrightarrow M^{+\cdot} + S$$

Since ionization takes place at atmospheric pressure, excess energy is released by impacts with nitrogen (nebulizing and auxiliary gas) which makes APCI a softer ionization method than chemical ionization under vacuum (especially since the negative mode is even softer and is therefore particularly suitable for the determination of the molecular mass) [47]. However, due to the high-energy processes during ionization, some fragmentation might occur, which sometimes can also be helpful for structural elucidation.

2.3. Relative Response in ESI and APCI. One of the major drawbacks of the atmospheric pressure techniques is their rather selective sensitivity with respect to certain analyte's characteristics. For example, ESI response of equimolar concentrations of different analytes in solution can vary by > 3 orders of magnitude [50]. Response depends on all analyte's, solvent's, and instrument's characteristics influencing the processes of ionization and ion desorption, e.g., solution and gas-phase basicity and chemistry, polarity (log P), the number of charge sites or different charge states in solution (pH), the susceptibility to oxidation/reduction, the tertiary structure and molecular size of the analyte (mainly for higher molecular weight compounds), vaporization energy, or surface affinity [51–59]. However, the reported findings depended on study design: for example, Zhou and Cook [60] found that signal intensities for caffeine and arginine were independent on the pH of the solution as a consequence of their basic character; these bases stay protonated till a pH of 10 (pKa of caffeine). Ehrmann et al. [61] did not find evidence for the importance of gas-phase basicity; very likely, these results were again influenced by the strong basic character of the compounds under investigation. Thus, Kiontke et al. found that the general importance of the fundamental parameters as there were compound basicity, polarity, and molecular surface, respectively, hold true to be factors indeed determining ESI sensitivity; their quantitative impact, however, is rather subject to interplay with other parameters such as solvent pH and instrumental configuration [15].

One of the most important compound characteristics known to determine the intensity of the $(M+H)^+$ signal in MS after electrospray ionization, is the extent of its protonation in solution, i.e., the solution basicity [61, 62]. The ability to attract a proton in solution is best described by the pKa of the respective compound that can be retrieved from public databases such as SciFinder and ChemAxon. Solution basicity is so closely related to the interplay of electron-donating and withdrawing effects in the structure of the analyte, i.e., the electron density of the investigated molecules, that these parameters cannot be separately assessed; ESI-response is determined to the same extent by solution basicity as by the structural effects that also account for the basicity of a compound [15].

The second important compound characteristic after basicity is compound polarity. The correlation between polarity and signal response is interacting with solvent pH *and* polarity which has not been extensively studied yet [61]. In the literature, different findings about this interaction were reported: polar analytes provided a higher relative ESI response at neutral pH, while nonpolar analytes appeared to be less sensitive to solvent pH [63], the log D for pH 10-14 was best correlated with ESI response at pH 7 [62], while Kiontke et al. [15] found the correlation of the response at pH 7 and 3 strongest with log D at pH 3, potentially related to the fact that in ESI positive mode electrochemical oxidation leads to acidification of the ESI solvent.

At acid pH, the polarity of the molecular surface becomes important but in dependency on the amount of the organic or aqueous, respectively, phase. Nonpolar targets particularly benefit from acidification of the aqueous solvent *only at*

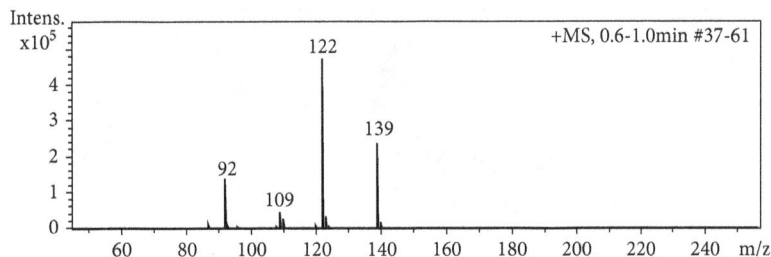

FIGURE 3: APCI mass spectrum of 10 μM 4-nitroaniline in methanol/H$_2$O 1:1 (v/v). m/z 139 is the molecular ion (M + H)$^+$.

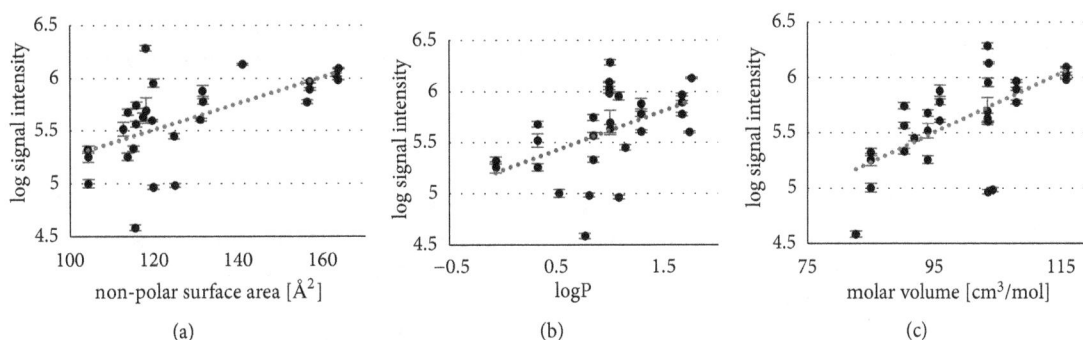

(a)

(b)

(c)

FIGURE 4: Log signal intensity (peak height) of APCI measurements in dependency on (a) the nonpolar surface area (SciFinder), (b) log P (ChemAxon), and (c) the molar volume.

low organic content. However, when decreasing the solvent pH, ion suppression by pH modifiers has to be considered which occurs by impaired desolvation due to decreased solvent volatility. In return, the volatility of a compound exerts its influence particularly upon signal enhancement by acidification suggesting that it is an additional advantage in competition to a pH modifier, since this effect was independent from compound basicity itself [15].

As to the research about sensitivity in APCI, it has been comparably less investigated. Sunner et al. grouped the compounds with respect to APCI response into three different classes [49, 64, 65]. One class, mostly nitrogen bases, are compounds with a gas phase basicity >~830 kJ/mol, which easily undergo gas-phase protonation. Concerning compounds with a gas-phase basicity below ~830 kJ/mol, for another group, often oxygen-containing bases, sensitivity increases with increasing basicity under thermodynamic control. The third group consists of substances forming gas-phase hydrates with a very low stability, which sensitivity is mostly influenced by the gas temperatures [49, 66–69]. Other, compound-independent parameter were suggested to be the reagent ion plasma density [70, 71], discharge current [72], space charge effects [73, 74], the residence time in the ion source [49], the distance between the discharge needle and the vacuum interface [72], and flow dynamics [75].

For a systematic comparison, we analyzed the relative signal response of the same 31 aromatic amino compounds that were analyzed with ESI [15] to assess the quantitative impact of the investigated molecular descriptors. Compared to ESI in agreement with [76], APCI produced more fragments and

less sodium adducts. As an example, Figure 3 shows the APCI spectrum of 4-nitroaniline.

The APCI spectrum of 4-nitroaniline is dominated by four species of which m/z 139 represents the [M + H]$^+$, the first cleavage [M+H-17]$^+$ most likely corresponds to the elimination of an OH-radical (confirmed by accurate mass, m/z = 122.0480) despite of suggestive NH$_3$-cleavage. Oxygen can also be protonated in the gas phase and an O-protonated species is formed [77, 78]. The presence of m/z 109 indicates a cleavage of NO, while m/z 92 corresponds to an HNO$_2$-cleavage. This example illustrates the considerably harder ionization conditions of APCI compared to ESI.

The logarithmized signal intensities of APCI measurements were then analyzed for Pearson's correlation with the values of the available molecular descriptors. The results are presented in Figure 4 (note that correlation strength was evaluated according to [79] (a) very weak: 0.00-0.019, (b) weak: 0.20-0.39, (c) moderate: 0.40-0.59, (d) strong: 0.60-0.79, and (e) very strong: 0.80-1.0).

A moderate positive correlation was found for the polarity descriptors molecular nonpolar surface area (Figure 4(a), data from SciFinder, and R = 0.59, p < 0.01) and log P (Figure 4(b), data from ChemAxon, R = 0.48, p < 0.01), and a strong positive correlation was found for the molar volume (Figure 4(c), R = 0.64, p < 0.001). Curiously, sensitivity was not enhanced with increasing gas-phase basicity. Possibly, since most of our target compounds were nitrogen bases with a gas phase basicity above the claimed threshold of ~830 kJ/mol [49, 64, 65], the quantitative impact of gas-phase protonation is exhausted for our compounds and here,

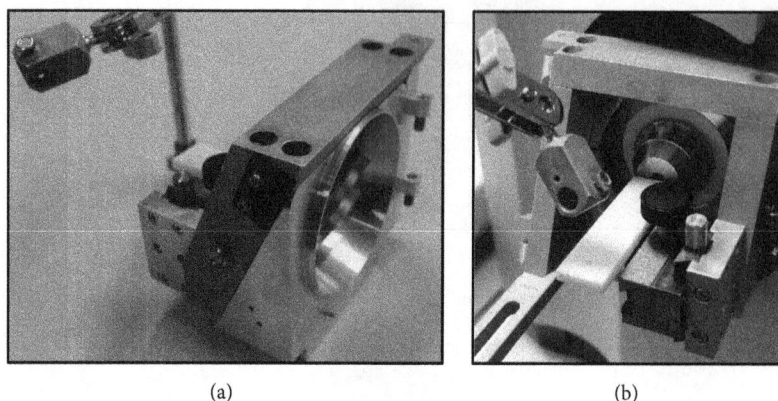

(a)   (b)

FIGURE 5: (a) Modular frame for mounting the plasma source to Bruker mass spectrometers and (b) completely mounted to a Bruker micrOTOF.

we observe the impact of other compound characteristics beyond that. However, concerning volatility, vaporization enthalpy also had no significant influence, which may be in agreement with Sunner et al. [49] who rather suggested the gas temperature to enhance signal response, indicating that *solvent* volatility may be more important than the *compound* volatility.

Thus, our results from analysis of the amines rather emphasize that desorption characteristics of the target compounds play an important role for this atmospheric pressure ionization technique. In conclusion, descriptors such as the molecular nonpolar surface area, log P and molar volume are very important for the surface activity of an analyte. The larger the nonpolar surface area and log P are, the higher the presence of the analytes at the liquid-air interface [23, 80] improving desorption, hence, ionization efficiency. In addition, the positive influence of the molar volume might attribute to the size of the molecule, which with increasing size not only stabilizes the protonated form in the gas phase [81] but also increases the probability of at least partial occupancy of the droplet exterior.

In summary, APCI appeared much less selective compared to ESI, and solution and gas-phase basicity, which are the determining compound characteristics in ESI, did not play the same crucial role for the compounds under investigation. Consequently, ion suppression in APCI-MS was often reported not as severe as in ESI-MS [82] which may be beneficial for methods employing less sample preparation steps. Instead, molecular polarity descriptors determining the surface affinity and desorption characteristics of an analyte, seem to play the major role here which in ESI become only more important mainly in situations where the protonation homeostasis is no longer limiting ionization, i.e., at low pH [15]. However, for the molar volume the existence of an optimal value is presumed since APCI is known less suitable for high molecular weight analytes >1500 u [83]; interestingly, ~1000 u was suggested as an approximate range for the change between the IEM to the CRM regime [84], which was suggested to be one of the reasons of this appearance [85].

3. Modern Ambient Ionization – Cold Atmospheric Plasma as an Easy and Particularly Promising Technique

3.1. Mechanism of Low Temperature Plasma Ionization. At the beginning of the 21st century, several new ionization techniques denoted as "ambient" were introduced beginning with *desorption electrospray ionization* (DESI) by Takats et al. 2004 [16] and *direct analysis in real time* (DART) by Cody et al. 2005 [17]. The term "ambient" introduced by Takats *et al.* for this type of ionization, is not strictly defined [86] and many similarities exist with the traditional methods APCI, APPI (atmospheric pressure photoionization) [87], AP-MALDI (atmospheric pressure matrix-assisted laser desorption/ionization) [88], or hybrid ionization techniques, for instance by coupling laser desorption to the traditional methods APCI [89–91] and ESI [92]. First of all, all operate at atmospheric pressure so that samples do not have to be introduced into a vacuum. Furthermore, in most of the ambient ionization methods, ESI- and APCI- processes are dominating with few limits to the new ambient methods. Since this is a very active field in MS research, various reviews already outlined requirements and possibilities of the ambient methods, e.g., Cooks et al. [93], Harris et al. [94], Weston [95], and others [96–98].

In contrast to the closed ionization chambers of traditional mass spectrometers, ionization with ambient methods occurs *outside* the instrument, so that the surface of very large or bulky objects can also be analysed. According to Harris et al. [94], ambient ion sources can be easily coupled to most differentially-pumped mass spectrometers, eventually with the help of special adapters as illustrated in Figure 5.

This feature promises easy implementation in mobile on-site MS-analysis [99, 100]. An enormous advantage and usually mentioned first when talking about AI-MS is the minimum or no sample preparation. This means that extraction, derivatization, desalting, dissolution, pre-concentration or separating techniques do not have to take place in advance, which can lead to enormous savings in time and resources. Furthermore, the ambient ionization methods should ionize

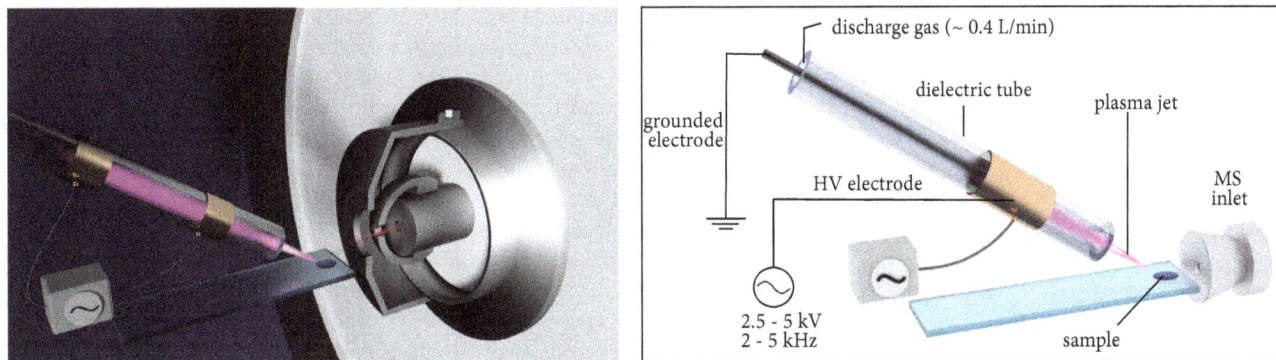

FIGURE 6: Physical appearance of the LTPI-source: schematic drawing of the plasma directed onto a target.

at least as softly as the traditional atmospheric pressure methods and should also maintain the native state and spatial integrity of the sample. Since the ambient ionization methods are essentially noninvasive, they are ideally suited for the examination of sensitive surfaces such as living biological tissues which makes these techniques particular interesting for *in situ* analysis in clinical diagnostics and surgery.

Within the many ambient methods introduced so far, plasma-based ambient ionization techniques are particularly fascinating thanks to their simple and at the same time inexpensive, yet usually robust construction. They are not dependent on high purity solvents, they generate mainly easy-to-interpret mass spectra since the ionization mechanism typically involves the protonation of the analyte by protonated water clusters that are formed by the interaction of the plasma with atmospheric water from the ambient air [101, 102]. In literature, numerous variations of plasma ionization techniques can be found. Differences exist, for example, in the voltage applied to generate the plasma, DC is used for DART, *atmospheric-pressure glow discharge*, APGD [103], and the corresponding *flowing atmospheric-pressure afterglow*, FAPA [104], while AC is applied in *dielectric-barrier discharge ionization*, DBDI [105], *low temperature plasma ionization*, LTPI [106], and *plasma-assisted desorption/ionization*, PADI [107]. Also the techniques differ in temperature; some are operated without additional heating (PADI, LTPI, DBDI), with Joule heating (APGD, FAPA) or, in the case of DART, the temperature is increased by additional heating to assist thermal desorption [97].

In comparison to all other techniques including DART, characteristic of the so-called *dielectric barrier discharge plasma* is that at least one electrode is covered with a dielectric, a nonconducting material (insulator, typically glass or ceramics) in which the charged particles or rather the charge cannot freely move in contrast to conductors. For this reason, DBD plasma must be operated with AC, as no further charge transport is possible with DC. Under AC, on the other hand, the dielectric acts as capacitor whose capacitance depends, among other things, on its thickness and permittivity [108]. Figure 6 illustrates the principal physical appearance of a DBD-LTPI source and its corresponding parts.

A dielectric barrier discharge (DBD) or "silent discharge" is a nonequilibrium plasma under atmospheric pressure [109]. The use of a dielectric between the electrodes and the plasma gas (e.g., helium) limits the current, resulting in nonequilibrium, low-temperature plasma (LTP) [109] enabling the direct ionization from a surface and subsequent MS analysis of compounds at a very low process gas flow rate, with high signal intensity and minimal fragmentation.

In LTP, statistically present electrons are accelerated and, impacting the surrounding gas, the electrons release their energy to collision partners producing more electrons, ions and excited species. If the particle density is low or the electric field is strong enough, the frequency of excitation is important because it determines the behaviour of electrons and ions. Due to their lower mass, the speed of the electrons on average will be higher than the speed of the gas molecules, atoms and/or ions. In this case it is called nonequilibrium plasma, or cold plasma [110]. If the particle density is so high that the mean free path of the electrons is small or the electric field is very low, the energy of the heavy gas particles will approach that of the electrons and all particles will have the same temperature. This is referred to as equilibrium plasma, called hot plasma.

The ionization mechanism was investigated in more detail for a helium-based LTP [101, 111] where the helium dimer $He_2^{+\cdot}$ formed near the plasma discharge [112–114] was found to be the predominant positive ion for charge transport and formation of $N_2^{+\cdot}$ through a charge transfer reaction, which, analogously to APCI, is then responsible for the formation of water clusters. $N_2^{+\cdot}$ is mainly present in the afterglow region of the He-LTP. To a lesser extent, the ionization of N_2 can also be caused by Penning ionization upon collision with excited helium. Other processes of formation are not further mentioned here due to their low probability.

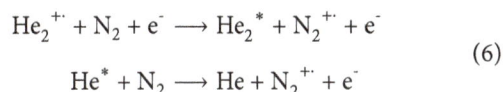

$$He_2^{+\cdot} + N_2 + e^- \longrightarrow He_2^* + N_2^{+\cdot} + e^-$$
$$He^* + N_2 \longrightarrow He + N_2^{+\cdot} + e^- \tag{6}$$

The occurrence of $N_2^{+\cdot}$ establishes a similarity with the APCI-mechanism. Indeed, with increasing concentration of N_2 in the He-plasma, the formation of $N_4^{+\cdot}$ increases due to conversion from $N_2^{+\cdot}$.

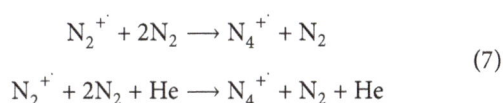

$$N_2^{+\cdot} + 2N_2 \longrightarrow N_4^{+\cdot} + N_2$$
$$N_2^{+\cdot} + 2N_2 + He \longrightarrow N_4^{+\cdot} + N_2 + He \tag{7}$$

Penning ionization of N_2 and the charge transfer to N_2^{+*} are relatively suppressed in an argon discharge, which on the other hand produces a strong OH response when analyzed with optical emission spectroscopy [115, 116]. The spectrum of the nitrogen plasma jet again has a series of NO_y lines. Here, the N_2^{+*} line is weaker than N_2 second positive system bands, which is quite different from that of a helium jet.

N_2^{+*} can transfer the charge to atmospheric water, which leads to protonated water clusters and finally to protonation of the sample [117].

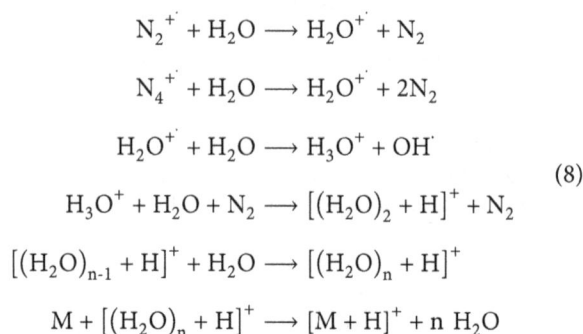

$$N_2^{+*} + H_2O \longrightarrow H_2O^{+*} + N_2$$

$$N_4^{+*} + H_2O \longrightarrow H_2O^{+*} + 2N_2$$

$$H_2O^{+*} + H_2O \longrightarrow H_3O^{+} + OH^{*}$$

$$H_3O^{+} + H_2O + N_2 \longrightarrow [(H_2O)_2 + H]^{+} + N_2 \qquad (8)$$

$$[(H_2O)_{n-1} + H]^{+} + H_2O \longrightarrow [(H_2O)_n + H]^{+}$$

$$M + [(H_2O)_n + H]^{+} \longrightarrow [M + H]^{+} + n\, H_2O$$

LTPI was already successfully applied for measurements under ambient conditions with superior performance. Moreover, it was also already used with a handheld low-temperature plasma source [99] and later for on-site analysis in combination with a miniature backpack mass spectrometer [100]. With this technology, explosives could be detected very quickly and in low quantities under ambient conditions from any surface [118] and even in mixtures [119]. Other promising applications were the screening of drugs of abuse [120, 121], agrochemicals in foods [122, 123], or fungicides in wine [124].

3.2. Relative Response with Low Temperature Plasma Ionization. In LTPI, analytes are typically detected as $[M + H]^{+}$, it is a relatively soft ionization method with nearly no fragmentation of the analytes. In contrast to the still more common electrospray ionization [95, 125], sensitivity with the different plasma-based techniques or in dependence on source parameters was hardly investigated yet with the exception of the used plasma gas, electrode spacing and the power consumption related to the distances of the electrodes [18, 126–128]. With respect to compound characteristics, a low vaporization enthalpy and low polarity (i.e., log P, large molecular nonpolar surface area and the molar volume) of the analyte as the most influential factors in LTPI are advantageous for achieving high signal intensities [129]. In addition, for substances with a boiling point beyond 200°C, the supply of additional energy, e.g., in form of heat, might be recommended in order to achieve improved signal intensities [117, 129]. In general, a lower vaporization enthalpy results in easier evaporation and thus, the number of desorbed analyte molecules available for ionization in the gas phase is enhanced. Indeed, the use of higher temperatures during LTPI has already been described in the literature to improve analytical sensitivity [121, 122, 130, 131]. While ionization of low-boiling and less polar substances is particularly favoured, signal responses show a negative linear correlation to surface

tension [129] which in return strongly correlates with the vaporization enthalpy in an inverse manner [132, 133].

Compared to the impact of the analyte's molecular characteristics such as volatility and polarity and in contrast to ESI, the solvent exerted much less impact with respect to relative and absolute signal intensity. In general, a better signal intensity of the analyte was obtained with higher boiling solvents; however, except water, most of the solvents appeared to be almost equally suitable when using LTPI for MS [129]. Nevertheless, signal response in different solvents also seemed to be determined rather by specific interactions between analytes and solvent, indicating that unpredictable matrix effects will interfere with signal response in applications of this technique. Indeed, our LTPI analysis of chlorpyrifos from the surface of several fruits suggested the occurrence of such matrix effects modulating signal response by more than two orders of magnitude (Figure 7).

Surprisingly and at a first glance in contrast to conclusions made by Kiontke et al. [129], who suggested a favorable matrix for detection of aromatic amines featuring a relatively low vapour pressure with low surface tension, it might be concluded that the waxy surface consisting of high boiling constituents as present in citrus fruits makes a particularly bad matrix for sensitive detection of chlorpyrifos. Thus, detailed investigations of the observed effect in the range of a magnitude are required to better understand the reasons for such behavior. Possibly, the abundant presence of other volatile compounds on the surface of citrus fruits might create a transient microenvironment (TME, [134]) responsible for the observed effect decreasing LTPI efficiency. Specific analyte-matrix effects as well as a high variation in replicate analyses still seem to hamper quantitative analyses and further investigation is required to address this bottleneck.

4. Differential Sensitivity of ESI, APCI and LTPI for the Aromatic Amines

4.1. Influence of Compound Solution Basicity. The relative response of the aromatic amino compounds was finally used to further explore differential relative sampling efficiencies with the three atmospheric pressure methods, namely, LTPI [129], APCI, and ESI [15]. For that, correlation analysis of peak signal intensities with physicochemical properties was performed and compared.

In accordance to the literature, the results suggest that only the ESI response from a solution pH = 7 correlates well with the pKa of the analyte (R = 0.51, p < 0.01) (Figure S1a in the Supplementary Material). In ESI for the used set of aromatic nitrogen-containing compounds, the protonation of the analyte and subsequent desorption from the droplet is essential and at pH 7 (without additives) the signal intensity primarily follows the solution basicity of the analytes, where basic analytes can easily take up a proton from the solvent and desorb from the droplet. Less basic analytes are more difficult to protonate and therefore show a lower signal intensity. The situation is different after solvent acidification, where protonation is no longer the limiting factor due to an excess of protons in solution and the analytes are protonated more easily. Thus for ESI pH = 3, but also with APCI and

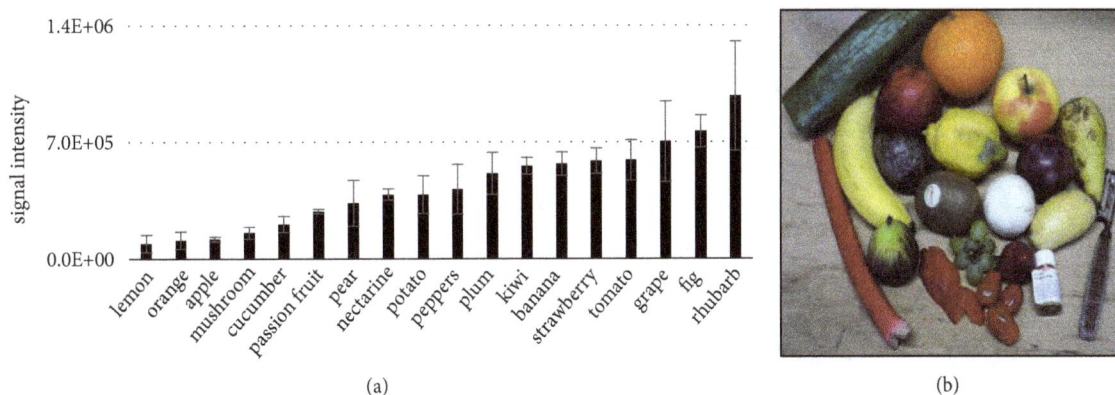

FIGURE 7: Signal intensity of LTPI measurements of chlorpyrifos as a function of the analyzed surface.

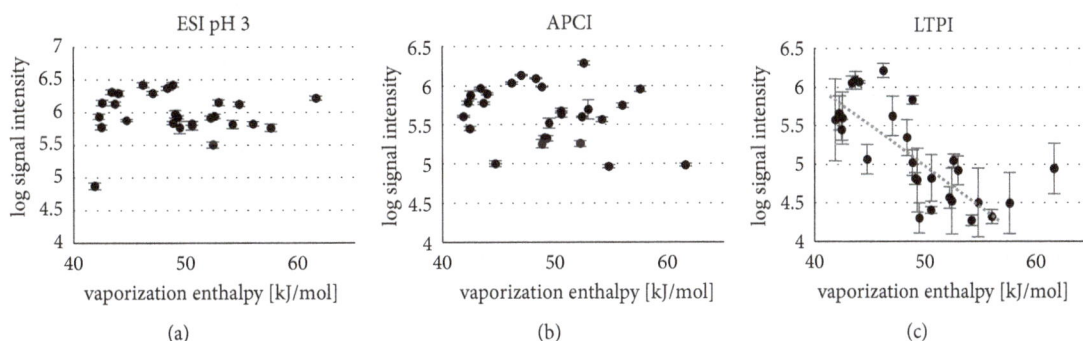

FIGURE 8: Log signal intensity (peak height) in dependency on the vaporization enthalpy for (a) ESI pH = 3, (b) APCI, and (c) LTPI.

LTPI (Figure S1b-d) no dependence of signal response on solution basicity was found. Moreover, also no correlation of the signal intensity with the proton affinity was found, which is somewhat surprising. According to the current understanding of APCI and LTPI ionization mechanism, the neutral analyte is vaporized and protonated solvent species transfer a proton to the amines in the gas phase depending on their proton affinity, which in turn corresponds to the acidity or basicity of a compound in the gas phase. Given that all of our analytes are nitrogen-bases with a gas-phase basicity > 830 kJ/mol providing a high proton affinity (as discussed before), other factors appear to be responsible for the observed sensitivity under the respective conditions.

4.2. Influence of Compound Volatility. Among the volatility descriptors, boiling point, vaporization enthalpy, vapor pressure, and surface tension all were tested for their influence on relative signal response. Since the effect of the vaporization enthalpy was most prominent with LTPI, the obtained relative signal intensities for the three different ionization techniques were plotted over this parameter (Figure 8).

The most eye-catching observation is the high variance of LTPI response compared to the other two methods. An average standard deviation of 46% for plasma ionization (Figure 8(c)) was observed over all analytes, compared to 19% for ESI at pH 7 (Figure S2a) and 9% for ESI at pH 3 (Figure 8(a)) and APCI (Figure 8(b)), respectively. ESI and

APCI are carried out in standardized, closed devices while the used plasma source has an open structure (Figure 6) and might therefore be more susceptible to variable environmental conditions such as convection, temperature, or humidity.

In accordance with the literature [131], for LTPI signal intensity a strong dependence on the analytes volatility was found based on negative correlation with the vaporization enthalpy (Figure 8(c), R = -0.64, p < 0.001), boiling point (not shown, R = -0.63, p < 0.001), and a moderate positive correlation with the vapour pressure (not shown, R = 0.55, p < 0.01). In general, a lower vaporization enthalpy improves the evaporation of the analytes making a larger number of analyte molecules amenable to ionization in the gas phase. In conclusion, this means that the supply of heat and the resulting improved evaporation and desorption of the analyte will result in a higher signal intensity. This behaviour has already been described in the literature to improve LODs with LTPI [121, 122, 130, 131]. However, no such correlation was observed for the other techniques except a positive correlation of signal enhancement in ESI response at pH 3 compared to pH 7 with the boiling point [15] which was also suggested to be a consequence of better desorption of more volatile compounds at adequate availability of charge carriers in solution, i.e., protons.

Furthermore, a very strong negative correlation (R = -0.86, p < 0.001) with the surface tension of the analytes was observed in LTPI (not shown). Surface tension is the

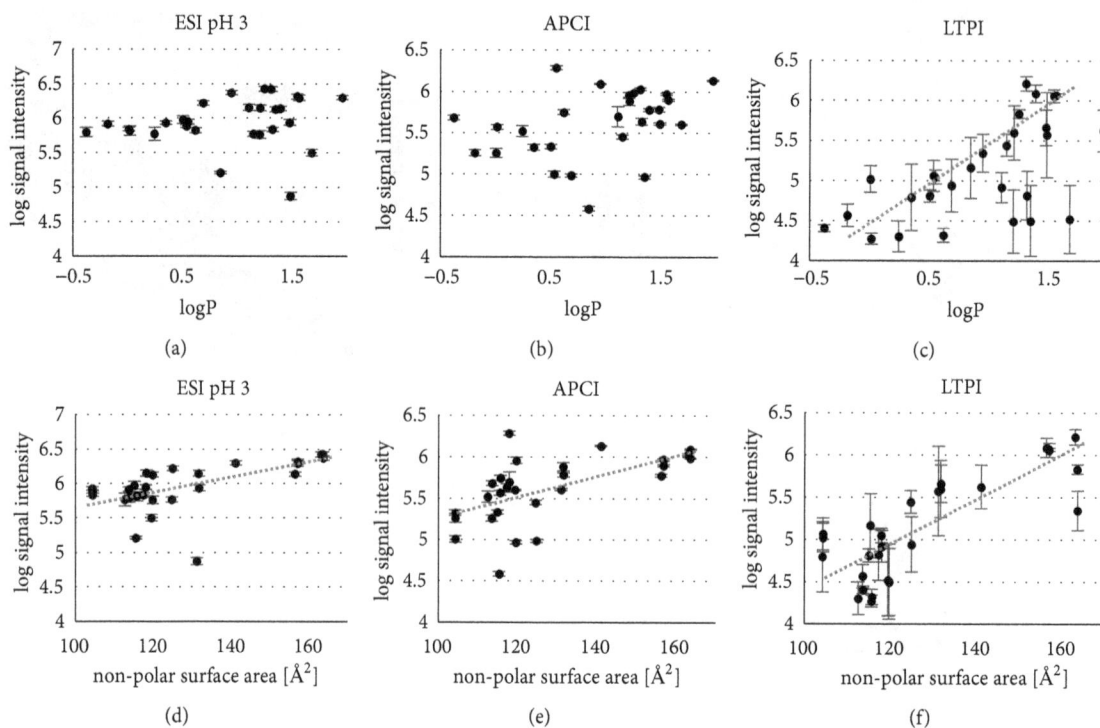

FIGURE 9: Log signal intensity (peak height) in dependency on the partition coefficient log P (SciFinder) for (a) ESI pH = 3, (b) APCI and (c) LTPI and the nonpolar surface area for (d) ESI pH = 3, (e) APCI, and (f) LTPI ((c) and (f).

result of increased cohesive intermolecular forces between molecules at the interfaces between air and liquid. For many liquids, the evaporation enthalpy changes linearly with the macroscopic surface tension [132, 133], and therefore both are dependent on each other. High surface tension of the analyte can suppress desorption and thus lead to a poor signal intensity. The fact that only LTPI is dependent on the volatility of the analytes tested here might be reasoned by the different temperature conditions applied during ionization; with APCI, sufficient energy was supplied by elevated temperatures through heated nitrogen streams (heater/auxiliary and dry gas at 250°C) to ensure complete vaporization.

While in ESI at pH 7 the influence of vaporization enthalpy seems overshadowed by the strong quantitative impact of solution basicity, at pH 3, when solution basicity loses its importance due to the enhanced availability of charge carriers, it is a low vaporization enthalpy leading to increased signal enhancement. However, at pH 3 other parameters than at pH 7 become important, in which impact was observed to be dependent on the particular instrument used, i.e., polarity or molecular size [15]. Thus, the influence of polarity was assessed for the three techniques; the results are illustrated in Figure 9.

Obviously, polarity of the analyte is a common parameter influencing the sampling efficiency of all three methods. Thus, for LTPI a strong correlation was found between the signal intensity and the log P (data from ChemAxon, Figure 9(c), R = 0.61, and p < 0.001) and an even very strong correlation with the nonpolar surface area (Figure 9(f), R = 0.81, and p < 0.001). The signal intensity in APCI also

shows a moderate correlation with the log P (R = 0.48, p < 0.01, and Figure 9(b)). Interestingly, for ESI at pH 7 polar compounds show an increased sampling efficiency (Figure S2b and c), while for APCI and LTPI the reversed behaviour was observed. This hints to the interplay between ionization and desorption; in ESI, ionization occurs *before* (ion) desorption, while for APCI and LTPI desorption is first and the desorbed analyte is *afterwards* ionized in the gas phase; thus, desorption of the *neutral* analyte would benefit from a low polarity favouring the droplets liquid/gas interface while for ESI, where chargeability *in solution* has the highest impact on sampling efficiency, a higher polarity is beneficial. Disappearance of this influence in ESI at pH 3, where chargeability is largely improved by the enhanced density of charge carrier, would support this perception.

A similar situation may apply to the size of the nonpolar surface area. In situations, where desorption is not limited by a required charge; i.e., for ESI at pH = 3 (Figure 9(a), R = 0.56, and p < 0.01) and APCI (Figure 9(b), R = 0.59, p < 0.01, a moderate correlation), a correlation of this parameter with the sampling efficiency was observed. Both descriptors, polarity and nonpolar surface area, have an impact on the surface activity of an analyte [23, 80] and help to improve desorption, which is necessary for a good signal intensity. Ions and molecules with a large nonpolar area have a high surface affinity, as they prefer the air-liquid interface before the aqueous bulk solution within the droplets.

Finally, Figure 10 illustrates the influence of the molar volume on the signal intensity obtained with the investigated ionization techniques.

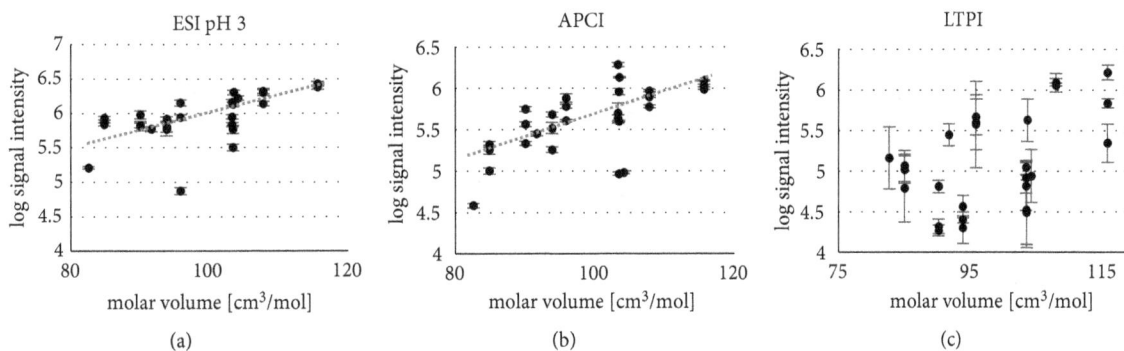

FIGURE 10: Log signal intensity (peak height) in dependency on the molar volume for (a) ESI pH = 3, (b) APCI, and (c) LTPI.

Within this context, the strong correlation of the molar volume with the signal intensities of ESI pH = 3 (Figure 10(a), R = 0.61, and p < 0.001), APCI (Figure 10(b), R = 0.64, and p < 0.001) and a moderate correlation for LTPI (Figure 10(c), R = 0.49 p < 0.01) also fits in well. In ESI, the competition for charge and desorption takes place on the surface of the droplet, but only charged species will be further transmitted. Furthermore, it is known that a (charged) analyte in ESI needs access to the droplet's surface for successful detection [135]. Again, at pH 7 the limited availability of charge carriers overshadows the impact of the molar volume (Figure S2d), but at pH = 3 it can be assumed that all analyte molecules are protonated in the droplet. The charged particles tend to spread over the surface of the droplets at maximal distance due to Coulombs repulsion. According to Wu et al., ions with larger molecular volumes can occupy a larger proportion of the droplet surface than ions with smaller volumes [136] enabling easier desorption from the droplet (note that the molecular volume strongly correlates with the molar volume). Furthermore, the larger the molecule, the better its protonated form is stabilized in the gas phase [81] which should exert its effect in all three, ESI, APCI, and plasma ionization.

Under the chosen conditions, desorption from the solvent seems to play a similarly strong role for LTPI and APCI as solution basicity for ESI at pH 7. This was initially unexpected for APCI, as complete evaporation should be ensured with the help of nebulizer, heated auxiliary gas and dry gas in accordance with the APCI mechanism. A better evaporation (choosing a higher temperature for the APCI heater and the dry gas, or an easier vaporizable solvent) may counter the observed dependence and uncover other influencing variables than the nonpolar surface area or the molar volume. Desorption and thus surface activity is also one of the key aspects in LTPI [129]. As no additional heat was added, the analytes have to first reach the surface of the solvent droplet and then evaporate. In this case, the nonpolar surface of a species is decisive for the deposition on the droplets surface, as is the vapour pressure, respectively the boiling temperature for the subsequent evaporation. However, in LTPI still a droplet-pick up mechanism seems to crucially enhance the sampling efficiency since the mere presence of a solvent was observed to enhance signal response significantly [129].

The detailed investigation on the quantitatively most impacting parameters of the three ionization techniques resulted in a rather complex and sophisticated picture which prompted us to analyze similarities in the pattern of obtained signal responses within the three ionization techniques. As a result, response pattern after ESI at pH 3 appeared to be almost equally similar to all other conditions, i.e., ESI pH 7 (R = 0.54, p = 0.003), APCI (R = 0.51, p = 0.001), and LTPI (R = 0.65, p < 0.001), which suggests that solvent acidification in ESI leads to a situation where sensitivity becomes dependent on similar analyte characteristics as in APCI and LTPI instead of the solution basicity. However, the correlation between the two ESI conditions greatly improved using the signal log values for analysis (R = 0.75, p < 0.001) but impaired the cobehaviour with APCI and LTPI. After log-transformation, the influence of extreme values in a linear correlation analysis is usually decreased. Therefore, we concluded that similarity of response patterns in ESI pH 3 with pH 7 might be based on the behaviour of different analytes than similarity with APCI and LTPI and analysed the cobehaviour of the target compounds. Indeed, we found a reverse response pattern of analytes with amino-, hydroxyl-, and methoxy-substituents as one group and a second comprising mainly pyridine, 2- and 3-fluoroanilines, and analytes with electron-withdrawing substituents in o-position (Figure 11).

While for the target compounds with electron-donating substituents, the highest relative intensity was observed with ESI pH 7 where solution basicity determines the response, the behavior is reversed for volatile analytes with electron-withdrawing and less polar analytes due to H-sharing of the substituent in o-position to the amino group [15]. Consequently, when we removed 2- and 3-fluoroaniline and pyridine, where this appearance was most developed, from correlation analysis of the log transformed data, we again obtained the strongest correlation between ESI pH 3 and LTPI instead (R = 0.70, p < 0.001 vs. R = 0.49, and p = 0.014 for ESI pH 3/ pH 7). Consequently, requirements for a good ESI response of polar, strong bases in acidified solvent are most similar to those in LTPI, while less polar weaker bases exhibit an impaired response in ESI pH 7 compared to the other analytes and ionization techniques leading to an (unwanted?) selectivity of ESI.

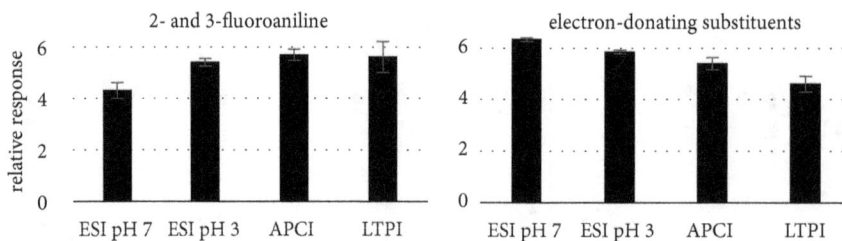

FIGURE 11: Response pattern of 2- and 3-fluoroanilines vs. electron-donating substituted compounds over the four ionization conditions reflect the fact that ESI pH 7 response is dominated by solution basicity while this advantage is lost when it comes to conditions where volatility of the compound matters.

TABLE 1: Summary of analyte and solvent characteristics in how they impact sampling efficiency of the three ionization methods, ESI, APCI, and LTPI, for MS.

	ESI pH 7	ESI pH 3	APCI	LTPI
Analyte				
surface tension	(disadvantageous)	disadvantageous	no influence	disadvantageous
polarity	beneficial	(beneficial)	disadvantageous	disadvantageous
volatility	no influence	beneficial	no influence	beneficial
basicity	beneficial	(beneficial)	no influence	no influence
fragmentation	hardly observed	hardly observed	observed	hardly observed
molar size	beneficial	beneficial	beneficial	beneficial
Solvent				
surface tension	disadvantageous	disadvantageous	not analyzed	disadvantageous
polarity	beneficial	no influence	not analyzed	no influence
volatility	beneficial	beneficial	not analyzed	disadvantageous
pH	contradictory*	contradictory*	not analyzed	not analyzed
Reproducibility	moderate	high	high	low

*Depending on the interaction between solvent evaporability, electrolyte, and instrumental configuration [15].

Interestingly, a comparison of the correlation factors obtained from analysis of response in relation to all molecular descriptors indicated the highest similarity of influencing factors for ESI pH 3 and APCI. Obviously, however, the quantitative impact of these molecular descriptors on signal response is different under the two conditions given the results detailed above.

5. Conclusions

In our comparison of two standard atmospheric pressure techniques, ESI and APCI, with the still quite new, ambient LTPI, we could show that generally different parameters of desorption properties, determined by the descriptors of volatility and surface affinity, seem to be decisive for the signal response of an analyte in a liquid matrix. Only at ESI pH 7 does the influence of basicity exceed the influence of desorption properties (Table 1).

According to this, APCI seems to be the least selective method against the aromatic amines, which would be a great advantage when it comes to multiselective analysis or screening methods. However, within this context the dynamic range of signal response (highest divided by least abundance) with ESI pH 3 actually suggested a weaker selectivity (factor 35 for ESI pH 3 vs. 50 for APCI contrasting ~700 for

ESI pH 7) and indicates that it is rather the fact that, in APCI a more complicated, but not too strong affecting interplay of compound characteristic, might determine its signal response than nonselective ionization. The partially large differences in the quantitative impact of the various molecular descriptors between the methods also illustrate the fact that there is still no universally optimal ionization method available for all analytical tasks. The requirement of ionization and subsequent transfer of the ion into the gas phase still is the main limitation in application of MS. Therefore, at present and for the foreseeable future, it is only possible to optimize the interface for a specific analytical challenge and a multitude of further developments may still be required towards an optimal, generally applicable ionization method for MS.

6. Materials and Methods

6.1. Chemicals. 3-aminophenol, 2-fluoroaniline, 3-fluoroaniline, 4-fluoroaniline, 2-methoxyaniline (*o*-anisidine), 3-methoxyaniline (*m*-anisidine), 4-methoxyaniline (*p*-anisidine), 2-nitroaniline, 3-nitroaniline, 4-nitroaniline, 3-methylaniline (*m*-toluidine), 3-aminoaniline (*m*-phenylenediamine), 4-aminoaniline (*p*-phenylenediamine), 2-aminobenzonitrile, 3-aminobenzonitrile, 4-aminobenzonitrile, and pyridine were purchased from Sigma Aldrich (Taufkirchen, Germany), besides 2-methylaniline (*o*-toluidine), 4-methylaniline (*p*-toluidine), chlorpyrifos from Fluka (Buchs, Switzerland), and aniline from Acros (Geel, Belgium). Acetonitrile (ACN) was purchased from VWR (Dresden, Germany). Methanol was purchased from Carl Roth (Karlsruhe, Germany), water from BIOSOLVE (Valkenswaard, Netherlands). 2-amino-aniline (*o*-phenylenediamine), 2-aminopyridine, 3-amino-pyridine, 4-aminopyridine, 2-aminophenol, 4-aminophenol, 2-aminobenzoic acid, 3-aminobenzoic acid, 4-aminobenzoic acid, sulfanilic acid, and 4-chloroaniline were kindly provided by Professor *em*. S. Berger (Institute of Analytical Chemistry, University of Leipzig, Germany).

6.2. Sample Preparation. A set of 31 anilines was prepared in ACN/H_2O 1:1 (*v/v*) for each aniline (10 μM for ESI and APCI, 1 mM for LTPI) and the signal intensity was determined with each of the three ionization methods, namely, atmospheric pressure chemical ionization (APCI), electrospray ionization (ESI), and low temperature plasma ionization (LTPI). Solvent blanks were run on a frequent basis to ensure the absence of cross-contamination. Prior to the measurement of the complete set of aromatic amines, dynamic behaviour was successfully confirmed using concentration series of the analytes [15, 129]. For detailed structures of all analytes see Kiontke et al. [129]. The analytes were selected for a systematic study of the influence of the molecular descriptors such as polarity and vapour pressure; they provide a very broad structural variety, importance in biological contexts and many compounds of interest contain structural units that are similar to our analytes. In addition, these analytes are already well characterized in the literature and publicly available databases.

6.2.1. APCI Response of Anilines. APCI measurements were performed on a Bruker Impact II QTOF MS (Bruker, Bremen, Germany) equipped with a Dionex Ultimate 3000 autosampler (Thermofisher, Dreieich Germany). The capillary voltage was at 4.0 kV, and 100 μL sample was injected at a flow rate of 50 μL/min with 2.5 bar nebulizer and 2.5 L/min dry gas flow rate (both nitrogen) at 250°C. Scan range was *m/z* 50-500 with 3 scans rolling average.

6.2.2. LTPI Response of Chlorpyrifos. To simulate the on-site analysis of pesticides directly from the surface using plasma ionization, fruit, vegetables and mushrooms were purchased from a local supermarket (REWE, Tarostraße, Leipzig, Germany). A strip of apple, banana, pear, mushroom, strawberry, fig, cucumber, potato, kiwi, nectarine, orange, pepper, passion fruit, plum, rhubarb, tomato, grape, and lemon peelers at least 1 cm long was taken from each peel to obtain the most reproducible layer thickness. Four circular cut-outs with a diameter of 4 mm were punched out of each of these peel strips. Subsequently, 1 μL chlorpyrifos solution (100 μM in ACN) was added to three cut-outs (the fourth served as blank) and the solvent was allowed to evaporate before the measurement.

An optimized home-built low-temperature plasma source was used in the experiments [18]. Briefly, it consists of an ignition transformer (EBI4 CM S, Danfoss, Nordborg, Denmark) and a glass tube (GC liner, Thermo Scientific, Waltham, MA, USA) with two surrounding outer electrodes made of copper foil tape (Noll GmbH, Wörrstadt, Germany) and isolated by a homemade Teflon housing. The flow was adjusted to 20 mL/min with an Ellutia 7000 GC Flowmeter (Ellutia Ltd, Ely, UK).

Optimized parameters of plasma configuration were used [18], e.g., dielectric thickness 2 mm, width and distance of the electrodes 10 mm each and distance of the electrode to the outlet 20 mm. Mass spectra were acquired on an Esquire 3000+ MS (Bruker, Bremen, Germany) with the following instrumental parameters: high voltage off, dry gas (nitrogen) 1.5 mL/min with a temperature set to 350, scan range: *m/z* 50-400, target mass: 350. The number of ions per scan was limited to 20,000 with a maximum accumulation time of 200 ms and a rolling average of three scans. After spotting 1 μL of the solutions on a paper target, data acquisition was immediately started at least for 2 minutes and the average response of each analyte was calculated from triplicate analysis.

6.3. Data Evaluation. The *m/z* peak signal intensities were averaged over 1 min analysis time using Bruker Data analysis software 4.2 and the corresponding signal intensities of triplicate analyses were used for data evaluation. The relative response of the anilines was assessed as the average intensity (cps, peak height) of the corresponding peak for the $[M + H]^+$ ion of the analyte of interest and for chloroaniline as sum of the two most abundant isotope peak signal intensities.

Characteristic chemical constants (pKa, molecular polar surface area, solvent accessible molecular surface area, log P, log D, proton affinity, gas phase basicity, boiling point, vapour pressure, vaporization enthalpy, and surface tension) were retrieved from public databases, namely, ChemSpider by the Royal Society of Chemistry, London, UK [http://www.chemspider.com/], chemicalize.org by ChemAxon, Budapest, Hungary [http://www.chemicalize.org/], Scifinder by the Chemical Abstracts Service, Columbus/Ohio, USA [https://scifinder.cas.org/], and the NIST Chemistry WebBook by the National Institute of Standards and Technology (NIST), Gaithersburg, USA [http://webbook.nist.gov/chemistry/]. The molecular volume was calculated using the Spartan software package (Spartan 14, Wavefunction Inc., Irvine, CA, USA). The settings for calculation were DFT (density functional theory) B3LYP with a 6–31G* basis set.

Correlation analysis of peak signal intensities with physicochemical characteristics (Pearson's product-moment correlation coefficient and significance) was carried out using the Analysis ToolPak in MS Excel 2013 (Microsoft Corp., Redmond, USA). Before correlation analysis, a visual inspection of appropriate data distribution was carried out using scatter plots.

Acknowledgments

This work was supported by the German Federal Environmental Foundation (Deutsche Bundesstiftung Umwelt, DBU) [Grant no. 20015/375]; the European Regional Development Fund (ERDF, Europäischer Fond für Regionale Entwicklung EFRE, "Europe funds Saxony") [Grant no. 100195374]; the Federal Ministry of Economy and Energy of Germany (Bundesministerium für Wirtschaft und Energie der Bundesrepublik Deutschland, BMWi, "Central Innovation Program for SME," "Zentrales Innovationsprogramm Mittelstand") [ZF4548701SL8]; and the University of Leipzig. The authors acknowledge Sebastian Piendl and Dr. Josef Heiland, University of Leipzig, for preparation of the figures. In addition, they thank Professor Detlev Belder and Professor *em.* Stefan Berger, at the University of Leipzig, for their kind, long-standing support and advice.

References

[1] S. Naylor and R. Kumar, "Emerging role of mass spectrometry in structural and functional proteomics," *Advances in Protein Chemistry and Structural Biology*, vol. 65, pp. 217–248, 2003.

[2] S. G. Villas-Bôas, S. Mas, M. Åkesson, J. Smedsgaard, and J. Nielsen, "Mass spectrometry in metabolome analysis," *Mass Spectrometry Reviews*, vol. 24, no. 5, pp. 613–646, 2005.

[3] T. Hu and J.-L. Zhang, "Mass-spectrometry-based lipidomics," *Journal of Separation Science*, vol. 41, no. 1, pp. 351–372, 2018.

[4] R. L. Foltz, D. M. Andrenyak, and D. J. Crouch, "Forensic Science, Applications of Mass Spectrometry," in *Encyclopedia of spectroscopy and spectrometry*, J. C. Lindon, G. E. Tranter, and D. W. Koppenaal, Eds., pp. 707–711, Elsevier AP Academic Press, Amsterdam, Boston, Heidelberg, 2017.

[5] N. Nakayama, Y. Bando, T. Fukuda et al., "Developments of mass spectrometry-based technologies for effective drug development linked with clinical proteomes," *Drug Metabolism and Pharmacokinetics*, vol. 31, no. 1, pp. 3–11, 2016.

[6] A. N. Gloess, C. Yeretzian, R. Knochenmuss, and M. Groessl, "On-line analysis of coffee roasting with ion mobility spectrometry–mass spectrometry (IMS–MS)," *International Journal of Mass Spectrometry*, vol. 424, pp. 49–57, 2018.

[7] T. A. Garrett, "Major roles for minor bacterial lipids identified by mass spectrometry," *Biochimica et Biophysica Acta (BBA) - Molecular and Cell Biology of Lipids*, vol. 1862, no. 11, pp. 1319–1324, 2017.

[8] J. A. R. Teodoro, H. V. Pereira, M. M. Sena, E. Piccin, J. J. Zacca, and R. Augusti, "Paper spray mass spectrometry and chemometric tools for a fast and reliable identification of counterfeit blended Scottish whiskies," *Food Chemistry*, vol. 237, pp. 1058–1064, 2017.

[9] D. J. Douglas, A. J. Frank, and D. Mao, "Linear ion traps in mass spectrometry," *Mass Spectrometry Reviews*, vol. 24, no. 1, pp. 1–29, 2005.

[10] B. A. Mamyrin, "Time-of-flight mass spectrometry (concepts, achievements, and prospects)," *International Journal of Mass Spectrometry*, vol. 206, no. 3, pp. 251–266, 2001.

[11] M. B. Comisarow and A. G. Marshall, "Fourier transform ion cyclotron resonance spectroscopy," *Chemical Physics Letters*, vol. 25, no. 2, pp. 282–283, 1974.

[12] A. Makarov, "Electrostatic axially harmonic orbital trapping: A high-performance technique of mass analysis," *Analytical Chemistry*, vol. 72, no. 6, pp. 1156–1162, 2000.

[13] K. H. Kingdon, "A method for the neutralization of electron space charge by positive ionization at very low gas pressures," *Physical Review A: Atomic, Molecular and Optical Physics*, vol. 21, no. 4, pp. 408–418, 1923.

[14] T. Kind and O. Fiehn, "Advances in structure elucidation of small molecules using mass spectrometry," *Bioanalytical Reviews*, vol. 2, no. 1, pp. 23–60, 2010.

[15] A. Kiontke, A. Oliveira-Birkmeier, A. Opitz, and C. Birkemeyer, "Electrospray ionization efficiency is dependent on different molecular descriptors with respect to solvent pH and instrumental configuration," *PLoS ONE*, vol. 11, no. 12, 2016.

[16] Z. Takáts, J. M. Wiseman, B. Gologan, and R. G. Cooks, "Mass spectrometry sampling under ambient conditions with desorption electrospray ionization," *Science*, vol. 306, no. 5695, pp. 471–473, 2004.

[17] R. B. Cody, J. A. Laramée, and H. D. Durst, "Versatile new ion source for the analysis of materials in open air under ambient conditions," *Analytical Chemistry*, vol. 77, no. 8, pp. 2297–2302, 2005.

[18] A. Kiontke, F. Holzer, D. Belder, and C. Birkemeyer, "The requirements for low-temperature plasma ionization support miniaturization of the ion source," *Analytical and Bioanalytical Chemistry*, vol. 410, no. 16, pp. 3715–3722, 2018.

[19] M. Dole, L. L. Mack, R. L. Hines et al., "Molecular beams of macroions," *The Journal of Chemical Physics*, vol. 49, no. 5, pp. 2240–2249, 1968.

[20] L. L. Mack, P. Kralik, A. Rheude, and M. Dole, "Molecular beams of macroions. II," *The Journal of Chemical Physics*, vol. 52, no. 10, pp. 4977–4986, 1970.

[21] J. B. Fenn, M. Mann, C. K. Meng, S. F. Wong, and C. M. Whitehouse, "Electrospray ionization for mass spectrometry of large biomolecules," *Science*, vol. 246, no. 4926, pp. 64–71, 1989.

[22] M. Yamashita and J. B. Fenn, "Electrospray ion source. Another variation on the free-jet theme," *The Journal of Physical Chemistry C*, vol. 88, no. 20, pp. 4451–4459, 1984.

[23] P. Kebarle and L. Tang, "From ions in solution to ions in the gas phase: the mechanism of electrospray mass spectrometry," *Analytical Chemistry*, vol. 65, no. 22, 1993.

[24] N. B. Cech, J. R. Krone, and C. G. Enke, "Predicting electrospray response from chromatographic retention time," *Analytical Chemistry*, vol. 73, no. 2, pp. 208–213, 2001.

[25] G. J. Van Berkel, "Electrolytic corrosion of a stainless-steel electrospray emitter monitored using an electrospray-photodiode array system," *Journal of Analytical Atomic Spectrometry*, vol. 13, no. 7, pp. 603–607, 1998.

[26] R. Abburi, S. Kalkhof, R. Oehme, A. Kiontke, and C. Birkemeyer, "Artifacts in amine analysis from anodic oxidation of organic solvents upon electrospray ionization for mass spectrometry," *European Journal of Mass Spectrometry*, vol. 18, no. 3, pp. 301–312, 2012.

[27] M. G. Ikonomou, A. T. Blades, and P. Kebarle, "Electrospray-Ion Spray: A Comparison of Mechanisms and Performance," *Analytical Chemistry*, vol. 63, no. 18, pp. 1989–1998, 1991.

[28] G. Taylor, "Disintegration of water drops in an electric field," *Proceedings of the Royal Society A Mathematical, Physical and*

Engineering Sciences, vol. 280, no. 1382, pp. 383–397, 1964.

[29] P. Nemes, I. Marginean, and A. Vertes, "Spraying mode effect on droplet formation and ion chemistry in electrosprays," *Analytical Chemistry*, vol. 79, no. 8, pp. 3105–3116, 2007.

[30] J. W. Strutt and L. Rayleigh, "On the instability of jets," *Proceedings of the London Mathematical Society*, vol. 10, no. 1, pp. 4–13, 1878.

[31] L. Rayleigh, " XX. ," *The London, Edinburgh, and Dublin Philosophical Magazine and Journal of Science*, vol. 14, no. 87, pp. 184–186, 2009.

[32] J. V. Iribarne and B. A. Thomson, "On the evaporation of small ions from charged droplets," *The Journal of Chemical Physics*, vol. 64, no. 6, pp. 2287–2294, 1976.

[33] L. Konermann, E. Ahadi, A. D. Rodriguez, and S. Vahidi, "Unraveling the mechanism of electrospray ionization," *Analytical Chemistry*, vol. 85, no. 1, pp. 2–9, 2013.

[34] L. Konermann, A. D. Rodriguez, and J. Liu, "On the formation of highly charged gaseous ions from unfolded proteins by electrospray ionization," *Analytical Chemistry*, vol. 84, no. 15, pp. 6798–6804, 2012.

[35] E. Ahadi and L. Konermann, "Modeling the behavior of coarse-grained polymer chains in charged water droplets: Implications for the mechanism of electrospray ionization," *The Journal of Physical Chemistry B*, vol. 116, no. 1, pp. 104–112, 2012.

[36] E. C. Horning, M. G. Horning, D. I. Carroll, I. Dzidic, and R. N. Stillwell, "New Picogram Detection System Based on a Mass Spectrometer with an External Ionization Source at Atmospheric Pressure," *Analytical Chemistry*, vol. 45, no. 6, pp. 936–943, 1973.

[37] D. I. Carroll, I. Dzidic, R. N. Stillwell, M. G. Horning, and E. C. Horning, "Subpicogram Detection System for Gas Phase Analysis Based upon Atmospheric Pressure Ionization (API) Mass Spectrometry," *Analytical Chemistry*, vol. 46, no. 6, pp. 706–710, 1974.

[38] I. Dzidic, D. I. Carroll, R. N. Stillwell, and E. C. Horning, "Comparison of Positive Ions Formed in Nickel-63 and Corona Discharge Ion Sources Using Nitrogen, Argon, Isobutane, Ammonia and Nitric Oxide as Reagents in Atmospheric Pressure Ionization Mass Spectrometry," *Analytical Chemistry*, vol. 48, no. 12, pp. 1763–1768, 1976.

[39] D. I. Carroll, I. Dzidic, R. N. Stillwell, K. D. Haegele, and E. C. Horning, "Atmospheric pressure ionization mass spectrometry. corona discharge ion source for use in a liquid chromatograph-mass spectrometer-computer analytical system," *Analytical Chemistry*, vol. 47, no. 14, pp. 2369–2373, 1975.

[40] J. Zhao, J. Zhu, and D. M. Lubman, "Liquid Sample Injection Using an Atmospheric Pressure Direct Current Glow Discharge Ionization Source," *Analytical Chemistry*, vol. 64, no. 13, pp. 1426–1433, 1992.

[41] W. C. Davis and R. K. Marcus, "An atmospheric pressure glow discharge optical emission source for the direct sampling of liquid media," *Journal of Analytical Atomic Spectrometry*, vol. 16, no. 9, pp. 931–937, 2001.

[42] F. Gosetti, E. Mazzucco, D. Zampieri, and M. C. Gennaro, "Signal suppression/enhancement in high-performance liquid chromatography tandem mass spectrometry," *Journal of Chromatography A*, vol. 1217, no. 25, pp. 3929–3937, 2010.

[43] M. Mann, C. K. Meng, and J. B. Fenn, "Interpreting Mass Spectra of Multiply Charged Ions," *Analytical Chemistry*, vol. 61, no. 15, pp. 1702–1708, 1989.

[44] W. M. A. Niessen, "State-of-the-art in liquid chromatography-mass spectrometry," *Journal of Chromatography A*, vol. 856, no. 1-2, pp. 179–197, 1999.

[45] A. Fridman, *Plasma chemistry*, Cambridge University Press, Cambridge, UK, 2012.

[46] L. C. Herrera, J. S. Grossert, and L. Ramaley, "Quantitative Aspects of and Ionization Mechanisms in Positive-Ion Atmospheric Pressure Chemical Ionization Mass Spectrometry," *Journal of The American Society for Mass Spectrometry*, vol. 19, no. 12, pp. 1926–1941, 2008.

[47] K. Hiraoka, *Fundamentals of Mass Spectrometry*, Springer New York, New York, NY, 2013.

[48] A. Vaikkinen, T. J. Kauppila, and R. Kostiainen, "Charge exchange reaction in dopant-assisted atmospheric pressure chemical ionization and atmospheric pressure photoionization," *Journal of The American Society for Mass Spectrometry*, vol. 27, no. 8, pp. 1291–1300, 2016.

[49] J. Sunner, G. Nicol, and P. Kebarle, "Factors Determining Relative Sensitivity of Analytes in Positive Mode Atmospheric Pressure Ionization Mass Spectrometry," *Analytical Chemistry*, vol. 60, no. 13, pp. 1300–1307, 1988.

[50] K. R. Chalcraft, R. Lee, C. Mills, and P. Britz-McKibbin, "Virtual quantification of metabolites by capillary electrophoresis- electrospray ionization-mass spectrometry: Predicting ionization efficiency without chemical standards," *Analytical Chemistry*, vol. 81, no. 7, pp. 2506–2515, 2009.

[51] I. Leito, K. Herodes, M. Huopolainen et al., "Towards the electrospray ionization mass spectrometry ionization efficiency scale of organic compounds," *Rapid Communications in Mass Spectrometry*, vol. 22, no. 3, pp. 379–384, 2008.

[52] T. L. Constantopoulos, G. S. Jackson, and C. G. Enke, "Effects of salt concentration on analyte response using electrospray ionization mass spectrometry," *Journal of The American Society for Mass Spectrometry*, vol. 10, no. 7, pp. 625–634, 1999.

[53] V. Gabelica, E. De Pauw, and M. Karas, "Influence of the capillary temperature and the source pressure on the internal energy distribution of electrosprayed ions," *International Journal of Mass Spectrometry*, vol. 231, no. 2-3, pp. 189–195, 2004.

[54] C. G. Enke, "A predictive model for matrix and analyte effects in electrospray ionization of singly-charged ionic analytes," *Analytical Chemistry*, vol. 69, no. 23, pp. 4885–4893, 1997.

[55] G. J. Van Berkel, S. A. McLuckey, and G. L. Glish, "Electrochemical Origin of Radical Cations Observed in Electrospray Ionization Mass Spectra," *Analytical Chemistry*, vol. 64, no. 14, pp. 1586–1593, 1992.

[56] N. B. Cech and C. G. Enke, "Practical implications of some recent studies in electrospray ionization fundamentals," *Mass Spectrometry Reviews*, vol. 20, no. 6, pp. 362–387, 2001.

[57] P. Kebarle, "A brief overview of the present status of the mechanisms involved in electrospray mass spectrometry," *Journal of Mass Spectrometry*, vol. 35, no. 7, pp. 804–817, 2000.

[58] C. M. Alymatiri, M. G. Kouskoura, and C. K. Markopoulou, "Decoding the signal response of steroids in electrospray ionization mode (ESI-MS)," *Analytical Methods*, vol. 7, no. 24, pp. 10433–10444, 2015.

[59] M. E. Sigman, P. A. Armstrong, J. M. MacInnis, and M. R. Williams, "Equilibrium partitioning model applied to RDX-halide adduct formation in electrospray ionization mass spectrometry," *Analytical Chemistry*, vol. 77, no. 22, pp. 7434–7441,

2005.

[60] S. Zhou and K. D. Cook, "Protonation in electrospray mass spectrometry: Wrong-way-round or right-way-round?" *Journal of The American Society for Mass Spectrometry*, vol. 11, no. 11, pp. 961–966, 2000.

[61] B. M. Ehrmann, T. Henriksen, and N. B. Cech, "Relative Importance of Basicity in the Gas Phase and in Solution for Determining Selectivity in Electrospray Ionization Mass Spectrometry," *Journal of The American Society for Mass Spectrometry*, vol. 19, no. 5, pp. 719–728, 2008.

[62] V. J. Mandra, M. G. Kouskoura, and C. K. Markopoulou, "Using the partial least squares method to model the electrospray ionization response produced by small pharmaceutical molecules in positive mode," *Rapid Communications in Mass Spectrometry*, vol. 29, no. 18, pp. 1661–1675, 2015.

[63] J. Liigand, A. Kruve, I. Leito, M. Girod, and R. Antoine, "Effect of mobile phase on electrospray ionization efficiency," *Journal of The American Society for Mass Spectrometry*, vol. 25, no. 11, pp. 1853–1861, 2014.

[64] J. Sunner, M. G. Ikonomou, and P. Kebarle, "Sensitivity Enhancements Obtained at High Temperatures in Atmospheric Pressure Ionization Mass Spectrometry," *Analytical Chemistry*, vol. 60, no. 13, pp. 1308–1313, 1988.

[65] G. Nicol, J. Sunner, and P. Kebarle, "Kinetics and thermodynamics of protonation reactions: H3O+ (H2O)h + B = BH+ (H2O)b + (h − b + 1) H2O, where B is a nitrogen, oxygen or carbon base," *International Journal of Mass Spectrometry*, vol. 84, no. 1-2, pp. 135–155, 1988.

[66] S. Kawasaki, H. Ueda, H. Itoh, and J. Tadano, "Screening of organophosphorus pesticides using liquid chromatography-atmospheric pressure chemical ionization mass spectrometry," *Journal of Chromatography A*, vol. 595, no. 1-2, pp. 193–202, 1992.

[67] K. A. Barnes, J. R. Startin, S. A. Thorpe, S. L. Reynolds, and R. J. Fussell, "Determination of the pesticide diflubenzuron in mushrooms by high-performance liquid chromatography-atmospheric pressure chemical ionisation mass spectrometry," *Journal of Chromatography A*, vol. 712, no. 1, pp. 85–93, 1995.

[68] S. Lacorte, C. Molina, and D. Barceló, "Temperature and extraction voltage effect on fragmentation of organophosphorus pesticides in liquid chromatography-atmospheric pressure chemical ionization mass spectrometry," *Journal of Chromatography A*, vol. 795, no. 1, pp. 13–26, 1998.

[69] P. Manini, R. Andreoli, M. Careri, L. Elviri, and M. Musci, "Atmospheric pressure chemical ionization liquid chromatography/mass spectrometry in cholesterol oxide determination and characterization," *Rapid Communications in Mass Spectrometry*, vol. 12, no. 13, pp. 883–889, 1998.

[70] S. N. Ketkar, S. M. Penn, and W. L. Fite, "Influence of Coexisting Analytes in Atmospheric Pressure Ionization Mass Spectrometry," *Analytical Chemistry*, vol. 63, no. 9, pp. 924-925, 1991.

[71] Y. Kato and Y. Numajiri, "Chloride attachment negative-ion mass spectra of sugars by combined liquid chromatography and atmospheric pressure chemical ionization mass spectrometry," *Journal of Chromatography B: Biomedical Sciences and Applications*, vol. 562, no. 1-2, pp. 81–97, 1991.

[72] J. A. Eiceman, J. K. Tofferi, and J. H. Kremer, "Quantitative Assessment of a Corona Discharge Ion Source in Atmospheric Pressure Ionization-Mass Spectrometry for Ambient Air Monitoring," *International Journal of Environmental Analytical Chemistry*, vol. 33, no. 3-4, pp. 161–183, 1988.

[73] M. Busman and J. Sunner, "Simulation method for potential and charge distributions in space charge dominated ion sources," *International Journal of Mass Spectrometry*, vol. 108, no. 2-3, pp. 165–178, 1991.

[74] M. Busman, J. Sunner, and C. R. Vogel, "Space-charge-dominated mass spectrometry ion sources: Modeling and sensitivity," *Journal of The American Society for Mass Spectrometry*, vol. 2, no. 1, pp. 1–10, 1991.

[75] I. M. Lazar, M. L. Lee, and E. D. Lee, "Design and optimization of a corona discharge ion source for supercritical fluid chromatography time-of-flight mass spectrometry," *Analytical Chemistry*, vol. 68, no. 11, pp. 1924–1932, 1996.

[76] C. Aguilar, I. Ferrer, F. Borrull, R. M. Marcé, and D. Barceló, "Comparison of automated on-line solid-phase extraction followed by liquid chromatography-mass spectrometry with atmospheric pressure chemical ionization and particle beam mass spectrometry for the determination of a priority group of pesticides in environmental waters," *Journal of Chromatography A*, vol. 794, no. 1-2, pp. 147–163, 1998.

[77] Y. Chai, G. Weng, S. Shen, C. Sun, and Y. Pan, "The protonation site of para-dimethylaminobenzoic acid using atmospheric pressure ionization methods," *Journal of The American Society for Mass Spectrometry*, vol. 26, no. 4, pp. 668–676, 2015.

[78] Y. Chai, N. Hu, and Y. Pan, "Kinetic and thermodynamic control of protonation in atmospheric pressure chemical ionization," *Journal of The American Society for Mass Spectrometry*, vol. 24, no. 7, pp. 1097–1101, 2013.

[79] J. D. Evans, *Straightforward statistics for the behavioral sciences*, Brooks/Cole Publishing Company, Pacific Grove, CA, USA, 1996.

[80] N. B. Cech and C. G. Enke, "Relating electrospray ionization response to nonpolar character of small peptides," *Analytical Chemistry*, vol. 72, no. 13, pp. 2717–2723, 2000.

[81] M. Oss, A. Kruve, K. Herodes, and I. Leito, "Electrospray ionization efficiency scale of organic compound," *Analytical Chemistry*, vol. 82, no. 7, pp. 2865–2872, 2010.

[82] J. T. Shelley and G. M. Hieftje, "Ionization matrix effects in plasma-based ambient mass spectrometry sources," *Journal of Analytical Atomic Spectrometry*, vol. 25, no. 3, pp. 345–350, 2010.

[83] A. Piccolo, M. Spiteller, and A. Nebbioso, "Effects of sample properties and mass spectroscopic parameters on electrospray ionization mass spectra of size-fractions from a soil humic acid," *Analytical and Bioanalytical Chemistry*, vol. 397, no. 7, pp. 3071–3078, 2010.

[84] M. Wilm, "Principles of electrospray ionization," *Molecular & cellular proteomics*, vol. 10, no. 7, 2011.

[85] M. Commisso, A. Anesi, S. Dal Santo, and F. Guzzo, "Performance comparison of electrospray ionization and atmospheric pressure chemical ionization in untargeted and targeted liquid chromatography/mass spectrometry based metabolomics analysis of grapeberry metabolites," *Rapid Communications in Mass Spectrometry*, vol. 31, no. 3, pp. 292–300, 2017.

[86] J. H. Gross, *Massenspektrometrie: Ein Lehrbuch*, Springer, Berlin, Heidelberg, Germany, 2013.

[87] D. B. Robb, T. R. Covey, and A. P. Bruins, "Atmospheric pressure photoionization: an ionization method for liquid chromatography - mass spectrometry," *Analytical Chemistry*, vol. 72, no. 15, pp. 3653–3659, 2000.

[88] V. V. Laiko, M. A. Baldwin, and A. L. Burlingame, "Atmospheric pressure matrix-assisted laser desorption/ionization mass spectrometry," *Analytical Chemistry*, vol. 72, no. 4, pp. 652–657, 2000.

[89] L. Kolaitis and D. M. Lubman, "Detection of Nonvolatile Species by Laser Desorption Atmospheric Pressure Mass Spectrometry," *Analytical Chemistry*, vol. 58, no. 11, pp. 2137–2142, 1986.

[90] S. D. Huang, L. Kolaitis, and D. M. Lubman, "Detection of explosives using laser desorption in ion mobility spectrometry/mass spectrometry," *Applied Spectroscopy*, vol. 41, no. 8, pp. 1371–1376, 1987.

[91] J. J. Coon and W. W. Harrison, "Laser desorption-atmospheric pressure chemical ionization mass spectrometry for the analysis of peptides from aqueous solutions," *Analytical Chemistry*, vol. 74, no. 21, pp. 5600–5605, 2002.

[92] J. S. Sampson, A. M. Hawkridge, and D. C. Muddiman, "Generation and Detection of Multiply-Charged Peptides and Proteins by Matrix-Assisted Laser Desorption Electrospray Ionization (MALDESI) Fourier Transform Ion Cyclotron Resonance Mass Spectrometry," *Journal of The American Society for Mass Spectrometry*, vol. 17, no. 12, pp. 1712–1716, 2006.

[93] R. G. Cooks, Z. Ouyang, Z. Takats, and J. M. Wiseman, "Ambient mass spectrometry," *Science*, vol. 311, no. 5767, pp. 1566–1570, 2006.

[94] G. A. Harris, A. S. Galhena, and F. M. Fernández, "Ambient sampling/ionization mass spectrometry: applications and current trends," *Analytical Chemistry*, vol. 83, no. 12, pp. 4508–4538, 2011.

[95] D. J. Weston, "Ambient ionization mass spectrometry: Current understanding of mechanistic theory; Analytical performance and application areas," *Analyst*, vol. 135, no. 4, pp. 661–668, 2010.

[96] T. J. Kauppila and R. Kostiainen, "Ambient mass spectrometry in the analysis of compounds of low polarity," *Analytical Methods*, vol. 9, no. 34, pp. 4936–4953, 2017.

[97] M.-Z. Huang, S.-C. Cheng, Y.-T. Cho, and J. Shiea, "Ambient ionization mass spectrometry: a tutorial," *Analytica Chimica Acta*, vol. 702, no. 1, pp. 1–15, 2011.

[98] G. J. Van Berkel, S. P. Pasilis, and O. Ovchinnikova, "Established and emerging atmospheric pressure surface sampling/ionization techniques for mass spectrometry," *Journal of Mass Spectrometry*, vol. 43, no. 9, pp. 1161–1180, 2008.

[99] J. S. Wiley, J. T. Shelley, and R. G. Cooks, "Handheld low-temperature plasma probe for portable "point-and- shoot" ambient ionization mass spectrometry," *Analytical Chemistry*, vol. 85, no. 14, pp. 6545–6552, 2013.

[100] P. I. Hendricks, J. K. Dalgleish, J. T. Shelley et al., "Autonomous in situ analysis and real-time chemical detection using a backpack miniature mass spectrometer: Concept, instrumentation development, and performance," *Analytical Chemistry*, vol. 86, no. 6, pp. 2900–2908, 2014.

[101] G. C.-Y. Chan, J. T. Shelley, J. S. Wiley et al., "Elucidation of reaction mechanisms responsible for afterglow and reagent-ion formation in the low-temperature plasma probe ambient ionization source," *Analytical Chemistry*, vol. 83, no. 10, pp. 3675–3686, 2011.

[102] S. B. Olenici-Craciunescu, A. Michels, C. Meyer et al., "Characterization of a capillary dielectric barrier plasma jet for use as a soft ionization source by optical emission and ion mobility spectrometry," *Spectrochimica Acta Part B: Atomic Spectroscopy*, vol. 64, no. 11-12, pp. 1253–1258, 2009.

[103] F. J. Andrade, W. C. Wetzel, G. C.-Y. Chan et al., "A new, versatile, direct-current helium atmospheric-pressure glow discharge," *Journal of Analytical Atomic Spectrometry*, vol. 21, no. 11, pp. 1175–1184, 2006.

[104] F. J. Andrade, J. T. Shelley, W. C. Wetzel et al., "Atmospheric pressure chemical ionization source. 2. Desorption-ionization for the direct analysis of solid compounds," *Analytical Chemistry*, vol. 80, no. 8, pp. 2654–2663, 2008.

[105] N. Na, M. Zhao, S. Zhang, C. Yang, and X. R. Zhang, "Development of a dielectric barrier discharge ion source for ambient mass spectrometry," *Journal of The American Society for Mass Spectrometry*, vol. 18, no. 10, pp. 1859–1862, 2007.

[106] J. D. Harper, N. A. Charipar, C. C. Mulligan, X. Zhang, R. G. Cooks, and Z. Ouyang, "Low-temperature plasma probe for ambient desorption ionization," *Analytical Chemistry*, vol. 80, no. 23, pp. 9097–9104, 2008.

[107] L. V. Ratcliffe, F. J. M. Rutten, D. A. Barrett et al., "Surface analysis under ambient conditions using plasma-assisted desorption/ionization mass spectrometry," *Analytical Chemistry*, vol. 79, no. 16, pp. 6094–6101, 2007.

[108] T. T. Grove, M. F. Masters, and R. E. Miers, "Determining dielectric constants using a parallel plate capacitor," *American Journal of Physics*, vol. 73, no. 1, pp. 52–56, 2005.

[109] U. Kogelschatz, "Dielectric-barrier discharges: their history, discharge physics, and industrial applications," *Plasma Chemistry and Plasma Processing*, vol. 23, no. 1, pp. 1–46, 2003.

[110] C. Tendero, C. Tixier, P. Tristant, J. Desmaison, and P. Leprince, "Atmospheric pressure plasmas: a review," *Spectrochimica Acta Part B: Atomic Spectroscopy*, vol. 61, no. 1, pp. 2–30, 2006.

[111] G. C.-Y. Chan, J. T. Shelley, A. U. Jackson et al., "Spectroscopic plasma diagnostics on a low-temperature plasma probe for ambient mass spectrometry," *Journal of Analytical Atomic Spectrometry*, vol. 26, no. 7, pp. 1434–1444, 2011.

[112] T. Martens, A. Bogaerts, W. J. Brok, and J. V. Dijk, "The dominant role of impurities in the composition of high pressure noble gas plasmas," *Applied Physics Letters*, vol. 92, no. 4, p. 041504, 2008.

[113] T. Martens, D. Mihailova, J. Van Dijk, and A. Bogaerts, "Theoretical characterization of an atmospheric pressure glow discharge used for analytical spectrometry," *Analytical Chemistry*, vol. 81, no. 21, pp. 9096–9108, 2009.

[114] L. Mangolini, C. Anderson, J. Heberlein, and U. Kortshagen, "Effects of current limitation through the dielectric in atmospheric pressure glows in helium," *Journal of Physics D: Applied Physics*, vol. 37, no. 7, pp. 1021–1030, 2004.

[115] H. M. Joh, S. J. Kim, T. H. Chung, and S. H. Leem, "Comparison of the characteristics of atmospheric pressure plasma jets using different working gases and applications to plasma-cancer cell interactions," *AIP Advances*, vol. 3, no. 9, p. 092128, 2013.

[116] X. Xiang, B. Kupczyk, J. Booske, and J. Scharer, "Diagnostics of fast formation of distributed plasma discharges using X-band microwaves," *Journal of Applied Physics*, vol. 115, no. 6, 2014.

[117] A. Albert and C. Engelhard, "Characteristics of low-temperature plasma ionization for ambient mass spectrometry compared to electrospray ionization and atmospheric pressure chemical ionization," *Analytical Chemistry*, vol. 84, no. 24, pp. 10657–10664, 2012.

[118] Y. Zhang, X. Ma, S. Zhang, C. Yang, Z. Ouyang, and X. Zhang, "Direct detection of explosives on solid surfaces by low temperature plasma desorption mass spectrometry," *Analyst*, vol. 134, no. 1, pp. 176–181, 2009.

[119] J. F. Garcia-Reyes, J. D. Harper, G. A. Salazar, N. A. Charipar, Z. Ouyang, and R. G. Cooks, "Detection of explosives and related compounds by low-temperature plasma ambient ionization mass spectrometry," *Analytical Chemistry*, vol. 83, no. 3, pp. 1084–1092, 2011.

[120] A. U. Jackson, J. F. Garcia-Reyes, J. D. Harper et al., "Analysis of drugs of abuse in biofluids by low temperature plasma (LTP) ionization mass spectrometry," *Analyst*, vol. 135, no. 5, pp. 927–933, 2010.

[121] J. K. Dalgleish, M. Wleklinski, J. T. Shelley, C. C. Mulligan, Z. Ouyang, and R. G. Cooks, "Arrays of low-temperature plasma probes for ambient ionization mass spectrometry," *Rapid Communications in Mass Spectrometry*, vol. 27, no. 1, pp. 135–142, 2013.

[122] J. S. Wiley, J. F. García-Reyes, J. D. Harper, N. A. Charipar, Z. Ouyang, and R. G. Cooks, "Screening of agrochemicals in foodstuffs using low-temperature plasma (LTP) ambient ionization mass spectrometry," *Analyst*, vol. 135, no. 5, pp. 971–979, 2010.

[123] S. Soparawalla, F. K. Tadjimukhamedov, J. S. Wiley, Z. Ouyang, and R. G. Cooks, "In situ analysis of agrochemical residues on fruit using ambient ionization on a handheld mass spectrometer," *Analyst*, vol. 136, no. 21, pp. 4392–4396, 2011.

[124] M. Beneito-Cambra, P. Pérez-Ortega, A. Molina-Díaz, and J. F. García-Reyes, "Rapid determination of multiclass fungicides in wine by low-temperature plasma (LTP) ambient ionization mass spectrometry," *Analytical Methods*, vol. 7, no. 17, pp. 7345–7351, 2015.

[125] A. Albert, J. T. Shelley, and C. Engelhard, "Plasma-based ambient desorption/ionization mass spectrometry: State-of-the-art in qualitative and quantitative analysis," *Analytical and Bioanalytical Chemistry*, vol. 406, no. 25, pp. 6111–6127, 2014.

[126] Y. Liu, M. Xiaoxiao, Z. Lin et al., "Imaging mass spectrometry with a low-temperature plasma probe for the analysis of works of art," *Angewandte Chemie International Edition*, vol. 49, no. 26, pp. 4435–4437, 2010.

[127] P. Nemes and A. Vertes, "Laser ablation electrospray ionization for atmospheric pressure, in vivo, and imaging mass spectrometry," *Analytical Chemistry*, vol. 79, no. 21, pp. 8098–8106, 2007.

[128] H. Hayen, A. Michels, and J. Franzke, "Dielectric barrier discharge ionization for liquid chromatography/mass spectrometry," *Analytical Chemistry*, vol. 81, no. 24, pp. 10239–10245, 2009.

[129] A. Kiontke, C. Engel, D. Belder, and C. Birkemeyer, "Analyte and matrix evaporability – key players of low-temperature plasma ionization for ambient mass spectrometry," *Analytical and Bioanalytical Chemistry*, vol. 410, no. 21, pp. 5123–5130, 2018.

[130] G. Huang, W. Xu, M. A. Visbal-Onufrak, Z. Ouyang, and R. G. Cooks, "Direct analysis of melamine in complex matrices using a handheld mass spectrometer," *Analyst*, vol. 135, no. 4, pp. 705–711, 2010.

[131] H. J. Lee, J. Oh, S. W. Heo et al., "Peltier Heating-Assisted Low Temperature Plasma Ionization for Ambient Mass Spectrometry," *Mass Spectrometry Letters*, vol. 6, no. 3, pp. 71–74, 2015.

[132] M. Keeney and J. Heicklen, "Surface tension and the heat of vaporization: A simple empirical correlation," *Journal of Inorganic and Nuclear Chemistry*, vol. 41, no. 12, pp. 1755–1758, 1979.

[133] D. S. Viswanath and N. R. Kuloor, "On Latent Heat of Vaporization, Surface Tension, and Temperature," *Journal of Chemical & Engineering Data*, vol. 11, no. 1, pp. 69–72, 1966.

[134] L. Song, S. C. Gibson, D. Bhandari, K. D. Cook, and J. E. Bartmess, "Ionization mechanism of positive-ion direct analysis in real time: A transient microenvironment concept," *Analytical Chemistry*, vol. 81, no. 24, pp. 10080–10088, 2009.

[135] J. V. Iribarne, P. J. Dziedzic, and B. A. Thomson, "Atmospheric pressure ion evaporation-mass spectrometry," *International Journal of Mass Spectrometry and Ion Physics*, vol. 50, no. 3, pp. 331–347, 1983.

[136] Z. Wu, W. Gao, M. A. Phelps, D. Wu, D. D. Miller, and J. T. Dalton, "Favorable Effects of Weak Acids on Negative-Ion Electrospray Ionization Mass Spectrometry," *Analytical Chemistry*, vol. 76, no. 3, pp. 839–847, 2004.

16

Development of a New Sequential Extraction Procedure of Nickel Species on Workplace Airborne Particulate Matter: Assessing the Occupational Exposure to Carcinogenic Metal Species

Catalani Simona,[1] Fostinelli Jacopo ⓘ,[1] Gilberti Maria Enrica,[1] Orlandi Francesca,[1] Magarini Riccardo,[2] Paganelli Matteo,[1] Madeo Egidio,[1] and De Palma Giuseppe[1]

[1]*Unit of Occupational Health and Industrial Hygiene, Department of Medical and Surgical Specialties, Radiological Sciences and Public Health, University of Brescia, Italy*
[2]*PerkinElmer (Italia), Milano, Italy*

Correspondence should be addressed to Fostinelli Jacopo; j.fostinelli@unibs.it

Guest Editor: Seyyed E. Moradi

Nickel (Ni) compounds and metallic Ni have many industrial and commercial applications, including their use in the manufacturing of stainless steel. Due to the specific toxicological properties of the different Ni species, there is a growing interest about the availability of analytical methods that allow specific risk assessment, particularly related to exposure to the Ni species classified as carcinogenic. In this paper, we described a speciation method of inorganic Ni compounds in airborne particulate matter, based on selective sequential extractions. The analytical method reported in this paper allows the determination of soluble, sulfidic, metallic, and oxide Ni by a simple sequential extraction procedure and analysis by Atomic Absorption Spectroscopy using small volumes of solutions and without long evaporation phases. The method has been initially set up on standard laboratory mixtures of known concentrations of different Ni salts. Then it has then been tested on airborne particulate matter (powder and filters) collected in different workstations of a large stainless steel production facility. The method has occurred effectively in the comparison of the obtained results with occupational exposure limit values set by the main international scientific and regulatory agencies for occupational safety and health, in order to prevent both toxic and carcinogenic effects in humans.

1. Introduction

Nickel (Ni) compounds and metallic Ni have many industrial and commercial applications, including their use in stainless steel production, in a large series of metal alloys, as catalysts, in batteries, pigments, and ceramics [1]. An industrial sector in which Ni exposure can be particularly relevant is the production of special stainless steel in secondary steel foundries: the workers engaged in this industry are potentially exposed to various forms of airborne Ni, in particular during the operations of melting and casting and at all the stages of the process characterized by the need for high temperatures [2, 3]. This kind of production has been widespread for decades in northern Italy, involving thousands of workers and

consequently arousing high interest on the related occupational and public health issues. The toxicological properties of Ni compounds yet represent an important challenge in terms of risk assessment and are also of great concern for the necessary enforcement of preventive measures in exposed workers [4, 5].

Exposure to Ni oxides and sulfides, which have low solubility in water, has been recognized as one of the prominent causes for occupational Ni-related lung and nasal cancer [6, 7].

The carcinogenic potential of water-soluble Ni compounds and Ni tetracarbonyl has been continuously discussed for decades [8]. Although there is no evidence that exposure to metallic Ni increases the risk of respiratory cancer, it is well known as the most important sensitizer among metal elements [9–12].

In 1990, the International Agency for Research on Cancer (IARC) concluded that there were sufficient evidences in humans for the carcinogenicity of Ni sulfate and of combinations of Ni sulfide and oxides in the Ni refining industry [13]. In 2012, with specific referral to the inhalator exposure route, IARC updated the evaluation classifying Ni compounds as "carcinogenic to humans-group 1", whereas metallic Ni and Ni alloys were categorized as "possibly carcinogenic to humans-group 2B", specifically for cancer of the lung and the nasal cavity [14].

The current understanding of the carcinogenic potential of the most prominent Ni species in sulfidic Ni (Ni subsulphide (Ni_3S_2), Ni oxide (NiO), Ni metal (Ni^0), and soluble Ni (primarily Ni sulfate, $NiSO_4$) has been determined through studies based on a combination of animal testing (of pure compounds) and human epidemiological data [15].

Due to the above-mentioned reasons, the speciation of Ni in workplaces' airborne particulate is of the utmost importance for the assessment of the respiratory health risks.

Regarding occupational exposure limits, different threshold levels for Ni and Ni compounds in workplaces and emissions are available (Table 1). In 1998 the American Conference of Governmental Industrial Hygienists (ACGIH) published separated threshold values for the organic and inorganic forms of Ni [16]. In doing so it was recognized that different Ni species had different toxic and carcinogenic properties [17].

In the "Recommendation from the Scientific Committee on Occupational Exposure Limits" for Ni and inorganic Ni compounds", by the European SCOEL, some occupational exposure limits (OELs) aiming to protect both from inflammatory effects in the lung and from cancer were published [18].

The determination of the total concentration of Ni, accordingly, gives no information about environmental risks or knowledge of the various forms, which makes monitoring of specific chemical species of Ni in environmental samples, such as airborne particulates, extremely important [19]. Consequently, the development of analytical techniques for the determination of various compounds of metallic elements in environmental samples, such as ambient aerosols, is presently one of the most challenging tasks for environmental analytical chemistry [20, 21].

Ni determination can be performed with various analytical techniques, including spectrophotometry, atomic absorption spectrometry (FAAS and ETAAS), inductively coupled argon plasma optical emission spectrometry (ICP-OES), inductively coupled plasma mass spectrometry (ICP-MS) and voltammetry. In the past two decades many techniques have been widely developed for the speciation of inorganic contaminants in environmental samples [22, 23]. Several of them make use of sequential extraction schemes to determine the metal distribution over different fractions, usually including species such as soluble, sulfidic, metallic, and oxide fractions. The application of sequential extraction procedures provides relevant environmental information.

Several of the sequential extraction procedures found in the literature are variants of the method proposed by Zatka et al. in the early 1990s [24] in which the solubility of the different fractions was utilized for the sequential determination of Ni ions.

The procedure of Zatka involves a sequential leaching of airborne dust from Ni production sites and Ni-using workplaces by using ammonium citrate, hydrogen peroxide/ammonium citrate, and bromine-methanol. Acceptable recoveries were obtained, for most species better than 95%.

Up to now, the sequential extraction procedure proposed by Zatka has mainly been applied for speciation of Ni in work-room air (for instance, in Ni refinery) present at mg/m^3. These levels however, obtaining correct results for the concentrations of trace elements in out-door air at ng/m^3 levels, are still a great concern, mainly due to the extremely small amounts of sample and analyte.

2. Aims

In consideration of the toxicological properties of Ni compounds, this paper describes a modified method which is mainly based on the fractionation proposed by Zatka et al. [25] but achieving both time optimization and greater sensitivity. The method has been first setup on a mix of different Ni species. Subsequently, the method has been applied on airborne particulate matter sampled in different departments of a steel production facility, in order to assess the airborne levels of different Ni species.

3. Materials and Methods

3.1. Reagents. All chemicals used were of analytical grade. The water used was bidistilled water, for inorganic trace analysis (Merck KgaA, Darmstadt, Germany).

The reagent in the solution for the elution of different Ni species from the filters was as follows:

(A) Ammonium citrate solution: 1.7% ammonium hydrogen citrate and 0.5% citric acid solution, the solution was prepared by dissolving 1.7 g of diammonium hydrogen citrate ((NH_4)2H-Cit)), CAS No. 3012-65-5, Carlo Erba, Milan, Italy) and 0.5 g of citric acid (99%, $C_6H_8O_7$, Sigma Aldrich, Saint Louis, Missouri, USA) in 100 ml of bidistilled water.

(B) H_2O_2 citrate: ammonium citrate 0.1M and hydrogen peroxide 30% (w/w) ratio 2:1 (H_2O_2, CAS No. 7722-84-1 Sigma Aldrich, Saint Louis, Missouri, USA).

(C) Methanol-bromine solution: methanol (CH_3OH, Chromasolv ≥99.9%, CAS No. 67-56-1, Honeywell, Thermo Fisher Scientific); Bromine (Br_2, CAS No. 7726-95-6 Sigma Aldrich, Saint Louis, Missouri, USA), 50:1.

(D) Nitric and hydrochloric acid solution: nitric acid (HNO_3 70%, CAS No. 7697-37-2 Sigma Aldrich, Saint Louis, Missouri, USA) and hydrochloric acid (HCl 37%, CAS No. 7647-01-0 Sigma Aldrich, Saint Louis, Missouri, USA) ratio 1:1.

The mixture of Ni compounds was prepared by several salts: Ni(II) sulfate hexahydrate ($NiSO_4$, PM 262.7, reagent plus® 99.99%, CAS No. 7786-81-4 Sigma Aldrich, Saint Louis,

TABLE 1: Ni species, with chemical formulas, solubility characteristics, 8h-TWA occupational exposure limits proposed by international agencies, and hazard statements assigned by the EU CLP Regulation, ACGIH, SCOEL, and OSHA.

Nickel Species		Solubility in water (g/100ml) [temperature]	CAS Number	ACGIH	SCOEL	OSHA	CLP
Nickel Sulfate	$NiSO_4$	65.5 [0°C]	7786-81-4	A4 0.1 mg/m^3 (Ni Sol)	0.01 mg/m^3 #	-	H351
Nickel subsulphide	Ni_3S_2	poorly soluble	12035-72-2	A1 0.1 mg/m^3	0.005 mg/m^3 § 0.01 mg/m^3 #	0.1 mg/m^3 (Ni insol)	H350i
Nickel monoxide	NiO	poorly soluble	1313-99-1	A1 0.2 mg/m^3 (Ni insol)	0.005 mg/m^3 § 0.01 mg/m^3 #	0.1 mg/m^3 (Ni insol)	H350i
Metallic nickel	Ni	poorly soluble	7440-02-0	A5 1.5 mg/m^3	0.005 mg/m^3 §	0.5 mg/m^3	H351

#Inhalable fraction; §respirable fraction.

TABLE 2: AAS instrumental parameters for determination of Ni

Operating conditions	
Primary source	Nickel Hollow Cathode lamp (Agilent Technologies)
Lamp current	5 mA
Analytical wavelength	232 nm
Background correction system	Zeeman effect based (Transversal)
Slit width	0.2 nm
Mode	Absorbance (peak height)
Graphite furnace operation	
Atomization tube	Partition tubes (coated)-GTA (Agilent Technologies)
Sheath/Purge gas	Argon (Ar) of 99.999% purity
Sample Injection (sample, μL)	30

Missouri, USA), Ni sulfide (Ni_3S_2, 99.7%, PM 240.1 CAS No. 12035-72-2 Sigma Aldrich, Saint Louis, Missouri, USA), Ni powder (PA 58.7, 100mesh, 99.999%, CAS No. 7440-02-0 Sigma Aldrich, Saint Louis, Missouri, USA), and Ni(II)oxide (NiO, PM 74.7, 99.999%, CAS No. 1313-99-1 Sigma Aldrich, Saint Louis, Missouri, USA).

The sequential leaching was carried out in an all hydrophilic Teflon filter holder (DigiFILTER 0.45 micron, SCP Science, Quebec, Canada); the holder was fitted with a 25mm (5μm) PVC filters (SKC Inc. 25mm, 5.0μm) (Figure 1). The filtrates are collected in a test tube (DigiTUBEs 50ml, SCP Science, Quebec, Canada).

A 0.45 micron Teflon membrane inserted in every Digi-FILTER guarantees 98% particle retention.

The DigiFILTERs were connected to a vacuum pump; the filtering system is set up so that mild suction can be turned on and off at short intervals if required.

3.2. Determination of Nickel. The determination of different fraction of Ni were performed by atomic absorption spectrometry (AAS Spectra 400 Varian, Medical Systems, Inc. Palo Alto, CA) equipped with a transversal Zeeman-effect background correction system and an auto sampler was used

FIGURE 1: Filter device utilized for sequential extraction of Ni's fractions (DigiFILTER 0.45 micron, SCP Science, Quebec, Canada).

for all measurements. The instrumental operating parameters of the AAS apparatus are reported in Tables 2 and 3.

TABLE 3: Temperature program of the AAS method for determination of nickel.

Step N°	Temperature °C	Time (sec)	Flow (L/min)	Type of gas
1	40	5.0	3.0	Argon
2*	150	35.0	3.0	Argon
3*	150	5.0	3.0	Argon
4$^\alpha$	900	10.0	3.0	Argon
5$^\alpha$	900	15.0	3.0	Argon
6$^\alpha$	900	2.0	0.0	
7$^\beta$	2400	1.0	0.0	
8$^\beta$	2400	2.0	0.0	
9	2500	1.0	3.0	Argon
10	2500	2.0	3.0	Argon

*Drying step, $^\alpha$pyrolysis step, and ^{B}atomizing step.

TABLE 4: Results of dissolution of different nickel fractions (N°=3).

Nickel Species	Solution	Treatment	Aspect	Ni expected(mg)	Ni determined(mg)	% extracted
NiSO$_4$	ammonium citrate (Sol A)	10mL, 37°C, 60min	clear	1.6	1.70±0.1	102%
	H$_2$O$_2$ citrate: ammonium citrate (Sol B)	10mL, ΔT, 60min	clear	2.3	2.1±0.2	91%
	solution methanol:Bromine (solution C)	10mL, ΔT, 2h	clear	1.8	1.6±0.1	89%
	Solution HCl: HNO$_3$ 1:1 (Sol D)	4mL, 70°C, 30min	clear	2.6	2.7±0.3	103%
Ni$_3$S$_2$	ammonium citrate (Sol A)	10mL, 37°C, 60min	residue	10.8	0.31±0.1	0.03%
	H$_2$O$_2$ citrate: ammonium citrate (Sol B)	10mL, ΔT, 60min	clear	7.6	7.4±1.1	96%
	solution methanol: Bromine (solution C)	10mL, ΔT, 2h	clear	8.0	8.3±0.8	103%
	Solution HCl: HNO$_3$ 1:1 (Sol D)	4mL, 70°C, 30min	clear	7.4	8.8±1.2	118%
Ni (0) metallic	ammonium citrate (Sol A)	10mL, 37°C, 60min	residue	4.6	/	/
	H$_2$O$_2$ citrate: ammonium citrate (Sol B)	10mL, ΔT, 60min	residue	12.2	0.5±0.1	4%
	solution methanol: Bromine (solution C)	10mL, ΔT, 2h	clear	8.7	9.2±1.1	105%
	Solution HCl: HNO$_3$ 1:1 (Sol D)	4mL, 70°C, 30min	clear	13.4	15.7±2.3	117%
NiO	ammonium citrate (Sol A)	10mL, 37°C, 60min	residue	7.0	/	/
	H$_2$O$_2$ citrate: ammonium citrate (Sol B)	10mL, ΔT, 60min	residue	13.4	0.1±0.1	1%
	solution methanol: Bromine (solution C)	10mL, ΔT, 2h	residue	7.4	/	/
	Solution HCl: HNO$_3$ 1:1 (Sol D)	4mL, 70°C, 30min	clear	10.4	11.8±3.2	114%

Ni stock standard solutions was prepared from 1 mg/mL (1000ppm) of standard solution (Ni(0) in 2% HNO$_3$, O2Si smart solution, Charleston, USA). Working solutions at 0.05; 0.1; and 0.5 mg/L were prepared by serial dilution in bidistilled water of the standard at 1000 mg/L solution (0.05ppm, 0.1ppm, and 0.5ppm).

The accuracy of the method was evaluated by analyzing certified reference materials (NIST 1643e-1643d trace elements in water for ultrafiltrate).

Instrumental limit of detection (LOD) of the total Ni, calculated as three standard deviations of the background signal obtained on 10 blank samples, was equal to 1 μg/L.

The limit of quantification (LOQ) of the total Ni, calculated as ten standard deviations of the background signal obtained on 5 blank samples, was equal to 3 μg/L.

The relative standard deviation (RSDs) of measurements of Ni solutions was between 5 and 10 %.

3.2.1. Extraction Tests. A little amount of all the Ni salts was treated in falcon with one leaching solution at a time and then the concentration of Ni was then determined.

To facilitate the dissolution the falcons were placed in an ultrasonic bath for 5 minutes.

Several tests have been carried out to evaluate different extraction conditions by using different volumes of solutions, times and temperatures (date not shown). The best conditions are reported in Table 4 which illustrates the composition of the solutions, conditions of the extractions, dissolution of the salts, and the recovery of Ni fraction.

The Ni soluble fraction was effectively extracted by all the solutions. The sulfidic fraction was not solubilized by solution A, while it was dissolved by other solutions. The metallic fraction was well extracted by solution C and D. The Ni oxide is solubilized only by a solution HCl:HNO$_3$ (1:1); the leaching with the other solutions does not show traces of Ni in solutions.

TABLE 5: Extraction of specific Ni's compounds in two mixtures of powders. For each Ni species the percentage of recovery with respect to the Ni salt added is reported.

Ni species	Mixture A			Mixture B		
	Ni added (mg)	Ni found (mg)	%	Ni added (mg)	Ni found (mg)	%
NiSO$_4$	0.016	0.013	81	0.054	0.049	91
Ni$_3$S$_2$	0.096	0.080	83	0.310	0.290	93
Ni(0)	0.114	0.11	96	0.367	0.371	101
NiO	0.064	0.065	102	0.207	0.220	106

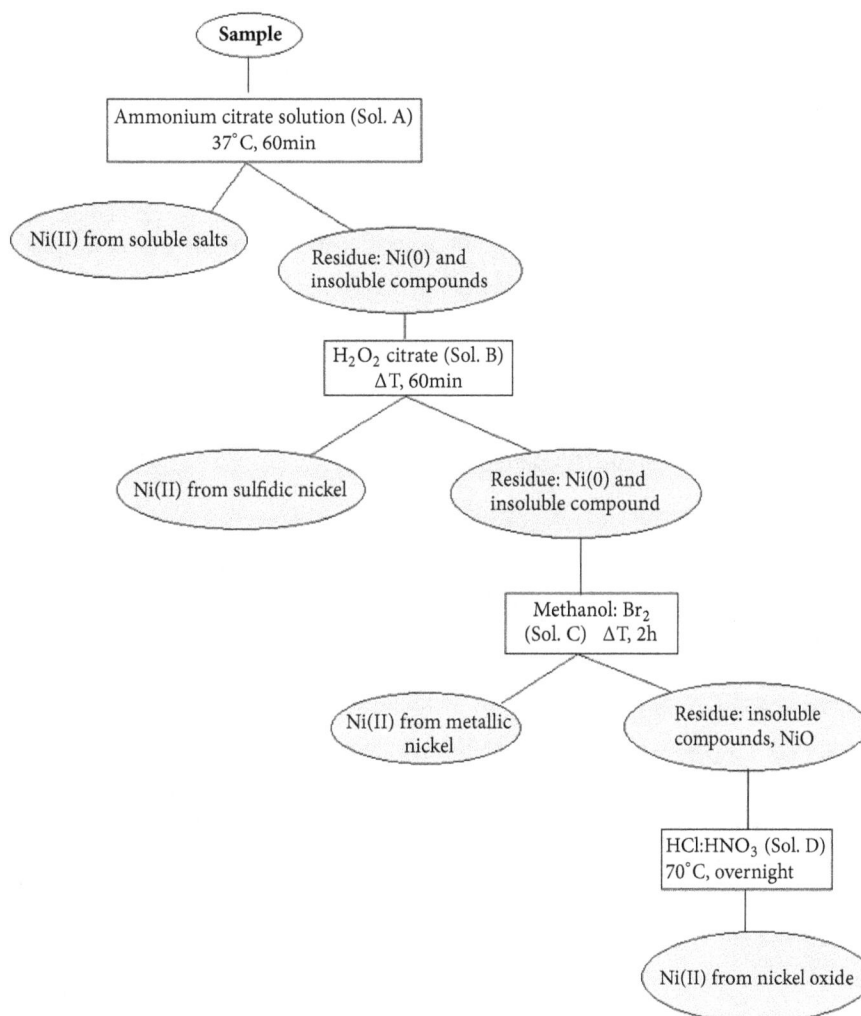

FIGURE 2: Selective sequential solubilization of inorganic Ni compounds: scheme of the procedure.

The same procedure was carried out on all the solutions (A, B, C, D) without the addition of Ni salts.

3.3. *Speciation Procedure.* In order to have Ni concentrations nearer to those found in real samples, each Ni compound was homogenously dispersed and grinded in an agate mortar.

The final mixed salts contained 10.4 mg NiSO$_4$ equal to 2.3 mg as Ni, 18.5 mg Ni$_3$S$_2$ equal to 113.3 mg as Ni, 15.8 mg Ni(0), and 11.3 mg NiO equal to 8.9 mg as Ni.

Quantities ranging from 1 to 2 mg of the mix weighed to the 5th decimal were deposited in a PVC filter placed on the DigiFILTER (Mix A = 0.4 mg; Mix B = 1.3 mg).

The sequential extraction procedures are illustrated in Figure 2 and the results in Table 5.

3.4. *Determination of Soluble Nickel.* Add 10mL of ammonium citrate solution (solution A) in DigiFILTER inserted on a falcon; place the DigiFILTER in the oven at 37°C for 60

TABLE 6: Determination of Ni's fraction in powders (μg/g) and environmental samples (μg/m^3) collected in the departments of the steel production facility.

Powder (μg/g)	NiSO$_4$	Ni$_3$S$_2$	Ni(0)	NiO	Σ	Ni Tot AAS
A (1.2 mg)	/	/	145	196	341	423
B (1.1 mg)	2727	395	745	6673	10540	11509
C (0.1mg)	/	/	323	356	679	695
D (1.0 mg)	580	232	2180	4440	7432	7538
Filters (μg/m^3)	NiSO$_4$	Ni$_3$S$_2$	Ni(0)	NiO	Σ	Ni Tot AAS
Filter A	0.32	0.07	0.64	1.5	2.53	2.81
Filter B	0.25	0.67	0.94	1.0	2.86	3.12
Filter C	0.76	0.28	1.6	3.5	6.14	6.44
Filter D	0.87	0.73	4.0	5.0	10.6	10.88
Filter E	0.89	0.24	0.55	1.3	2.98	3.50

minutes. With a vacuum pump, draw the solution into the falcon and keep it for the determination of the soluble Ni fraction. Insert a new falcon in the DigiFILTER.

3.5. Determination of Sulfidic Nickel. To the filter from which soluble Ni phases have been leached out add 10mL of the solution H$_2$O$_2$ citrate: ammonium citrate (solution B). Keep at room temperature for 60 minutes under a hood.

With a vacuum pump, draw the solution into the falcon and keep it for the determination of the sulfidic Ni fraction. Insert a new falcon in the DigiFILTER.

3.6. Determination of Metallic Nickel. To the filter from which soluble and sulfidic Ni phases have been leached out add 10mL of the solution methanol:bromine (solution C). Keep at room temperature for 2 hours. With a vacuum pump, draw the solution into the falcon and keep it for the determination of the metallic Ni. Insert a new falcon in the DigiFILTER.

3.7. Determination of Oxide Nickel. To the filter from which the first Ni's fraction have been leached, add 5mL of solution HCl:HNO$_3$ (solution D) and keep at room temperature overnight under the hood. With a vacuum pump, draw the solution in the new falcon. Transfer the filter from the support into the falcon and add another 5 mL of HCl:HNO$_3$ and heat in a water bath at 70°C for 15 minutes.

4. Application of Sequential Leaching to Real Samples

4.1. Sampling Site and Equipment. The sampling was performed in a steel foundry plant specialized in stainless steels for naval and aerospace industry; the production cycle is based on an electric arc furnace with subsequent casting in a continuous plant. The melting is essentially performed in a three-phase furnace equipped with three graphite electrodes. The different qualities of steel are obtained by mixing recycled scrap with chromium, Ni, and other raw materials. Through the refining in the ladle furnace, specific compositions and quality of the steel are reached (the exact composition of the alloy is proprietary information).

In order to characterize occupational exposure to airborne Ni compounds, we carried out a characterization of Ni species collected on two different types of substrate:

(1) Particulate collected through an IOM (Institute of Occupational Medicine) sampler (SKC Inc.) on PVC membrane filters (diameter: 25 mm; porosity 0.5 μm), according to Italian standards [26]. The sampling time of the inhalable fraction ranged from three to six hours in each location.

(2) Samples of deposition powders generated by industrial processes.

All samples were collected in production areas subjected to possible Ni airborne exposure, i.e., the ladle furnace, the continuous casting area, and the electric arc furnace.

5. Leaching and Results

The speciation procedure described in this paper was applied to real samples of environmental dust and filters.

A small amount of collected powder (in the order of a few mg) was placed on a PVC filter and treated with the sequential leaching described above. The results of speciation are reported in Table 6.

The analysis was also carried out on a PVC filter clean to control any forms of contamination.

In all the samples, the most represented species was NiO (35-63%), followed by metallic Ni (7-48%); the soluble and sulfidic fractions were equally distributed (8-30% and 3-23%, respectively). In two samples of powders (A and C) the soluble and sulfidic fractions were not detectable.

In all remove samples, the sum of the different Ni fractions was comparable with the amount of total Ni. The sum of the fraction was slightly lower than total Ni measured in AAS, and the difference was comprised between 1.4 and 19.4%.

At each analysis set, a sample of weighed salts mix is extracted in parallel to verify the quality and effectiveness of the extraction. The quality of determination of total Ni was checked with certified material (NIST 1640, metallic elements in water).

6. Discussion

The speciation of Ni in environmental air samples has been debated for several years, since there are threshold limits values relative to the single species but the analytical techniques for determining them are not always available or easy to use. The speciation of Ni can be obtained through chemical leaching and subsequent analysis of the solutes with traditional methods of Ni determinations or by X-ray determination [27]. In the literature several authors have tried to standardize the leaching process and the contrasting results demonstrate the high variability of the method.

The first available method was published by Zatka et al. [28], which, through several washings and dry evaporations, determines the soluble Ni fraction, the sulfidic Ni, the metallic Ni, and the oxide in sequence with a relative standard deviation (RSD) ranged from 3.2 to 3.7%.

Over the years, other authors have tried to replace by using different solutions. Bolt et al. [28] have developed an inline system for extracting soluble Ni compounds (with ammonium citrate buffer), Ni sulfide (with ammonium citrate and hydrogen peroxide), and metal Ni (with CuCl2 / KCl). The final digestion of the residues on the membrane with HNO_3/HCl leads to the determination of the Ni oxide fraction. The method is based on the one developed by Zatka but requires much less time in execution. In 2003, Profumo et al. [29] reported a method of speciation based on sequential extraction. The authors succeeded in speciating Ni metal and soluble Ni compounds, such as sulfate and chloride and among from the insoluble Ni oxide and sulfide. The Ni recoveries of the different species were in the range 94/99%.

Also Conard et al. [30] tried to improve the sequence of Zatka especially in the separation step of the Ni sulfide/metal phases with leaching with ammonium citrate and hydrogen peroxide increasing the volume of the solution, time, and percolation methods. Despite the studies and tests to date, there is no standardized extraction method to carry out the chemical speciation of Ni, and the proposed methods are difficult to apply on environmental samples.

Our study allowed us to overcome some of the prominent methodological limits of the previously described speciation techniques: the excessive length of the extraction phase and the powder leak during the leaching process.

This method is based on the different solubility based on the different chemical-physical properties of the inorganic Ni species which allow a sequential extraction of the fraction and a determination of Ni.

The test on all the solutions of each single fraction shows how the solutions selectively dissolve the inorganic Ni species.

$NiSO_4$, more soluble compound, is dissolved by all four solutions with an efficacy ranging from 89 to 103%, while the less soluble Ni dissolves only in a solution of HNO_3 and HCl 1:1.

Compared to the previously published methods we reduced the overall volume of the solvents used to only 10 mL, therefore avoiding the evaporation phase and the issues related to the retaining of the solutes. Previous tests on the same sequence of leaching conducted on a filter placed on a simple vacuum extraction device led to very low retaining rates in the order of 50% (date not shown); the adoption of DigiFILTER devices allowed to improve the retaining rate (reaching 94-99%) probably preventing the loss of small particles independently from the different Ni species.

The determination of species of Ni in real powder or filter allowed identifying the different species associated with the different samples.

In all remove samples, the sum of the different Ni fractions was comparable with the amount of total Ni. The sum of the fraction was slightly lower than total Ni measured in AAS; the difference was comprised between 1.4 and 19.4%.

Moreover, taking into account the great progress that has been made in the identification of new health related particle size exposure assessment, the methods described in this paper could also be applied in analyzes carried out on particulate ultrafine material [31]. The value of inhalable Ni aerosols monitoring by species has been widely demonstrated in a variety of working environments as a fundamental tool in assessing respiratory cancer risk [32, 33]. A species-specific approach to setting occupational exposure limits guarantees that the best available health and exposure data will be used. Future research may lead to consideration of setting species-specific OELs on the basis of certain subfractions of inhalable Ni aerosols.

Better Ni speciation techniques will improve the capability of assessing the occupational exposure to carcinogens together with both the measurement of the inhalable fraction and the particle size distribution, as expected and hoped in previous studies [34].

7. Conclusion

The analytical method described in this paper allows the determination of soluble, sulfidic, metallic, and oxide Ni by a simple sequential extraction procedure and determination by Atomic Absorption Spectroscopy using small volumes of solutions and without long evaporation phases. For the purpose of assessing occupational exposure to carcinogens, the method developed allowed the separation and speciation of the different Ni species on the inhalable fraction of the airborne particulate sampled in different working environments.

The speciation of Ni in real environmental samples allowed us to compare our results with the TLVs proposed by the different agencies and regulations, which is not possible with the sole determination of the total metallic Ni.

Acknowledgments

The study was supported by University of Brescia.

References

[1] C. Klein and M. Costa, "Nickel," in *Handbook on the Toxicology of Metals*, Academic Press, Waltham, MA, USA, 4th edition, 2014.

[2] M. A. Riaz, A. B. T. Akhtar, A. Riaz, G. Mujtaba, M. Ali, and B. Ijaz, "Heavy metals identification & exposure at workplace environment its extent of accumulation in blood of iron & steel recycling foundry workers of Lahore, Pakistan," *Pakistan Journal of Pharmaceutical Sciences*, vol. 30, no. 4, pp. 1233–1238, 2017.

[3] NiDI (Nickel Development Institute), *Safe Use of Nickel in the Workplace*, Nickel Development Institute, Toronto, Canada, 2nd edition, 1997.

[4] H. Lu, X. Shi, M. Costa, and C. Huang, "Carcinogenic effect of nickel compounds," *Molecular and Cellular Biochemistry*, vol. 279, no. 1-2, pp. 45–67, 2005.

[5] M. Costa, "Molecular mechanisms of nickel carcinogenesis," *Biological Chemistry*, vol. 383, no. 6, pp. 961–967, 2002.

[6] D. Beyersmann and A. Hartwig, "Carcinogenic metal compounds: recent insight into molecular and cellular mechanisms," *Archives of Toxicology*, vol. 82, no. 8, pp. 493–512, 2008.

[7] D. G. Barceloux, "Nickel," *Journal of Toxicology - Clinical Toxicology*, vol. 37, no. 2, pp. 239–258, 1999.

[8] L. T. Haber, L. Erdreicht, G. L. Diamond et al., "Hazard identification and dose response of inhaled nickel-soluble salts," *Regulatory Toxicology and Pharmacology*, vol. 31, no. 2, pp. 210–230, 2000.

[9] M. G. Ahlström, J. P. Thyssen, T. Menné, and J. D. Johansen, "Prevalence of nickel allergy in Europe following the EU Nickel Directive – a review," *Contact Dermatitis*, vol. 77, no. 4, pp. 193–200, 2017.

[10] M. Saito, R. Arakaki, A. Yamada, T. Tsunematsu, Y. Kudo, and N. Ishimaru, "Molecular mechanisms of nickel allergy," *International Journal of Molecular Sciences*, vol. 17, no. 2, article no. 202, 2016.

[11] P. Apostoli and S. Catalani, "Mechanisms of action for metallic elements and their species classified carcinogen R 45 and R 49 by EU," *Giornale Italiano di Medicina del Lavoro ed Ergonomia*, vol. 30, no. 4, pp. 382–391, 2008.

[12] J. Zhao, X. Shi, V. Castranova, and M. Ding, "Occupational toxicology of nickel and nickel compounds," *Journal of Environmental Pathology, Toxicology and Oncology*, vol. 28, no. 3, pp. 177–208, 2009.

[13] IARC, "Chromium, Nickel, and Welding," in *Monographs on the Evaluation of Carcinogenic Risks to Humans*, vol. 49, IARC, Lyon, France, 1990.

[14] IARC, "Arsenic, Metals, Fibers and dust," in *Monographs on the Evaluation of Carcinogenic Risks to Humans*, vol. 100, IARC, Lyon, France, 2009.

[15] G. G. Fletcher, F. E. Rossetto, J. D. Turnbull, and E. Nieboer, "Toxicity, uptake, and mutagenicity of particulate and soluble nickel compounds," *Environmental Health Perspectives*, vol. 102, no. 3, pp. 69–79, 1994.

[16] ACGIH, "Threshold Limit Values and Biological Exposure Indices for Chemical Substances and Physical Agents," in *Proceedings of the American Conference of Governmental Industrial Hygienists*, Cincinnati, OH, USA, 1998.

[17] K. S. Kasprzak, F. W. Sunderman Jr., and K. Salnikow, "Nickel carcinogenesis," *Mutation Research - Fundamental and Molecular Mechanisms of Mutagenesis*, vol. 533, no. 1-2, pp. 67–97, 2003.

[18] SCOEL, "Recommendation from the Scientific Committee on Occupational Exposure Limits for nickel and inorganic nickel compounds," SCOEL/SUM/85, 2011.

[19] P. Apostoli, "The role of element speciation in environmental and occupational medicine," *Fresenius' Journal of Analytical Chemistry*, vol. 363, no. 5-6, pp. 499–504, 1999.

[20] J. Szpunar and R. Łobiński, "Speciation in the environmental field - Trends in analytical chemistry," *Fresenius' Journal of Analytical Chemistry*, vol. 363, no. 5-6, pp. 550–557, 1999.

[21] G. W. Hughson, K. S. Galea, and K. E. Heim, "Characterization and assessment of dermal and inhalable nickel exposures in nickel production and primary user industries," *Annals of Occupational Hygiene*, vol. 54, no. 1, pp. 8–22, 2010.

[22] L. Füchtjohann, N. Jakubowski, D. Gladtke, D. Klockow, and J. A. C. Broekaert, "Speciation of nickel in airborne particulate matter by means of sequential extraction in a micro flow system and determination by graphite furnace atomic absorption spectrometry and inductively coupled plasma mass spectrometry," *Journal of Environmental Monitoring*, vol. 3, no. 6, pp. 681–687, 2001.

[23] D. Schaumlöffel, "Nickel species: analysis and toxic effects," *Journal of Trace Elements in Medicine and Biology*, vol. 26, no. 1, pp. 1–6, 2012.

[24] V. J. Zatka, J. Stuart Warner, and M. David, "Chemical Speciation of Nickel in Airborne Dusts: Analytical Method and Results of An Interlaboratory Test Program," *Environmental Science & Technology*, vol. 26, no. 1, pp. 138–144, 1992.

[25] OSHA—Occupational Health and Safety Administration, "Permissible Exposure Limits - Annotated Tables Z-1," https://www.osha.gov/dsg/annotated-pels/index.html, (accessed on 3rd July 2018).

[26] UNICHIM—Associazione Per l'Unificazione Nel Settore Dell'industria Chimica, "Ambienti di Lavoro—Determinazione Della Frazione Inalabile Delle Particelle Aerodisperse, Metodo Gravimetrico," http://www.uni.com/index.php?option=com_content&view=article&id=454&Itemid=2448&lang=it, (accessed on 26 June 2018).

[27] L. L. Van Loon, C. Throssell, and M. D. Dutton, "Comparison of nickel speciation in workplace aerosol samples using sequential extraction analysis and X-ray absorption near-edge structure spectroscopy," *Environmental Science: Processes & Impacts*, vol. 17, no. 5, pp. 922–931, 2015.

[28] H. M. Bolt, C. Noldes, and M. Blaszkewicz, "Fractionation of nickel species from airborne aerosols: Practical improvements and industrial applications," *International Archives of Occupational and Environmental Health*, vol. 73, no. 3, pp. 156–162, 2000.

[29] A. Profumo, G. Spini, L. Cucca, and M. Pesavento, "Determination of inorganic nickel compounds in the particulate matter of emissions and workplace air by selective sequential dissolutions," *Talanta*, vol. 61, no. 4, pp. 465–472, 2003.

[30] B. R. Conard, N. Zelding, and G. T. Bradley, "Speciation/fractionation of nickel in airborne particulate matter: Improvements in the Zatka sequential leaching procedure," *Journal of Environmental Monitoring*, vol. 10, no. 4, pp. 532–540, 2008.

[31] G. Marcias, J. Fostinelli, S. Catalani et al., "Composition of Metallic Elements and Size Distribution of Fine and Ultrafine Particles in a Steelmaking Factory," *International Journal of Environmental Research and Public Health*, vol. 15, no. 6, 2018.

[32] D. J. Sivulka, "Assessment of respiratory carcinogenicity associated with exposure to metallic nickel: A review," *Regulatory Toxicology and Pharmacology*, vol. 43, no. 2, pp. 117–133, 2005.

[33] D. J. Sivulka and S. K. Seilkop, "Reconstruction of historical exposures in the US nickel alloy industry and the implications for carcinogenic hazard and risk assessments," *Regulatory Toxicology and Pharmacology*, vol. 53, no. 3, pp. 174–185, 2009.

17

Determination of Pyrethroids in Tea Brew by GC-MS Combined with SPME with Multiwalled Carbon Nanotube Coated Fiber

Dongxia Ren,[1] Chengjun Sun,[1,2] Guanqun Ma,[3] Danni Yang,[1] Chen Zhou,[1] Jiayu Xie,[1] and Yongxin Li🆔[1,2]

[1]*West China School of Public Health, Sichuan University, Chengdu 610041, China*
[2]*Provincial Key Laboratory for Food Safety Monitoring and Risk Assessment of Sichuan, Chengdu 610041, China*
[3]*College of Life and Environmental Sciences, Shanghai Normal University, Shanghai, China*

Correspondence should be addressed to Yongxin Li; lyxlee2008@hotmail.com

Academic Editor: Alberto Chisvert

A new method has been developed to simultaneously determine 7 pyrethroid residues in tea brew using gas chromatography-mass spectrometry (GC-MS) combined with solid phase microextraction (SPME) with multiwalled carbon nanotubes (MWCNTs) coated fiber. The MWCNTs coated fiber of SPME was homemade by using stainless steel wire as coating carrier and polyacrylonitrile (PAN) solution as adhesive glue. Under the optimized conditions, a good linearity was shown for bifenthrin, fenpropathrin, permethrin, and cyfluthrin in 1–50 ng mL^{-1} and for cypermethrin, fenvalerate, and deltamethrin in 5–50 ng mL^{-1}. The correlation coefficients were in the range of 0.9948–0.9999. The average recoveries of 7 pyrethroids were 94.2%–107.3% and the relative standard deviations (RSDs) were less than 15%. The detection limit of the method ranged from 0.12 to 1.65 ng mL^{-1}. The tea brew samples made from some commercial tea samples were analyzed. Among them, bifenthrin, fenpropathrin, and permethrin were found. The results show that the method is rapid and sensitive and requires low organic reagent consumption, which can be well used for the detection of the pyrethroids in tea brew.

1. Introduction

Synthetic pyrethroids are a class of bionic synthetic ester pesticides whose structural or biological activity is similar to natural pyrethroids, with the features of high efficiency, broad spectrum, and less residue. Despite their low toxicity, long-term exposure and ingestion of these pesticides may result in endocrine disorders and affect nervous system function [1]. China is the homeland of tea, and tea has several beneficial functions, such as refreshing and relieving restlessness. Nowadays, people not only require better quality of tea, but also pay particular attention to the harmful substances in it [2]. Since tea is susceptible to suffer from pests and diseases, farmers usually spray pesticides to reduce the adverse effects, which results in pesticide residues [3–5]. The level of pesticide residues in tea brew is directly related to the intake amount of human, while the determination of these 7 pyrethroids in tea brew has rarely been reported

[6, 7]. Therefore, it is necessary to develop a rapid and efficient method for the detection of these 7 pyrethroid pesticides in tea brew.

Recently, it is in tea that most researches are mainly focused on the detection of pyrethroid residues [2, 8]. Such methods as solid phase extraction, gel permeation chromatography, and matrix solid phase dispersion are commonly used as sample preparation techniques for analysis of pesticide residues in tea [9, 10]. However, organic solvents are employed during the extraction procedure, which poses a threat to the environment and human health. SPME is a new method of microextraction without solvent based on solid phase extraction, which has the advantages of simultaneous sampling, extraction and concentration, and low consumption of sample and reagents, as well as ease of operation and automation [8, 11]. Extraction phase is the core of SPME, whose performance directly determines the SPME efficiency, thus affecting the sensitivity of the method and the reliability of the results. However, the commercial fibers tend to swell off, and most of them use fused silica as the fiber carrier, which is expensive, is easy to break, and has short service life. In addition, the assortment of commercially available extraction phases is quite limited, which also limits their application. In order to improve the SPME performance, many researchers are committed to the research on new fibers.

In recent years, the emergence of many kinds of materials has brought new opportunities for the development of SPME fiber coating. Wang et al. used dispersive solid phase extraction with polyaniline-coated magnetic particles to determine 5 pyrethroids in tea drinks [12]. Sun et al. determined the pollutants in the aquatic environment by SPME coupled with surface enhanced Raman spectroscopy, in which the SPME was coated with ZnO nanorods (ZnO NRs) decorated with Au@4-ATP@Ag core–shell nanoparticles (NPs) [13]. Wang et al. synthesized a core–shell TiO_2@C fiber for SPME to detect polycyclic aromatic hydrocarbons in river water samples [14]. Wu et al. prepared a single-walled carbon nanotubes coated SPME fiber for extraction of several pyrethroids [15]. Among these nanomaterials, MWCNTs, a new kind of carbon nanomaterials, have attracted much attention in many fields. As new adsorbents, MWCNTs present a promising future in the preconcentration of environmental contaminants, due to their unique hydrophobic structure and large specific surface area, as well as strong reaction with a variety of organic compounds [16, 17].

Up to now, the detection of these 7 pyrethroids in tea brew is rarely reported [6, 7]. In this study, a SPME with MWCNTs coated fiber was developed using stainless steel wire as the coating carrier, which in combination with GC-MS was applied to the detection of the target pyrethroids in tea brew with satisfactory results.

2. Experimental

2.1. Chemicals and Materials. The standards of bifenthrin, fenpropathrin, permethrin, cyfluthrin, cypermethrin, fenvalerate, and deltamethrin were purchased from National Standard Material Center of China. n-Hexane, methanol, and acetone were of HPLC grade and provided by Tianjin Branch Miou Chemical Reagent Co., Ltd. MWCNTs (10–30 μm length, 10–20 nm outer diameter) were purchased from Chinese Academy of Sciences Chengdu Organic Chemistry Co., Ltd. N,N-Dimethylformamide (DMF) was obtained from Shanghai Guoyao Chemical Reagent Co., Ltd. Polyacrylonitrile (PAN) was purchased from Sigma (St. Louis, MO, USA). Water used throughout the experiments was obtained from a Millipore water purification system (18.25 M Ω cm, Millipore, USA).

Stainless steel wires (12 cm × 0.15 mm) were obtained from Shenzhen Santk Metal Material Co., Ltd. SPME needles (SPME-S-02) were purchased from Shanghai New Extension Analytical Instruments Technology Co., Ltd. SPME handle (57330-U) was obtained from Supelco.

2.2. Preparation of the Standard Solutions. The stock solutions of 100 μg mL^{-1} were prepared with pyrethroid standards

containing bifenthrin, fenpropathrin, permethrin, cyfluthrin, cypermethrin, fenvalerate, and deltamethrin in n-hexane and stored at −20°C. Five levels of calibration scales were obtained by diluting the stock standard solutions with n-hexane, which were stored at 4°C until used.

2.3. Preparation of MWCTs Coated Fiber for SPME. The stainless steel wires were cleaned with methanol, acetone, and n-hexane in order to remove the surface contaminants and then dried in the air at room temperature. Meanwhile, 0.5 mg of PAN was dissolved in 7.5 mL of DMF, and the mixture was kept at 90°C for 1 h. Five mg of MWCNTs was taken into a centrifuge tube and mixed with PAN solution by ultrasound. Subsequently, the cleaned stainless steel wire was dipped into the coating suspension with a height of 1.5 cm for one minute [18–20] and then withdrew at a constant rate to ensure the formation of a smooth coating on the wire surface. After this, the fiber was dried at 180°C for 2 min. This above procedure from "dipped into the coating suspension" was repeated 3 times. Prior to use, the proposed SPME was conditioned at 280°C for 2 h.

2.4. Sample Preparation. The tea samples were randomly purchased from the local markets. After fully mixed, 3 g (accurate to 0.001 g) of each was weighed into a 250 mL beaker and 100 mL of boiled water was added and then kept for 3 h. 10 mL of the above-mentioned supernatant in a vial with cap was placed on a magnetic stirrer and extracted with the proposed SPME with MWCNTs coated fiber at 750 rpm for 20 min. The target compounds were desorbed at 280°C for 8 min and determined by GC-MS.

2.5. GC-MS Analysis. GC-MS analysis was performed by Agilent Technologies 7890A GC with a 5975C MS, equipped with a DB-5 MS fused silica capillary column (30 m × 0.25 mm × 0.25 μm) (Agilent Technologies, Santa Clara, CA, USA). Helium (purity 99.999%) served as carrier gas at a flow rate of 1 mL min^{-1}. To ensure rapid and good separation of all the target compounds, a programmed temperature was employed. The initial column temperature was held at 60°C, and then to 150°C at 30°C min^{-1}, and finally increased to 290°C at 10°C min^{-1}, with hold time of 6 min. The injection port temperature was set at 280°C and the injection mode was in splitless mode.

In this study, the mass spectrometer was performed with an EI source in selected ion monitoring (SIM) mode. The electron energy was 70 eV with the temperature of the ion source and transfer line at 230°C and 280°C, respectively. The solvent delay was setting as 7 min. In the NIST search library, higher abundance of sub-ions was selected as qualitative ions and quantitative ions (Table 1).

3. Results and Discussion

3.1. Surface Structure of the Fiber. Carbon nanotubes have the advantages of large specific surface area, good chemical stability, superior mechanical strength, and high heat resistance and can set off π-π interactions [21, 22]. Therefore, they

TABLE 1: The retention time, monitoring ions, and quantitative ions of 7 pyrethroids.

Peak number	Pesticides	Retention time (min)	Quantitative ions (m/z)	Monitoring ions (m/z)
(1)	Bifenthrin	13.25	181.0	166.0, 182.0, 165.0
(2)	Fenpropathrin	13.45	349.0	141.0, 265.0, 209.0
(3)	Permethrin	15.10	182.0	163.0, 91.0, 165.0
(4)	Cyfluthrin	15.89	163.0	206.1, 226.1, 165.0
(5)	Cypermethrin	16.40	181.0	163.0, 209.0, 91.0
(6)	Fenvalerate	17.6, 18.02	125.0	167.1, 181.0, 225.0
(7)	Deltamethrin	19.07	253.0	251.0, 255.0, 181.0

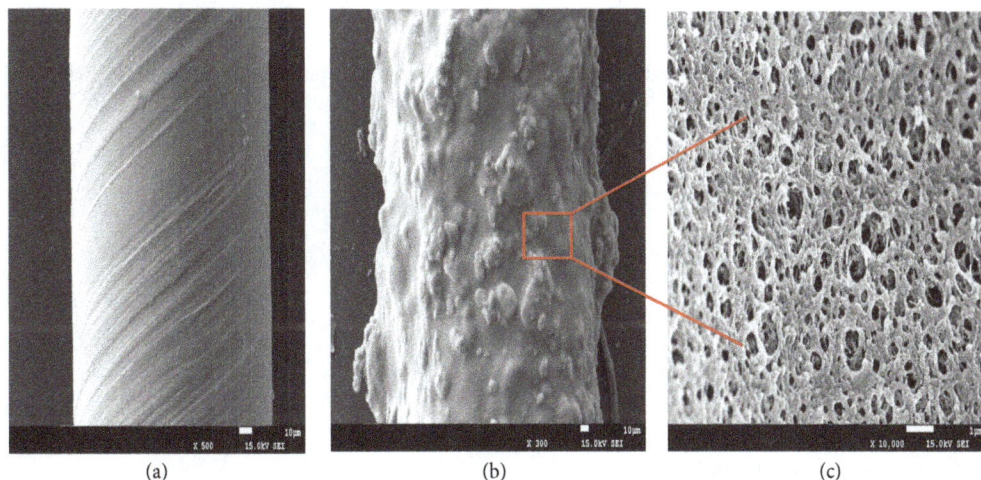

(a)　　　　　　　　　　(b)　　　　　　　　　　(c)

FIGURE 1: SEM images of the MWCNTs coated fiber, (a) the image of the stainless steel wire, (b) the image of the coated fiber's surface, and (c) the image of the coated fiber's inner structure.

have a promising application potential in the extraction field, for example, as SPME fiber. In order to investigate the surface properties of MWCNTs coated SPME fibers, seven pyrethroid pesticides were extracted with the proposed fibers and the uncoated stainless steel wires under the same experimental conditions. The results showed that the MWCNTs coated SPME fibers have a strong adsorption ability. Additionally, scanning electron microscopy (SEM) was employed to further explore the surface characteristics of the proposed fiber, as shown in Figure 1. It could be seen from Figure 1(b) that the multiwalled carbon nanotubes were uniformly attached on the surface of the stainless steel wire and were semiexposed. Figure 1(c) showed that the typical porous structure of MWCNTs coating was formed on the fiber surface, which resulted in high surface area of the coating, thereby offering extraordinary adsorption capacity. And the thickness of the coating is about 70 μm.

3.2. Optimization of GC-MS Conditions.
Tea has a challenging matrix, containing pigments, caffeine, and other impurities, which would interfere with the detection of the target compounds. In order to improve the accuracy and sensitivity of the method, it is necessary to select the appropriate mass spectrometry conditions. Under optimized GC-MS conditions, the chromatogram of the seven pyrethroids was shown in Figure 2.

3.3. Optimization of SPME Conditions.
There are several main factors affecting SPME extraction and desorption efficiency, including stir rate, desorption temperature, and time, as well as extraction time. In this study, these factors were optimized as follows: the blank sample was spiked with the stock solution to obtain a final concentration of 20 ng mL^{-1}. The analytes were extracted under different SPME conditions and analyzed by GC-MS. The optimal conditions were selected according to the peak area of the analytes.

3.3.1. Optimization of Extraction Time.
According to the digital model of direct SPME and the dynamic adsorption model of nonbalance solid phase microextraction, the extraction efficiency is proportional to concentration of the target compound when other conditions are constant [23, 24]. In this study, analytes were extracted for 10, 15, 20, 25, and 30 min, respectively, with a stirring rate of 750 rpm and desorbing for 5 min at 280°C. The peak area of the analytes was used as the reference index. Each experiment was repeated for three times. As shown in Figure 3, the amount of the target pyrethroids increased with the increase of the extraction time. When the extraction time was 20 min, the extraction amount of the analytes reached the maximum and tended to be balanced. Therefore, taking into account the extraction efficiency and the analytical time, the extraction time of 20 min was selected.

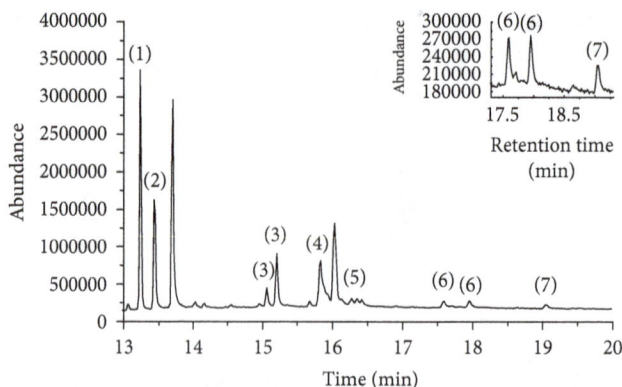

FIGURE 2: GC-MS chromatogram of the mixed standard solution of 7 pyrethroids. (1) bifenthrin, (2) fenpropathrin, (3) permethrin, (4) cyfluthrin, (5) cypermethrin, (6) fenvalerate, and (7) deltamethrin.

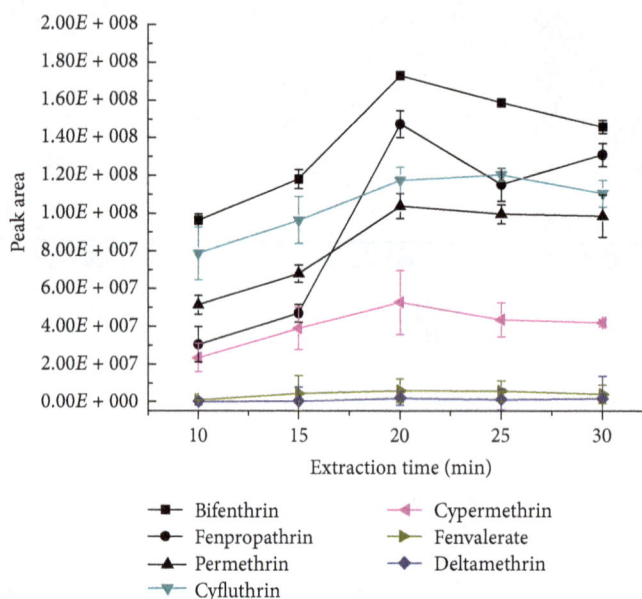

FIGURE 3: The effect of the extraction time on the extraction efficiency ($n = 3$). The standard mixture concentration: 20 ng mL^{-1}, stirring rate: 750 rpm, desorption temperature: 280°C, and desorption time: 5 min.

3.3.2. Optimization of Stirring Rate. The essence of SPME is to extract the analytes from the sample solution to the stationary phase. During the extraction process, the analytes in the solution diffused from the solid-liquid layer to the surface of the solid phase coating. Thus, the appropriate agitation facilitates the diffusion of molecules in the solution to improve the extraction efficiency [25]. The extraction efficiency of 7 pyrethroids was studied with stirring rates ranging from 0 to 1250 rpm when the other conditions were fixed, including extraction time of 20 min and desorption at 280°C for 5 min. Each experiment was repeated for three times. As shown in Figure 4, the peak area of the analytes increased as the stirring rate increased. Until 750 rpm, the peak area of the analytes reached the maximum. If the stirring rate continued to rise, it was prone to depression in the center of the liquid surface, which not only affected the normal operation of the experiments, but also led to a decrease of the

extraction efficiency. Thus, 750 rpm was chosen as the stirring rate for subsequent experiments.

3.3.3. Optimization of Desorption Temperature. In SPME-GC-MS detection, if the desorption temperature is too low, the analytes cannot be effectively desorbed, and the carry-over would exhibit an adverse effect on both the results and the follow-up experiments, while if the temperature is too high, it might make the analytes decomposed and at the same time also affect the service life of SPME fibers [15, 26]. The influence of desorption temperature on the desorption efficiency for 7 pyrethroids was investigated in the range of 220–300°C. The other conditions were as follows: the stir rate was 750 rpm and the extraction time was 20 min with desorption for 5 min. Each experiment was repeated for three times. As Figure 5 shows, the elevated temperature facilitated the effective desorption of the analytes. The peak area of

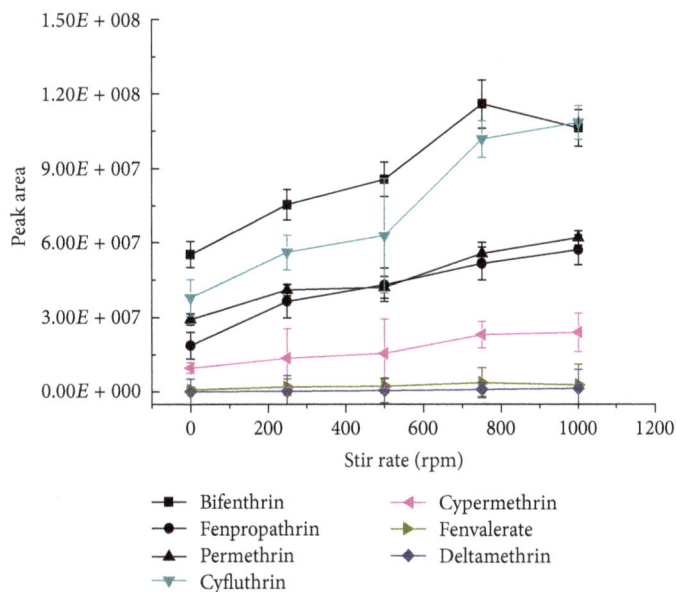

FIGURE 4: The effect of the stirring rate on the extraction efficiency ($n = 3$). The standard mixture concentration: 20 ng mL^{-1}, extraction time: 20 min, desorption temperature: 280°C, and desorption time: 5 min.

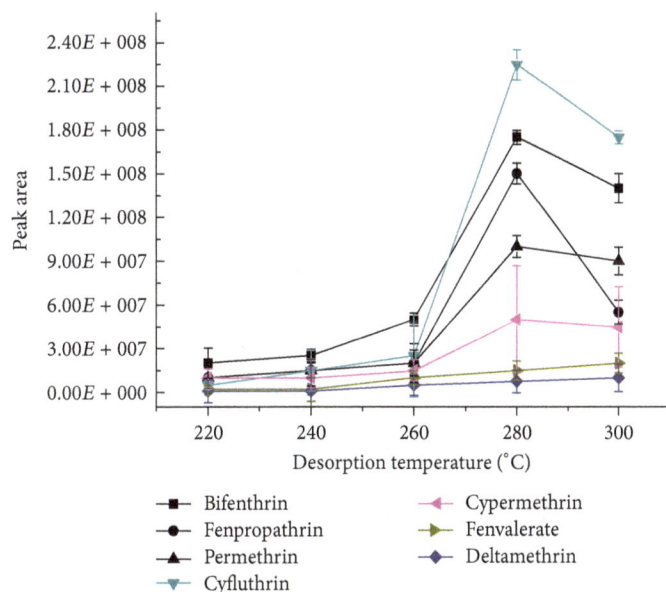

FIGURE 5: The effect of the temperature on the desorption efficiency ($n = 3$). The standard mixture concentration: 20 ng mL^{-1}, stirring rate: 750 rpm, extraction time: 20 min, and desorption time: 5 min.

each analyte increased with the increase of the desorption temperature when it was in the range of 220–280°C. When the desorption temperature was increased to 300°C, the peak area of some of the analytes did not increase significantly and that of others decreased considerably, indicating that the analytes had been completely desorbed at 280°C. Therefore, 280°C was selected as the desorption temperature in subsequent experiments.

3.3.4. *Optimization of Desorption Time.* In order to investigate the effect of desorption time, the analytes were determined when the desorption time was 2, 5, 8, and 11 min, respectively, under the conditions of extraction for 20 min with a stir rate of 750 rpm and desorption at 280°C. Each experiment was repeated for three times. As shown in Figure 6, when the desorption time was 2–8 min, the peak area of the analytes increased with the desorption time. However, when the desorption time continued to increase to 11 min, the peak area of most analytes did not change significantly. Therefore, 8 min was selected as the desorption time.

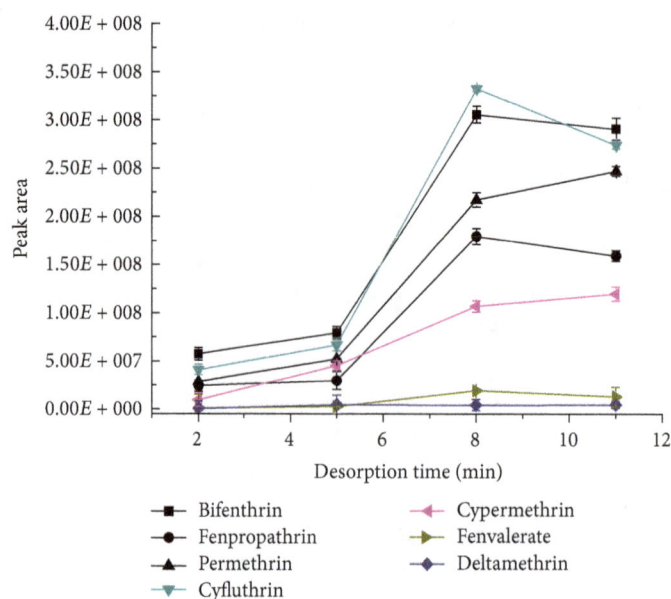

FIGURE 6: The effect of the desorption time on the desorption efficiency ($n = 3$). The standard mixture concentration: 20 ng mL^{-1}, stirring rate: 750 rpm, extraction time: 20 min, and desorption temperature: 280°C.

TABLE 2: The linear ranges, correlation coefficients, LODs, and LOQs of the method.

Pesticides	Linear range/(ng mL^{-1})	Regression equation	Correlation coefficient	LOD/(ng mL^{-1})	LOQ/(ng mL^{-1})
Bifenthrin	1–50	$Y = (2 \times 10^6 \pm 5 \times 10^4)x + (4 \times 10^6 \pm 2 \times 10^5)$	0.9978	0.12	0.36
Fenpropathrin	1–50	$Y = (9 \times 10^5 \pm 5 \times 10^4)x + (2 \times 10^6 \pm 2 \times 10^5)$	0.9997	0.30	0.90
Permethrin	1–50	$Y = (8 \times 10^5 \pm 6 \times 10^4)x + (1 \times 10^6 \pm 1 \times 10^5)$	0.9999	0.33	0.99
Cyfluthrin	1–50	$Y = (4 \times 10^6 \pm 2 \times 10^5)x - (4 \times 10^6 \pm 3 \times 10^5)$	0.9999	0.24	0.72
Cypermethrin	5–50	$Y = (4 \times 10^5 \pm 4 \times 10^4)x + (9 \times 10^5 \pm 1 \times 10^5)$	0.9991	0.60	1.80
Fenvalerate	5–50	$Y = (6 \times 10^4 \pm 4 \times 10^3)x + (1 \times 10^5 \pm 1 \times 10^4)$	0.9997	0.99	2.97
Deltamethrin	5–50	$Y = (9 \times 10^4 \pm 6 \times 10^3)x + (1 \times 10^5 \pm 1 \times 10^4)$	0.9948	1.65	4.95

3.4. Method Validation

3.4.1. Regression Equation, Linear Range, and Detection Limits.
A standard series of 1, 5, 10, 30, and 50 ng mL^{-1} was prepared in the blank tea sample, and the assay was carried out in accordance with the procedure of 2.4. The standard curves were plotted with the peak area as the ordinate (y) and the mass concentration of the standard solution as the abscissa (x, ng mL^{-1}). The limits of detection (LODs) and the limits of quantification (LOQs) are calculated on the basis of the 3-fold and 10-fold baseline noise, respectively. The results were presented in Table 2. There was a good linear relationship for bifenthrin, fenpropathrin, permethrin, and cyfluthrin in the range of 1 to 50 ng mL^{-1} and for cypermethrin, fenvalerate,

and deltamethrin in the range of 5 to 50 ng mL^{-1}, with the correlation coefficients (r) in 0.9948–0.9999. The detection limits were 0.12–1.65 ng mL^{-1}, and the limits of quantification were 0.36–4.95 ng mL^{-1}.

3.4.2. Recovery and Precision of the Method.
Under the optimized pretreatment and chromatographic conditions, the blank sample was spiked with the stock standard mixture to obtain a final concentration of 5, 10, and 50 ng mL^{-1}, respectively, and then treated with MWCNTs coated SPME and analyzed by GC-MS. The recoveries were calculated according to the ratio of the determined concentration and the actual value. Each experiment was determined for 6 times. The results were shown in Table 3. The average recoveries

TABLE 3: The recovery and RSDs of the method.

Pesticides	Background/(ng mL^{-1})	Add/(ng mL^{-1})	Found/(ng mL^{-1})	Recovery/(%)	RSD/(%)
Bifenthrin	ND[a]	5.00	6.29	126	9.9
		10.00	10.02	100	2.5
		50.00	47.8	95.6	2.1
Fenpropathrin	ND	5.00	5.60	112	12.1
		10.00	7.93	79.3	13.9
		50.00	45.6	91.2	6.2
Permethrin	ND	5.00	5.89	118	12.6
		10.00	10.27	103	4.8
		50.00	50.28	101	2.0
Cyfluthrin	ND	5.00	4.73	94.6	14.4
		10.00	9.75	97.6	4.1
		50.00	51.96	104	2.9
Cypermethrin	ND	5.00	5.61	112	7.1
		10.00	10.83	108	3.5
		50.00	49.80	99.6	5.7
Fenvalerate	ND	5.00	4.60	92.1	8.5
		10.00	10.25	102	11.5
		50.00	48.41	96.8	1.3
Deltamethrin	ND	5.00	5.53	110	10.6
		10.00	9.76	97.6	7.2
		50.00	50.96	102	6.6

[a]Not detected.

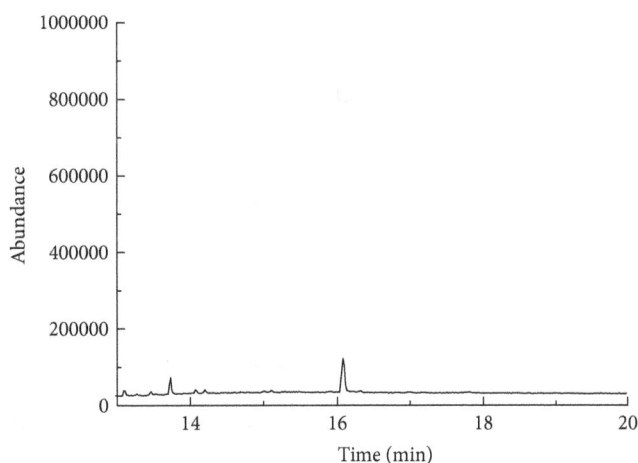

FIGURE 7: The chromatograms of the Maofeng Tea.

ranged from 79.3% to 126% and the RSDs were less than 15%, indicating the favorable potential of the proposed method to be implemented in real application.

3.4.3. Determination of Commercial Samples. The proposed method was employed for the determination of 7 pyrethroid residues in tea brew samples made from different brands of tea randomly purchased from the local markets, and the results were shown in Figures 7–9. Bifenthrin, fenpropathrin, and permethrin were detected, although none reached the limit of quantification (the former two were found in jasmine tea and green tea, respectively, and the third one was only found in jasmine tea). Kang [27] determined 16 pesticide residues including fenpropathrin, cypermethrin, deltamethrin, fenvalerate, and lambda-cyhalothrin in tea brew by gas chromatography-electron capture detector (GC-ECD) after extracted by ethyl acetate for 3 times combining with LC-Florisil SPE column cleanup. The linear ranges and the recoveries of the method for the pyrethroids were in the range of 10 ng mL^{-1}–1000 ng mL^{-1} and 63.1%–105.2%,

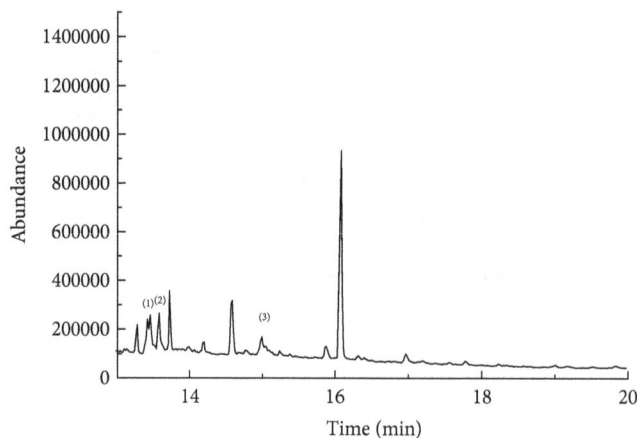

FIGURE 8: The chromatograms of the jasmine tea, (1) bifenthrin, (2) fenpropathrin, and (3) permethrin.

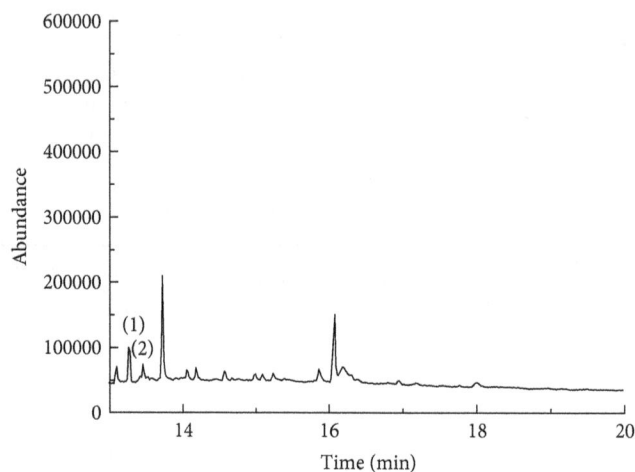

FIGURE 9: The chromatograms of the green tea, (1) bifenthrin and (2) fenpropathrin.

respectively. When compared with this method, our method is more sensitive, is easier to operate, and consumes less organic reagents.

4. Conclusions

A multiwalled carbon nanotube coated SPME fiber was prepared by using stainless steel wire as the coating carrier and PAN as the adhesive. The SPME sampling combined with GC-MS analysis was employed to determine seven pyrethroids in tea brew samples. The proposed MWCNTs SPME not only has the advantages of fastness, efficiency, and convenience, but also can overcome the defects of the commercial fiber, such as ease of being broken and short service life as well as expensive price. The method is simple, rapid, and sensitive and with low consumption of organic reagents, which is suitable for the detection of pyrethroid pesticides in tea brew.

Acknowledgments

This work was supported by Project of Sichuan Health Bureau of China (no. 150111) and Project of Chengdu Science and Technology Bureau of China (2015-HM01-00182-SF).

References

[1] Y. B. Liu, L. H. Jia, R. F. Xue, and M. X. Zhang, "Rapid GC-MS Determination of Residual Pyrethroid Pesticides in Vegetables and Fruits," *Physical Testing And Chemical Analysis Part B Chemical Analysis*, vol. 42, no. 8, pp. 637–643, 2006.

[2] H. P. Chen, X. Liu, Q. H. Wang, and Y. Jiang, "Simultaneous Determination of 72 Pesticide Residues in Tea by Gas Chromatography-Mass Spectrometry," *Food Science*, vol. 32, no. 6, pp. 159–164, 2011.

[3] P. Schreinemachers and P. Tipraqsa, "Agricultural pesticides and land use intensification in high, middle and low income countries," *Food Policy*, vol. 37, no. 6, pp. 616–626, 2012.

[4] J. P. Dos Anjos and J. B. De Andrade, "Determination of nineteen pesticides residues (organophosphates, organochlorine, pyrethroids, carbamate, thiocarbamate and strobilurin) in coconut water by SDME/GC-MS," *Microchemical Journal*, vol. 112, pp. 119–126, 2014.

[5] M. Tankiewicz, C. Morrison, and M. Biziuk, "Multi-residue method for the determination of 16 recently used pesticides from various chemical groups in aqueous samples by using DI-SPME coupled with GC–MS," *Talanta*, vol. 107, pp. 1–10, 2013.

[6] X. Li, Z. Zhang, P. Li, Q. Zhang, W. Zhang, and X. Ding, "Determination for major chemical contaminants in tea (Camellia sinensis) matrices: A review," *Food Research International*, vol. 53, no. 2, pp. 649–658, 2013.

[7] M. Gupta and A. Shanker, "Persistence of acetamiprid in tea and its transfer from made tea to infusion," *Food Chemistry*, vol. 111, no. 4, pp. 805–810, 2008.

[8] Q. Zhou, J. Xiao, and W. Wang, "Using multi-walled carbon nanotubes as solid phase extraction adsorbents to determine dichlorodiphenyltrichloroethane and its metabolites at trace level in water samples by high performance liquid chromatography with UV detection," *Journal of Chromatography A*, vol. 1125, no. 2, pp. 152–158, 2006.

[9] G. Min, S. Wang, H. Zhu, G. Fang, and Y. Zhang, "Multi-walled carbon nanotubes as solid-phase extraction adsorbents for determination of atrazine and its principal metabolites in water and soil samples by gas chromatography-mass spectrometry," *Science of the Total Environment*, vol. 396, no. 1, pp. 79–85, 2008.

[10] P. Liu, Y. Chen, C. Zhao, and L. Tian, "Determination of ten photoinitiators in fruit juices and tea beverages by solid-phase micro-extraction coupled with gas chromatography/mass spectrometry," *Chinese Journal of Chromatography*, vol. 31, no. 12, pp. 1232–1239, 2013.

[11] F. Guo, "Progress of Sample Pretreatment Techniques in Determination of Pesticide Residue in Tea," *Studies of Trace Elements and Health*, vol. 29, no. 1, pp. 61–64, 2012.

[12] Y. Wang, Y. Sun, Y. Gao et al., "Determination of five pyrethroids in tea drinks by dispersive solid phase extraction with polyaniline-coated magnetic particles," *Talanta*, vol. 119, pp. 268–275, 2014.

[13] L. Sun, M. Zhang, V. Natarajan, X. Yu, X. Zhang, and J. Zhan, "Au@Ag core-shell nanoparticles with a hidden internal reference promoted quantitative solid phase microextraction-surface enhanced Raman spectroscopy detection," *RSC Advances*, vol. 7, no. 38, pp. 23866–23874, 2017.

[14] F. Wang, J. Zheng, J. Qiu et al., "In Situ Hydrothermally Grown TiO2@ C CoreShell Nanowire Coating for Highly Sensitive Solid Phase Microextraction of Polycyclic Aromatic Hydrocarbons," *ACS applied materials interfaces*, vol. 9, no. 2, pp. 1840–1846, 2017.

[15] F. Wu, W. Lu, J. Chen, W. Liu, and L. Zhang, "Single-walled carbon nanotubes coated fibers for solid-phase microextraction and gas chromatography-mass spectrometric determination of pesticides in Tea samples," *Talanta*, vol. 82, no. 3, pp. 1038–1043, 2010.

[16] G. Wu, X. Bao, S. Zhao, J. Wu, A. Han, and Q. Ye, "Analysis of multi-pesticide residues in the foods of animal origin by GC-MS coupled with accelerated solvent extraction and gel permeation chromatography cleanup," *Food Chemistry*, vol. 126, no. 2, pp. 646–654, 2011.

[17] Y. Li, "SPE-gas chromatography/mass spectrometry determination of pesticide residues in tea," *Chinese Journal of Health Laboratory Technology*, vol. 20, no. 6, pp. 1271–1274, 2010.

[18] J. Lü, J. Liu, Y. Wei et al., "Preparation of single-walled carbon nanotube fiber coating for solid-phase microextraction of organochlorine pesticides in lake water and wastewater," *Journal of Separation Science*, vol. 30, no. 13, pp. 2138–2143, 2007.

[19] F. Wu, J. Chen, W. Lu, Z. Lin, and L. Zhang, "Solid-phase microextraction using poly (1-hexadecene-co-trimethylolpropane trimethacrylate) coating for analysis of pesticides in tea samples by high-performance liquid chromatography," *Current Analytical Chemistry*, vol. 7, no. 4, pp. 341–348, 2011.

[20] L. Chen, J. Zeng, C. Ma et al., "Preparation of multi-walled carbon nanotubes coated solid-phase microextraction fiber and its application on the analysis of polybromobiphenyls in seawater," *Scientia SinicaChimica*, vol. 39, no. 12, pp. 1652–1657, 2009.

[21] T. Fukushima, A. Kosaka, Y. Ishimura et al., "Molecular ordering of organic molten salts triggered by single-walled carbon nanotubes," *Science*, vol. 300, no. 5628, pp. 2072–2074, 2003.

[22] D.-Q. Yang, B. Hennequin, and E. Sacher, "XPS demonstration of π-π interaction between benzyl mercaptan and multiwalled carbon nanotubes and their use in the adhesion of Pt nanoparticles," *Chemistry of Materials*, vol. 18, no. 21, pp. 5033–5038, 2006.

[23] D. Louch, S. Motlagh, and J. Pawliszyn, "Dynamics of organic compound extraction from water using liquid-coated fused silica fibers," *Analytical Chemistry*, vol. 64, no. 10, pp. 1187–1199, 1992.

[24] J. Ai, "Solid Phase Microextraction for Quantitative Analysis in Nonequilibrium Situations," *Analytical Chemistry*, vol. 69, no. 6, pp. 1230–1236, 1997.

[25] L. Chen, "Determination of pyrethrins in tea brew by using solid phase microextraction," *Journal ofMinjiang University*, vol. 31, no. 5, pp. 103–107, 2010.

[26] L. Chen, W. Chen, C. Ma, D. Du, and X. Chen, "Electropolymerized multiwalled carbon nanotubes/polypyrrole fiber for solid-phase microextraction and its applications in the determination of pyrethroids," *Talanta*, vol. 84, no. 1, pp. 104–108, 2011.

[27] W. B. Kang, "GC determination of organophosphorus, organochlorine and pyrethroid pesticide residues in tea drinker," *Fujian Analysis Testing*, vol. 16, no. 4, pp. 47–49, 2007.

18

Performance Evaluation of STARPAM Polymer and Application in High Temperature and Salinity Reservoir

Chengli Zhang,[1] **Peng Wang** ⓘ**,**[1] **and Guoliang Song** ⓘ[2]

[1]College of Petroleum Engineering, Northeast Petroleum University, Daqing, Heilongjiang 163318, China
[2]College of Mathematics and Statistics, Northeast Petroleum University, Daqing, Heilongjiang 163318, China

Correspondence should be addressed to Peng Wang; 2537298882@qq.com

Academic Editor: Valentina Venuti

Based on the properties of high temperature and salinity reservoir, the water-soluble polymer with good heat resistance and salt tolerance can be obtained through copolymerization between 2-acrylamide-2-methyl sulfonate monomer (AMPSN) and acrylamide monomer (AM) in water. The star shaped stable complexes (STARPAM) with the star nucleus of β-CD are prepared by living radical polymerization, which can improve the viscosity and change the percolation characteristics of the polymer in porous media. In the article, the performance of the STARPAM (star-shaped polymer) with heat resistance and salt tolerance was evaluated by comparing the viscosification property, heat and salt resistance, calcium and magnesium tolerance, and long-term thermal stability of STARPAM (star-shaped polymer) with those of HPAM (partially hydrolyzed polyacrylamide) and MO-4000 (linear polymer). The results of physical simulation experiment showed that the viscosity of the STARPAM is 3.3 times that of MO-4000 and 4 times that of HPAM under the conditions of mineralization degree of 20000 mg/L, concentration of 1500 mg/L, and 75°C, which indicated that heat resistance and salt tolerance of the STARPAM are excellent. Oil displacement experiments showed that STARPAM can enhance oil recovery by 20.53% after water flooding, and the effect of oil displacement is excellent. At present, 19 wells were effective with a ratio of 95.2%. Compared with before treatment, the daily liquid production increased by 136 m^3, daily oil production increased by 44.6 t, water cut decreased by 4.67 percentage points, and flow pressure decreased by 1.15 MPa.

1. Introduction

The heterogeneity of the reservoir and the unfavorable mobility ratio are two important factors that affect the sweep efficiency and oil recovery of the water flooding. In the middle and later stage of water flooding, there is a problem of development that the injected water moves along the high permeability layers, while the utilization degree of the low permeability layers reduced, which could result in low recovery. HPAM is a linear water-soluble polymer, which is one of the most widely used water-soluble polymers. At present, it is widely used in oil field for tertiary recovery. The practice of field development has proved that polymer flooding is an effective method of improving oil recovery (EOR) and has become an important part of oil production in middle and later stage [1–3]. In actual development, it is found that, with the increase of temperature and salinity of the reservoir (Table 1), the electrical properties of the sodium

carboxyl group in HPAM molecules are shielded, and the HPAM molecules are curly so that the ability of increasing viscosity declines [4–6]. When the content of Ca^{2+} and Mg^{2+} is higher, and the degree of hydrolysis of polyacrylamide is more than 40%, molecules of HPAM will combine with Ca^{2+}, Mg^{2+}, and other polyvalent ions, resulting in flocculation and sedimentation. The stability of the polymer is very important due to the long period of tertiary oil recovery. Therefore, hydrolysis degree of the polymer molecules used for tertiary recovery must be less than 40%. Only in this way can the polymer have the characteristics of heat resistance and salt tolerance in oil field application. However, the hydrolysis reaction of an amide group in HPAM is very rapid under acid and alkaline conditions, and the rate of hydrolysis under neutral conditions is accelerated with the increase of temperature, which makes HPAM do not have the characteristics of heat resistance and salt tolerance [7–9].

TABLE 1: Standard of oil layer division in an oil field.

Category	Screening criteria		
	Original formation temperature / °C	Formation water salinity / mg/L	Ca^{2+} and Mg^{2+} content / mg/L
Class I	≤70	≤10000	≤200
Class II	70-80	10000-30000	200-400
Class III	80-93	30000-1000000	≥400
Class IV	Oil layer is sanding seriously and has poor connectivity of the oil layer		

The polymers with heat resistance and salt tolerance for tertiary oil recovery have been developed at home and abroad, including HPAM of super-high molecular weight, amphoteric ion polymer, monomer copolymer, hydrophobically associating polymer, multiple composite polymer, comb polymer, and star-shaped polymer. By analyzing the mechanism of heat resistance and salt tolerance, it is considered that star-shaped polymers are the most promising, if the star-shaped polymers with function of heat resistance and salt tolerance are developed, which can provide theoretical basis and technical support for tertiary recovery to enhance oil recovery [10, 11].

Flory P J [12] put forward the concept of star-shaped polymer in 1948; unlike linear straight chain polymers, it is a type of polymer that several or more polymer chains can be produced from one fulcrum or nucleus. In 1950s, Flory P J [13] proposed the idea of synthesizing highly branched polymers with star structures with ABn monomers and predicted the parameter of relative molecular mass distribution of polymers and so on. In 1956, Morton M et al. [14, 15] synthesized four-arms star-shaped polymers with tetrachlorosilane as the core and straight chain polystyrene as the arm. Because of their branching structure and dispersion, this kind of chemical agent has special properties and functions, which has become a research focus of star-shaped polymers. In 1985, Tomalia DA and Newkome XS et al. [16] published research achievements on the dendritic supramolecules synthesized by diffusion outward from a core, which opened a new field for the research of star-shaped polymers. At present, the commonly used polymerization methods for preparing star-shaped polymers include atom transfer radical polymerization, reversible addition fragmentation chain transfer polymerization, and ring opening polymerization [17, 18]. Wei Ding [19] and Ying Sun et al. [20] synthesized star-shaped polyacrylamide through the methods of "COREFIRST" and single electron transfer radical polymerization. Fuxiao Wang [21] synthesized a series of star-shaped hydrophobically associating polyacrylamide with different content of ODAC by photoinitiated radical polymerization.

β-CD is a kind of natural macromolecules linked by glucan; large amounts of hydroxyl at both ends can be directly modified and used as macromolecular initiators [22]. The star-shaped polymer with core of β-CD has more abundant properties and applications than ordinary cross-linked star-shaped polymers. The molecular structure of β-CD includes polyhydroxy and hydrophobic cavities, and the properties of polyhydroxyl groups determine that β-CD can be used as nuclei of the star-shaped polymers. What

is more important is that the hydrophobic cavities of β-CD have supramolecular inclusion for a wide range of guest molecules, which can change the seepage characteristics of the star-shaped polymers in reservoirs [22–24]. Through the active free radical polymerization, a stable star-shaped complex with β-CD as the core is formed, which can improve the viscosity and change the seepage characteristics of the polymer in porous media. The existence of multifunctional monomers in STARPAM polymer ensures that the hydrolysis of star-shaped polymer can be limited under the conditions of high temperature and high salinity, and the star-shaped structure can increase the structural regularity of polymer molecular cluster and be with temperature and salt tolerance characteristics.

2. Mechanism of Synthesis of a New Type of Heat Resistance and Salt Tolerance Polymer (STARPAM)

2.1. The Method of Synthesis of STARPAM.
The water-soluble polymer with good characteristics of heat resistance and salt tolerance can be obtained by aqueous copolymerization of 2-acrylamide-2-methyl sulfonate monomer (AMPSN) and acrylamide monomer (AM). β-CD is modified as a seven-membered functional initiator by 2,2,6,6-tetramethylpiperidinyloxy-TEMPO [25]. The stable star-shaped complex (STARPAM) with the nucleus of β-CD (β-cyclodextrin) is formed through TEMPO-mediated reactive radical polymerization; on this basis, the viscosity of polymer is increased and the percolation characteristics of the polymer in porous media are changed, as shown in Figure 1.

2.2. Steps of the Synthesis of STARPAM.
(1) The quantitative 2-acrylamide-2-methylsulfonate monomer (AMPSN) was added to wide-mouth bottle, which was dissolved with a moderate amount of deionized water, neutralizing with sodium hydroxide in the ice water bath until to pH of 7~8, and slowly stirring to completely dissolve.

(2) The PAM was precipitated with anhydrous alcohol for several times and then dried in vacuum. A certain amount of PAM was accurately weighed and dissolved in appropriate amount of deionized water, when the solution reached the expected value and was placed in a wide-mouth bottle.

(3) Input the nitrogen after the above solution was sealed, and the weighed sodium bisulfite and potassium persulfate solution are added in this order using a pipette

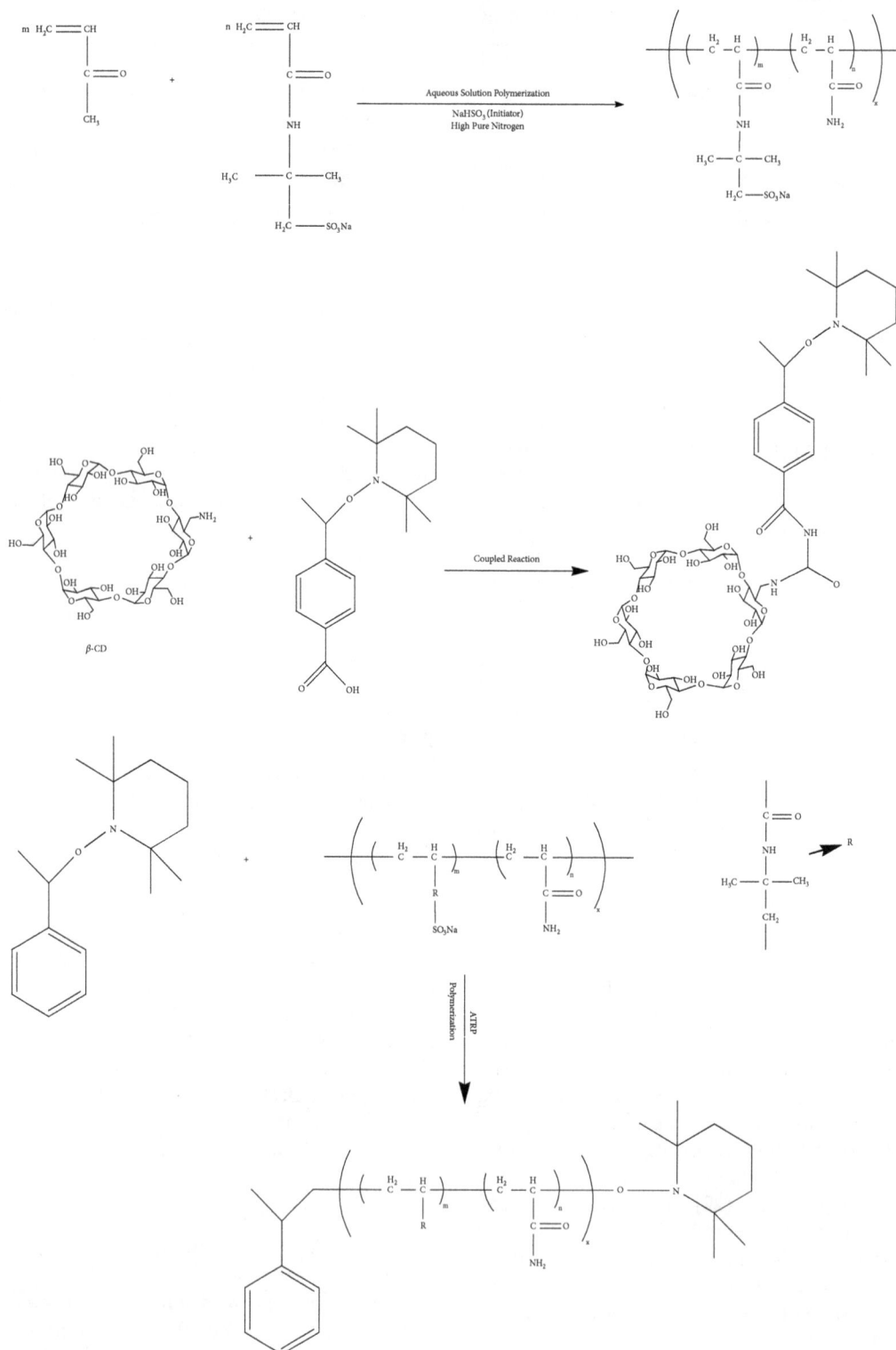

FIGURE 1: Synthesis of STARPAM with a nucleation of β-CD.

(respectively, prepare the deionized water solution of a certain concentration in advance).

(4) Sealing and the nitrogen were continued input, and then the wide-mouth bottle was put in the water bath with constant temperature, and the reaction time is specified. After

the end of the reaction, the product of colloidal sulfonated polyacrylamide was obtained.

(5) 2,2,6,6-Tetramethylpiperidinyloxy benzoic acid (3.4 g, 9.4 mmol) was coupled with β-CD (β-cyclodextrin 8.6 g, 7.0 mmol) using dicyclohexylcarbodiimide (DCC 2.1 g,

TABLE 2: Comparison of adding viscosity.

Concentration /(mg/L)	Apparent viscosity/(mPa·s)		
	STARPAM	MO-4000	HPAM
250	2.60	0.43	1.30
500	5.63	1.73	1.73
750	12.12	3.89	2.60
1000	21.20	7.36	5.19
1250	32.45	9.95	7.36
1500	45.00	13.85	11.25
1750	58.41	21.20	15.58
2000	75.00	29.40	21.63

9.5 mmol) in DMF (dimethylformamide 100 mL). To couple the carboxylic acid exhaustively with the amino groups, the reaction was performed in the presence of N-hydroxybenzotriazole (HBT 2.0 g, 12.1 mmol) and triethylamine (Et3N 1.01 g, 9.0 mmol).

(6) Initiator (2.2 g, 0.70 μmol) was dissolved in sulfonate polyacrylamide (15 g, 120 mmol). Oxygen was removed from the solution by freezing in liquid nitrogen, evacuating the flask, warming to room temperature, and flushing the flask with argon gas. This procedure was repeated three times. The mixture was then stirred at 120°C for 6 h. After cooling in liquid nitrogen, the mixture was diluted with chloroform (25 mL) and then poured into methanol (1.5 L). The precipitate was filtered off and purified by reprecipitation with chloroform-methanol and dried in vacuo to give polymer as a white powder.

3. Heat Resistance and Salt Tolerance Performance Evaluation of STARPAM

3.1. Experimental Equipment and Reagents. (1) SNB-2 Digital Viscometer (Shanghai Jingke day U.S. Trade Co. Ltd.); speed is 6 r/min.

(2) HPAM, relative molecular mass 12 million, Karamay Xinke chemical (Group) Co., Ltd.; the degree of hydrolysis is 22.3%.

(3) MO-4000, relative molecular mass > 20 million, Karamay Xinke chemical (Group) Co., Ltd.; the degree of hydrolysis is 22.8%.

(4) STARPAM, relative molecular mass > 25 million, the degree of hydrolysis is 32.5%, and the content of β-CD (mass fraction) is 0.065%.

3.2. The Performance of Viscosity Increasing and Viscosity-Temperature. At 75°C, the concentration of polymers was changed with 250 mg/L as the concentration step, and the effect of solution concentration on the viscosity of polymers was observed. The results were shown in Table 2 and Figure 2.

It can be seen from Table 2 and Figure 2 that the viscosity of polymer solution increases with the increase of the concentration. When the concentration reaches 1500 mg/L, the viscosity of STARPAM solution is 3.3 times of HPAM polymer solution, and the performance of viscosity increasing is obviously superior to the other two linear polymers.

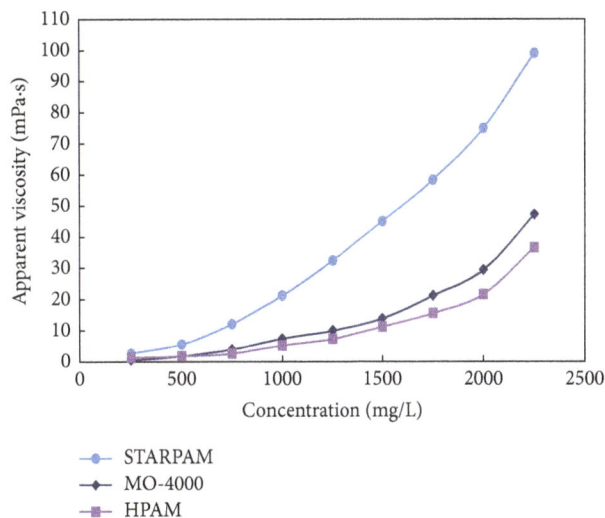

FIGURE 2: Comparison of adding viscosity.

When reaching the same viscosity, the content of STARPAM polymer is lower, which can reduce the cost of oil production. The STARPAM contains more than 7 long chain arms, with large viscosity and molecular weight of that is more than 25 million, which can increase the volume of the hydrodynamics of the molecular chain. The viscosity of the polymer solution is relatively higher under the same polymer concentration.

When the concentration of polymer solution remained of 1500 mg/L and the salinity remained of 20000 mg/L, changing the temperature on the basis of the temperature step with 10°C, the effect of temperature on the viscosity of polymer solution was observed. The results were shown in Table 3 and Figure 3.

It can be seen from Table 3 and Figure 3 that the viscosity of polymer solution gradually decreases with the increase of temperature, and the heat resistance of STARPAM polymer solution is the best, which indicated that the displacement viscosity of the leading edge of the series of polymer is high. The main chains of polymer molecules changing into stars can effectively increase the rigidity of molecular chain and the regularity of molecular structure, which makes the polymer molecular chain crimp difficult and the hydraulic radius

TABLE 3: Comparison of heat resistance.

Temperature /(°C)	Apparent viscosity/(mPa·s)		
	STARPAM	MO-4000	HPAM
30	61.62	25.00	22.08
40	57.01	23.00	20.00
50	53.23	20.42	16.67
60	50.00	18.33	14.58
70	48.30	16.67	12.08
80	44.85	13.01	9.40
90	37.31	11.60	8.75

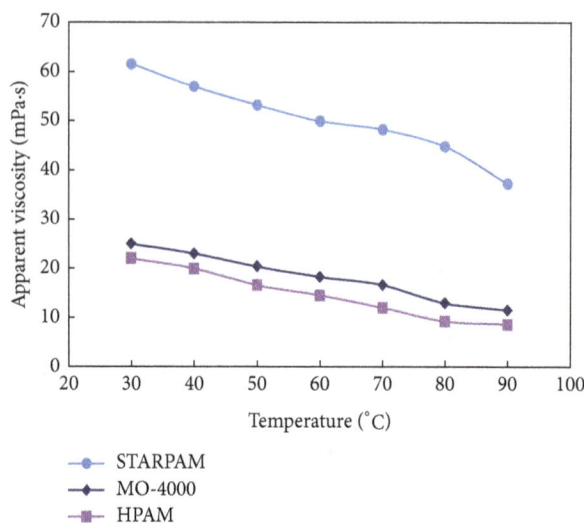

FIGURE 3: Comparison of heat resistance.

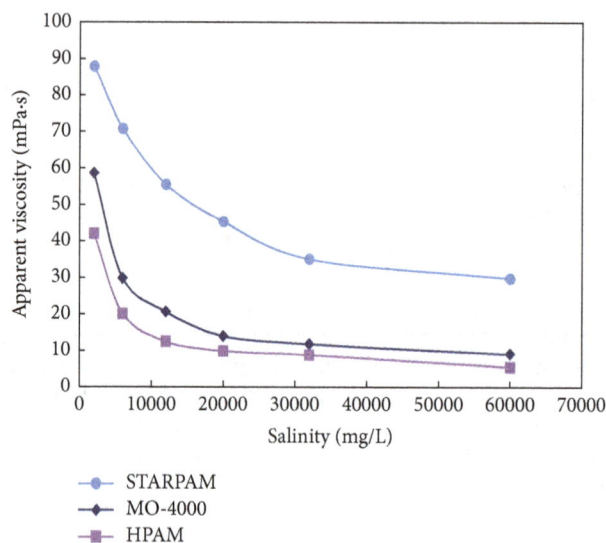

FIGURE 4: Comparison of the influence of polymers viscosity on the salinity.

of molecular chain rotation increase. Therefore, viscosity of STARPAM is high under high temperature.

The adding of sulfonic acid groups on polyacrylamide molecules can enhance the polarity of molecules and form hydrogen bonds between macromolecules, and the long chain of carbons in the group also can enhance the rigidity of the molecule; thus the heat resistance is increased.

3.3. The Performance of Ca^{2+}, Mg^{2+}, and Salt Tolerance. Three kinds of polymer solutions with different mineralization degree were prepared respectively, and the effects of different salinity on the viscosity of polymer solution were observed to verify the salt resistance. The results were shown in Table 4 and Figure 4.

It can be seen from Table 4 and Figure 4 that the viscosity of polymer solution decreases with the increase of salinity. When the salinity is less than 12000 mg/L, the viscosity of STARPAM solution is slower than other two solutions, and the salt sensitivity is relatively lower. When the salinity is more than 12000 mg/L, the viscosity of STARPAM solution is higher, which indicated that the viscosity of the three polymer solutions decreases slowly, and the performance of salt resistance of STARPAM polymer is well.

The polymer solutions with different concentration of Ca^{2+} were prepared. The effect of Ca^{2+} concentration on the viscosity of polymer solutions was observed. The results were shown in Table 5 and Figure 5.

As seen from Table 5 and Figure 5, with the increase of Ca^{2+} concentration in polymer solutions, the viscosity of polymer solutions gradually decreases, and the performance of Ca^{2+} resistance of STARPAM polymer solution is the best. After introducing the strong anion groups in STARPAM chain, the structure contains strong anionic, water-soluble sulfonic groups, screened amido groups, and unsaturated double bond, so that it has excellent performance. The sulfonic groups can effectively inhibit the hydrolysis of the amide groups and has a good tolerance to the two valence cations, which will not react with the two valence ions to produce precipitation and enhance the salt resistance.

3.4. The Performance of Stability. The relationship between the apparent viscosity of the three polymers and the aging time was recorded under the conditions of high purity nitrogen and vacuum deoxidization at 75°C, seen as Table 6

TABLE 4: Comparison of the influence of polymers viscosity on the salinity.

Salinity /(mg/ L)	Apparent viscosity /(mPa·s)		
	STARPAM	MO-4000	HPAM
2000	88.00	58.72	42.01
6000	70.84	29.83	20.20
12000	55.60	20.66	12.56
20000	45.41	14.02	10.00
32000	35.17	11.88	8.90
60000	29.89	9.13	5.59

TABLE 5: Comparison of the influence of polymers viscosity on the Ca^{2+}.

Ca^{2+} concentration /(mg/L)	Apparent viscosity /(mPa·s)		
	STARPAM	MO-4000	HPAM
0	45.06	13.66	11.05
200	31.74	11.39	8.77
400	27.14	10.28	7.37
600	24.29	9.17	6.55
800	21.72	8.35	5.74
1000	19.74	7.24	5.21
1200	45.06	13.66	11.05

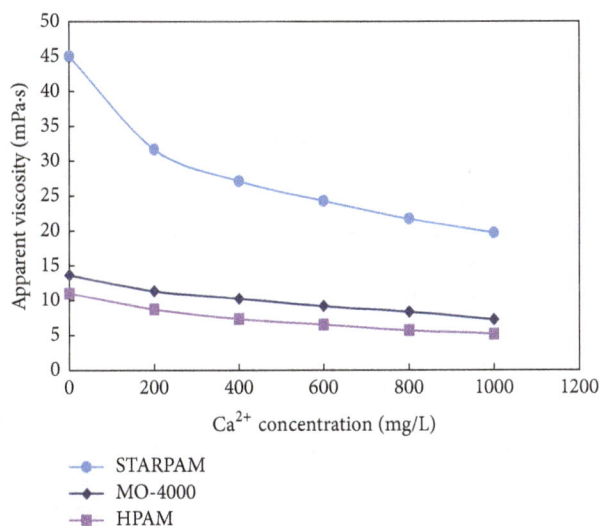

FIGURE 5: Comparison of the influence of polymers viscosity on the Ca^{2+}.

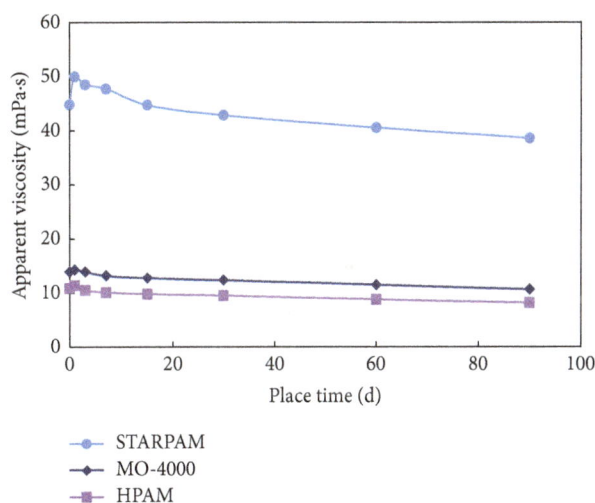

FIGURE 6: The relationship between three kinds of polymer viscosity and aging time (75°C, high purity nitrogen and vacuum deoxidization).

and Figure 6, and the retention rate of viscosity was calculated as shown in Table 7 and Figure 7.

The viscosity of the three kinds of polymer solution decreases with the increase of aging time. It is because that polymer chain breaks down and degrades due to high temperature aging, and the interaction between polymer molecules is weakened and the viscosity is reduced. The shielding effect of salt makes the STARPAM polymer chain curl, which effectively protects the main chain when aging at high temperature, thus slowing down the degradation.

The viscosity retention rate of the three polymers is over 90% in the first 30 days. The viscosity retention rate of

STARPAM polymer solution is 7.4% higher than HPAM polymer and is 9.2% higher than MO-4000 polymer, indicating that the stability of STARPAM polymer solution is better.

4. Percolation Characteristics Evaluation of STARPAM

The core that gas measurement permeability 1450~1500 × $10^{-3} \mu m^2$, length 5.00~8.00 cm, and diameter 2.50 cm was selected. According to "standard Q/ SDY 1119-2003 implementation rule 6.19", under the conditions of the salinity

TABLE 6: The relationship between three kinds of polymer viscosity and aging time.

Place time /(d)	Apparent viscosity /(mPa·s)		
	STARPAM	MO-4000	HPAM
0	44.91	13.99	10.98
1	50.06	14.35	11.52
3	48.58	13.98	10.59
7	47.83	13.23	10.20
15	44.85	12.83	9.90
30	42.96	12.40	9.60
60	40.64	11.56	8.90

TABLE 7: The relationship between viscosity retention rate and aging time of three polymers.

Place time /(d)	Viscosity retention rate /(%)		
	STARPAM	MO-4000	HPAM
0	100.00	100.00	100.00
1	111.47	102.62	104.96
3	108.18	99.92	96.49
7	106.50	94.56	92.93
15	99.88	91.69	90.20
30	95.65	88.68	87.47
60	90.49	82.63	81.09

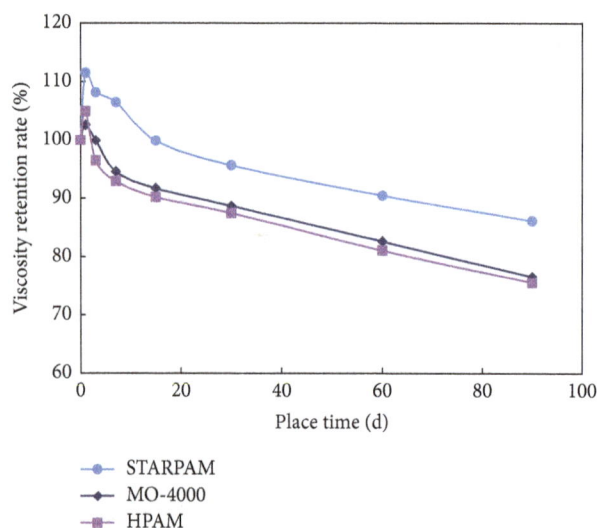

FIGURE 7: The relationship between viscosity retention rate and aging time of three polymers (75°C, high purity nitrogen and vacuum deoxidization).

of compound salt water 19334 mg/L, the concentration of polymer, respectively, 800 mg/L, 1000 mg/L, and 1500 mg/L, temperature of 75°C, and injection rate of 0.50 mL/min, the resistance coefficient and residual resistance coefficient of the new polymer samples were measured.

Experimental steps are as follows: (1) injecting 19334 mg/L compound salt water until the pressure is stable; (2) injecting new polymer solution to flood oil until the pressure is stable (the new polymer solution was filtered by a sand core funnel of G2 type before injection); (3) injecting 19334 mg/L

compound salt water until stable pressure. The experimental process is shown in Figure 8 and data are presented in Table 8 and Figures 9 and 10.

From Table 8 and Figures 9-10, it is shown that the resistance coefficient and the residual resistance coefficient increase with the increase of the concentration of the polymer solution for the same molecular weight, and whatever the core and the concentration, which of STARPAM polymers are higher than those of other two kinds of polymers.

The resistance coefficient and residual resistance coefficient are generated through the capture and retention of polymer solution flowing in porous media. The trapped polymer molecules are strongly resistance to water and relatively weakly resistance to oil. Therefore, the original percolation characteristics and flow channels of the reservoir will be greatly changed, and the permeability will be reduced.

5. Oil Displacement Performance Evaluation of STARPAM

(1) Cemented core: permeability of core is $1450 \sim 1500 \times 10^{-3} um^2$.

(2) Simulated oil: the mixture of oil and kerosene in SL oil field with the ratio of 5:1, the viscosity is 24.2 mPa·s at 75°C, and the viscosity of crude oil is 97.7 mPa·s at 75°C.

(3) Experimental steps: ① measuring permeability of core with water and saturating simulated oil; ② water flooding to water cut of 98%; ③ polymer flooding (0.2 PV); ④ following up water flooding (water cut 98%). The experimental data are presented in Table 9 and Figure 11.

The effect of three kinds of polymer flooding is shown in Table 9. The oil displacement efficiency of STARPAM

TABLE 8: Resistance coefficient and residual resistance coefficient of three polymers.

| Sample | Core parameters | | | | | | | R_F | R_K | C / mg/L |
	Number	L / cm	R / cm	V / cm^3	ϕ / %	K / $10^{-3} \mu m^2$	μ / mPa·s			
STARPAM	S-1	5.92	2.50	8.25	28.39	1476	18.6	22.35	2.36	800
STARPAM	S-2	5.99	2.51	8.33	28.16	1506	22.6	29.36	3.28	1000
STARPAM	S-3	5.92	2.50	8.19	27.89	1466	45.2	38.56	4.67	1500
MO-4000	S-1	5.92	2.50	8.25	28.39	1476	9.6	12.32	1.45	800
MO-4000	S-2	5.99	2.51	8.33	28.16	1506	10.3	16.23	2.23	1000
MO-4000	S-3	5.92	2.50	8.19	27.89	1466	14.8	25.60	3.60	1500
HAMP	S-1	5.92	2.50	8.25	28.39	1476	8.2	5.66	0.89	800
HAMP	S-2	5.99	2.51	8.33	28.16	1506	9.6	12.32	1.22	1000
HAMP	S-3	5.92	2.50	8.19	27.89	1466	12.6	20.13	2.63	1500

Remarks: L: length, R: diameter, V: pore volume, ϕ: porosity, K: permeability, μ: viscosity, R_F: resistance coefficient, R_K: residual resistance coefficient, and C: concentration.

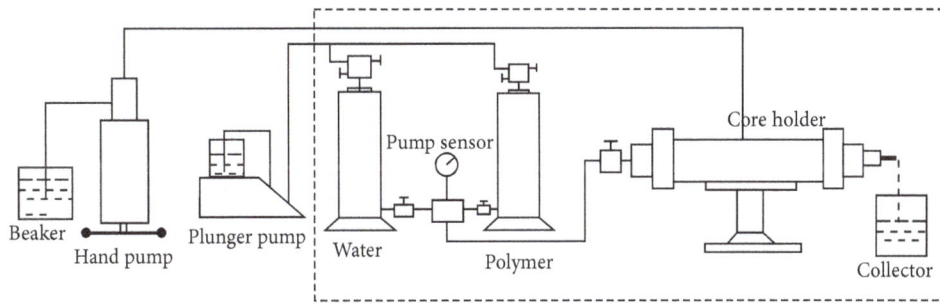

FIGURE 8: The flow characteristics evaluation process.

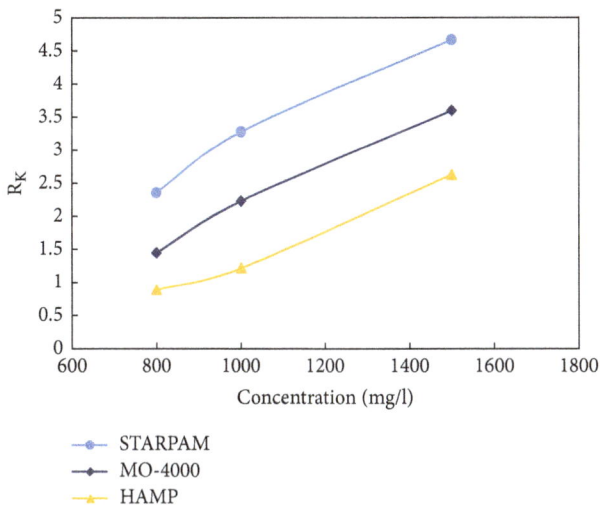

FIGURE 9: Resistance coefficient of three polymers.

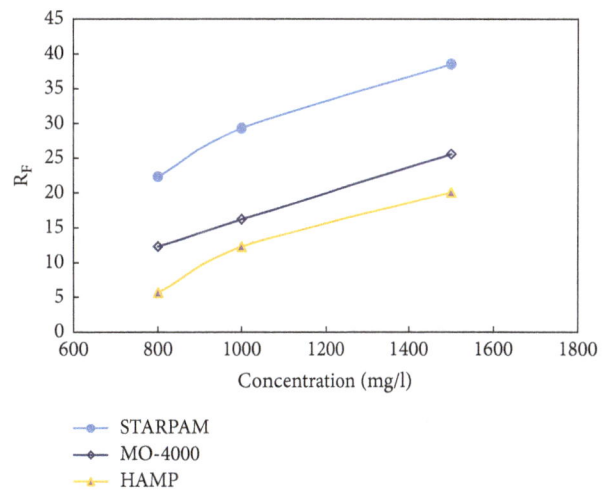

FIGURE 10: Residual resistance coefficient of three polymers.

polymer is relatively higher, and the recovery rate is 20.53%, the other two kinds of polymer flooding efficiency are lower, and the recovery rate is 17.13% and 12.09%, respectively. From Figure 11, we can see that the oil recovery at the end stage of water flooding is almost constant, but increasing after polymer flooding, of which the STARPAM polymer is obviously the best. The efficiency of polymer flooding depends on the difference of viscosity. According to the

analysis of the performance of the three kinds of polymer, the viscosity of STARPAM polymer is the highest, so that the effect of oil displacement is the best.

It can be seen from the experimental results that, for every kind of polymers, the injection pressure increases with the increase of polymer injection volume and then be stability. Although the molecular weight of star-shaped polymers is much higher than that of the others, the injection pressure is similar to that of MO-4000 linear polymer, and the

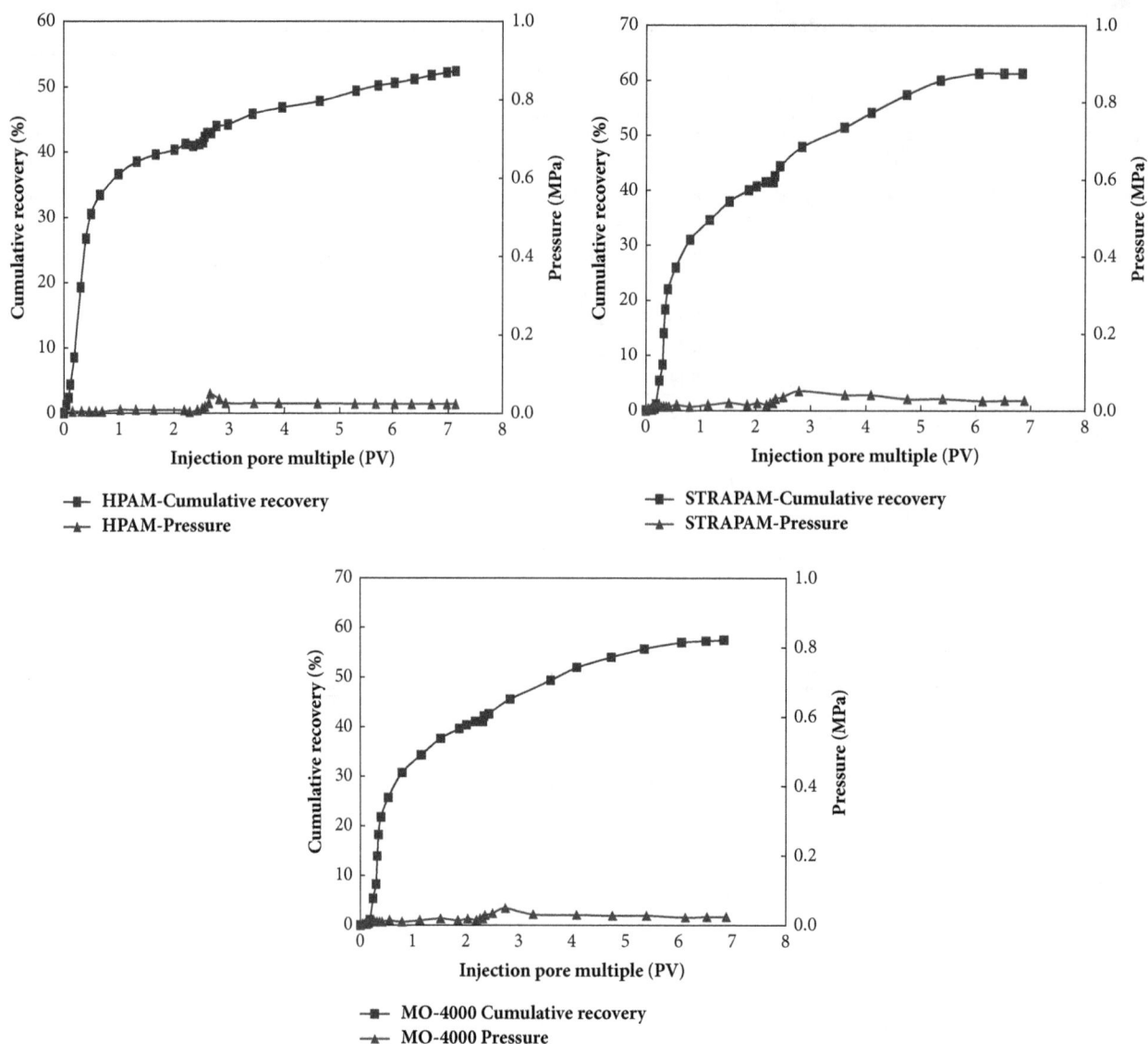

FIGURE 11: Polymer flooding curve.

injection ability is good. It is because the conical structure of star nuclear β-CD is hydrophilic to the outer cavity and hydrophobic to the inner cavity. The hydrogen bond formed between the hydroxyl group on the β-CD and the water molecules makes the β-CD water-soluble. Unique cavity structure of β-CD can make matching between subject and object. Through action of noncovalent bonding, a stable complex with hydrophobicity, certain shape, and suitable size can be formed to affect the behavior of polymer solution and change its seepage characteristics in porous media.

6. Field Application

6.1. Block Overview. The area of SL-A block in SL high temperature and salinity oilfield is $0.48 \, \text{km}^2$, and the underground pore volume is $102.3 \times 10^4 \, \text{m}^3$. The geological reserve of target stratum is 65.2×10^4 t, the average shooting

sandstone thickness in single well is 12.32 m, the effective thickness is 10.14 m, and the average effective permeability is $230.6 \times 10^{-3} \, \mu\text{m}^2$. Five-point well pattern is adopted in A block, and there are 33 wells, including 13 injection wells, 20 production wells; the distance between injection and production wells is 120 m. The average temperature of the reservoir is 75°C, and the total salinity of formation water is 15167 mg/L. SL-A block is a typically reservoir with high temperature and salinity, but it has potential for development.

6.2. Analysis of Field Application Effect. In order to reduce formation temperature and ion interference, blank water flooding was conduct of SL-A block in January 2015. Star-shaped polymer injection system was adopted in the initial stage of polymer flooding in August 2016, and the speed of the high and low concentration of polymer injection was kept at 0.18 ~ 0.19 PV/a; the injection speed remains around 0.20

Polymer Flooding
SLA-A-PRO SI-A.irf

FIGURE 12: SL-A block comprehensive water cut curve.

Polymer Flooding
SLA-Oil Rate of Field. irf

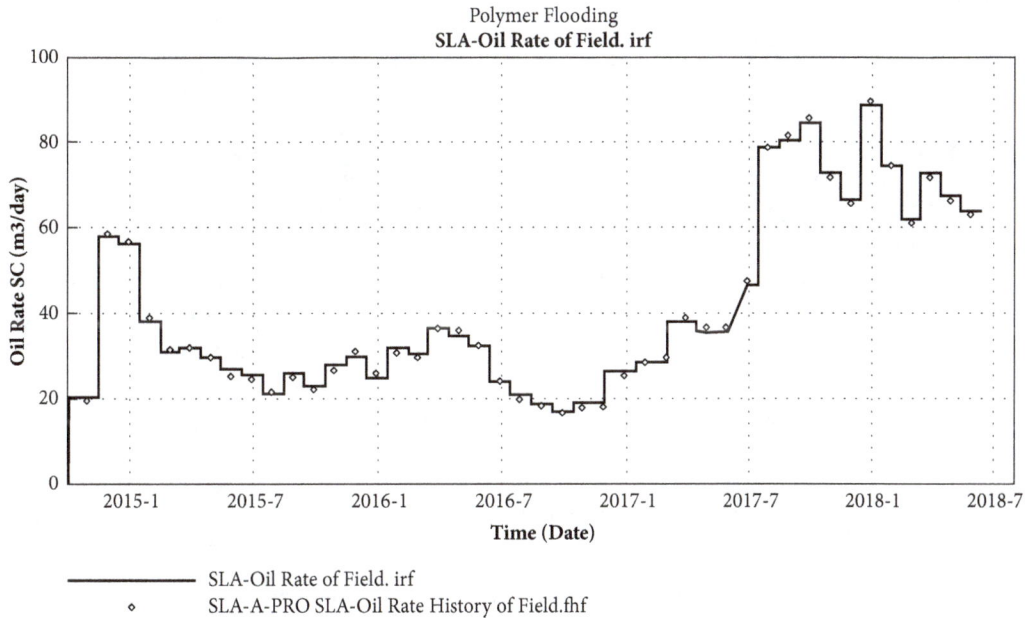

FIGURE 13: SL-A block daily oil production curve.

~ 0.22PV/a in the concentration reduction and acceleration stage, the average injection concentration was 1596 mg/L, and the total amount of polymer was 1050 mg/ L·PV.

Block SL-A entered the low-water-cut period in August 2017 and water cut reached the lowest value of 87.8% in November 2017. At present, 19 wells have been effective with a ratio of 95.2%. Compared with before, daily liquid production increased by 136 m^3, daily oil production increased by 44.6 t, water cut decreased by 4.67 percentage points, and flow pressure decreased by 1.15 MPa. The water cut and daily oil production for SL-I layers of SL-A block are shown as Figures

12 and 13. The polymer concentration and residual oil are presented in Figures 14 and 15. These two indicators are tested every six months, and the time points are October 2016, April 2017, October 2017, and April 2018, respectively.

As the polymer-slug gradually approaching the oil wells, the polymer concentration in the polymer injection wells in the SL-I layers became lower, and the polymer accumulated between the oil wells and the polymer injection wells, and the remaining oil was driven to the bottom of the oil wells along the main line. It is proved that the percolation performance of the STARPAM polymer is good and the sweeping range

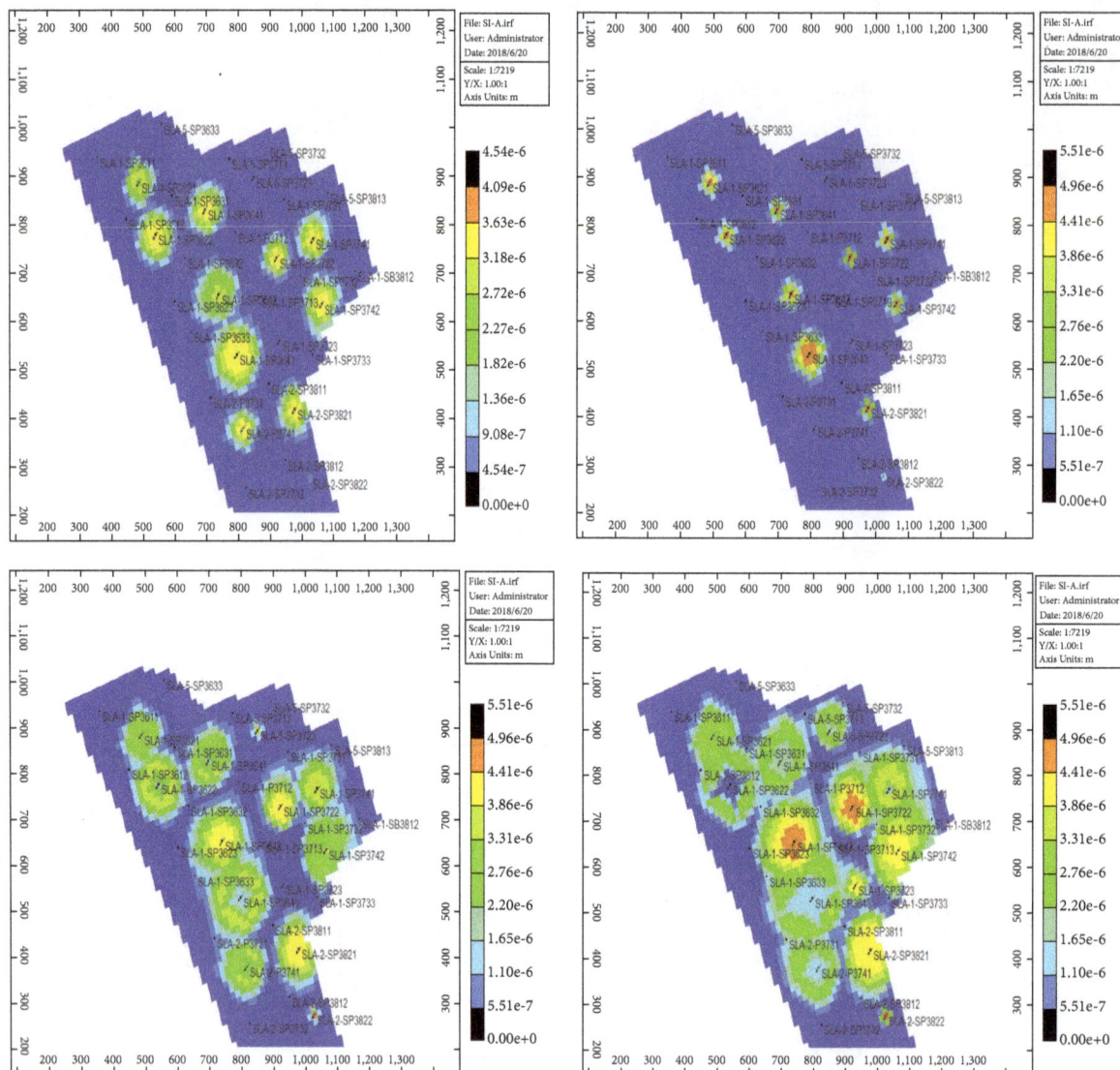

FIGURE 14: Distribution map of polymer concentration field in SL-I layer.

TABLE 9: Comparison of oil displacement effect.

K / $10^{-3} \mu m^2$	Sample	μ / mPa·s	S_O / %	R_W / %	Polymer flooding 1500 mg/L, 0.2PV	
					R_P / %	R_E / %
1050	STARPAM	45.1	67.64	40.77	61.30	20.53
1039	MO-4000	14.5	66.67	40.37	57.50	17.13
1145	HPAM	11.3	66.79	40.31	52.40	12.09

Remarks: K: permeability, μ: viscosity, S_O: oil saturation, R_W: recovery of water flooding, R_P: cumulative recovery, and R_E: enhanced oil recovery of polymer flooding.

is wide. It can be used in the high temperature and salinity reservoirs to enhance oil recovery.

7. Conclusions

(1) Under the condition of high temperature and salinity, the rupture of the molecular chain will reduce the polymer degradation viscosity; the main chains of polymer molecules changing into stars can effectively increase the rigidity of molecular chain and the regularity of molecular structure, which makes the of polymer molecular chain crimp difficult and the hydraulic radius of molecular chain rotation increase. Therefore, viscosity of STARPAM is high under high temperature.

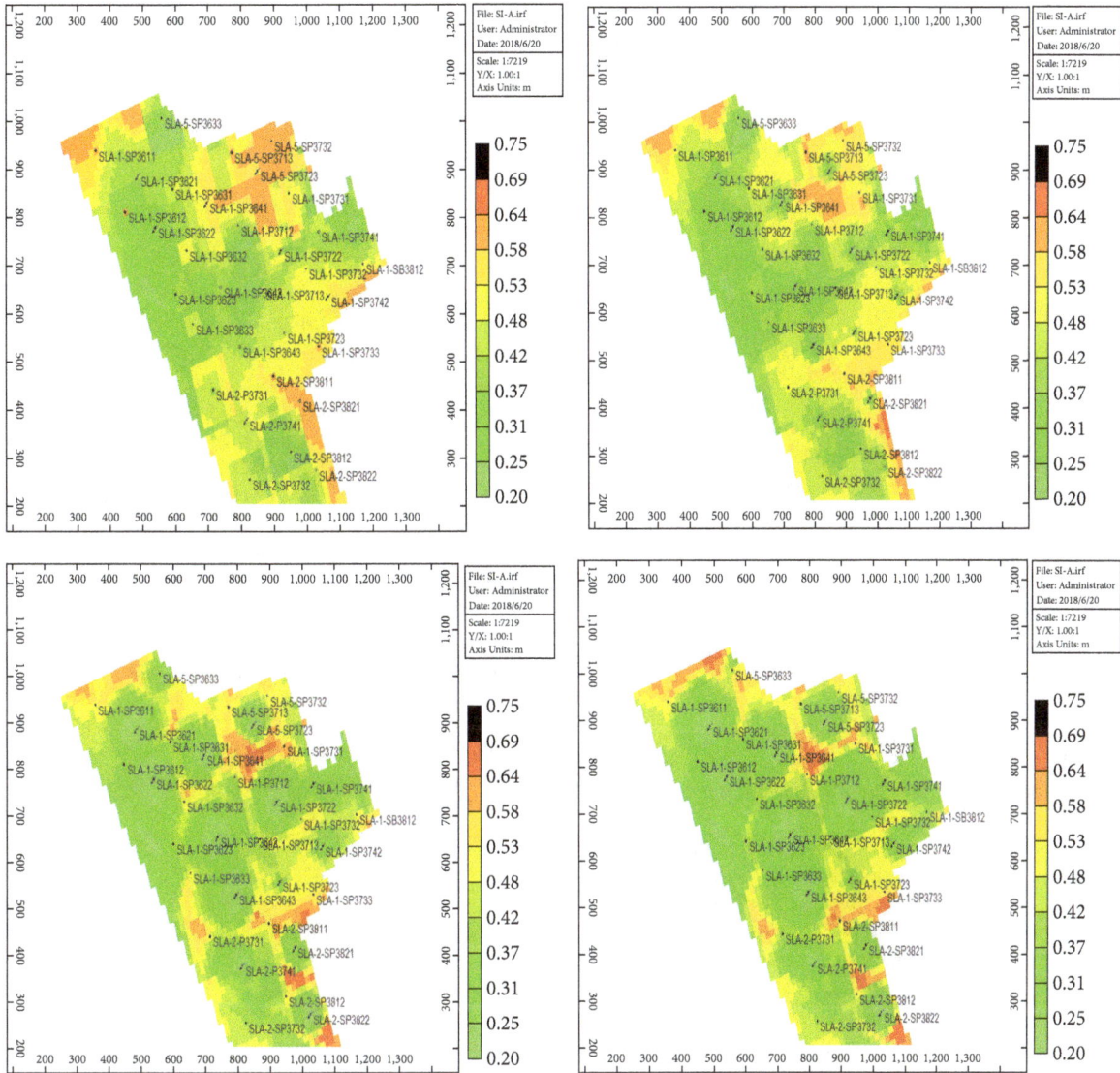

FIGURE 15: Distribution map of residual oil in SL-I layer.

(2) The stable star-shaped complex (STARPAM) with β-CD (β-cyclodextrin) as the star nucleus is formed by polymerization of active radical, which can affect the behavior of polymer solution and change its seepage characteristics in porous media.

(3) Through a series of experiments, it is proved that the viscosity of STARPAM polymer solution is relatively higher under the same polymer concentration, the oil displacement efficiency is the best and the oil recovery is higher after water flooding.

(4) STARPAM polymer solution was injected to block SL-A firstly in August 2016, the period of low water cut was appeared in August 2017, and the smallest value of water cut reached 87.8% in November 2017. At present, 19 wells have been effective with a ratio of 95.2%. Compared with before, daily liquid production increased by 136 m^3, daily oil production increased by 44.6 t, water cut decreased by 4.67 percentage points, and flow pressure decreased by 1.15 MPa.

Acknowledgments

This work is financially supported by the National Natural Science Foundation of China under Grant no. 51504069, China Postdoctoral Science Foundation under Grant no. 2017M621241, Heilongjiang Postdoctoral Science Foundation under Grant no. LBH-Z16039, and the Natural Science Foundation of Heilongjiang Province no. E2016014. These foundations provide us with many financial help, such as the cost of experimental materials, the layout of the articles, and so on.

References

[1] S. Dai, L. Wang, and H. Quan, "Synthesis and Solution Properties of Dendritic Side Based Polymers," *Fine Chemicals*, vol. 30, no. 6, pp. 691–695, 2013.

[2] J. J. Sheng, B. Leonhardt, and N. Azri, "Status of polymer-flooding technology," *Journal of Canadian Petroleum Technology*, vol. 54, no. 2, pp. 116–126, 2015.

[3] Lei. Zhang, *Study on Degradation and Stability of Polymer Flooding [D]*, Shandong University, 2011.

[4] L. Luowen, D. Han, L. Wei, L. Qingxia, and F. Jian, "Synthesis and Performance Evaluation of Anti Salinization Star Polymers," *Petroleum Exploration and Development*, vol. 37, no. 4, pp. 477–482, 2010.

[5] F.-G. Wang, J.-R. Hou, F.-L. Zhao, H.-D. Hao, and Z.-H. Lou, "Feasibility study of comb copolymer flooding in high temperature reservoir," *Oil-Field Chemistry*, vol. 31, no. 4, pp. 559–563, 2014.

[6] Y. Zhu, Z. Wang, K. Wu, Q. Hou, and H. Long, "Enhanced Oil Recovery by Chemical Flooding from the Biostromal Carbonate Reservoir," in *Proceedings of the SPE Enhanced Oil Recovery Conference*, Kuala Lumpur, Malaysia, 2013.

[7] Y. Zhu, M. Lei, and Z. Zhu, "Development and Performance of Salt-Resistant Polymers for Chemical Flooding," in *Proceedings of the SPE Middle East Oil & Gas Show and Conference*, Manama, Bahrain, 2015.

[8] Y. Wu, A. Mahmoudkhani, P. Watson, T. R. Fenderson, and M. Nair, "Development of New Polymers with Better Performance under Conditions of High Temperature and High Salinity," in *Proceedings of the SPE EOR Conference at Oil and Gas West Asia*, Muscat, Oman, 2012.

[9] D.-D. Yin, Y.-Q. Li, B. Chen et al., "Study on compatibility of polymer hydrodynamic size and pore throat size for Honggang reservoir," *International Journal of Polymer Science*, vol. 2014, Article ID 729426, 7 pages, 2014.

[10] Y. Li, J. Gao, D. Yin, J. Li, and H. Liu, "Study on the matching relationship between polymer hydrodynamic characteristic size and pore throat radius of target block s based on the microporous membrane filtration method," *Journal of Chemistry*, vol. 2014, Article ID 569126, 7 pages, 2014.

[11] R. Kumar, "A Review on Epoxy and Polyester Based Polymer Concrete and Exploration of Polyfurfuryl Alcohol as Polymer Concrete," *Journal of Polymers*, vol. 2016, Article ID 7249743, 13 pages, 2016.

[12] J. R. Schaefgen and P. J. Flory, "Synthesis of Multichain Polymers and Investigation of their Viscosities," *Journal of the American Chemical Society*, vol. 70, no. 8, pp. 2709–2718, 1948.

[13] P. J. Flory, " Molecular Size Distribution in Three Dimensional Polymers. VI. Branched Polymers Containing A—R—B ," *Journal of the American Chemical Society*, vol. 74, no. 11, pp. 2718–2723, 1952.

[14] M. Morton, T. E. Helminiak, S. D. Gadkary, and F. Bueche, "Preparation and properties of monodisperse branched polystyrene," *Journal of Polymer Science*, vol. 57, no. 165, pp. 471–482.

[15] D. A. Tomalia, H. Baker, J. Dewald et al., "A new class of polymers: starburst-dendric macromolecules," *Polymer Journal*, vol. 17, no. 1, pp. 117–132, 1985.

[16] X.-S. Feng and C.-Y. Pan, "Synthesis and characterization of star polymers initiated by hexafunctional discotic initiator through atom transfer radical polymerization," *Journal of Polymer Science Part A: Polymer Chemistry*, vol. 39, no. 13, pp. 2233–2243, 2001.

[17] Y. Zhao, Y. Chen, C. Chen, and F. Xi, "Synthesis of well-defined star polymers and star block copolymers from dendrimer initiators by atom transfer radical polymerization," *Polymer Journal*, vol. 46, no. 15, pp. 5808–5819, 2005.

[18] Y. H. Kim, W. T. Ford, and T. H. Mourey, "Branched poly(styrene-b-terf-butyl aerylate) and poly (styrene-b-acrylic acid) by ATRP from a dendritic poly(propylene imine)(NH2) 64 core," *Journal of Polymer Science Part A: Polymer Chemistry*, vol. 45, no. 20, pp. 4623–4634, 2007.

[19] W. Ding, Y. Sun, and C. Lv, "Single electron transfer active free radical polymerization in ionic liquids for the preparation of star polyacrylamide," *Applied Chemistry*, vol. 28, no. 10, pp. 1148–1154, 2011.

[20] Y. Sun, *Preparation of water-soluble polymer by single electron transfer living radical polymerization*, Northeast Petroleum University, Daqing, China, 2011.

[21] F. Wang, M. Duan, and S. Fang, "Study on the properties of star type hydrophobic associating polyacrylamide solution," *Petrochemicals*, vol. 39, no. 5, pp. 537–541, 2010.

[22] L. Zhang, C. Jing, J. Liu, and K. Nasir, "A Study on a Copolymer Gelant With High Temperature Resistance for Conformance Control," *Journal of Energy Resource Technology*, vol. 140, no. 3, p. 032907, 2018.

[23] J. Yubao, L. Xiangguo, L. Jinxiang, Y. Qin, W. Zijian, and G. song, "Transport capacity and mechanism of hydrophobically associating polymers," *Petroleum and Chemical Industry*, vol. 46, no. 5, pp. 600–607, 2017.

[24] R. Askarinezhad, D. G. Hatzignatiou, and A. Stavland, "Core-Based Evaluation of Associative Polymers as Enhanced Oil Recovery Agents in Oil-Wet Formations," *Journal of Energy Resource Technology*, vol. 140, no. 3, p. 032915, 2018.

[25] T. Kakuchi and A. Narum, "Glycoconjugated polymer.4. Synthesis and aggregation property of well-defined end-mnctionalized polysterene with β-cyclodextrin," *Mascromolecules*, vol. 36, pp. 3909–3913, 2003.

19

Rapid Screening of Volatile Organic Compounds from *Aframomum danielli* Seeds using Headspace Solid Phase Microextraction Coupled to Gas Chromatography Mass Spectrometry

Mosotho J. George [1,2] **Patrick B. Njobeh,**[3] **Sefater Gbashi,**[3] **Gabriel O. Adegoke,**[3,4] **Ian A. Dubery,**[2] **and Ntakadzeni E. Madala** [2]

[1] *Department of Chemistry and Chemical Technology, National University of Lesotho, Roma 180, Lesotho*
[2] *Department of Biochemistry, University of Johannesburg, P.O. Box 524, Auckland Park 2006, South Africa*
[3] *Department of Biotechnology and Food Technology, University of Johannesburg, Doornfontein Campus, P.O. Box 17011, Johannesburg 2028, South Africa*
[4] *Department of Food Technology, University of Ibadan, Ibadan, Nigeria*

Correspondence should be addressed to Mosotho J. George; jm.george@nul.ls

Academic Editor: Jan Åke Jönsson

Volatile organic compounds (VOCs) derived from plants have been used in the fragrance industry since time immemorial. Herein we report on the rapid screening of VOCs from seeds of ripe *Aframomum danielli* (family, Zingiberaceae) using a polydimethylsiloxane fibre headspace solid phase microextraction coupled to a gas chromatography mass spectrometry (SPME-GC/MS) instrument. Portions of 0.25, 0.35, and 0.50 g of ground sample were weighed and extraction of volatile organic compounds (VOCs) was achieved using a 100 μm polydimethylsiloxane solid phase microextraction (PDMS SPME) fibre, with the equilibrium time of 40 minutes and extraction temperature of 50°C; the following compounds with their respective relative abundances were obtained as the top ten most abundant and annotated ones using NIST, Wiley, and Fragrances Libraries: eucalyptol (58%); β-pinene (22%); α-pinene (7.5%); α-terpineol (4%), α-terpinyl acetate (2%); α-bergamotene (1%); pinocarveol (0.39%); α-copaene (0.35%); caryophyllene (0.34); and β-bisabolene (0.31%). These compounds have been reported elsewhere in the literature and listed in the Fragrances Library, incorporated into the Saturn QP2020 GCMS Solution® software used for their analysis.

1. Introduction

Plants have always been part of human life where they do serve as food source not only for humans but for animals as well. There are a number of benefits that can be derived from plants other than nutritional value. Many different plants have been used in traditional medicine since time immemorial. With the increasing frequency of degenerative diseases occurrence, wild plants that have been traditionally ignored are now receiving considerable attention owing to their potential in antioxidant activity and other medical benefits thereof. As such, there is much curiosity in understanding the phytochemical composition and chemical characteristics of these herbal plants.

Aframomum danielli, a plant that grows widely in West Africa, is an underutilized plant species known to contain an enormous variety of interesting phytochemicals [1–3]. There are a number of reports where the *Aframomum* species demonstrated some medicinal effects such as anticancer, antiplasmodial, antiulcer, antimicrobial [4], and antifungal [5]. This plant can also act as food preservative when added to packaged foods [6]. For example, it improved the postharvest storage shelf-life of tomato [7] and has been found to stabilise the refined peanut oil more effectively than the synthetic antioxidants such as butylated hydroxytoluene and α-tocopherol [1]. Although different parts of this plant (flower, leaf, stem, root, and seeds) have been investigated [2, 3, 8], the seeds in particular have demonstrated very potent pharmacoactive and sensory (flavour) properties [8–10]. The seeds of *A. danielli* are smooth, shiny, and olive brown in colour [11] and upon crushing produce a very strong aromatic smell that resembles eucalyptus leaves, which suggests an abundance of VOCs and essential oils such as those found in *Eucalyptus* trees. Essentially, the main chemical classes of VOCs produced by plants include terpenoids, benzenoids and phenylpropanoids, alkanes, alkenes, alcohols, esters, and various derivatives of fatty acids and amino acids [12, 13].

Essentially, VOCs from plant sources are widely used in the pharmaceutical, antiseptic, flavouring, fragrance, and other cosmetic and pharmacological industries, and their analysis has been well established [14–17]. Solvent extraction and hydrodistillation are the two major conventional ways to extract VOCs from plants [17], although various other extraction methods for VOCs in different plant matrices have been reviewed in literature [17, 18]. However, there are eminent disadvantages associated with these methods, such as low recovery, destruction of sample matrix, and use of nonenvironmentally friendly organic solvents [17, 19]. Headspace solid phase microextraction (HS-SPME) is a solvent-free, nondestructive, and easy approach for collecting VOCs emitted from plants [20, 21]. This approach can be practiced even using live plants without harvesting them. Moreover, HS-SPME coupled with gas chromatography (GC) has been shown to be very efficacious, collecting considerable amounts of volatile compounds [21–24].

Herein we report the development of a screening method for VOCs using HS-SPME-GC-MS analysis from the crushed ripe seeds of *A. danielli* with the view of using some of these volatiles for possible agricultural and pharmacological applications. Different parameters, namely, temperature, amount of sample, and extraction time, were optimised followed by a qualitative and semiquantitative analysis of the most abundant VOCs obtained under the optimum conditions.

2. Experimental

2.1. Sample Collection and Preparation. Mature seeds of *A. danielli* plant used in this study were collected from the Southern region of Nigeria. The collected seeds were crushed to powder (≤0.5 mm) using a quartz mortar and pestle.

2.2. Sample Extraction Using HS-SPME. Different parameters amenable to headspace sampling were investigated, namely, temperature, amount of sample material used, and sampling time. SPME extraction was achieved using a 100 μm polydimethylsiloxane solid phase microextraction (PDMS SPME) fibre, preconditioned for 30 minutes in a GC injection port at 200°C. Different masses (0.25, 0.35, and 0.50 g) of the ground *A. danielli* seeds were introduced into a 2 mL GC vial and the fibre was introduced 5 mm above the sample contained in the GC vial and incubated at set temperatures (20, 35, 40, and 50°C, resp.) in a water bath (Pierce, Rockford,

Illinois, USA) equipped with a multivial heating unit. The incubation period was varied between 10, 20, 30, 40, and 50 minutes. After each extraction time, the fibre was retracted into the needle and introduced into the GC injection port for desorption of analytes, chromatographic separation, and subsequent detection of VOCs via mass spectrometry. The different parameters, namely, temperature, mass of sample, and the extraction time, were optimised in a univariate manner. Each extraction and GC analysis were performed in triplicate ($n = 3$).

2.3. Annotation of Volatiles. Tentative annotation of the analytes was achieved through the mass comparisons of the mass spectra of individual compounds and compared with the NIST 2008, Wiley 2009, and Fragrances Libraries interfaced in the GCMS Solution Software running the instrument.

2.4. Instrumentation. Analysis of VOCs present in *A. danielli* seeds was performed using a Shimadzu QP 2010 gas chromatograph with mass spectrometer (Kyoto, Japan) fitted with a Restek Rtx-5ms (5% phenyl-95% dimethyl-polysiloxane) capillary column with the dimensions 30 m × 0.25 mm × 0.25 μm. The injection port temperature was set at 200°C and the optimised oven temperature programme began at 50°C held for 2 minutes, ramped to 170°C at a rate of 10°C/minutes, and then ramped to 250°C at a rate of 25°C/minutes and held for 2 minutes. Sample injection mode was splitless with a sampling time of 2 minutes followed by a split ratio of 1 : 10 using Helium (UHP Helium, Afrox, South Africa) as carrier gas pumped through the column at a constant flow rate of 1 mL/minute. The MS a transfer line temperature was set at 250°C, and ion source temperature was 200°C, with a scanning mode mass range of 50–500 amu.

3. Results and Discussion

3.1. Profiling and Annotation of Volatiles from the Seeds of A. danielli. Figure 1 shows the chromatogram of the extracted volatiles from the seeds of *A. danielli* using a headspace SPME with extraction conditions of 50°C, 0.25 g of sample, and 40-minute extraction. The number annotations on the chromatogram indicates the peak indexes which correlates with their mass spectral data shown in Table 1.

It can be seen that peak (3) had the highest intensity followed by peak (2) and then peak (4). Peak (3) was annotated as eucalyptol (also known as 1,8-cineole) following confirmation on three libraries (NIST, Wiley, and Fragrances) all interfaced with the GCMS Solution software. However, a close inspection of this peak revealed that it could be composed of several compounds and not just one as shown by the small spikes at the apex of the peak as well as the different mass spectra detected at different positions of the same peak.

From Table 1 it can be seen that Wiley shows the highest match for all the compounds with NIST showing the lowest matches. However, the differences are only a few percentage points from one another. The listing of the compounds in the Flavour and Fragrances Library (FFRSC) indicates that such compounds have been used in the fragrances. Hence,

TABLE 1: Some chromatographic data and tentative annotation of the top ten most abundant VOCs obtained from the headspace extraction of the ground *A. danielli* seeds.

Peak	Retention time (min)	Compound name	RMM	Ref Ion	Qual Ion	Different library matches (%)		
						NIST	FFSC	Wiley
(1)	5.087	α -Pinene	136	93	77	96	96	97
(2)	6.086	β-Pinene	136	93	69	96	96	97
(3)	7.259	Eucalyptol	154	81	154	91	89	91
(4)	10.082	α-Terpineol	154	59	136	94	94	96
(5)	12.434	α-Terpinyl acetate	196	121	93	93	95	96
(6)	13.515	Bergamotene	204	119	93	95	95	96
(7)	14.445	β-Bisabolene	204	69	93	90	95	96

Key. RMM: relative molecular mass; Ref Ion: reference ion; Qual Ion: qualifier ion.

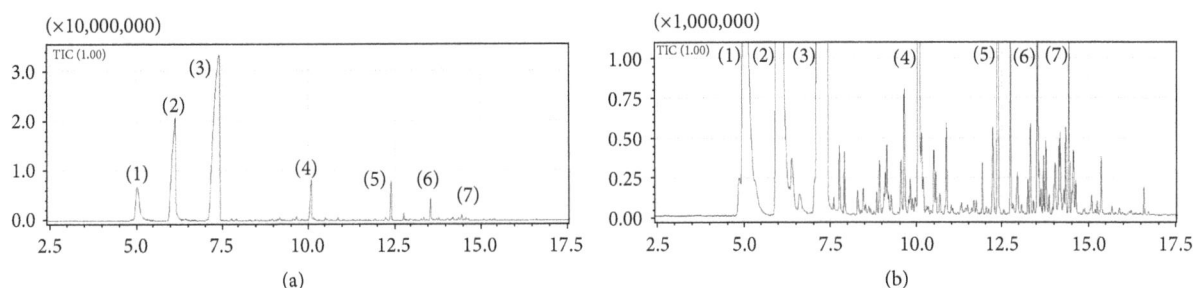

FIGURE 1: A chromatogram of the VOCs following a 40-minute extraction at 50°C: (a) depicts the chromatogram standardised against the largest peak (at about 7.5 min) expanded to cover the region between 2.5 and 17.5 minutes while (b) depicts 10 times magnified chromatogram showing several tens of peaks that are not visible in (a) due to their low relative abundances compared to the seven peaks visible in (a).

this makes this plant a good candidate for use in fragrances industry.

3.2. The Effect of Mass of the Sample on the Production and Extraction of the VOCs.

Logically the increase in the amount of sample should increase the production of the VOCs in a fixed unit volume. To optimise the mass of the sample required to yield the highest amount of the VOCs, different masses of the ground seeds were used and the amount of the VOCs produced is presented in Figure 2, plotted relative to 0.25 g sample for $n = 3$ replicates.

As can be seen, the amount of the VOCs increased with the amount of the sample used. However, the increase is not linear as the data yielded the coefficient of determination, $R^2 \leq 0.9126$ for terpinyl acetate although visually this was the most nonlinear curve (data not shown pictorially). The extraction seems to level off beyond 0.35 g of the sample while there is also a drop in repeatability as shown by the drop in relative standard deviations from the average of 7.7% for 0.25 g sample to 18.6% for 0.5 g samples, with the abundance of the VOCs increasing to an average of 134% using 0.5 g samples relative to 0.25 g samples. The compounds that resulted in the highest VOCs production (about 140%) were α-pinene and the two terpenoids while β-pinene and eucalyptol demonstrated the lowest increase (about 120%). The levelling off could be attributed to the rate of uptake of the

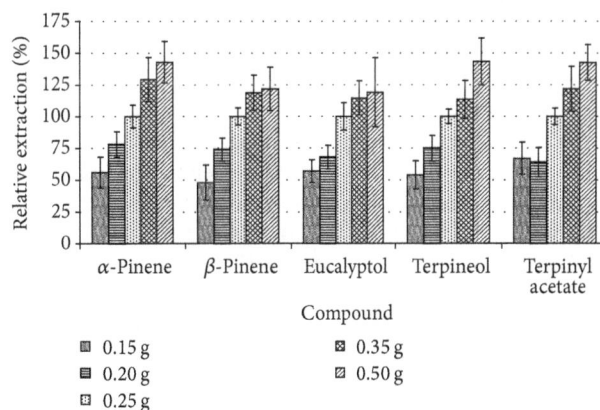

FIGURE 2: The effect of varying the amount of sample on the production of the VOCs.

analytes into the fibre which in this case becomes the limiting factor. Another factor could be the saturation of the fibre; however, this is unlikely since peak areas varied considerably. If this was due to saturation then the peak areas would be almost equal at all times. This, however, cannot be argued confidently given that the analysis was restricted to a few compounds (10, although only 5 are shown on the charts for ease of visualisation), yet from Figure 1 it can be seen that there are quite a number of compounds.

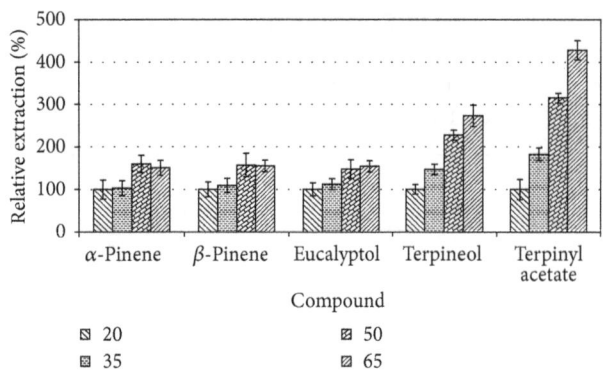

FIGURE 3: The effect of varying temperature on the extraction of volatiles from *A. danielli* seeds.

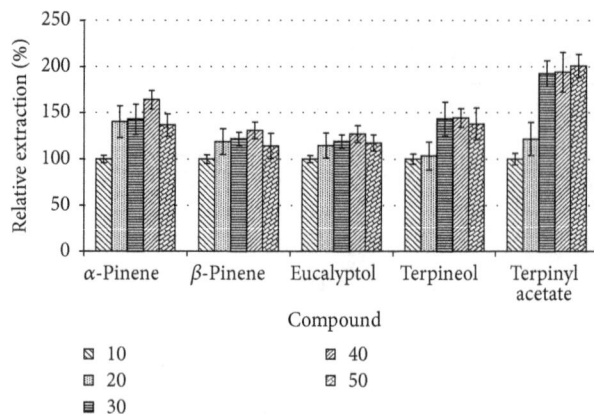

FIGURE 4: The effect of varying extraction time of volatiles of *A. danielli*.

Another important aspect is the increase in the deviations as the amount of the sample was increased. This could be attributed to the reduced headspace volume that resulted in the fibre touching the sample rather than being suspended in the headspace, especially with the 0.5 g samples. This negates the benefit of increasing the amount of sample since the sample particles act as variable barriers for the passage of the analytes onto the fibre surface. This is so because the particle shape and size of the ground samples were not uniform, hence having different barriers at different times.

Due to the above factors, the amount of 0.25 g was selected as the most ideal sample amount and was used for further optimisation experiments.

3.3. The Effect of Temperature on Extraction Efficiency of Volatiles.

Temperature is one of the most universal parameters that affect the efficiency of extraction through transfer of analytes to the headspace. Recently, a solvent-assisted headspace sampling was performed using organic solvents in driving analytes into the headspace; thus considerably reducing sampling time was reported albeit in aqueous matrices [25]. To assess the effect of temperature on extraction efficiency, samples in the GC-vials in this present study were incubated at different temperatures (25–65°C) and extracted for a fixed period of 20 minutes.

Figure 3 shows the extraction efficiency relative to room temperature (20°C) and only five compounds were shown for ease of visualisation chosen because of their abundance.

As can be seen in Figure 3, extraction efficiency of the VOCs increased 1.5-fold for the two pinene types and eucalyptol while those of terpineol and terpinyl acetate increased about 2.5 to 4.5 at 65°C compared to the 20°C (room temperature). The other interesting observation is that the latter two compounds showed the continued increase with temperature while the first three levelled off after 50°C, after which time different behaviours were noticed. This is an indication of the dynamism of SPME extraction and the increased fluidity of the polymer with increase in temperature increasing the exchange of the analytes between the fibre and the headspace volume; therefore, 50°C was chosen for further extractions. Besides, the first three compounds eluting much earlier than the last two (terpineol and terpinyl acetate)

indicated that their vapour pressure is much higher and they easily saturate the headspace. The latter compounds only enrich the headspace as the temperature is increased due to their relatively lower vapour pressure.

3.4. The Effect of Extraction Time on Extraction of the Volatiles.

Given the solid nature and surface area of the fibre, SPME is characterised by slower extraction kinetics taking as long as one hour than its counterpart liquid-based methods such as single-drop microextraction that typically become saturated within about 20 minutes [26]. To assess the effect of temperature and the ideal time required for the highest extraction, different 0.25 g samples of *A. danielli* were incubated at 50°C and extracted after different time periods between 10 and 50 minutes. Figure 4 illustrates the effect of varying extraction times on the extraction of 5 most abundant volatiles recovered from the analysed spice.

As can be seen from Figure 4, the extraction of the lower volatile compounds seems to increase continuously with the increase in time (about twice that of 10-minute extraction time at 40 minutes), while for the most volatile compounds the extraction peaks and levels off at 40 minutes of extraction yielding about 1.25- to 1.75-fold that of 10-minute extraction time at 40 minutes. This trend can still be explained by the dynamism of the fibre allowing the compounds to be exchanged between the fibre and the headspace volume as alluded to in the earlier section. However, the extraction time of 40 minutes was selected as the ideal time for the extraction of these VOCs from the crushed seeds.

3.5. Characterisation of the VOCs by Abundance and Comparison with Reported Literature on *A. danielli* and Other Plants.

Following the optimised extraction time and temperature, a semiquantitative analysis was carried out on extracts obtained at optimal temperature of 50°C for 40 minutes. Table 2 provides a list of top 30 compounds selected on the basis of their relative abundance when extracted under optimised conditions as well as comparison with those reported in literature from the same plant (seeds, etc.).

TABLE 2: Top 30 most abundant VOCs extracted from *A. danielli* seeds annotated using the three libraries listed in Table 1.

Retention time (min)	Percentage abundance	Annotations*	Literature reported (*Aframomum* spp.)	Literature reported (other plant species)
4.02	0.15	α-Thujene	[27, 28]	
4.134	7.58	α-Pinene	[28, 29]	
4.898	22.11	β-Pinene oxide		[30]
5.886	58.10	Eucalyptol	[29]	
6.173	0.08	γ-Terpinene	[31]	
6.294	0.09	Sabinene hydrate ⟨E⟩δ	[28, 32]	
6.768	0.08	Sabinene hydrate ⟨Z⟩	[28]	
7.187	0.12	α-Campholenal	[31]	
7.388	0.39	Pinocarveol ⟨Z⟩		[33]
7.743	0.11	Pinocarvone	[28, 34]	
7.823	0.28	δ-Terpineol	[31]	
7.976	0.11	Terpinen-4-ol	[34]	
8.235	4.08	α-Terpineol	[28, 34]	
8.594	0.21	Carveol ⟨Z⟩	[28]	
8.941	0.18	Carvone	[28]	
10.271	0.21	Carvyl acetate ⟨Z⟩		[35]
10.419	1.83	α-Terpinyl acetate	[28]	
10.788	0.35	α-Copaene	[28, 29]	
10.988	0.11	β-Elemene	[28, 36]	
11.359	0.12	Caryophyllene ⟨(E)⟩		[37, 38]
11.553	1.17	α-Bergamotene ⟨Z⟩	[34]	
11.73	0.09	Valerena-4,7(11)-diene		[39]
11.789	0.12	β-Bergamotene ⟨E⟩	[28, 32]	
12.05	0.15	Aciphyllene#		[40–42]
12.168	0.34	Caryophyllene (Z)		[38]
12.356	0.21	Isodaucene		[43]
12.446	0.31	β-Bisabolene	[29, 34]	
12.571	0.19	β-Selinene	[29, 34]	
13.365	0.13	Caryophyllene oxide	[34]	
14.493	0.07	α-Bergamotol ⟨(Z), E⟩		[44]

*Annotation was made with the match of ≥85% from the NIST Library, with the lowest match of 85% obtained for the last three entries (sabinene hydrate ⟨E⟩, γ-terpinene, and α-bergamotol) with percentage abundances between 0.07 and 0.08. δThe E isomer has been reported elsewhere to elute earlier than the "Z" isomer [28, 32]. #Also reported in bacteria.

Clearly, it can be seen that the volatiles are dominated by compound depicted as peak (3) assigned as eucalyptol accounting for about 60% of the total peak areas for 30 peaks integrated automatically using autointegration software. This is followed by β-pinene oxide at 22%, α-pinene at 7.5%, and terpineol at 4% as well as terpinyl acetate and bergamotene that accounted for about 2 and 1%, respectively, with all the 26 remaining compounds, each with less than a percentage point collectively taking the remaining 6.5%. However, it must be noted that the percentage indicated for each compound is relatively skewed due to the abundance of the first three compounds; most of these compounds are still significantly abundant as can be depicted in Figure 1.

4. Conclusions

The presented work demonstrated the effectiveness of the proposed method for screening the VOCs from the *Aframomum danielli* which in essence demonstrated the applicability to other plant seeds with rich content of volatile oils. In the current study, the amount of seeds material, time of extraction, and temperature of extraction were found to affect the outcomes of the extracted metabolites, both qualitatively and quantitatively. Thus, 0.25 g of finely ground ripe seeds, extraction time of 40 minutes, and a temperature of 50°C were found to be optimal for the extraction of the reported metabolites. Our results suggest that *A. danielli* seeds contain eucalyptol as a dominant VOC (about 60% of all the VOCs detected) with only six (6) compounds accounting for almost 95% of the total VOCs produced. Overall, A. *danielli* can be regarded as a potent source of eucalyptol oil; hence this plant can be used as a potent source for commercial production of this eucalyptol oil.

Acknowledgments

The authors gratefully acknowledge the University of Johannesburg for supporting this work. Mosotho J. George further acknowledges National University of Lesotho (RCC Grant P116-9008) and OPCW (Grant no. L/ICA/ICB/204905/16) for the research fellowship.

References

[1] G. O. Adegoke and A. G. Gopala Krishna, "Extraction and identification of antioxidants from the spice Aframomum danielli," *Journal of the American Oil Chemists' Society*, vol. 75, no. 8, pp. 1047–1052, 1998.

[2] S. O. Rotimi, O. A. Rotimi, T. Bisi-Adeniyi et al., "Gas chromatography mass spectrometry identification of antiangiogenic phytochemicals in Aframomum danielli K. Schum: An in silico study," *International Journal of Pharmacognosy and Phytochemical Research*, vol. 7, no. 6, pp. 1194–1201, 2015.

[3] M. O. Afolabi, G. O. Adegoke, and F. M. Mathooko, "Phytochemical characterization of the extracts of Aframomum danielli flower, leaf, stem and root," *African Journal of Agricultural Research*, vol. 6, no. 1, pp. 97–101, 2011.

[4] A. M. El-Halawany, R. S. E. Dine, N. S. El Sayed, and M. Hattori, "Protective effect of Aframomum melegueta phenolics against CCl 4 -induced rat hepatocytes damage; Role of apoptosis and pro-inflammatory cytokines inhibition," *Scientific Reports*, vol. 4, article no. 5880, 2014.

[5] G. O. Adegoke and B. J. Skura, "Nutritional profile and antimicrobial spectrum of the spice Aframomum danielli K. Schum," *Plant Foods for Human Nutrition*, vol. 45, no. 2, pp. 175–182, 1994.

[6] A. Dauda and G. Adegoke, "Preservation of Some Physico-Chemical Properties of Soymilk-Based Juice with *Aframomum Danielli* Spice Powder," *American Journal of Food Science and Technology*, vol. 2, no. 4, pp. 116–121, 2014.

[7] G. O. Babarinde, G. O. Adegoke, and R. Akinoso, "Effect of Aframomum danielli extract on some chemical and antioxidant components of roma tomato variety during storage," *American Journal of Food Technology*, vol. 9, no. 1, pp. 28–38, 2014.

[8] S. B. Fasoyiro and G. O. Adegoke, "Phytochemical characterization and the antimicrobial property of *Aframomum danielli* extract," *African Journal of Agricultural Research*, vol. 2, no. 3, pp. 76–79, 2007.

[9] I. A. Ajayi, I. O. Ademola, and S. V. Okotie, "Larvicidal effects of *Aframomum danieli* seed extracts against gastrointestinal nematode of sheep: in vitro studies," *African Journal of Traditional, Complementary and Alternative Medicines*, vol. 5, no. 3, pp. 244–246, 2008.

[10] M. O. Afolabi and G. O. Adegoke, "Antioxidative and flavouring effects of Aframomum danielli on biscuits," *African Journal of Food Science*, vol. 8, no. 4, pp. 200–203, 2014.

[11] F. R. Irvine, "The Useful Plants of West Tropical Africa. By J. D. Dalziel, M.D., B.Sc, F.L.S. Being an Appendix to The Flora of West Tropical Africa, by Hutchinson and Dalziel. London: Crown Agents for the Colonies. 1937. Pp. 612. 18s.," *Africa*, vol. 11, no. 02, pp. 254-255, 1938.

[12] M. D'Alessandro and T. C. J. Turlings, "Advances and challenges in the identification of volatiles that mediate interactions among plants and arthropods," *Analyst*, vol. 131, no. 1, pp. 24–32, 2006.

[13] J. Kesselmeier and M. Staudt, "Biogenic volatile organic compounds (VOC): an overview on emission, physiology and ecology," *Journal of Atmospheric Chemistry*, vol. 33, no. 1, pp. 23–88, 1999.

[14] A. C. Steinemann, I. C. MacGregor, S. M. Gordon et al., "Fragranced consumer products: Chemicals emitted, ingredients unlisted," *Environmental Impact Assessment Review*, vol. 31, no. 3, pp. 328–333, 2011.

[15] R. Gyawali and K. S. Kim, "Volatile organic compounds of medicinal values from Nepalese Acorus calamus L , Kathmandu University Journal of Science," *Journal of Engineering and Technology*, vol. 5, no. 2, pp. 51–65, 2009.

[16] F. Manzoor, S. A. Malik, K. Naz, K. J. Cheema, and S. Naz, "Potential of antitermitic activities of eucalyptus oil," *Pakistan Journal Of Zoology*, vol. 44, no. 2, pp. 335–339, 2012.

[17] E. Ormeño, A. Goldstein, and Ü. Niinemets, "Extracting and trapping biogenic volatile organic compounds stored in plant species," *TrAC - Trends in Analytical Chemistry*, vol. 30, no. 7, pp. 978–989, 2011.

[18] D. Tholl, W. Boland, A. Hansel, F. Loreto, U. S. R. Röse, and J.-P. Schnitzler, "Practical approaches to plant volatile analysis," *The Plant Journal*, vol. 45, no. 4, pp. 540–560, 2006.

[19] C. C. Teo, S. N. Tan, J. W. H. Yong, C. S. Hew, and E. S. Ong, "Pressurized hot water extraction (PHWE)," *Journal of Chromatography A*, vol. 1217, no. 16, pp. 2484–2494, 2010.

[20] C. Bicchi, C. Cordero, E. Liberto, B. Sgorbini, and P. Rubiolo, "Headspace sampling of the volatile fraction of vegetable matrices," *Journal of Chromatography A*, vol. 1184, no. 1-2, pp. 220–233, 2008.

[21] M. Kusano, M. Kobayashi, Y. Iizuka, A. Fukushima, and K. Saito, "Unbiased profiling of volatile organic compounds in the headspace of Allium plants using an in-tube extraction device," *BMC Research Notes*, vol. 9, no. 1, article no. 133, 2016.

[22] D. Materić, D. Bruhn, C. Turner, G. Morgan, N. Mason, and V. Gauci, "Methods in plant foliar volatile organic compounds research," *Applications in Plant Sciences*, vol. 3, no. 12, 2015.

[23] N. P. Brunton, D. A. Cronin, F. J. Monahan, and R. Durcan, "A comparison of solid-phase microextraction (SPME) fibres for measurement of hexanal and pentanal in cooked turkey," *Food Chemistry*, vol. 68, no. 3, pp. 339–345, 2000.

[24] Q. L. Ma, N. Hamid, A. E. D. Bekhit, J. Robertson, and T. F. Law, "Optimization of headspace solid phase microextraction (HS-SPME) for gas chromatography mass spectrometry (GC-MS) analysis of aroma compounds in cooked beef using response surface methodology," *Microchemical Journal*, vol. 111, pp. 16–24, 2013.

[25] M. J. George, L. Marjanovic, and D. B. G. Williams, "Solvent-Assisted Headspace Sampling Using Solid Phase Microextraction for the Analysis of Phenols in Water," *Analytical Chemistry*, vol. 87, no. 19, pp. 9559–9562, 2015.

[26] D. B. G. Williams, M. J. George, R. Meyer, and L. Marjanovic, "Bubbles in solvent microextraction: The influence of intentionally introduced bubbles on extraction efficiency," *Analytical Chemistry*, vol. 83, no. 17, pp. 6713–6716, 2011.

[27] A. C. Ngakegni-Limbili, B. Zebib, M. Cerny et al., "Aframomum stipulatum (gagnep) k. schum and aframomum giganteum (oliv. & hanb) k. schum as aroma tincto oleo crops resources: Essential oil, fatty acids, sterols, tocopherols, and tocotrienols composition of different fruit parts of congo varieties," *Journal of the Science of Food and Agriculture*, vol. 93, no. 1, pp. 67–75, 2013.

[28] G. O. Adegoke, L. Jagan Mohan Rao, and N. B. Shankaracharya, "A comparison of the essential oils of Aframomum daniellii (Hook, f.) K, Schum, and Amomum subulatum Roxb," *Flavour and Fragrance Journal*, vol. 13, no. 5, pp. 349–352, 1998.

[29] G. D. Diomandé, A. M. Koffi, Z. F. Tonzibo, G. Bedi, and G. Figueredo, "GC and GC/MS analysis of essential oil of five *Aframomum species* from Côte Divoire," *Middle-East Journal of Scientific Research*, vol. 11, no. 6, pp. 808–813, 2012.

[30] E. Salminen, P. Mäki-Arvela, P. Virtanen, T. Salmi, J. Wärnå, and J.-P. Mikkola, "Kinetics upon isomerization of α,β-pinene oxides over supported ionic liquid catalysts containing lewis acids," *Industrial & Engineering Chemistry Research*, vol. 53, no. 52, pp. 20107–20115, 2014.

[31] M. A. Couppé de K. Martin, H. Joseph, S. Bercion, and C. Menut, "Chemical composition of essential oils from aerial parts of Aframomum exscapum (Sims) Hepper collected in Guadeloupe, French West Indies," *Flavour and Fragrance Journal*, vol. 21, no. 6, pp. 902–905, 2006.

[32] P. H. Amvam Zollo, R. Abondo, L. Biyiti, C. Menut, and J. M. Bessière, "Aromatic plants of Tropical Central Africa XXXVIII: Chemical composition of the essential oils from four Aframomum species collected in Cameroon (1)," *Journal of Essential Oil Research*, vol. 14, no. 2, pp. 95–98, 2002.

[33] M. Couladis, V. Tsortanidou, J. Francisco-Ortega, A. Santos-Guerra, and C. Harvala, "Composition of the essential oils of Argyranthemum species growing in the Canary Islands," *Flavour and Fragrance Journal*, vol. 16, no. 2, pp. 103–106, 2001.

[34] E. Adjalian, F. T. D. Bothon, B. Yehouenou et al., "GC/MS and GC/FID analysis and evaluation of antimicrobial performance of Aframomum sceptrumessential oils of Benin," *World Journal of Pharmaceutical Sciences*, vol. 2, no. 8, pp. 787–792, 2014.

[35] A. Nezhadali and M. Parsa, "Study of the Volatile Compounds in Artemisia Sagebrush from Iran using HS/SPME/GC/MS," *International Journal of Environmental Science and Development*, pp. 287–289, 2010.

[36] S. Eyob, M. Appelgren, J. Rohloff, A. Tsegaye, and G. Messele, "Traditional medicinal uses and essential oil composition of leaves and rhizomes of korarima (Aframomum corrorima (Braun) P.C.M. Jansen) from southern Ethiopia," *South African Journal of Botany*, vol. 74, no. 2, pp. 181–185, 2008.

[37] L. C. Salanta, M. Tofana, S. A. Socaci, C. L. Pop, D. Michiu, and A. Farcas, "Determination of the volatile compounds from hop and hop products using ITEX/GC-MS technique," *Journal of Agroalimentary Processes and Technologies*, vol. 18, no. 2, pp. 110–115, 2012.

[38] S. Oluwafemi, T. J. Bruce, J. A. Pickett, J. Ton, and M. A. Birkett, "Behavioral Responses of the Leafhopper, Cicadulina storeyi China, a Major Vector of Maize Streak Virus, to Volatile Cues from Intact and Leafhopper-Damaged Maize," *Journal of Chemical Ecology*, vol. 37, no. 1, pp. 40–48, 2011.

[39] B. W. Pyle, H. T. Tran, B. Pickel et al., "Enzymatic synthesis of valerena-4,7(11)-diene by a unique sesquiterpene synthase from the valerian plant (Valeriana officinalis)," *FEBS Journal*, vol. 279, no. 17, pp. 3136–3146, 2012.

[40] T. Nawrath, G. F. Mgode, B. Weetjens, S. H. E. Kaufmann, and S. Schulz, "The volatiles of pathogenic and nonpathogenic mycobacteria and related bacteria," *Beilstein Journal of Organic Chemistry*, vol. 8, pp. 290–299, 2012.

[41] C. M. Buré and N. M. Sellier, "analysis of the essential oil of indonesian patchouli (pogostemon cabin benth.) using GC/MS (EI/CI)," *Journal of Essential Oil Research*, vol. 16, no. 1, pp. 17–19, 2004.

[42] M. A. Hossain, M. J. Kabir, S. M. Salehuddin et al., "Antibacterial properties of essential oils and methanol extracts of sweet basil Ocimum basilicum occurring in Bangladesh," *Pharmaceutical Biology*, vol. 48, no. 5, pp. 504–511, 2010.

[43] V. Mazzoni, F. Tomi, and J. Casanova, "A daucane-type sesquiterpene from Daucus carota seed oil," *Flavour and Fragrance Journal*, vol. 14, no. 5, pp. 268–272, 1999.

[44] L. Jirovetz, G. Buchbauer, M. P. Shafi, and M. M. Kaniampady, "Chemotaxonomical analysis of the essential oil aroma compounds of four different Ocimum species from southern India," *European Food Research and Technology*, vol. 217, no. 2, pp. 120–124, 2003.

20

A Study on DSLM Transporting the Rare Earth Metal La (III) with a Carrier of PC-88A

Shibao Lu ⓘ,[1] **Yan Wang ⓘ,**[2] **Liang Pei,**[3] **and Wei Li ⓘ**[1]

[1]*School of Public Administration, Zhejiang University of Finance and Economics, Hang Zhou 310018, China*
[2]*Development and Planning Division, Tongji University, 1239 Siping Road, Shanghai, China*
[3]*Key Laboratory of Water Cycle and Related Land Surface Processes, Institute of Geographic Sciences and Natural Resources Research, Chinese Academy of Sciences, Beijing 100101, China*

Correspondence should be addressed to Yan Wang; wylady@tongji.edu.cn

Academic Editor: Adil Denizli

This paper studies transmission behavior of La (III) in dispersed supported liquid membrane (DSLM) of dispersed phase constituted by dispersed supported liquid membrane solution and HCl solution with polyvinylidene fluoride membrane (PVDF) as support and kerosene as membrane solvent, with 2-ethyl hexyl phosphonic acid-single-2-ethyl hexyl ester (PC-88A) and two-(2-ethyl hexyl) phosphoric acid (D2EHPA) as mobile carrier. It also investigates the influence of La (III) transmission by the material liquid acidity, initial concentration of La (III), HCl concentration, membrane solution, and HCl solution volume ratio, resolving agent and carrier concentration, as well as concluding that the optimal transmission and separation conditions are dispersed phase of 4.00 mol/L HCl concentration, 30:30 volume ratio of membrane solution, and HCl solution, within 0.160 mol/L controlled carrier concentration and 4.00 pH value of material liquid. Under the optimal conditions, the La (III) initial concentration of material liquid phase is $8.00 \times 10{-}5$ mol/L mol/L, 125 min, and 93.9% migration rate. Under the condition of unchanged acidity of resolving phase, HCL, H_2SO4, and HNO_3 as resolving agent, at 125th min, the migration rates of La (III) are 93.9%, 94.0%, and 87.8%, respectively. HCl solution, H_2SO_4 solution, and HNO_3 solution have a certain effect on the La (III) resolution, of which 4.00 mol/L HCl solution and 2.00 mol/L H_2SO_4 solution are better. The effect of HNO_3 is slightly lower than HCl and H_2SO_4.

1. Introduction

Rare earth metals have wide usage; they cannot only be used alone but also are used in the form of mixed rare earth. Adding a moderate amount of rare earth metals or their compounds in alloy can greatly improve the performance of the alloy; thus, the rare earth elements are also known as vitamin of metallurgical industry. For example, adding some rare earth elements in steel can increase the plasticity, toughness, wear resistance, heat resistance, oxidation resistance, and corrosion resistance [1–4]. Another example is that rare earth metals can be used as a pyrophoric alloy, permanent magnetic materials, superconducting materials, dyeing materials, light-emitting materials, and trace element fertilizer [5–7]. As a result, in addition to being widely used in metallurgy, petrochemical industry, glass ceramic, fluorescent materials, electronic materials, pharmaceutical, and

agricultural sectors, the rare earth metal also goes gradually into many areas of modern science and technology. With the wide application of rare earth elements in production and living, it is necessary to separate and enrich rare earth elements, at which a lot of foreign and domestic institutes have been studying recently. The characteristics of liquid membrane extracting rare earth metals comprise a short process, high speed, large enrichment ratio, less reagents, low cost, and wide industrial application prospect. In the system of liquid membrane extracting rare earth metal ion, generally, the organic solvent adopts kerosene or sulfonated kerosene, the carrier adopts LA, P_{204}, P_{507}, etc., and the internal phase adopts HCI and HNO_3 [8–11]. Mother liquor leached by the rare earth can be grouped, purified, and separated, etc. according to the need.

In order to overcome these above difficulties in the conventional LM systems, a new liquid membrane technique, namely, dispersion supported liquid membrane (DSLM) [12], was proposed. The DSLM technique is based upon surface renewal, diffusion theory, and our previous work, which also integrates the advantages of fiber membrane extraction process, liquid film transport process, and most of other liquid membrane systems [13]. This is a new type of LM process with several advantages such as increased stability of the membrane, reduced costs, increased simplicity of operation, extremely efficient stripping of the target species from the organic phase by obtaining a high flux, and a higher concentration of the recovered target species in the stripping solution. L. et al. [14] studied the separation of cobalt (II) and lithium (I) in the supported liquid membrane system. He also studied the stirring speed, pH value of material liquid phase, concentration of membrane phase carrier Cyanex272, the concentration of cobalt (II) and lithium (I) in the material liquid phase, and the influence of resolving agent on the two kinds of metal transmission flux. R. et al. [9]. utilized the technology of hollow fiber supporting liquid membrane to extract cadmium in seawater and used ultrasensitive graphite furnace atomic absorption spectrum as detection means to realize selective extraction of cadmium from the water sample. C. J. [15] et al. studied transmission separation of copper, zinc, diamond, and nickel in the supported liquid membrane system, respectively, and used L1X84I, YOPS-99, and Cyanex272 as supported liquid membrane flow carrier, to separate the mixture of copper, zinc, diamond, nickel. Singh et al. [16] reported transmission model of zinc ions in PC-88A-kerosene supporting liquid membrane system. Under the different experimental conditions, a large number of experiments are done to predict the transmission degree of zinc in the supporting liquid membrane system. J. V. et al. [17] reported the transmission of trivalent chromium in the liquid membrane system with the carrier of two (2-ethyl hexyl) phosphonic acid. T. [18] researched the transmission of divalent and trivalent metal ions as selective migration in the supported liquid membrane system with carrier of new organic phosphonic acid. The research results show that, by using different extraction agents, the mixed ions Cu (I), Co (II), Ni (II), Pb (II), Fe(III), and Cd (II) can be separated well. Sha m si pur et al. studied the transmission separation of silver and mercury in the supported liquid membrane system, improved the ordinary supported liquid membrane system, and set up the system of two film and three chambers, which realized the fast separation of the silver ions and mercury ion. China is rich in rare earth production, many kinds of which need to be separated, purified, and recovered in large quantities; hence the task is very difficult.

The operation of SLM system is simple, which does not require the introduction of expensive surface active agent [10, 19–21], but the membrane solution (organic solvents, extracting agent, and modifying agent) dissolves in water which decreases the stability of membrane and high transmission flux. To settle the existing problem of SLM, some home and abroad studies have been exploring new liquid membrane configuration recently, intending to keep the separation of liquid membrane and overcome the shortcomings [22–24]. So, the terms of "combined liquid brain" and "combined technology" are proposed, which combines a solid film

with the liquid film during all kinds of chemical process, effectively overcome the carrier of the supported liquid membrane leaking from membrane phase, and extend the life of the membrane. This work will combine the extraction dispersion technology with the supported liquid membrane and put forward the concept of dispersed supported liquid membrane.

At present, the literatures on dispersed supported liquid membrane separating and migrating rare earth metal are rare. This work mainly discusses and studies the feasibility of dispersed supported liquid membranes separating and migrating La (III), utilizes membrane module design, carrier optimization, and migration rate control to realize the migration and separation of rare earth metals, researches the migration process, and establishes new methods and a new system of dispersed supported liquid membranes migrating and separating rare earth metals, which is expected to be a breakthrough in the industrial application.

2. Experiment

2.1. Experimental Facility. The homemade migration pool of DSLM consists of the material liquid pool, the dispersed pool, and the supported pool, where the capacities of the material liquid pool and the dispersed pool are both 80 mL, equipped with adjustable speed electric blender. The supporting body is hydrophobic porous polyvinylidene fluoride membrane PVDF (Shanghai Yadong Nuclear Grade Resin Co., Ltd.); the diameter of the pore is 0.22 mm, the film thickness 65 mm, the porosity ε 75%, the tortuosity factor τ 1.67, and the effective area 12 cm2. The experimental apparatus is shown in Figure 1.

UV-1200 type spectrophotometer (Shanghai HuiPuDa Instrument Factory), JJ-1 type precise timed electric blender (Jintan City DanYangMen Quartz Glass Factory), UV-2102PC type ultraviolet-visible spectrophotometer (Unico(Shanghai) Instrument Co., Ltd.), AY120 type electronic scales (Shimadzu), 520MPT atomic emission spectrometer (ChangChun JiLin University Little Swan Instruments Co., Ltd.).

2.2. Experimental Method

2.2.1. Preparation of the Solution. They are 1.00 mol/L HAc~ NaAc buffer solution; 1.00 mol/L NaH2PO4~Na2HPO4 buffer solution; 6.00 mol/L HCl; 4.00 mol/L H_2SO_4; 1.00×10^{-2} mol/L arsenazo III (C22H18As2O14N4S2). Except that the concentration of La (III) is diluted to 1.00×10^{-2} mol/L by 1.00 mol/L H_2SO_4, the concentration of the other standard solutions of various rare earth metal ions should be diluted to 1.00×10^{-2} mol/L with 1.00 mol/L HCl.

Membrane solution is made of the flow carrier of PC - 88 A, the concentration of which is diluted to 0.230 mol/L with kerosene.

2.2.2. Operating Steps. The experiment adopts the homemade liquid film transmission device, puts the supported PVDF membrane into the membrane solution to leach and absorb for a certain time (about 3 to 4 hours), then uses the filter

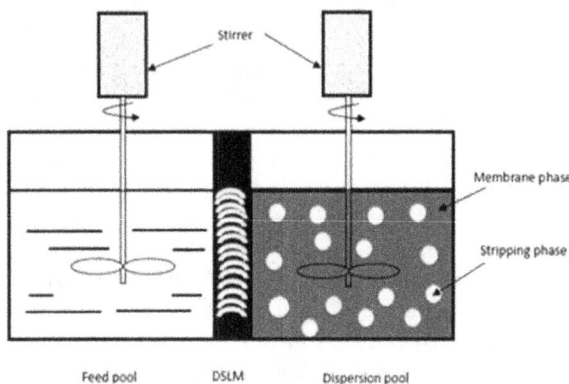

FIGURE 1: Schematic diagram of DSLM apparatus.

paper to suck up liquid on the surface of the membrane, and fixes it in DSLM migration pool. There are two-phase solution in the two slots, respectively, with the isolation of PVDF membrane; one is the material liquid, the other the dispersed phase. A certain amount ($5.00 \sim 10.0$ mL) of 1.00×10^{-3} mol/L solution of rare earth metals and buffer solution (total 60.0 mL) is added to the material liquid phase, while the 60.0 mL mixture of membrane solution and HCl solution are added to the dispersed phase. Start the blender and time it and then take samples of 1.00 mL to 10.0 mL the colorimetric tube from the material liquid phase at a certain amount of time.

2.2.3. Analysis of Samples.
Add a moderate amount of the buffer solution and a certain amount of the $1.00 \times 10{-4}$ mol/L chromogenic agent arsenazo III (C22H18As2O14N4S2) into the taken samples and dilute it with the deionized water to 10 mL; after 10 min past the chromogenic reaction, use UV-1200 spectrophotometer to measure the absorbance of La (III) at 653 nm.

2.2.4. Analysis of the Results.
According to the relationship curve of the absorbance value and the concentration of rare earth metals, the mobility (such as formula (1)) can be calculated. Then the mobility is analyzed quantitatively based on the relationship curve of the absorbance and the concentration of metal. The computation formula is as follows:

$$\eta = \frac{(c_0 - c_t)}{c_0} \times 100\% = \frac{(A_0 - A_t)}{A_0} \times 100\% \quad (1)$$

where η is for mobility; c0 for metal ion concentration in the starting material liquid (unit: mol/L); ct for metal ion concentration in material liquid phase at time of t (unit: mol/L); A0 for the starting absorbance; At for the absorbance at time of t.

2.2.5. Treatment of Membrane Solution.
At the end of the experiment, it is necessary to treat the membrane solution and remove residual rare earth metals in the solution. This experiment adopts 4.00 mol/L H2SO4 to resolve it, and the membrane solution can be recycled.

2.3. Experimental Principle.
The processes of reaction and the migration of metal ions in the dispersed supported liquid membrane system are as follows.

The La (III) ions in material liquid phase diffused through the water between the material liquid phase and the membrane phase. On the interface of the water phase and the membrane phase, the metal ions La (III) will have the following complex reaction with the carrier of PC–88A (abbreviated to HR):

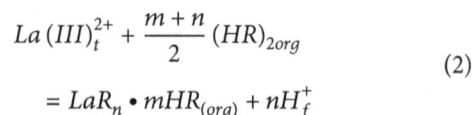

$$La\,(III)_t^{2+} + \frac{m+n}{2}\,(HR)_{2org}$$
$$= LaR_n \bullet mHR_{(org)} + nH_f^+ \quad (2)$$

where the subscript f stands for the water phase, the subscript org for the membrane phase, and (HR) 2 for the extraction agent in the form of dimers in the nonpolar oil.

The metal ions with carrier complex generated by the reaction diffuse from the interface between the material phase and the membrane phase to the inside of the membrane [25] and then diffuse in the membrane phase. At the interface of the membrane phase and the dispersed phase, it takes the following resolving reaction with the resolving agent:

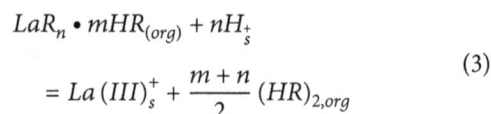

$$LaR_n \bullet mHR_{(org)} + nH_s^+$$
$$= La\,(III)_s^+ + \frac{m+n}{2}\,(HR)_{2,org} \quad (3)$$

where the right subscript S stands for resolving phase. Due to the fact that the stirring effect causes metal ion with carrier complexes to contact the resolving agent fully, this ensures the continuous extraction and the reverse extraction process and improves the transmission speed of the metal ions and the stability of liquid membrane system. Changing the volume ratio of the resolving agent and the liquid membrane solution can get the resolving solution containing a high concentration of metal ions. Stop stirring; let it stand. The resolving agent contains the high concentration of metal ions and the membrane phase will layer automatically, which is convenient to treat the concentration.

By [6], we know

$$\ln\frac{c_{f(t)}}{c_{f(0)}} = -\frac{A}{V_f}P_c t \quad (4)$$

where $c_{f(t)}$ and $c_{f(0)}$ stand for metal ion concentration at time of t and the initial material liquid phase, respectively; A for the effective area of membrane; V for the material liquid volume; Pc for the permeability coefficient of metal ions; t for the transmission time. By measuring the concentration of the metal ions under different conditions, draw relation figure between $-\ln(c/c0)$ with t and analyze the influence degree of various factors on the migration rate from the straight slope of the line. The permeability coefficient of the metal ions in the membrane Pc can be indicated by formula (4)

$$P_c = \frac{J}{[M]_f} \quad (5)$$

where J stands for the mass transfer flux of the metal ions; $[M]_f$ for the metal ion concentration of the material liquid phase.

When the material liquid phase contains two kinds of ions, if the permeability coefficients of the two kinds of ions in the membrane are different, the ions can be separated by the liquid membrane system. The separation factor is defined as

$$\beta = \frac{P_{c1}}{P_{c2}} = \frac{J_{M1}/[M_1]_f}{J_{M2}/[M_2]_f} \qquad (6)$$

3. Results and Discussions

3.1. The Influence of pH Value of Material Liquid Phase on the Rare Earth Metal. Known from the mass transfer mechanism of rare earth metal ions in the DSLM, differential concentration of H+ in the material liquid phase and the dispersed phase is mass transfer power of rare earth metals in DSLM [26]. As a result, the higher the pH value of the material liquid, the more conducive to the migration of rare earth metal. But the resolving agent used in dispersed phase is strong acid. So, when the pH value of material liquid is increased to a certain extent, the high strength of H+ differential concentration between the two phases speeds up permeation of the dispersed phase H+ through membrane phase, which seriously affects not only the stability of liquid membrane, but also the migration rate of rare earth metal in the supported liquid membrane. Therefore, the acidity difference between material liquid phase and the dispersed phase is one of the key factors influencing the mass transfer rate of rare earth metals.

What is more, the pH value of material liquid phase can influence the existence of the rare earth metal ions; under the proper pH value, the rare earth metal ions can form complex carrier with the membrane carrier and enter into the liquid film. If the metal ions are transported, the separation effect is well; otherwise, the separation effect is poor. If the pH value becomes too low, the acidity difference between the material liquid phase and the dispersed phase is too subtle and the migration effect will not be satisfying; If the pH of the material liquid becomes too high, it may cause the rare earth metal ion hydrolyse or form hydroxy complex, which affects the migration rate [27]. So, the selection of the material liquid pH plays an important role in metal ion migration.

The value ratio of the membrane and HCL solution in the dispersed phase of La (III) in DSLM migration system is selected as 30:30; the HCl concentration of the dispersed phase is 4.00 mol/L; the initial concentration of La (III) is $1.00 \times 10\text{-}4$ mol/L; the concentration of PC–88A in membrane solution is 0.160 mol/L. Under the condition, study the influence of material liquid pH on the migration behavior of La (III) in the DSLM; the experimental results are shown in Figure 2 and in Table 1.

In Figure 2, when the pH values are 3.00, 3.30, 3.60, 4.00, and 4.30, at 125th min, the migration rate of La (III) can reach 4.91%, 38.3%, 75.4%, 81.2% and 82.0%, respectively. When the pH of material liquid is less than 3.00, the difference of the concentrations of H+ in the material liquid phase and the dispersed phase is minor and the migration effect of La (III)

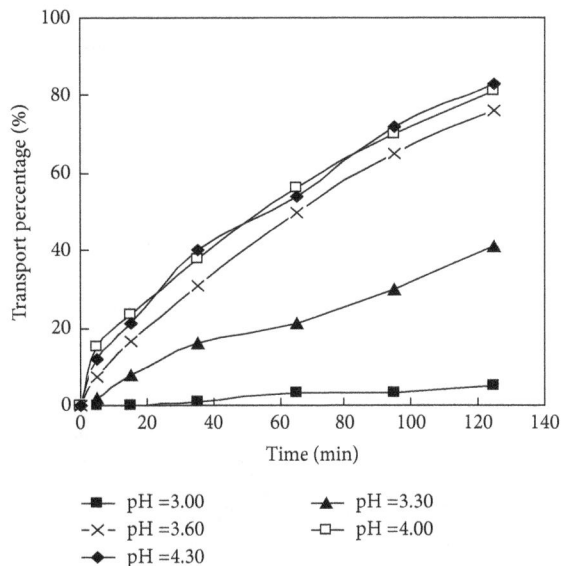

FIGURE 2: Effect of pH in feed phase on transport of La (III).

is not obvious. When the material liquid pH achieves 3.00, the difference between the concentrations of H+ in the two phases is still small and the migration rate of La (III) is only 4.91%. When the pH value climbs to 3.60, the migration rate increases obviously, twice as much as that of pH 3.30. When the pH value is 4.00, the migration rate increases obviously more than that at pH 3.60; when the pH value is 4.30, the migration rate of La (III) only increases over 0.80% than that of pH 4.00. When pH becomes greater than 4.30, the material liquid acidity is low and high strength of H+ concentration difference between the two phases accelerates the permeation of the dispersed phase H+ through the membrane phase, which influences the stability of liquid membrane and the migration rate of La (III) in DSLM.

Continue to reduce liquid H+ concentration and the La (III) of material liquid hydrolyses, and the solution becomes muddy. From Table 1 we can see that when pH is 3.00, the permeability coefficient of La (III) is 4.47×10^{-7} m/s; when the pH increases to 3.60, the permeability coefficient is increased to 1.25×10^{-5} m/s, 30 times more than that of 3.00 pH; when pH is 4.00, the permeability coefficient increases to 1.49×10^{-5} m/s; on this basis to increase the pH to 4.30, the permeability coefficient is only increased by 3.00×10^{-7} m/s when pH is 4.00. Obviously, the best condition is when the liquid pH control is within 4.00.

During the migration process of La (III), the best material liquid pH is selected as 4.00, at 125 min, and the migration rate of rare earth metals in selected condition is 75.2%.

3.2. The Effect of HCl Concentration in Dispersed Phase on Rare Earth Metal Migration. The concentration difference of H+ in material liquid phase and dispersed phase is the mass transfer dynamic in DSLM for rare earth metal. We can also change the mass transfer dynamic in DSLM for the rare earth metal via changing the concentration of resolution agent in the dispersed phase based on the determination of the pH

TABLE 1: Effect of pH in feed phase on transport of rare earths.

Rare earth metal	(min) Migration Time	Item	Results				
La(III)	125	pH	3.00	3.30	3.60	4.00	4.30
		$-\ln c_t/c_0$	0.0503	0.483	1.40	1.67	1.71
		P_c (m/s)	4.47×10^{-7}	4.29×10^{-6}	1.25×10^{-5}	1.49×10^{-5}	1.52×10^{-5}

Note. Ct and C0 express concentration at time of t and the initial concentration of rare earth metal, respectively, unit: mol/L; Pc expresses permeability coefficient, the unit: m/s.

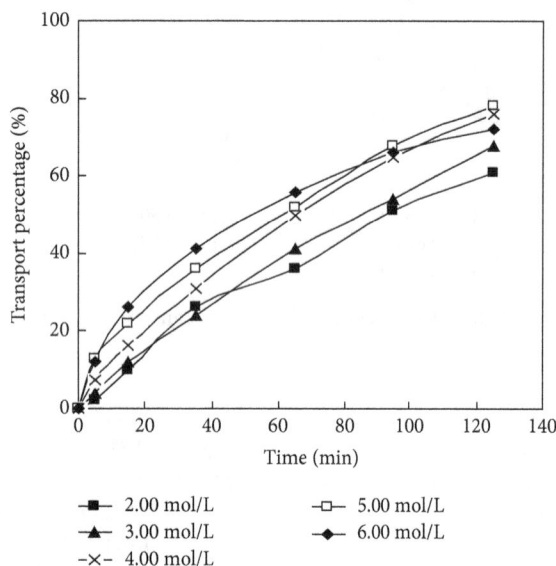

FIGURE 3: Effect of HCl concentration in dispersion phase on transport of La (III).

value in the material liquid phase. If the concentration of the resolving agent increases, resolving rate increases and the mobility will increase, too. But when the resolving agent concentration increases to a certain extent, the difference of H+ concentrations between the dispersed phase and the material liquid phase becomes too large that the corresponding osmotic pressure difference enlarges; therefore, it is possible that H+ would osmose from the dispersed phase to the material liquid phase. In this case, the membrane solution on the support body and the carrier will run off due to this process, resulting in the phenomenon of decrease in the mobility or instability in the membrane phase. Therefore, we need to study the impact of the concentration of the resolving agent HCl in the dispersed phase on the migration of the rare earth metal.

The material liquid phase pH of La (III) in DSLM migration system was selected as 3.6, the volume ratio of membrane solution and HCl in dispersed phase was 30: 30, and the concentration of carrier PC-88A in membrane solution was 0.160 mol/L. The initial concentration of La (III) was 1.00×10^{-4} mol/L. In the research on the impact of the concentration of HCl solution in dispersed phase of migration behavior of La (III) in DSLM under this condition, the experimental results are shown in Figure 3 and Table 2.

From Figure 3 we can find that when the HCl concentration of the resolving agent in the dispersed phase increases

from 2.00 mol/L to 5.00 mol/L, the migration rate tends to increase, which shows that the more the concentration of the agent becomes, the greater the resolving rate will be, and the migration rate will also be higher accordingly.

From Figure 3 we can see that when the concentrations of HCl are 2.00 mol/L, 3.00 mol/L, 4.00 mol/L, 5.00 mol/L, and 6.00 mol/L, at the time of 125 min, the migration rates of La (III) are 60.3%, 68.1%, 75.2%, 76.0%, and 71.7%, respectively. When the HCl concentration of the dispersed phase is 6.00 mol/L, from 0 to 85 minutes, the migration rate of La (III) is higher than 5.00 mol/L. But after 85 min, the migration rate began to decline, and at 125 min it decreased by 3.50% more than 4.00 mol/L, which attributes to the high acidity of the dispersed phase. Since the large concentration difference of H+ between dispersed phase and material liquid phase leads to large osmotic pressure difference between the dispersed phase and the material liquid phase, the H+ of dispersed phase is likely to permeate to the liquid phase. Meanwhile, the membrane solution on the supporting body is running off. Subsequently the carrier flows away, which results in the reduction of migration rate or instability of membrane phase. When the HCl concentration is 4.00 mol/L, the migration rate significantly presents higher than that of the 2.00 mol/L and 3.00 mol/L of HCl concentration. But when continuing to increase the concentration by 5.00 mol/L, the migration rate increases by only 0.80%. Tables 3-2 also show that the 4.00 mol/L and 5.00 mol/L HCl concentration and the permeability coefficient of the La (III) in DSLM are 1.24×10^{-5} m/s and 1.27×10^{-5} m/s, with only 3.00×10^{-7} m/s difference from each other. From the consideration of acidity control, the appropriate concentration of HCl in the dispersed phase should be 4.00 mol/L.

During migration of La (III), the required best HCl concentration is 4.00 mol/L, respectively, at 125 min, 75 min, 95 min, 130 min, 95 min, and 155 min and the migration rates of six kinds of rare earth metals in the selected condition are 75.2%, 91.2%, 73.5%, 80.6%, 73.5%, and 67.9%.

3.3. The Effect of Volume Ratio of the Membrane Solution over the HCl Solution on Rare Earth Metals Migration. Because the dispersed phase is the solution of HCl being dispersed uniformly in the membrane solution, the volume ratio between the membrane solution and the HCl solution directly affects the extraction and resolution rates of rare earth metals [28]. When the total volume of the dispersed phase and the concentration of the carrier and HCl are constant, the higher the ratio of HCl in dispersed phase is, the more unstable the dispersion becomes, which is unfavorable for the migration of rare earth metals. Moreover, with the

TABLE 2: Effect of HCl concentration in the dispersion phase on transport of rare earths.

Rare earth metal	(min) Migration n time	item	Results				
		HCl (mol/L) HCl concentration	2.00	3.00	4.00	5.00	6.00
La(III)	125	$-\ln c_t/c_0$	0.924	1.14	1.39	1.42	1.26
		P_c (m/s)	8.21×10^{-6}	1.02×10^{-5}	1.24×10^{-5}	1.27×10^{-5}	1.12×10^{-5}

Note. C_t and C_0 express concentration at time of t and the initial concentration of rare earth metal, respectively, unit: mol/L; Pc expresses permeability coefficient, the unit: m/s.

TABLE 3: Effect of volume ratio of membrane solution and HCl solution on transport of rare earth.

rare earth metal	(min) Migration Time (min)	item	Data results				
		volume ratio	10:50	20:40	30:30	40:20	50:10
La(III)	125	$-\ln c_t/c_0$	0.660	0.814	1.39	1.41	1.46
		P_c (m/s)	5.87×10^{-6}	7.24×10^{-6}	1.24×10^{-5}	1.26×10^{-5}	1.30×10^{-5}

Note. Ct and C0 express concentration at time of t and the initial concentration of rare earth metal, respectively, unit: mol/L; Pc expresses permeability coefficient, the unit: m/s.

increase of the HCl solution volume ratio, the membrane solution volume is reduced which means that the carrier number also decreases, so the extraction rate decreases, the resolution rate increases, and the mobility of rare earth metals decreases. When the volume ratio of the HCl solution decreases, the membrane solution volume will increase which means that the carrier number increases, so the extraction reaction rate increases, the resolution rate decreases, and the mobility of rare earth metals increases. When the volume ratio of HCl solution is reduced to a certain extent and then continually reduced, the complex rate of rare earth metals in the material liquid phase is reduced because of the fewer carrier number, as well as the decreased mobility of rare earth metals. The appropriate volume ratio of the membrane solution and HCl solution is the key to improve the mobility.

The material liquid phase pH of La (III) in DSLM migration system was selected as 3.6. The initial concentration of La (III) was 1.00×10^{-4} mol/L. HCl concentration in dispersed phase was 4.00 mol/L. The concentrations of carrier PC-88A in membrane solution were 0.160 mol/L, 0.160 mol/L, 0.100 mol/L, 0.160 mol/L, 0.100 mol/L, and 0.160 mol/L, respectively. Researching the impact of the volume ratio of the membrane solution over the HCl solution on the migration behavior of La (III) in DSLM under this condition, the experimental results are shown in Figure 4 and Table 3.

It can be seen from Figure 4 that La (III) mobility decreased in turn when the volume ratio of the membrane solution and HCl solution in the dispersed phase changed from 50:10 to 10: 50, and the La (III) mobility were 76.9%, 76.3%, and 75.8% when the volume ratios were 50: 10, 40: 20, and 30: 30, respectively, at 125 min. The mobility was only 55.7% and 48.3%, respectively, when the volume ratios were 20:40 and 10:50. Table 3 also shows that the permeability coefficient of La (III) in DSLM increases by only 6.00×10^{-7} m/s when the volume ratio increased from 30:30 to 50:10. And the permeability coefficient decreases significantly when the volume ratios are 20:40 and 10:50. It is the traditional SLM system, if the volume ratio was 0:60, the equivalence of the only resolution phase without the membrane solution

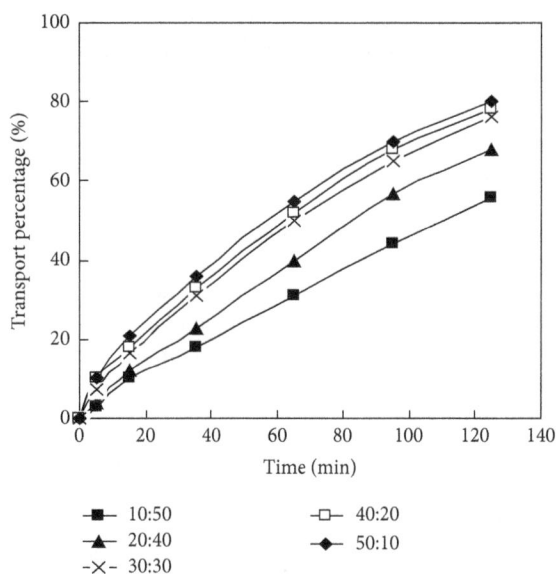

FIGURE 4: Effect of volume ratio of membrane solution and HCl solution on transport of La (III).

in the dispersed phase. It can be seen from the results that using DSLM system, when the membrane solution and HCl solution volume ratio in dispersed phase increases from 10:50 to 50:10, the mobility increases, which demonstrates the use of the dispersed phase instead of the traditional SLM resolution phase which helps to improve the mobility of SLM and which proves the superiority of DSLM. The appropriate volume ratio of the membrane solution and the HCl solution should be selected at about 30:30 from economical consideration.

By studying the impact of the volume ratio between the membrane solution and HCl solution in the dispersed phase on the migration behavior of rare earth metals in DSLM system, we can understand the following: the migration of rare earth metals in DSLM system is codetermined by the chemical reaction and the diffusion dynamics and is a dynamic equilibrium process. The entire process is controlled

TABLE 4: Effect of initial concentrations on transport of rare earth.

rare earth metal	(min) Migration Time (min)	item	Data results				
La(III)	125	(mol/L) initial concentration (mol/L)	5.00×10^{-5}	8.00×10^{-5}	1.00×10^{-4}	1.50×10^{-4}	2.00×10^{-4}
		$-\ln c_t/c_0$	~	2.80	1.67	1.12	0.751
		P_c (m/s)	~	2.49×10^{-5}	1.49×10^{-5}	9.97×10^{-6}	6.68×10^{-6}

Description: "~" represents undetectable, namely, full migration.

by the chemical reaction, that is, extraction reaction and resolution reaction, when the volume ratio between the membrane solution and HCl solution in the dispersed phase is relatively tiny. According to the principle of the chemical equilibrium, increasing the volume ratio of the membrane solution and HCl solution favors the formation of the carrier complex; therefore the mobility of rare earth metal increases rapidly; but when the volume ratio reaches a certain level, the concentrations of the carrier, rare earth metal, and complex at interface close to saturation. The diffusion process will play a decisive role, so the increase of rare earth metals mobility gradually slows down with the increase of the volume ratio. If the volume ratio continues to increase, the proportion of the resolution agent reduces. And resolution rate will inevitably decline; hence the migration rate also reduces.

In DSLM system and PC-88A as carrier, the best volume ratio of the membrane solution over HCl solution was 30:30 in La (III) migration process and the mobility of rare earth metal La (III) was 75.8% at 125 min under the selected conditions.

3.4. The Effect of the Initial Concentration on Rare Earth Metal Migration.

In certain DSLM system, if the initial concentration of rare earth ions is too large, the rare earth metal is not fully migrated within a certain period of time [25]; if the initial concentration of rare earth metal ions is too small, the contact rate of metal ions with membrane is very low. These will affect the migration rate. And the measurement range of the instrument for certain elements concentrations should be taken into consideration. Therefore, the initial concentration of material liquid phase has some influence on the migration behavior of rare earth metals.

From (1), the rare earth metal La (III) forms a complex through chemical reaction with the carrier PC-88A at the interface between the material liquid phase and the membrane phase. When the concentration of rare earth metals is relatively low, the balance shifts left and leads to a reduced mobility. Increasing the concentration of rare earth metals, the balance shifts to the right and the mobility increases. But the rare earth metal mobility is also affected by the carrier concentration and the membrane area. When the carrier concentration and membrane area are constant, the number of rare earth metal ions migrating per unit time is certain. So the rare earth metals mobility does not increase with the initial concentration increasing infinitively. When the rare earth metal concentration increases to a certain extent and

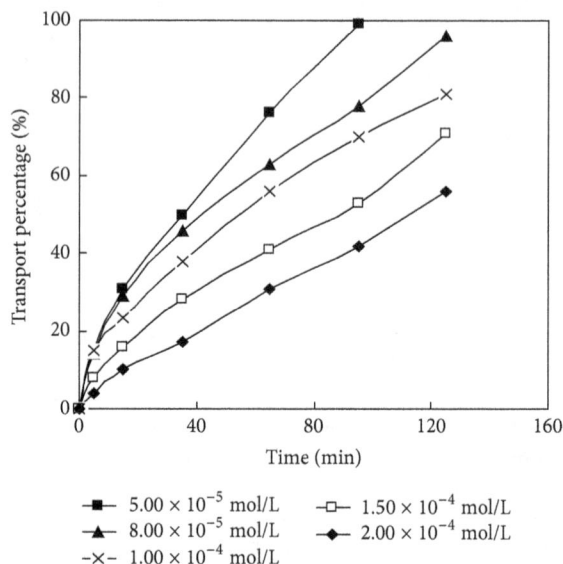

FIGURE 5: Effect of initial concentrations on transport of La(III).

then as the initial concentration increases, the mobility begins to decline.

The volume ratios of the membrane solution and HCl in dispersed phase of La (III) in DSLM migration system were selected as 30: 30, 40: 20, 30: 30, 30: 30, 40: 20, and 40:20. The HCl concentration in the dispersed phase is 4.00 mol/L. The concentration values of carrier PC-88A in the membrane solution were 0.160 mol/L, 0.160 mol/L, 0.100 mol/L, 0.160 mol/L, 0.100 mol/L, and 0.160 mol/L, respectively. The pH values of the material liquid phase were 4.00, 1.00, 5.20, 4.20, 5.00, and 5.10 respectively. Studying the impact of the initial concentration of rare earth metal in the material liquid phase on the migration behavior of La (III) in DSLM under this circumstances, the experimental results are shown in Figure 5 and Table 4.

It can be seen from Figure 5 that La (III) mobility was 93.9%, 81.2% and 67.4%, and 52.8%, respectively, when its initial concentrations were 8.00×10^{-5} mol/L, 1.00×10^{-4} mol/L, 1.50×10^{-4} mol/L, and 2.00×10^{-4} mol/L at 125 min. At this time, with the decrease of the initial concentration, the mobility and permeability coefficient are increased. When the initial concentration of La (III) is 5.00×10^{-5} mol/L, its migration rate reaches 99.8% at 95 min and no La (III) can be detected at 125 min which has fully migrated. But when La (III) concentration continues to decrease, the balance

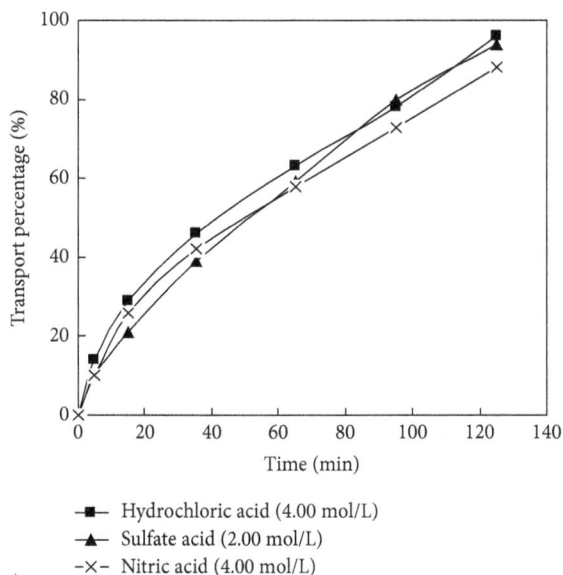

FIGURE 6: Effect of different stripping agents on transport of La(III).

shifts left and leads to reduced mobility and permeability coefficient.

3.5. The Effect of Different Analytical Agents on the Rare Earth Metals Migration. From the mass transfer mechanism of DSLM, the concentration difference of H+ in material liquid phase and dispersed phase is the mass transfer power for metal ions in DSLM [26]. Therefore, when the acidity of material liquid phase and analytical phase are determined, the mass transfer power is generally stable. However, its nature necessarily differs for different types of analytical agents and the ionic environment of analytical phase varies, so the mass transfer power will be slightly different. In this study, the analytical agents are strong acids. HCl, H2SO4, and HNO3 are the strong analytical agents commonly owning different natures. HCl becomes easily volatile and its Cl– is easy to form complex ion with metal ion; SO42– in H2SO4 is easy to form insoluble salts with metal ion; HNO_3 is oxidizing acid and easy to oxidize organic complexes when its concentrations are high and unfavorable for complex reaction between metal ions and carrier in an oxidizing environment. Therefore, the study of the impact of different analytic agents in dispersed phase on the migration behavior of rare earth is of certain significance.

The volume ratio of the membrane solution and HCl in dispersed phase of La (III) in DSLM migration system was selected as 30: 30. The H+ concentration in the dispersed phase was 4.00 mol / L. The concentration of carrier PC-88A in the membrane solution was 0.160 mol/L. The initial concentration was 8.00×10^{-5} mol/L. The pH value of the material liquid phase was 4.00. Researching the impact of different analytical agents in the dispersed phase on the migration behavior of La (III) in DSLM under this condition, the experimental results are shown in Figure 6.

The effects of HCl solution, H2SO4 solution, and HNO_3 solution on La(III) migration were studied, respectively, in this experiment, keeping the acidity of the analytic phase in dispersed phase stable.

As shown in Figure 6, La (III) mobility values were 93.9%, 94.0%, and 87.8% with HCl, H2SO4, and HNO3 as analytic agent, respectively, at 125 min. It can be seen that the HCl solution, the H2SO4 solution, and the HNO3 solution all have a certain effect on La (III) resolution, in which the 4.00 mol/L HCl solution and the 2.00 mol/L H2SO4 solution were better and the HNO_3 solution comes the third. This is because of the formation of a stable complex $LaCln_3$-n between Cl– in dispersed phase and La (III). Therefore, the 4.00 mol /L HCl was chosen as the analytic agent.

3.6. The Effect of the Carrier Concentration on the Rare Earth Metal Migration. The process of DSLM migration of rare earth metals is jointly nominated by the chemical reaction between rare earth metal complexes and the carrier and its diffusion process after bonding [28]. When the carrier is of a low concentration, the mobility is primarily controlled by the chemical reaction between the complex with the carrier. According to the principle of the chemical equilibrium, increasing the concentration of reactants favors complex formation of the carrier material, and therefore the rapider the mobility of metal ions increases, the higher the carrier concentration becomes and the more the adequate response takes place, the higher the mobility turns out. But when the concentration reaches a certain value, the interface concentration is close to saturation, increasing of the carrier concentration contributing to the increasing of the rare earth metal ions mobility levels off gradually. If the carrier concentration is too high and the membrane carrier concentration is saturated, the mobility is primarily controlled by the diffusion of complexes and carrier, which will clog the membrane pores to some extent and lead to the mobility dropping down. Thus, studying the effect of carrier concentration on the rare earth migration is necessary.

The volume ratios of the membrane solution and HCl in dispersed phase of La (III) in DSLM migration system were selected as 30: 30, 40: 20, 30: 30, 30: 30, 40: 20, and 40: 20; the HCl concentration in the dispersed phase is 4.00 mol / L; the initial concentrations were 8.00×10^{-5} mol/L, 7.00×10^{-5} mol/L, 1.00×10^{-4} mol/L, 8.00×10^{-5} mol/ L, 8.00×10^{-5} mol/L, and 1.00×10^{-4} mol/L, respectively. The pH values of material liquid phase are 4.00, 1.00, 5.20, 4.20, 5.00, and 5.10, respectively. Researching the impact of the carrier concentration in dispersed phase on the migration behavior of La (III) in DSLM under this condition, the experimental results are shown in Figure 7 and Table 5.

As can be seen from Figure 7, when the concentrations of PC-88A were 0.036 mol/L, 0.065 mol/L, 0.100 mol/L, 0.160 mol/L, and 0.230 mol/L, respectively, and when the initial concentration of La (III) was 8.00×10–5mol / L, the mobility was up to 73.7%, 86.2%, 90.3%, 93.9%, and 94.1% at 125 min. The PC-88A concentration increased; then La (III) mobility increased. When the PC-88A concentration increased from 0.036 mol/L to 0.065 mol/L, the La (III) mobility increased by 12.5%; when the PC-88A concentration increased from 0.065 mol/L to 0.100 mol /L, La (III) mobility increased 4.1%.

TABLE 5: Effect of carrier concentration on transport of rare earths.

rare earth metal	(min) Migration time (min)	item	Data results				
La(III)	125	(mol/L) carrier concentration (mol/L)	0.036	0.065	0.100	0.160	0.230
		$-\ln c_t/c_0$	1.34	1.98	2.33	2.80	2.83
		P_c (m/s)	1.19×10^{-5}	1.76×10^{-5}	2.16×10^{-5}	2.49×10^{-5}	2.52×10^{-5}

Note. Ct and C0 express concentration at time of t and the initial concentration of rare earth metal, respectively, unit: mol/L; Pc expresses permeability coefficient, the unit: m/s.

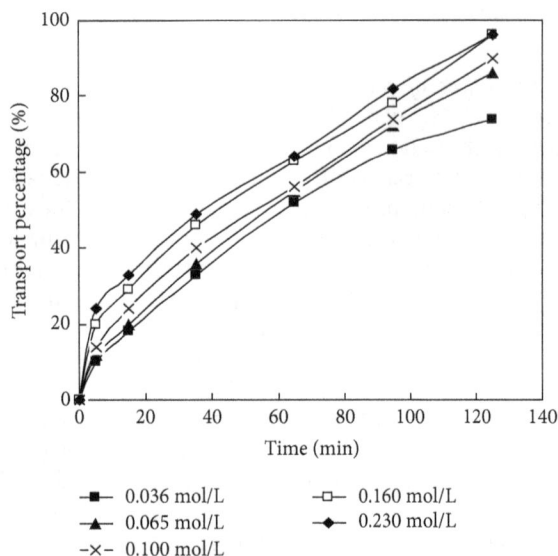

FIGURE 7: Effect of different carrier concentration on migration of La (III).

When PC-88A concentration increased from 0.100 mol /L to 0.160 mol/L, La (III) mobility increased by only 3.6 %. When the PC-88A concentration increased back to 0.230 mol/L, the mobility was 94.1% and only 0.2% higher than that of 0.160 mol /L. When the PC-88A concentration exceeded 0.160 mol/L, the La (III) mobility increased relatively smoothly and was tending towards stability. The La (III) mobility increased rapidly when there was increased carrier concentration in a low concentration range, because the whole process was controlled by the chemical reaction. According to the principle of the chemical equilibrium, increasing the concentration of reactants favors the formation of carrier complex; thus, rare earth metals mobility increases rapidly. But when the concentration reaches a certain level, the interface concentration closes to saturation. And increasing the carrier concentration makes the increasing of mobility of rare earth metals gradually level off. It also can be seen from Table 5 that when the PC-88A concentration increases from 0.036 mol/L to 0.065 mol/L, the osmotic coefficient of La (III) in DSLM increases from 1.19×10^{-5} m/s to 1.76×10^{-5} m/s by 5.70×10^{-6} m/s. When the PC-88A concentration increases from 0.065 mol/L to 0.100 mol/L, the osmotic coefficient increases 4.00×10-6 m/s. When the PC-88A concentration increases to 0.160 mol/L, the osmotic coefficient increases by only $3.30 \times$

10-6 m/s. So with the increasing of the concentration of PC-88A, the increasing of osmotic coefficient tends to level off gradually. On this basis, increasing the PC-88A concentration to 0.230 mol/L, the osmotic coefficient becomes 2.52×10^{-5} m/s, only 3.00×10^{-7} m/s higher than that of 0.160 mol/L. Thus, the concentration of the flow carrier PC-88A is selected as 0.160 mol/L.

Hence, the selected optimum carrier concentration of the dispersed phase in La (III) migration process is 0.160 mol/L and the mobility of rare earth metal La (III) is 93.9% under the selected conditions at 125 min.

4. Conclusions

(1) The experiments show that DSLM system of (PC-88A-) kerosene—HCl—has a significant role in enrichment and transmission of La (III). The acidity of the material liquid phase, the initial concentration of La (III), the concentration of HCl in the dispersed phase, and the volume ratio of HCI and the membrane solution will affect the transmission of La (III). During the rare earth metal migration process, the most appropriate analytical agent is HCl and the ionic strength of the feed phase has little effect on the migration behavior of rare earth metals in DSLM.

(2) The optimum mass transfer conditions for La (III) are that the concentration of HCl in dispersed phase is 4.00 mol / L, the volume ratio of membrane solution and HCl solution 30:30, the carrier concentration 0.160 mol / L, and the pH value of the feed phase 4.00. Under the optimal conditions, the migration rate reaches 93.9% after 125 min when the initial concentration of La (III) in the material liquid phase is 8.00×10–5 mol/L.

(3) Maintaining the dispersed phase acidity under the same premise, La (III) mobility is 93.9%, 94.0%, and 87.8%, respectively, after 125 min using HCl, H2SO4, and HNO3 as parsing agents. The HCl solution, H2SO4 solution, and HNO3 solution have some effect on La (III) resolving, in which the 4.00 mol/L HCl solution and 2.00 mol/L H2SO4 solution are better for resolving and then comes HNO3.

(4) La (III) is the best transmission condition in the separation of La (III) experiment when the concentration of the flow carrier PC-88A is selected at 4.00 mol/L. It proves that the best concentration value of HCl is 4.00

mol/L during the La (III) migration. At 125th min, 75th min, 95th min, 130th min, 95th min, and 155th min, the mobility of rare earth metal La (III) is 75.2%, 91.2%, 73.5%, 80.6%, 70.1%, and 67.9%, respectively, under the selected circumstances.

Acknowledgments

The research was funded by the National Natural Science Foundation of China (Grant nos. 51379219, 41371187) and Zhejiang province Funds for Distinguished Young Scientists (Grant no. LR15E090002).

References

[1] F. Valenzuela, C. Fonseca, C. Basualto, O. Correa, C. Tapia, and J. Sapag, "Removal of copper ions from a waste mine water by a liquid emulsion membrane method," *Minerals Engineering*, vol. 18, no. 1, pp. 33–40, 2005.

[2] L. G. Wang, C. J. Gao, and L. Wang, "Membrane integration technologyfor treatment of copper-bearing industrial wastewater," *Chin. Water Wastewater*, vol. 21, no. 3, pp. 83–85, 2005.

[3] L. Jelinek, Y. Wei, and M. Kumagai, "Adsorption of Ce(IV) anionic nitrato complexes onto anion exchangers and its application for Ce(IV) separation from rare earths(III)," *Journal of Rare Earths*, vol. 24, no. 4, pp. 385–391, 2006.

[4] A. A. Khan and A. Khan, "Electrical conductivity and cation exchange kinetic studies on poly-o-toluidine Th(IV) phosphate nano-composite cation exchange material," *Talanta*, vol. 73, no. 5, pp. 850–856, 2007.

[5] H. Abdul Aziz, A. H. Kamaruddin, and Z. M. Abu Bakar, "Process optimization studies on solvent extraction with naphthalene-2-boronic acid ion-pairing with trioctylmethylammonium chloride in sugar purification using design of experiments," *Separation and Purification Technology*, vol. 60, no. 2, pp. 190–197, 2008.

[6] "boronic acid ion-pairing with trioctylmethylammoniumchloride in sugar purification using design of experiments," Tech. Rep.

[7] Z. Jie, S. Chuanmin, Y. Guofeng, and X. Fei, "Separation and Enrichment of Rare Earth Elements in Phosphorite in Xinhua, Zhijin, Guizhou," *Journal of Rare Earths*, vol. 24, no. 1, pp. 413-414, 2006.

[8] D. He, S. Gu, and M. Ma, "Simultaneous removal and recovery of cadmium (II) and CN- from simulated electroplating rinse wastewater by a strip dispersion hybrid liquid membrane (SDHLM) containing double carrier," *Journal of Membrane Science*, vol. 305, no. 1-2, pp. 36–47, 2007.

[9] L. Pei, B. Yao, and C. Zhang, "Transport of Tm(III) through dispersion supported liquid membrane containing PC-88A in kerosene as the carrier," *Separation and Purification Technology*, vol. 65, no. 2, pp. 220–227, 2009.

[10] L. Pei, B. Yao, L. Wang, N. Zhao, and M. Liu, "Transport of Tb3+ in dispersion supported liquid membrane system with carrier P507," *Chinese Journal of Chemistry*, vol. 28, no. 5, pp. 839–846, 2010.

[11] J. V. Sonawane, A. K. Pabby, and A. M. Sastre, "Au(I) extraction by LIX-79/n-heptane using the pseudo-emulsion-based hollow-fiber strip dispersion (PEHFSD) technique," *Journal of Membrane Science*, vol. 300, no. 1-2, pp. 147–155, 2007.

[12] S. Gu, Y. Yu, D. He, and M. Ma, "Comparison of transport and separation of Cd(II) between strip dispersion hybrid liquid membrane (SDHLM) and supported liquid membrane (SLM) using tri-n-octylamine as carrier," *Separation and Purification Technology*, vol. 51, no. 3, pp. 277–284, 2006.

[13] D. He, X. Luo, C. Yang, M. Ma, and Y. Wan, "Study of transport and separation of Zn(II) by a combined supported liquid membrane/strip dispersion process containing D2EHPA in kerosene as the carrier," *Desalination*, vol. 194, no. 1-3, pp. 40–51, 2006.

[14] B. C. Zhang, G. Gozzelino, and G. Baldi, "State of art of the research on supported liquid membranes," *Membr. Sci.Technol*, vol. 20, no. 2, pp. 46–54, 2000.

[15] R. Guo, C. Hu, B. Li, and Z. Jiang, "Pervaporation separation of ethylene glycol/water mixtures through surface crosslinked PVA membranes: Coupling effect and separation performance analysis," *Journal of Membrane Science*, vol. 289, no. 1-2, pp. 191–198, 2007.

[16] C. J. Frizzell, M. In Het Panhuis, D. H. Coutinho et al., "Reinforcement of macroscopic carbon nanotube structures by polymer intercalation: The role of polymer molecular weight and chain conformation," *Physical Review B: Condensed Matter and Materials Physics*, vol. 72, no. 24, Article ID 245420, 2005.

[17] T. Singh and A. Kumar, "Fluorescence behavior and specific interactions of an ionic liquid in ethylene glycol derivatives," *The Journal of Physical Chemistry B*, vol. 112, no. 13, pp. 4079–4086, 2008.

[18] S. Lu and L. Pei, "A study on phenol migration by coupling the liquid membrane in the ionic liquid," *International Journal of Hydrogen Energy*, vol. 41, no. 35, pp. 15724–15732, 2016.

[19] L. PEI, B. YAO, and X. FU, "Study on transport of Dy(III) by dispersion supported liquid membrane," *Journal of Rare Earths*, vol. 27, no. 3, pp. 447–456, 2009.

[20] T. He, "Towards stabilization of supported liquid membranes: preparation and characterization of polysulfone support and sulfonated poly (ether ether ketone) coated composite hollow fiber membranes," *Desalination*, vol. 225, no. 1–3, pp. 82–94, 2008.

[21] S. Heitmann, J. Krings, P. Kreis et al., "Recovery of n-butanol using ionic liquid-based pervaporation membranes," *Separation and Purification Technology*, vol. 97, pp. 108–114, 2012.

[22] A. Plaza, G. Merlet, A. Hasanoglu, M. Isaacs, J. Sanchez, and J. Romero, "Separation of butanol from ABE mixtures by sweep gas pervaporation using a supported gelled ionic liquid membrane: Analysis of transport phenomena and selectivity," *Journal of Membrane Science*, vol. 444, pp. 201–212, 2013.

[23] A. B. Beltran, G. M. Nisola, E. L. Vivas, W. Cho, and W.-J. Chung, "Poly(octylmethylsiloxane)/oleyl alcohol supported liquid membrane for the pervaporative recovery of 1-butanol from aqueous and ABE model solutions," *Journal of Industrial and Engineering Chemistry*, vol. 19, no. 1, pp. 182–189, 2013.

[24] A. I. López-Lorente, B. M. Simonet, and M. Valcárcel, "The potential of carbon nanotube membranes for analytical separations," *Analytical Chemistry*, vol. 82, no. 13, pp. 5399–5407, 2010.

[25] K. F. Yee, Y. T. Ong, A. R. Mohamed, and S. H. Tan, "Novel MWCNT-buckypaper/polyvinyl alcohol asymmetric membrane for dehydration of etherification reaction mixture:

Fabrication, characterisation and application," *Journal of Membrane Science*, vol. 453, pp. 546–555, 2014.

[26] S. Trivedi and S. Pandey, "Interactions within a [ionic liquid + poly(ethylene glycol)] mixture revealed by temperature-dependent synergistic dynamic viscosity and probe-reported microviscosity," *The Journal of Physical Chemistry B*, vol. 115, no. 22, pp. 7405–7416, 2011.

[27] S. Y. Hu, Y. Zhang, D. Lawless, and X. Feng, "Composite membranes comprising of polyvinylamine-poly(vinyl alcohol) incorporated with carbon nanotubes for dehydration of ethylene glycol by pervaporation," *Journal of Membrane Science*, vol. 417-418, pp. 34–44, 2012.

[28] M. Shahverdi, B. Baheri, M. Rezakazemi, E. Motaee, and T. Mohammadi, "Pervaporation study of ethylene glycol dehydration through synthesized (PVA-4A)/polypropylene mixed matrix composite membranes," *Polymer Engineering & Science*, vol. 53, no. 7, pp. 1487–1493, 2013.

21

Simultaneous Determination of Cr, As, Se, and other Trace Metal Elements in Seawater by ICP-MS with Hybrid Simultaneous Preconcentration Combining Iron Hydroxide Coprecipitation and Solid Phase Extraction using Chelating Resin

Akihide Itoh [ID],[1] **Masato Ono,**[2] **Kota Suzuki,**[1] **Takumi Yasuda,**[1] **Kazuhiko Nakano,**[1] **Kimika Kaneshima,**[1] **and Kazuho Inaba**[1]

[1]*Department of Environmental Science, School of Life and Environmental Science, Azabu University, 1-17-71, Fuchinobe, Chuo-ku, Sagamihara-shi, Kanagawa 252-5201, Japan*
[2]*GL Science Inc., Shinjuku Square Tower 30F, 6-22-1 Nishi Shinjuku, Shinjuku-ku, Tokyo 163-1130, Japan*

Correspondence should be addressed to Akihide Itoh; a-ito@azabu-u.ac.jp

Guest Editor: Marcela Z. Corazza

In the present study, ICP-MS with a new hybrid simultaneous preconcentration combining solid phase extraction using chelating resin and iron hydroxide coprecipitation in one batch at a single pH adjustment (pH 6.0) were developed for multielement determination of trace metal ions in seawater. In multielement determination, the present method makes it possible to determine Cr(III), As(V), Se (IV), and other 14 trace metal elements (Ti, V, Co, Ni, Cu, Zn, Zr, Ge, Cd, Sb, Sn, W, Pb, and U) in seawater. Moreover, for speciation analyses of Cr, As, and Se, the pH dependence on recovery for the different chemical forms of Cr, As, and Se was investigated. In speciation analyses, Cr, As, and Se were determined as the total of Cr (III) and a part of Cr (VI), total of As (III) and As (V), and Se(IV), respectively. Determination of total of Se and Cr(VI) remains as future task to improve. Nevertheless, the present method would have possibility to develop as the analytical method to determine comprehensively most metal elements in all standard and guideline values in quality standard in environmental water in Japan, that is, most toxic metal elements in environmental water.

1. Introduction

ICP-MS has excellent analytical features such as simultaneous multielement capability, extremely high sensitivity, and wide linear dynamic range for most metal elements [1–3]. So, ICP-MS makes it possible to determine comprehensively almost all heavy metals, whose standard values or guideline values were established in water quality standards for human health relating to water pollution in Japan, without any special preconcentration. However, the measurement of trace metals such as heavy metals in seawater is difficult even using ICP-MS, because the salt contents in seawater are approximately 3.5% and they cause not only matrix effect and spectral interference but also the clogging of the torch top and the orifice of cone in ICP-MS [3]. These days, a high matrix introduction (HMI) unit permits the direct introduction of seawater into ICP-MS. In addition, novel ICP-MS with tandem quadrupole mass spectrometer (QMS/QMS) as well as with an octapole reaction cell (ORC) have become commercially available, which provides efficient removal of spectral interferences due to oxide species [4]. However, ICP-MS with preconcentration and desalting remains the most efficient method for the simultaneous and sensitive determination of trace metal elements in seawater without spectral interference and matrix effect.

The chelating resin preconcentration method has excellent analytical features of nonselective multielement determination for many trace elements in seawater, along with efficient removal of matrix elements such as Na, K, Ca, and Mg [5–10]. However, it is found that the chelating resin provided poor recoveries for some oxoanion-forming elements, such as As, Se, and Cr, which are toxic and important in environmental sciences. Coprecipitation methods [11] such as lanthanum hydroxide [12–14], iron hydroxide [15–17], yttrium hydroxide [18], and magnesium hydroxide [19] are also effective as other preconcentration methods to complement chelating resin technique, because both oxoanion-forming elements and cation-forming trace elements can be concentrated using this method. However, the coprecipitation methods do not allow one to analyze toxic metal elements comprehensively, although they provided good recoveries for some oxoanion-forming elements and/or a part of transition metals. In addition, coprecipitation carrier results in high concentration of matrix components. Accordingly, performing both chelating resin preconcentration and coprecipitation complementally under control of matrix components is effective to determine many trace metal elements including toxic ones simultaneously. Yabutani et al. developed the tandem preconcentration method based on chelating resin adsorption and lanthanum hydroxide coprecipitation method that was continuously used to determine the oxoanion-forming elements and other trace elements [20]. However, this tandem method was time-consuming, because a series of preconcentration procedures including pH adjustment need to be performed for the coprecipitation after the chelating resin preconcentration. In contrast, a proposed new hybrid simultaneous preconcentration method was examined as a batch method that uses a single pH adjustment to achieve solid phase extraction using chelating resin and iron hydroxide coprecipitation. Thus, the potentials of ICP-MS with the present hybrid preconcentration method were investigated for simultaneous preconcentration and determination of the oxoanion-forming elements such as Cr, As, Se, and other trace metal elements, whose standard and guideline values were established in environmental quality standards for water pollution in Japan, even in seawater containing high concentrations of salts.

2. Materials and Methods

2.1. Instruments. An ICP-MS instrument (Agilent 7700x, Agilent Technologies Co., Tokyo, Japan), equipped with a quadrupole mass spectrometer and an octapole reaction cell (ORC), was used for the determination of trace metals in preconcentration solution of seawater. The operating conditions for the ICP-MS instrument were summarized in Table 1. In the ICP-MS measurement, the internal standard correction was performed using Be, In, and Tl as internal standard elements to correct matrix effects due to major elements [21]. The purified water (18.2 MΩ cm) used throughout the present experiment was prepared by a Milli Q SP-TOC system (Nihon Millipore Kogyo, Tokyo, Japan).

2.2. Chemicals. The standard solutions for making the calibration curves in the ICP-MS measurements were prepared

TABLE 1: Operating conditions for the ICP-MS instrument.

ICP-MS: Agilent 7700x	
Plasma conditions:	
RF power	1.55 kW
Plasma gas flow rate	15.0 L min^{-1} Ar
Auxiliary gas flow rate	0.90 L min^{-1} Ar
Makeup gas flow rate	0 L min^{-1} Ar
Carrier gas flow rate	1.05 L min^{-1} Ar
Sampling depth (mm from load coil)	8.0 mm
Cell gas	He mode:4.3 mL min^{-1}
	H$_2$ mode:6.0 mL min^{-1}
Nebulizer	Micro Mist
Sample uptake rate	0.45 mL min^{-1}
Data acquisition:	
Accumulation time	0.3-1.0 s / point
Data point	3 points / peak
Repetition	3 times

by diluting commercial multielement standard stock solutions (XSTC-622, 35 elements, 10 mg L^{-1} each), which were purchased from SPEX (Metuchen, NJ, USA). As, Cr, and Se were involved as As(V), Cr(III), and Se(IV) in XSTC-622, respectively. Nitric acid, hydrochloric acid, acetic acid, and aqueous ammonia solution were of electronics industry grade (Kanto Chemical Co., Tokyo, Japan). A single-element standard stock solution of Fe 10000 mg L^{-1} for general tests in the Japanese pharmacopoeia (Wako Pure Chemical Industries Inc., Osaka, Japan) was used as the iron solution for iron hydroxide coprecipitation.

The added standard solutions to investigate the recovery values were prepared as follows. The standard stock solutions for Cr(III), Cr(VI), and As(V) were prepared by diluting chromium(III) standard for ICP, chromium(VI) standard for ICP, and Arsenic (V) standard solution (1000 mg L^{-1} each, Merck, Darmstadt, Germany), respectively. The standard stock solutions for As(III) and Se(IV) were prepared by diluting standard solution of arsenic (III) and selenium (IV) for chemical analysis (1000 mg L^{-1} each, Kanto Chemical Co.), respectively. The standard solution for Se(VI) was prepared by extra grade of dissolving sodium selenite (Wako Pure Chemical Industries Inc.) in ultrapure water.

The chelating resin particles (InertSep ME2, 60-70 μm in diameter, GL Science Inc., Tokyo, Japan) have iminodiacetic acid (pK_a = 2.98) and dimethylamino (pK_a =10.77) groups on methacrylate resin. This resin was beforehand conditioned with ethanol, 2 M HNO$_3$, purified water, and 0.1 M ammonium acetate solution, which was used for chelating resin preconcentration of seawater samples. The ammonium acetate solution (pH 6) used for the pH adjustment was prepared by mixing equivalent molar amounts of acetic acid and ammonia solution.

The artificial seawater was prepared as follows using some reagents of extra grade purchased from Wako Pure

Chemical Industries Inc.: 28.5 g of sodium chloride, 6.82 g of magnesium sulfate heptahydrate, 5.16 g of magnesium chloride hexahydrate, 1.47 g of calcium chloride dehydrate, 0.725g of potassium chloride, 0.084 g of sodium bromide, and 0.0273 g of boric acid were dissolved in ultrapure water. Then, the volume of the solution was adjusted to be 1 L with ultrapure water.

2.3. Procedure of the Hybrid Preconcentration Combing Iron Hydroxide Coprecipitation and Solid Phase Extraction Using Chelating Resin. In the preconcentration procedure, 50 mL of a sample solution was initially taken in a 50 mL plastic bottle (DigeTUBEs, SCP SCIENCE, Canada, Montreal) and then 250 mg of the chelating resin particles, 50 μL of 10000 mg L^{-1} iron standard solution, 1 mL of 1.0 M ammonium acetate (buffer solution), and 100 μL of 400 μg L^{-1} methyl red solution (pH indicator) were added into the sample solution. The pH of the sample solution was adjusted to 6.0±0.1 using 6 M ammonia solution. As described in Results and Discussion, the optimal pH value for seawater samples in the iron hydroxide precipitation was pH 6.0, which was also appropriate for solid phase extraction using chelating resin. In the process of hybrid preconcentration procedure, it is considered that some cation-forming trace metals were complexed on the surface of the chelating resin particles and some oxoanion-forming elements such as Cr, As, and Se were adsorbed on and/or occluded in iron hydroxides coprecipitation. Changes in color were monitored during the pH adjustment for a preliminary estimation, but the final pH of the sample solution was confirmed using with a compact pH meter (Twin pH meter, Horiba Ltd., Kyoto, Japan). Next, the iron hydroxide coprecipitations formed were ripened for 2 h at 70°C in a water bath. Then, all the solutions were filtered using ϕ25 mm the glass fiber filter (Digi Filter, SCP SCIENCE, Canada, Montreal) with pore size of 1 μm, and the chelating resin particles and iron hydroxide coprecipitates were collected on the filter simultaneously. They were then washed with 50 mL purified water several times. Trace metals from the chelating resin particles and iron hydroxide coprecipitates were eluded with an acid solution. In the elution process, 5 mL of 2.6 M nitric acid solutions was added divisionally twice (2 and 3 mL) into chelating resin particles and iron hydroxide coprecipitates on the filter. When the first 2 mL of 2.6 M nitric acid solution was added, the solution was maintained for 10 min before removing to dissolve adequately the iron hydroxide coprecipitates. Subsequently, 5 mL of ultrapure water and 3 mL of the 2.6 M nitric acid solution were added as elutants and aspirated by vacuum pump (MDA-05, ULVAC, Chigasaki, Japan). Totally, 10 mL of elutant was obtained per each sample. Hence, ca. 5-fold preconcentration was achieved for the analysis solution in the present procedure, which was subjected to the simultaneous multielement determination of Cr, As, and Se and other trace metals using ICP-MS.

In the recovery test for investigation for the added amount of Fe^{3+} and analysis of coastal seawater, 50 μL of a mixed standard solution (XSTC-622, SPEX) containing 35 trace metals (1.0 mg L^{-1} each), in which Cr, As, and Se were contained as As(V), Cr(III), and Se(IV), was added into 50 mL of an artificial seawater. Then, the preconcentration procedure described above was carried out for the spiked and the unspiked artificial seawaters to estimate the recovery values. The recovery values for analyte elements were obtained as the percentages of the differences between the amounts of analyte elements in the spiked and unspiked artificial seawater samples, which was measured by ICP-MS after the present hybrid simultaneous preconcentration procedure, to the amounts added to the artificial seawater (50 ng each). In recovery test for speciation analysis of Cr, As, and Se, two kinds of recovery tests were separately performed using the stock solution for lower oxidation state of Cr(III), As(III), and Se(IV) and that for higher oxidation state of Cr(VI), As(V), and Se(VI) to avoid the oxidation-reduction reaction among these elements. 50 μL of the mixed standard stock solution of Cr(III), As(III), and Se(IV) (1.0 mg L^{-1} each) was added to the artificial seawater sample to prepare 50 mL of the spiked test solution for the lower oxidation state, and those of Cr(VI), As(V), and Se(VI) (1.0 ngL^{-1} each) were added to another sample of the artificial seawater to prepare 50 mL of the spiked test solution for the higher oxidation state. These two kinds of the spiked test solutions and unspiked artificial seawater without addition of any standard solution were analyzed by ICP-MS with the present hybrid simultaneous preconcentration to investigate the pH dependence on recovery of each oxidation state of Cr, As, and Se. The recovery for each chemical form of Cr, As, and Se was calculated in a similar manner to the recovery test for investigation for the added amount of Fe^{3+} and analysis of coastal seawater described above.

2.4. Seawater Samples. Seawater samples were collected at the Senzu coast in Izu-Oshima Island, Tokyo, Japan. Collected samples were filtered through the membrane filters of ϕ47 mm with a pore size 0.45 μm (Omnipore filter, Millipore, Bedford, MA, USA) immediately after sampling. The dissolved samples filtered with the membrane filters were acidified to pH 1 by adding concentrated HNO$_3$ (EL grade, Kanto Chemical Co.) and then subjected to hybrid preconcentration.

3. Results and Discussion

3.1. Investigation for Added Amounts of Fe^{3+}. In the present study, iron hydroxide coprecipitation was employed along with solid phase extraction using chelating resin to develop a hybrid simultaneous preconcentration method, because Fe(OH)$_3$ precipitates have a positive charge at pH 4-8 [22]. Moreover, Fe(OH)$_3$ precipitates can form at pH 5-6 [16]: this acidic condition is also optimal for solid phase extraction using chelating resin. In iron hydroxide coprecipitation, the added amounts of Fe^{3+} are generally 1.5-50 mg for 50 mL of each seawater sample [15, 16]. In the present study, however, it was set as 0.5 mg for 50 mL samples, because the added

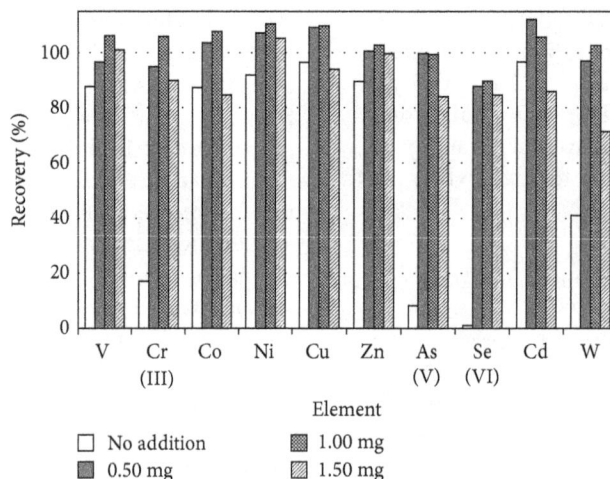

FIGURE 1: Comparison of recoveries of V, Cr(III), Co, Ni, Cu, Zn, As(V), Se(IV), Cd, and W in changing the added amount of Fe^{3+} in the hybrid preconcentration.

amount of Fe^{3+} as coprecipitation carrier should be kept minimal to decrease total matrix concentration and not to block the performance of chelating resin particles. Thus, the optimal added amount of Fe^{3+} was investigated in the present hybrid simultaneous preconcentration. When 0 mg, 0.50 mg, 1.0 mg, and 1.5 mg of Fe^{3+} were added to 50 mL of seawater sample with 250 mg of powdered chelating resin particles, respectively, and adjusted to pH 6.0, the recoveries of V, Cr(III), Co, Ni, Cu, Zn, As(V), Se(IV), Cd, and W are shown in Figure 1. As can be seen in Figure 1, the recoveries of most analyte elements were found to be higher at 0.50 mg and 1.0 mg. From this result, it was determined that the optimal amount of added Fe^{3+} was 0.50 mg, which resulted in the smaller total matrix and provided the smaller blank values. Then, Fe concentration was maintained below 50 mg L^{-1} in the 5-fold concentrated solutions and total residual concentrations of major ions such as Mg^{2+}, Ca^{2+}, Na^+, and K^+ were below 20 mg L^{-1}. Thus, the matrix effect caused by the added Fe and the residual concentration of major ions was so small to be corrected using the internal standard method [21].

3.2. Comparison of Recoveries in the Hybrid Preconcentration with Those in Solid Phase Extraction Using Chelating Resin. The recovery of each element in the developed hybrid preconcentration was investigated. The results are shown in Figure 2 with those in a single solid phase extraction using chelating resin. As seen in Figure 2, for single solid phase extraction using chelating resin, the recoveries of Cr(III), Ge, As(V), Se(IV), Zr, and Sb were very low below 10%, and those of Sn and W were below 60%, whereas those of V, Co, Ni, Cu, Zn, Cd, Pb, and U were over 80% and high enough for determination of trace metals in seawater. As reported in previous studies [14], it is considered that oxoanion-forming elements such as Cr, Ge, As, Se, Sb, and W provided poor recoveries in single chelating resin preconcentration, which may be ascribed to their low adsorption on the chelating resin with cation-exchange functional groups.

However, in the present hybrid preconcentration, the recoveries of oxoanion-forming elements such as As(V), Cr(III), Se(IV), Ge, Sb, and W were remarkably higher than those in single solid phase extraction using chelating resin, as seen in Figure 2. Particularly, the recoveries of As(III), Cr(III), Se(IV), and W were over 80%. Moreover, those of Ti, Cd, Zr, and Sn were 70-95% and become remarkably higher than those in single chelating resin preconcentration. The precisions of the recoveries were almost below 5% for the present hybrid method. Recovery values and precisions obtained were high enough to determine simultaneously the oxoanion-forming elements and other trace metals.

3.3. pH Dependence on the Recovery of Different Chemical Forms of Cr, As, and Se for Speciation Analysis. Cr, As, and Se are present as two different oxidation states in seawater and environmental water. Cr exists as either Cr^{3+} or CrO_4^{2-} (VI) in seawater, As as either AsO_3^{3-} (III) or AsO_4^{3-} (V), and Se as either SeO_3^{2-} (IV) or SeO_4^{2-} (VI) [23]. Because the toxicity and bioavailability of these elements to aquatic animals and plants depend on the oxidation state, speciation analysis for these elements was very important to evaluate the effects on the aquatic ecosystem. Thus, the pH dependence on recovery of each oxidation state of Cr, As, and Se in seawater samples was investigated for speciation analysis. The results are shown in Figure 3. It can be seen from Figure 3 that the optimal pH to recover Se (IV), As (III), As(V), and Cr(III) simultaneously was pH 6.0. At this pH, the recovery of As was over 80%, regardless of the oxidation state. On the other hand, the recoveries of Cr(VI) and Se(VI) were ca.30% and 2%, respectively, whereas those of Cr(III) and Se (IV) were 100% and 80%, respectively. Moreover, even in different pH value, the recovery of each chemical form was not enough for its determination. Therefore, the optimal pH is also 6.0 for the speciation analyses that allows determining separately each chemical form of Cr, As, and Se. Under these conditions, the total concentrations of As, the sum of As(III) and As (V), and the concentration of Se(IV) were determined.

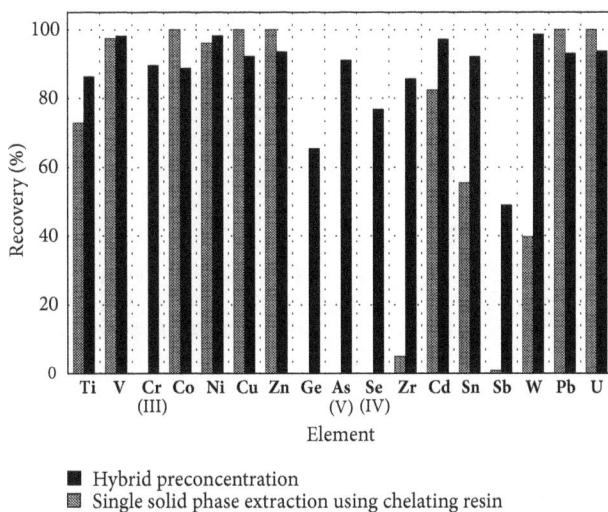

FIGURE 2: Comparison of the recoveries of Cr, As, Se, and other trace metal elements by the hybrid preconcentration with those by single preconcentration using chelating resin.

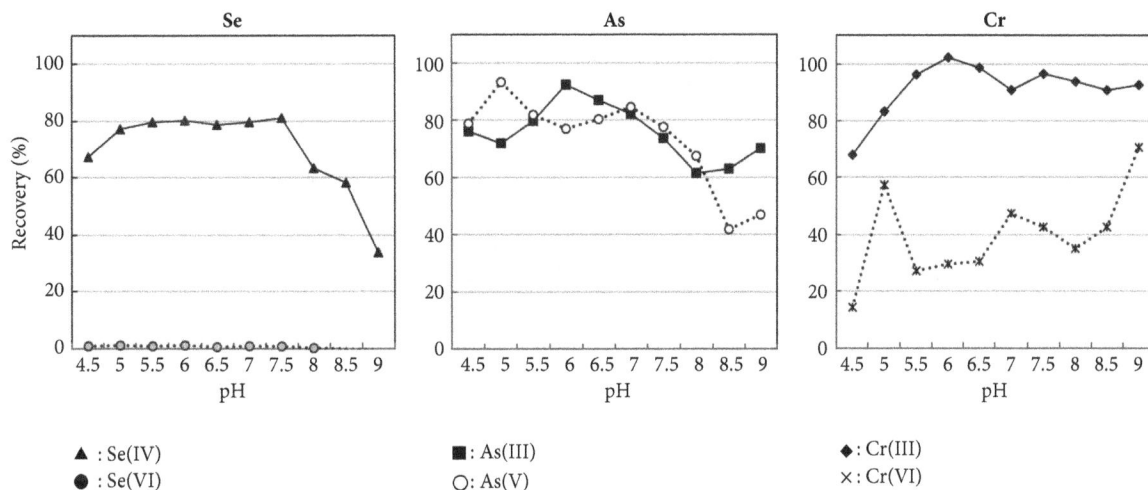

FIGURE 3: pH dependence on recovery of different chemical forms of Cr, As, and Se.

However, it was difficult to determine separately Cr(III) and Cr(IV).

3.4. Simultaneous Determination of Cr, As, and Se and Other Trace Metals in Coastal Seawater.
Analytical results for the oxoanion-forming elements and other trace metals dissolved in coastal seawater collected at Izu-Oshima Island are shown in Table 2 with the recovery values, blank values, instrumental detection limits (DL_{instru}), and analytical detection limits (DL_{anal}). The recovery values were calculated as the percentages of analyte element amounts recovered after preconcentration to those added before preconcentration (50 ng each), as described in "Experimental" section. The recovery of As in Table 2 was obtained for As (V). However, as the recovery of As(III) was as high as that of As(V) at pH 6.0 in Figure 3, the analytical result of As in coastal seawater is shown as As (III+V) in Table 2. The DL_{instru} of the analyte elements was obtained at the concentrations corresponding

to 3-fold the standard deviation (3σ) of the background signal intensities for the blank solution (2 M HNO_3), where the standard deviation (σ) was calculated from 10-times repeated measurement at each mass number. As it is confirmed that the linearities of the calibration curves for all analyte elements ranged from DL_{inst} to over 100 ng L^{-1}, all analyte elements in the concentrated sample solutions were measured within the liner range of the calibration curves. The DL_{anal} was estimated from the instrumental detection limits, taking into consideration the concentration factors and recovery values. The concentrations for 17 trace metals, which were corrected by the recovery values, concentration factors, and blank values, are shown in Table 2. The observed values and relative standard deviations (RSDs) were estimated from mean values and standard deviation (σ) of the independent 3-times analyses. As can be seen in Table 2, the concentrations of Cr (III), As (III+V), Se (IV), and other trace metals (Ti, V, Co, Ni, Cu, Zn, Ge, Zr, Cd, Sb, Sn, W, Pb, and U) were in the

TABLE 2: Analytical results for Cr, As, Se and other trace metals in coastal seawater (Izu Oshima, the senzu coast) determined by ICP-MS with hybrid preconcentration combining iron hydroxide coprecipitation and solid phase extraction using chelating resin.

Element	Cell gas mode	m/z	Concentration Observed [a] / μgL^{-1}	RSD/%	Recovery Mean [a] / %	RSD/%	Blank value/ μgL^{-1}	DL_{instru} [b] / μgL^{-1}	DL_{anal} [c] / μgL^{-1}
Ti	He	47	0.027 ± 0.009	32.7	96.9 ± 0.85	0.88	n.d. [d]	0.072	0.0074
V	He	51	1.87 ± 0.08	4.0	100.5 ± 0.99	0.99	0.001	0.0024	0.00024
Cr (III)	He	52	0.24 ± 0.02	7.4	96.3 ± 0.5	0.56	0.018	0.013	0.0013
Co	He	59	$0.012* \pm 0.001$	7.1	88.6 ± 0.57	0.64	0.006	0.0012	0.00014
Ni	He	60	0.22 ± 0.01	5.8	105.0 ± 3.2	3.3	0.012	0.0031	0.0031
Cu	He	63	0.17 ± 0.02	10.3	104.7 ± 1.9	2.0	0.034	0.021	0.0021
Zn	He	66	1.85 ± 0.03	1.8	86.8 ± 1.9	2.2	0.077	0.058	0.0067
Ge	He	72	$0.72* \pm 0.04$	5.2	65.5 ± 1.9	2.8	0.31	0.0018	0.00027
As (III + V)	He	75	1.34 ± 0.05	4.0	90.8 ± 0.84	0.92	0.0012	0.0024	0.00026
Se (IV)	H_2	78	0.0324 ± 0.0002	0.8	71.1 ± 0.17	0.24	n.d. [d]	0.0021	0.00029
Zr	He	90	0.055 ± 0.004	6.9	85.8 ± 2.3	2.6	n.d. [d]	0.0072	0.00084
Cd	He	111	0.013 ± 0.002	11.7	91.5 ± 0.53	0.58	n.d. [d]	0.0006	0.00007
Sn	He	118	0.065 ± 0.008	12.7	99.1 ± 0.57	0.57	0.0061	0.011	0.0011
Sb	He	121	0.23 ± 0.03	11.5	49.1 ± 0.39	0.80	0.0008	0.0003	0.00006
W	He	182	0.020 ± 0.001	5.1	98.6 ± 0.16	0.17	0.002	0.0029	0.00029
Pb	He	208	0.087 ± 0.01	10.4	98.9 ± 3.8	3.9	0.0023	0.0055	0.00056
U	He	238	3.0 ± 0.2	5.2	103.0 ± 0.010	0.010	n.d. [d]	0.0011	0.00011

(a) Mean $\pm \sigma$(standard deviation),n=3, The observed values with asterisk were corrected by the blank values over 10% of the observed values. (c) DL_{anal}: analytical detection limit. (d) Not detected.

(b) DL_{instru}: instrumental detection limit.

range of 3.0 μg L^{-1} for U to 0.012 μg L^{-1} for Co, which were determined with low RSDs below ca. 10% except for Ti. The recovery values for most elements in Table 1 were large (more than 85%) enough to obtain reliable analytical data, whereas those for Ge, Se(IV), and Sb were below 70%. However, they were employed for correction of the determined values, because the RSDs for these recoveries were below 3% and the precisions were very high. The blank values for most elements were below sub μg L^{-1} and low enough to correct the determined values. However, those for Co and Ge were over 10% of the observed values and relatively high. Therefore, these determined values were showed with asterisk in Table 2, as they may be less reliable than those for other elements.

4. Conclusion

In the present study, ICP-MS with a new hybrid simultaneous preconcentration combining the solid phase extraction using chelating resin and iron hydroxide coprecipitation in one batch at a single pH adjustment (pH 6.0) were developed for multielement determination of trace metal ions in seawater. In multielement determination, the present method made it possible to determine Cr(III), As(V), Se(IV), and 14 other trace metal elements (Ti, V, Co, Ni, Cu, Zn, Ge, Zr, Cd, Sb, Sn, W, Pb, and U). However, in speciation analyses, Cr, As, and Se were determined as the total of Cr (III) and a part of Cr (VI), total of As (III) and As (V), and Se(IV), respectively. Determination of total of Se and Cr (VI) remains as future task to improve. Nevertheless, the present method would have possibility to develop as the analytical method to determine comprehensively most metal elements in all standard and guideline values in quality standard in environmental water in Japan, that is, most toxic metal elements in environmental water.

Acknowledgments

We would like to thank Mr. Kunihisa Yamaguchi and his colleagues in Metropolitan Islands Area Research and Development Center of Agriculture, Forestry and Fisheries for sampling of coastal seawater in Izu-Oshima Island, Japan.

References

[1] M. Dziewatkoski, *Inductively Coupled Plasma Mass Spectrometry*, A. Montaser, Ed., Wiley-VCH, New York, NY, USA, 1998.

[2] H. E. Taylor, *Inductively Coupled Plasma-Mass Spectrometry*, Academic Press, San Diego, Calif, USA, 2001.

[3] H. Haraguchi, "Multielement Profiling Analyses of Biological, Geochemical, and Environmental Samples as Studied by Analytical Atomic Spectrometry," *Bulletin of the Chemical Society of Japan*, vol. 72, pp. 1163–1186, 1999.

[4] Y. Zhu and A. Itoh, "Direct Determination of Cadmium in Seawater by Standard Addition ICP-QMS/QMS with an ORC," *Analytical Sciences*, vol. 32, pp. 1301–1305, 2016.

[5] T. Yabutani, S. Ji, F. Mouri et al., "Multielement Determination of Trace Elements in Coastal Seawater by Inductively Coupled Plasma Mass Spectrometry with Aid of Chelating Resin Preconcentration," *Bulletin of the Chemical Society of Japan*, vol. 72, p. 2253, 1999.

[6] H. Sawatari, T. Toda, T. Saizuka, C. Kimata, A. Itoh, and H. Haraguchi, "Multielement Determination of Rare Earth Elements in Coastal Seawater by Inductively Coupled Plasma Mass Spectrometry after Preconcentration Using Chelating Resin," *Bulletin of the Chemical Society of Japan*, vol. 68, pp. 3065–3070, 1995.

[7] Y. Zhu, A. Itoh, and H. Haraguchi, "Multielement Determination of Trace Metals in Seawater by ICP-MS Using a Chelating Resin-Packed Minicolumn for Preconcentration," *Bulletin of the Chemical Society of Japan*, vol. 78, pp. 107–115, 2005.

[8] Y. Zhu, A. Itoh, E. Fujimori, T. Umemura, and H. Haraguchi, "Multielement Determination of Trace Metals in Seawater by Inductively Coupled Plasma Mass Spectrometry after Tandem Preconcentration Using a Chelating Resin," *Bulletin of the Chemical Society of Japan*, vol. 78, no. 4, pp. 659–667, 2005.

[9] A. Itoh, T. Ishigaki, T. Arakaki, A. Yamada, M. Yamaguchi, and N. Kabe, "Determination of trace metals in coastal seawater around Okinawa and its multielement profiling analysis," *Bunseki Kagaku*, vol. 58, no. 4, pp. 257–263, 2009.

[10] A. Itoh, T. Kodani, M. Ono et al., "Potential anthropogenic pollution by Eu as well as Gd observed in river water around urban area," *Chemistry Letters*, vol. 46, no. 9, pp. 1327–1329, 2017.

[11] S. Kagaya, "Rapid coprecipitation technique for the separation and preconcentration of trace elements," *Bunseki Kagaku*, vol. 65, no. 1, pp. 13–23, 2016.

[12] T. Yabutani, S. Ji, F. Mouri, A. Itoh, and H. Haraguchi, "Simultaneous multielement determination of hydride- and oxoanion-forming elements in seawater by inductively coupled plasma mass spectrometry after lanthanum coprecipitation," *Bulletin of the Chemical Society of Japan*, vol. 73, p. 895, 2000.

[13] D. Rahmi, Y. Zhu, E. Fujimori et al., "An in-syringe La-coprecipitation Method for the Preconcentration of Oxo-anion Forming Elements in Seawater Prior to an ICP-MS Measurement," *Analytical Sciences*, vol. 24, pp. 1189–1192, 2008.

[14] T. Arakaki, T. Ishigaki, M. Yamaguchi, and A. Itoh, "Characteristics of concentrations and chemical forms of trace elements in deep seawater near kume island in Okinawa prefecture as studied by multielement profiling analysis," *Bunseki Kagaki*, vol. 58, p. 707, 2009.

[15] H. Daidoji, S. Tamura, and M. Natsubara, "Determination of rare earth elements and thorium in sea water by inductively coupled plasma atomic emission spectrometry," *Bunseki Kagaku*, vol. 34, p. 340, 1985.

[16] T. Fujinaga, M. Koyama, M. Kawashima, K. Izutsu, and S. Himeno, "Coprecipitation of Arsenic (V), (III) and Antimony (III) on Iron (III) and Aluminium Hydroxides," *Nippon Kagaku Kaishi*, vol. 1974, no. 8, pp. 1489–1493, 1974.

[17] M. Raso, P. Censi, and F. Saiano, "Simultaneous determinations of zirconium, hafnium, yttrium and lanthanides in seawater according to a co-precipitation technique onto iron-hydroxide," *Talanta*, vol. 116, pp. 1085–1090, 2013.

[18] S. Kagaya, S. Miwa, T. Mizuno, and K. Tohda, "Rapid Coprecipitation Technique Using Yttrium Hydroxide for the Preconcentration and Separation of Trace Elements in Saline Water Prior to Their ICP-AES Determination," *Analytical Sciences*, vol. 23, pp. 1021–1024, 2007.

[19] Z. Arslan, T. Oymak, and J. White, "Triethylamine-assisted Mg(OH) 2 coprecipitation/preconcentration for determination of trace metals and rare earth elements in seawater by inductively coupled plasma mass spectrometry (ICP-MS)," *Analytica Chimica Acta*, vol. 1008, pp. 18–28, 2018.

[20] T. Yabutani, K. Chiba, and H. Haraguchi, "Multielement Determination of Trace Elements in Seawater by Inductively Coupled Plasma Mass Spectrometry after Tandem Preconcentration with Cooperation of Chelating Resin Adsorption and Lanthanum Coprecipitation," *Bulletin of the Chemical Society of Japan*, vol. 74, pp. 31–38, 2001.

[21] H. Sawatari, E. Fujimori, and H. Haraguchi, "Multi-Element Determination of Trace Elements in Seawater by Gallium Coprecipitation and Inductively Coupled Plasma Mass Spectrometry," *Analytical Sciences*, vol. 11, pp. 369–374, 1995.

[22] M. Sugiyama, "Ocaeanochemistry, 1998".

[23] Y. Nozaki, "EOS Trans. AGU, 1997".

Permissions

The contributors of this book come from diverse backgrounds, making this book a truly international effort. This book will bring forth new frontiers with its revolutionizing research information and detailed analysis of the nascent developments around the world.

We would like to thank all the contributing authors for lending their expertise to make the book truly unique. They have played a crucial role in the development of this book. Without their invaluable contributions this book wouldn't have been possible. They have made vital efforts to compile up to date information on the varied aspects of this subject to make this book a valuable addition to the collection of many professionals and students.

This book was conceptualized with the vision of imparting up-to-date information and advanced data in this field. To ensure the same, a matchless editorial board was set up. Every individual on the board went through rigorous rounds of assessment to prove their worth. After which they invested a large part of their time researching and compiling the most relevant data for our readers.

The editorial board has been involved in producing this book since its inception. They have spent rigorous hours researching and exploring the diverse topics which have resulted in the successful publishing of this book. They have passed on their knowledge of decades through this book. To expedite this challenging task, the publisher supported the team at every step. A small team of assistant editors was also appointed to further simplify the editing procedure and attain best results for the readers.

Apart from the editorial board, the designing team has also invested a significant amount of their time in understanding the subject and creating the most relevant covers. They scrutinized every image to scout for the most suitable representation of the subject and create an appropriate cover for the book.

The publishing team has been an ardent support to the editorial, designing and production team. Their endless efforts to recruit the best for this project, has resulted in the accomplishment of this book. They are a veteran in the field of academics and their pool of knowledge is as vast as their experience in printing. Their expertise and guidance has proved useful at every step. Their uncompromising quality standards have made this book an exceptional effort. Their encouragement from time to time has been an inspiration for everyone.

The publisher and the editorial board hope that this book will prove to be a valuable piece of knowledge for researchers, students, practitioners and scholars across the globe.

List of Contributors

Karen C. Bedin, Edson Y. Mitsuyasu, Amanda Ronix, André L. Cazetta, Osvaldo Pezoti and Vitor C. Almeida
Department of Chemistry, State University of Maringá, Av. Colombo 5790, CEP 87020-900 Maringá, PR, Brazil

Shizhuang Weng, Mengqing Qiu, Fang Wang, Jinling Zhao, Linsheng Huang and Dongyan Zhang
Anhui Engineering Laboratory of Agro-Ecological Big Data, Anhui University, 111 Jiulong Road, Hefei 230601, China

Shizhuang Weng, Mengqing Qiu, Fang Wang, Jinling Zhao, Linsheng Huang and Dongyan Zhang
Science and Technology on Communication Networks Laboratory, Shijiazhuang 050000, China

Ronglu Dong
Hefei Institute of Physical Science, Chinese Academy of Sciences, 350 Shushanhu Road, Hefei 230031, China

Md. Hasan Zahir
Center of Research Excellence in Renewable Energy, Research Institute, King Fahd University of Petroleum and Minerals, Dhahran 31261, Saudi Arabia

Shakhawat Chowdhury
Department of Civil and Environmental Engineering, Water Research Group, King Fahd University of Petroleum and Minerals, Dhahran 31261, Saudi Arabia

Md. Abdul Aziz
Center of Research Excellence in Nanotechnology, King Fahd University of Petroleum and Minerals, Dhahran 31261, Saudi Arabia

Mohammad Mizanur Rahman
Center of Research Excellence in Corrosion, King Fahd University of Petroleum and Minerals, Dhahran 31261, Saudi Arabia

Margarita Cid-Hernández, Luis Javier González-Ortiz, María Judith Sánchez-Peña and Fermín Paul Pacheco-Moisés
Departamento de Química, Centro Universitario de Ciencias Exactas e Ingenierías, Universidad de Guadalajara, Blvd. Marcelino García Barragán 1421, 44430 Guadalajara, Jalisco, Mexico

Fernando Antonio López Dellamary-Torals
Departamento de Madera, Celulosa y Papel, Universidad de Guadalajara, km 15.5 de la Carretera Guadalajara-Nogales, 45220 Zapopan, Jalisco, Mexico

Yong Wang and He Li
Beijing Advanced Innovation Center for Food Nutrition and Human Health, Beijing Technology & Business University, 11 Fuchenglu, Beijing 100048, China

Yong Wang
Beijing Key Laboratory of Nutrition & Health and Food Safety, COFCO Nutrition & Health Research Institute, No.4 Road, Future Science & Technology Park, Beijing 102209, China

Ping Sun
Processing Technology Research Center for Tomato, COFCO Tunhe, Changji, Xinjiang 831100, China

Benu P. Adhikari
School of Applied Sciences, RMIT University, City Campus, Melbourne, VIC 3001, Australia

Dong Li
College of Engineering, China Agricultural University, Beijing 100083, China

Tong Chen, Xinyu Chen, Daoli Lu and Bin Chen
School of Food and Biological Engineering, Jiangsu University, Zhenjiang 212013, China

Naser F. Al-Tannak
Department of Pharmaceutical Chemistry, Faculty of Pharmacy, Kuwait University, AlJabriyah, Kuwait

Ahmed Hemdan
Department of Pharmaceutical Analytical Chemistry, Faculty of Pharmacy, Ahram Canadian University, Giza, Egypt
Institute of Clinical Chemistry, UniversityMedical Center Hamburg-Eppendorf (UKE), Martinistraße 52, 20246 Hamburg, Germany

Maya S. Eissa
Department of Pharmaceutical Analytical Chemistry, Faculty of Pharmacy, Egyptian Russian University, Cairo, Egypt

Nathan W. Bower, Murphy G. Brasuel and Eli Fahrenkrug
Chemistry and Biochemistry Department, Colorado College, Colorado Springs, CO 80903, USA

Matthew D. Cooney
GIS Technical Director, Colorado College, Colorado Springs, CO 80903, USA

Maja Hadzieva Gigovska, Ana Petkovska, Packa Antovska and Sonja Ugarkovic
Research & Development, Alkaloid AD, Blvd. AleksandarMakedonski 12, 1000 Skopje, Macedonia

Jelena Acevska, Natalija Nakov and Aneta Dimitrovska
Faculty of Pharmacy, University "Ss Cyril and Methodius", Mother Theresa 47, 1000 Skopje, Macedonia

Haiyan Lei and Wei Cao
Department of Food Science and Engineering, School of Chemical Engineering, Northwest University, Xi'an, Shaanxi 710069, China

Jianbo Guo, Zhuo Lv and Xiaohong Zhu
Shaanxi Institute for Food and Drug Control, Xi'an, Shaanxi 710069, China

Xiaofeng Xue and Liming Wu
Institute of Apiculture Research, Chinese Academy of Agricultural Sciences, Beijing 100093, China

Hector Henrique Ferreira Koolen, Marcos Nogueira Eberlin and Giovana Anceski Bataglion
ThoMSon Mass Spectrometry Laboratory, Institute of Chemistry, University of Campinas (UNICAMP), 13083-970, Campinas, SP, Brazil

Hector Henrique Ferreira Koolen
Metabolomics and Mass Spectrometry Research Group, Amazonas State University (UEA), 69065-001, Manaus, AM, Brazil

Clécio Fernando Klitzke, Joe Binkley and Jeffrey Patrick
LECO Corporation, 49085, St. Joseph, MI, USA

Ana Cec-lia Rizatti de Albergaria-Barbosa
Laboratory of Marine Geochemistry, Geoscience Institute, Federal University of Bahia (UFBA), 40170-020, Salvador, BA, Brazil

Rolf Roland Weber and Márcia Caruso B-cego
Marine Organic Chemistry Laboratory, Oceanographic Institute, University of São Paulo (USP), 05508-120, São Paulo, SP, Brazil

Giovana Anceski Bataglion
Department of Chemistry, Federal University of Amazonas (UFAM), 69077-000, Manaus, AM, Brazil

Hui Chen, Chao Tan and Zan Lin
Key Lab of Process Analysis and Control of Sichuan Universities, Yibin University, Yibin, Sichuan 644000, China

Hui Chen
Hospital, Yibin University, Yibin, Sichuan 644000, China

Zan Lin
The First Affiliated Hospital, Chongqing Medical University, Chongqing 400016, China

Fernando F. Sodré
Institute of Chemistry, University of Brasília, Brasília 70910-000, Brazil

Cínthia M. P. Cavalcanti
Environmental Sanitation Company of the Federal District (CAESB), Águas Claras 71928-720, Brazil

Mohamed E. I. Badawy, Mahmoud A. M. El-Nouby and Abd El-Salam M. Marei
Department of Pesticide Chemistry and Technology, Faculty of Agriculture, Alexandria University, El-Shatby, Alexandria 21545, Egypt

Andreas Kiontke, Susan Billig and Claudia Birkemeyer
Research Group of Mass Spectrometry at the Faculty of Chemistry and Mineralogy, University of Leipzig, Linnéstr. 3, 04103 Leipzig, Germany

Catalani Simona, Fostinelli Jacopo, Gilberti Maria Enrica, Orlandi Francesca, Paganelli Matteo, Madeo Egidio and De Palma Giuseppe
Unit of Occupational Health and Industrial Hygiene, Department of Medical and Surgical Specialties, Radiological Sciences and Public Health, University of Brescia, Italy

Magarini Riccardo
PerkinElmer (Italia), Milano, Italy

Dongxia Ren, Chengjun Sun, Danni Yang, Chen Zhou, Jiayu Xie and Yongxin Li
West China School of Public Health, Sichuan University, Chengdu 610041, China

Chengjun Sun and Chengjun Sun
Provincial Key Laboratory for Food Safety Monitoring and Risk Assessment of Sichuan, Chengdu 610041, China

Guanqun Ma
College of Life and Environmental Sciences, Shanghai Normal University, Shanghai, China

Chengli Zhang and Peng Wang
College of Petroleum Engineering, Northeast Petroleum University, Daqing, Heilongjiang163318, China

Guoliang Song
College of Mathematics and Statistics, Northeast Petroleum University, Daqing, Heilongjiang163318, China

Mosotho J. George
Department of Chemistry and Chemical Technology, National University of Lesotho, Roma 180, Lesotho

Mosotho J. George, Ian A. Dubery and Ntakadzeni E. Madala
Department of Biochemistry, University of Johannesburg, Auckland Park 2006, South Africa

Patrick B. Njobeh, Sefater Gbashi and Gabriel O. Adegoke
Department of Biotechnology and Food Technology, University of Johannesburg, Doornfontein Campus, Johannesburg 2028, South Africa

Gabriel O. Adegoke
Department of Food Technology, University of Ibadan, Ibadan, Nigeria

Shibao Lu and Wei Li
School of Public Administration, Zhejiang University of Finance and Economics, Hang Zhou 310018, China

Yan Wang
Development and Planning Division, Tongji University, 1239 Siping Road, Shanghai, China

Liang Pei
Key Laboratory of Water Cycle and Related Land Surface Processes, Institute of Geographic Sciences and Natural Resources Research, Chinese Academy of Sciences, Beijing 100101, China

Akihide Itoh, Kota Suzuki, Takumi Yasuda, Kazuhiko Nakano, Kimika Kaneshima and Kazuho Inaba
Department of Environmental Science, School of Life and Environmental Science, Azabu University, 1-17-71, Fuchinobe, Chuo-ku, Sagamihara-shi, Kanagawa 252-5201, Japan

Masato Ono
GL Science Inc., Shinjuku Square Tower 30 F, 6-22-1 Nishi Shinjuku, Shinjuku-ku, Tokyo163-1130, Japan

Index

www.ingramcontent.com/pod-product-compliance
Lightning Source LLC
Chambersburg PA
CBHW061241190326
41458CB00011B/3550